D1274584

X-RAY
SPECTROSCOPY

The late F. K. Richtmyer was Consulting Editor of the series from its inception in 1929 to his death in 1939. Lee A. DuBridge was Consulting Editor from 1939 to 1946; and G. P. Harnwell from 1947 to 1954. Leonard I. Schiff served as consultant from 1954 until his death in 1971.

**McGRAW-HILL
BOOK COMPANY**
New York
St. Louis
San Francisco
Düsseldorf
Johannesburg
Kuala Lumpur
London
Mexico
Montreal
New Delhi
Panama
Rio de Janeiro
Singapore
Sydney
Toronto

Edited by
LEONID V. AZÁROFF
Professor of Physics and Director
Institute of Materials Science
University of Connecticut

X-Ray
Spectroscopy

This book was set in Times New Roman.
The editors were Jack L. Farnsworth and James W. Bradley;
the production supervisor was Sam Ratkewitch.
The drawings were done by John Cordes, J & R Technical Services, Inc.
The printer and binder was The Maple Press Company.

BELMONT COLLEGE LIBRARY

80339

Library of Congress Cataloging in Publication Data

Azároff, Leonid V.
X-ray spectroscopy.

(International series in pure & applied physics)
Includes bibliographies.
1. X-ray spectroscopy.
QC482.S6A9   544'.66   73-6612
ISBN 0-07-002674-2

**X-RAY
SPECTROSCOPY**

Copyright © 1974 by McGraw-Hill, Inc. All rights reserved.
Printed in the United States of America. No part of this publication may be reproduced,
stored in a retrieval system, or transmitted, in any form or by any means,
electronic, mechanical, photocopying, recording, or otherwise,
without the prior written permission of the publisher.

1234567890MAMM79876543

QC
482
S6
A9

CONTENTS

LIST OF CONTRIBUTORS

LEONID V. AZÁROFF, Physics Department and Institute of Materials Science, University of Connecticut, Storrs, Connecticut

W. L. BAUN, Air Force Materials Laboratory, Wright Patterson AFB, Dayton, Ohio

P. E. BEST, Physics Department and Institute of Materials Science, University of Connecticut, Storrs, Connecticut

JOHN R. CUTHILL, Alloy Physics Section, National Bureau of Standards, Washington, D.C.

CHARLES S. FADLEY, Department of Physics, University of Dar es Salaam, Dar es Salaam, Tanzania

STIG B. M. HAGSTRÖM, Department of Physics, Linköpping University, Linköpping, Sweden

LARS HEDIN, Institute of Theoretical Physics, University of Lund, Lund, Sweden

ROBERT P. MADDEN, Far-UV Physics Section, National Bureau of Standards, Washington, D.C.

D. J. NAGEL, Naval Research Laboratory, Washington, D.C.

DOUGLAS M. PEASE, Physics Department and Institute of Materials Science, University of Connecticut, Storrs, Connecticut

G. A. ROOKE, Ferranti, Ltd., Bracknell, Berks, Great Britain

JOHN S. THOMSEN, Department of Physics, The Johns Hopkins University, Baltimore, Maryland

EDITOR'S PREFACE

The field of x-ray spectroscopy grew directly from the discovery of x radiation by Wilhelm-Konrad Röntgen in 1895. The importance of this discovery was acknowledged in 1901 when Röntgen received the first Nobel prize awarded in physics. The role that x-ray spectroscopy played subsequently in the development of modern physics is attested to by the many Nobel prizes awarded to x-ray physicists in later years. By the mid-1930s most of this pioneering work had been completed so that Compton and Allison were able to present a fairly complete description of the accomplishments in their monumental book *X-rays in theory and experiment* (2d ed., D. Van Nostrand Company, Inc., Princeton, N.J., 1935) that has lasted to the present time as the authoritative statement on this subject. In fact, it can be said that the experimental developments by 1935 had outdistanced the abilities of theoretical physicists to provide quantitative interpretations for the experimental observations. Consequently, activity in this field gradually abated as most physicists turned their attention to other problems.

It is interesting to note that major developments in science, particularly important new theories, are often stimulated as much by advances in technology as by historical reasons. Thus the theory of relativity was developed by Einstein in response to the experimentally discovered limitations of Newtonian mechanics. The technology developed by scientists early in the twentieth century, in turn, enabled the verification of the developments of Einstein's fertile imagination. Without such interplay, interpretation of experimental results is subject to qualitative speculation, while hypotheses that cannot be tested experimentally belong more properly to the realm of philosophy.

By the second half of the twentieth century, technological advances in ionization detectors and x-ray spectrometry produced a renaissance of interest in x-ray spectroscopy as an analytical tool in chemical analysis and as a tool for studying the solid state in general. The parallel development of computational methods in atomic and solid-state physics and chemistry, abetted by modern high-speed computers, enabled theoreticians to return to the many unsolved problems in x-ray spectroscopy with a realistic hope of finding their solutions. Thus the last decades have seen a rebirth of effort in expanding the methods of x-ray spectroscopy as well as its application to a growing list of studies in various fields of science. The present book has been compiled in order to help this development along. It represents an attempt to present the state of the science by collecting in one place tutorial discussions of most of the important areas. The high degree of activity and the many advances made almost daily militate against an attempt at the present time at a synthesis of the kind carried out by Compton and Allison. Instead, an up-to-date description is presented that should enable the reader to learn what is already known and to discover where many interesting problems still remain.

Each chapter in this book has been written by one of the leading exponents of that particular field of activity. Following an introductory review of x-ray spectroscopy in Chap. 1, the next two chapters discuss modern methodologies and analyses involving two-crystal spectrometry and grating spectrometry. The modern theory of soft x-ray spectroscopy is developed in Chap. 4, emphasizing emission spectra. This is followed by a discussion of many-body effects in Chap. 5, since these play an important role in modern theoretical developments. The theoretical and experimental results of x-ray absorption spectroscopy are presented in Chap. 6, while the utilization of synchrotron radiation in x-ray absorption spectroscopy is discussed in Chap. 7. The field of x-ray photoelectron spectroscopy is considered in Chap. 8, and an extensive review of the utilization and application of x-ray spectroscopy to chemistry, metallurgy, and solid-state physics is presented in Chap. 9. A companion volume containing a synthesis of the modern advances in the field of x-ray diffraction is being published concurrently.

The compilation of this work has been made possible by moneys awarded to the University of Connecticut[1] by the U.S. Air Force Materials Laboratory through the personal efforts of Mr. William Baun. The material contained in Chaps. 3 and 7 was prepared with the support of the National Bureau of Standards and, consistent with its policies, is exempted from the copyright protecting the rest of this book. I trust that the readers of this book will share with me a deep appreciation of the efforts of the contributors as well as those of the typists and others who made the final book a reality.

LEONID V. AZÁROFF

[1]The preparation of the manuscript for this book was supported, in part, by Contract F33615-68-C-1602 issued by U.S.A.F. Air Force Systems Command, Aeronautical Systems Division, Wright Patterson AFB, Dayton, Ohio.

ERRATA

Page 55.   In Fig. 2-7, the angle $180° - \theta_n$ should be $180° - 2\theta_n$.

Page 150.   The legend for Fig. 3-14 should read:
             Steps in production of NPL laminar grating. (*Bennett, 1971.*)

Page 154.   The legend for Fig. 3-16 should read:
             Wafer-type photoelectron multiplier showing longitudinal cross section through focused-mesh multiplier below photograph of wafer.

Page 167.   The legend for Fig. 3-29 should read:
             Aluminum L_{III} emission spectrum from $AuAl_2$. (Curve *A* after Switendick.)

1

X-RAY SPECTRA

P. E. Best

I. INTRODUCTION

The subject of x-ray spectra, with its origin in early atomic physics, is today a topic of interest to a number of fields: solid-state and chemical physics, astrophysics, atomic and plasma physics—indeed any field that involves interactions of matter and electromagnetic energy in the range of about 500 eV to many GeV. Likewise, there is diversity in the types of processes which can give rise to x-ray spectra. For emission spectra, for example, these processes include characteristic atomic x-ray emission, Bremsstrahlung, synchrotron radiation, inverse Compton scattering, muonic x-ray emission, positron annihilation, and nuclear decays. The discussion in this chapter is limited to x-ray emission and absorption spectra which arise from processes directly involving a vacancy in an inner electron shell of an atom. The limited number of examples used to illustrate the discussion are drawn only from the fields of atomic, solid-state, and chemical physics.

Traditionally, the experimental x-ray field has been divided into short- and long-wavelength regions (Parratt, 1959). The division is not only experimental, however. The gross features of most x-ray spectra in the short-wavelength (<20 A) region can be accounted for on a hydrogenic model. It is on the details of these spectra that much study has been expended to explain features arising from many-electron and time-dependent effects. This is less so in the long-wavelength region, where even the gross features of the absorption spectra cannot be understood with a hydrogenic model. The emphasis in this chapter is on the fine-feature phenomena which have attracted the most attention in the last decade. Some mention is also made of the longer wavelength phenomena, but these have been reviewed in detail by Fano and Cooper (1968).

We shall describe, in a general way, the types of states which have been shown to participate in transitions giving rise to x-ray spectra, and the manner in which the transitions occur. In doing this we shall devote a relatively small amount of space to treating transitions which involve only one electron changing quantum numbers between singly ionized states, but which are otherwise unexcited. Yet these transitions, in a general way, account for the greater part of all observed x-ray intensities, and are the basis on which most information is extracted for studies of electronic structures of interest to chemical and solid-state physics. However, an understanding of the processes giving rise to x-ray spectra requires knowledge of all the types of states that can occur, as well as the important single-vacancy ones.

Often the same states are involved in processes monitored by a number of different experiments: x-ray spectra, (x-ray) photoelectron emission, Auger electron emission, and the many optical experiments which probe the structure of valence and conduction bands. Those other experiments more directly related to x-ray spectroscopy will be mentioned where appropriate. X-ray photoemission is discussed in detail in Chap. 8.

II. HISTORICAL NOTE

The early contributions to atomic physics which arose out of studies of x-ray spectra are well documented and need not be mentioned here (Siegbahn, 1931; Compton and Allison, 1935). Subsequent to about 1925 there was a time of considerable activity in areas of current interest: solid-state and chemical effects, satellite emission, and various absorption phenomena. In many instances the gross spectral features recorded in those days match those observed by state-of-the-art techniques today, and many of the early qualitative interpretations of the gaseous and solid-state spectra have remained unchanged. It is in the details of the spectra that today's results are superior to those of forty or so years ago, and it is in the quantitative descriptions that today's predictions are better.

Subsequent to about 1945, x-ray spectroscopy was continued in laboratories in about 10 countries throughout the world, as partly indicated by the books and review articles of the period (Cauchois, 1948; Shaw, 1956; Blokhin, 1957; Sandström, 1957;

Tomboulian, 1957; Parratt, 1959; Sawada et al., 1959). In this time, however, the field was not a part of the then current trends in physics. Without theoretical stimulus, progress was not rapid. Nevertheless, the unsolved major problems in the field were well defined in this period. These had to do with experimental techniques, with the many-electron nature of the transitions, and with the time scales of importance to the processes (Parratt, 1959). In the last decade progress toward solutions of these problems has been made. Indeed, today the field is again very active, with contributions coming from a wider array of laboratories and disciplines than ever before. A large part of the rejuvenation is due to the computer-given ability to make quantitative predictions about the various types of x-ray states and transitions. There have also been infusions from the complementary fields of x-ray photoelectron spectroscopy and x-ray chemical analysis.

Improvements in experimental techniques of this relatively old field have been evolutionary rather than revolutionary, with the possible exception of the recent utilization of high-intensity continuum sources in the long-wavelength region (Chap. 7). These sources have made absorption measurements much more accessible. For the short-wavelength region the advent of high-resolution solid-state detectors has greatly facilitated measurements (Goulding and Stone, 1970). On the evolutionary side, a notable recent improvement has been the availability of crystals with ever larger $2d$ spacings, extending now to 97 A (Ruderman, Ness, and Lindsay, 1965). In addition, there has been an increase in the use of polarization techniques in x-ray spectroscopy (Schnopper, 1966; Watanabe et al., 1971). These changes, together with improvements of electronic controls and automation (Gregory and Best, 1971), constitute the main advances in the field of spectral measurements over the last decade.

Various types of spectrometers are being used to gather data today: energy-dispersive detectors, two-crystal x-ray spectrometers, single plane-crystal spectrometers, bent-crystal spectrometers, and grating spectrometers. Each instrument has its uses, the choice depending on various factors such as the wavelength region and the requirements of resolution and transmission which are determined by the experiment itself. Readers are referred to the literature for discussions of the merits of each instrument (see also Chap. 3). For high-resolution work in the wavelength region of interest here, the two-crystal spectrometer is unique in that a good approximation to the instrumental window function can be obtained (Compton and Allison, 1935; Porteus, 1962), allowing for correction (by the unfolding technique) of the instrumental resolving power.

The catalog of experimental difficulties is not significantly different from that published a decade ago (Parratt, 1959). The situation with regard to the extraction of "true" spectra from measured data has recently seen essentially two changes, both being in the area of the study of valence emission spectra from certain types of metals and small-band-gap materials. In many spectra the presence of unrecognized satellites

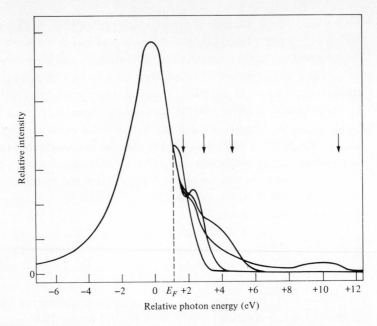

FIGURE 1-1
Shape of the threshold level iron $L\alpha$ line as a function of the incident electron energies indicated by the arrows. The spectra were recorded on a two-crystal vacuum spectrometer and have been corrected for background, detector dead time, system response, underlying continuum, and instrumental resolution. (*After Hanzely and Liefeld*, 1969.)

distorts the shape from that expected of the simplest of models (Parratt, 1959). The most satisfactory way to eliminate these is to record spectra at exciting electron-beam energies below threshold for the disturbing features (Liefeld, 1968). But even before this problem could be recognized and effectively resolved, the subject of target self-absorption had to be dealt with (Hanson and Herrera, 1957). Again, by keeping the incident electron energy low, and by measuring the spectra at close-to-normal take-off angle, spectra relatively free from these self-absorption perturbations can be obtained (Liefeld, 1968). In detailed studies of this type still other perturbations to the spectra were found, apparently due in part to emission arising from the transition of a bound-ejected electron to the inner hole (Hanzely and Liefeld, 1969) (Fig. 1-1) and in part to an enhanced continuum emission associated with the elastic component of the incident electron beam (Liefeld, 1971). As discussed later, there are some reasons for believing that the simple valence emission spectra of many materials are not subject to all the perturbations mentioned above; however, the systematic studies required to support these arguments have not been made.

III.   THE SPECTRA

A.   Absorption

For practical purposes composite data encompassing all x-ray absorption is collected (Hubbell et al., 1973). Among the many purposes it serves, this type of data is used to make approximate corrections to data in x-ray crystallography and chemical analysis by x-ray emission. Broad spectral data are also used in studies related to theories of continuum absorption (Manson and Cooper, 1968; McGuire, 1968), and in investigations of atomic-sum rules of the form

$$\sum_{s}^{\text{discrete}} f_s + \int_{I}^{\infty} \frac{df}{dE}\, dE = N$$

where N is the number of electrons in the atom, f_s is the oscillator strength to discrete states, and df/dE is the oscillator-strength density integrated from the ionization threshold I to continuum states (Piech and Levinger, 1964; Fano and Cooper, 1968).

In this chapter we are concerned with the processes that are responsible for details of the absorption spectra in the vicinity of absorption edges.[1] It can also be pointed out that the spectral dependence of x-ray absorption coefficients can also be found, approximately, from measurements of the energy-loss spectra of high-energy electrons at very small scattering angles (Lassettre and Francis, 1964; Swanson and Powell, 1968).

B.   Emission

An extensive literature exists on the formal quantum-mechanical interpretation of the general features of the spectra including the systematics of emission spectra, including line energies, doublet splittings, and relative intensities (Blokhin, 1957; Sandström, 1957). Despite the extensiveness of the literature, these subjects are by no means closed. Realistic calculations of relative line intensities have been made only recently, for example. Such calculations for atoms of low atomic number Z will be described later. For atoms of large Z, relative intensities have been calculated in the active-electron approximation, including the effects of retardation, utilizing wavefunctions obtained by self-consistent relativisitic Hartree-Fock methods with a Slater approximation to the exchange term (Scofield, 1969; Rosner and Bhalla, 1970). Experimenters

---

[1] An *absorption edge* is the step in the absorption coefficient which occurs as the photon energy increases across the threshold for exciting an electron in the atomic subshell in question. In older literature the term "edge" refers to the first main inflection point in the experimental absorption curve, without regard to origin. More recently there has been a tendency to reserve the term for the spectral feature associated with the onset of photoionization, a feature not easily located for most spectra.

in this area have measured relative intensities of lines arising from electric-dipole, quadrupole, and magnetic-dipole transitions (Smither, Freedman, and Porter, 1970; Rao, Palms, and Wood, 1971). In each case careful correction of the raw data for self-absorption effects, and for instrumental response, must be made. Many "forbidden" lines, i.e., those arising from nonelectric-dipole transitions, have been recorded on film, without accurate intensity measurements (Padalia and Rao, 1969). The angular correlation of x-ray emission has been considered from the point of view of admixing magnetic-quadrupole with electric-dipole radiation (Catz, 1970).

We also mention the subject of line widths at this time, restricting ourselves to lines arising from simple transitions directly involving only inner-shell electrons. The prime feature in the width of a line is the sum of the lifetimes of the initial and final states. The lifetime width of an inner-shell vacancy is proportional to the probability for decay, both radiative and nonradiative (i.e., Auger electron emission) decay processes being taken into account (Blokhin, 1957; Parratt, 1959). The ratio of the probability for radiative decay to that for decay by all modes is the fluorescent yield (Kostroun, Chen, and Craseman, 1971; McGuire, 1970).

The most important nonlifetime factor observed in line widths is multiplet structure unresolved from the particular line in question. An example is the structure which arises from the interaction of an unpaired inner-shell electron with an incomplete outer shell. This occurs particularly in the $K\alpha$ spectra from first transition elements, where the $L_{2,3}$ levels are split by this type of interaction (Blokhin and Nikiforov, 1964; see also Bonnelle and Jørgensen, 1964). In these same atoms, multiplet splittings in $M_{2,3}$ levels are sometimes so large as to give rise to distinct line structure in the $K\beta_{1,3}$ region of the spectrum, $K \rightarrow M_{2,3}$ (Tsutsumi and Nakamori, 1968; Ekstig et al., 1970).[1] This same structure is observed in corresponding photoelectron measurements, where the final atomic states are the same as the final states of the $K\beta_{1,3}$ emission process (Fadley and Shirley, 1970).

The unpaired outer electrons referred to above need not necessarily belong to the lowest energy state of the excited atom, but may be formed by a multiple-electron transition caused by the exciting radiation. Such valence-excited configurations (VEC states) (Parratt, 1959) can be identified in certain experiments which will be mentioned later.

In solids, an additional source of broadening of inner levels results from lattice vibrations (Overhauser, 1959; Parratt, 1959). The theory has been applied to x-ray spectra for both metals (Cuthill, McAlister, and Williams, 1967) and ionic crystals

---

[1] The notation "$K \rightarrow M_{2,3}$" is equivalent to $1s^{-1} \rightarrow 3p^{-1}_{1/2,3/2}$, both of which indicate the transition of the hole. Conventional x-ray nomenclature, of the type $K\beta_{1,3}$, is adequate for emission lines arising from one-electron transitions involving only singly ionized inner atomic levels. For emission involving outer electron levels of molecules or solids, this nomenclature is generally useful only for indicating the spectral region in which the emission occurs.

(Best, 1971). Widths of lines due to transitions between inner vacancies of molecules have been discussed in terms of the Franck-Condon principle (Best, 1968; Gilberg, 1970).

Various theoretical and experimental studies of the comparison between primary (incident-electron beam) and secondary (incident-x-ray beam) excitation of x-ray emission spectra have been made (Graeffe et al., 1969; Krause et al., 1970). Many studies have also been made of the excitation of x-ray spectra by atomic or ionic bombardment (Burch, Richard, and Blake, 1971; Knudson et al., 1971), although consideration of these would take us too far afield. In these cases the emission may be used as a probe of the exciting mechanism.

IV. STATES (ELECTRON CONFIGURATIONS)

While the features of most x-ray spectra can be described by a one-electron-jump model, an adequate description of x-ray *processes* in general must take into account the manner in which the relatively inactive electrons are affected by the change of potential brought about either by the production of the initial state or by its decay. To do this, some knowledge of the many-electron wavefunctions of the initial and final states is required. The most accurate common representation of such wavefunctions is that of a Slater determinant of orthonormal spin orbitals

$$\Psi = \frac{1}{N!} \det \left[\phi_{1s\alpha}(1), \phi_{1s\beta}(2), \phi_{2s\alpha}(3), \dots, \phi_{\eta\lambda\beta}(N) \right] \qquad (1\text{-}1)$$

Minimizing the total energy of the product leads to the well-known Hartree-Fock equations, which can be solved to yield the ϕ's that are good approximations to the spatial wavefunctions. For any transition in which an electron is ejected from the system, the decision to include the continuum orbital inside (outside) of the determinant depends on whether the inner-vacancy lifetime is longer (shorter) than the exchange period of the ejected electron with the residue.

In some cases, the quantitative interpretation of x-ray spectra is not possible using only states which are each described as a definite configuration of one-electron quantum numbers. Especially when states of the same parity and with the same J value are close in energy, electrostatic interaction between them will lead to configuration mixing in which the actual states are linear combinations of simple configurations of the type described by expression (1-1) (Kuhn, 1969; Weiss, 1969). The mixing coefficients can be determined by the variational principle.

To calculate state energies, Bagus (1965) has used wavefunctions of the type in equation (1-1). The correlation energy—the difference between the experimentally determined ionization energy and the corrected Hartree-Fock ionization energy—is of the order of 1 eV in 3000 (Bagus, 1965; Aberg, 1967). Calculations of this type for inner-vacancy states are not available for many elements. Then, Koopmans' theorem is often employed, in which the one-electron orbital energy obtained from the Hartree-Fock equations is used as the ionization energy (Koopmans, 1934). This is equivalent to assuming that the "inactive" electron orbitals are identical in the initial and final states. While it is not a good description of either the state or its energy (Bagus, 1965), the method is apparently accurate in estimating *changes* of inner ionization energies with changes of outer atomic electron configuration (Barinskii and Nefedov, 1969; Siegbahn et al., 1967; Leonhardt and Meisel, 1970).

It has been shown theoretically that inner-shell ionization energies calculated from Koopmans' theorem are related to a weighted-average energy for singly and multiply excited states which are formed by a sudden ionization process (Manne and Aberg, 1970). In this process the inactive electrons are spatially frozen, their original configuration being projected onto the orbitals which are eigenstates of the final potential, i.e., after ionization.

An atom with a $1s$-shell vacancy is referred to as being in the K state and an atom with simultaneous vacancies in the $1s$ and $3p_{1/2}$ shells in the KM_2 state, etc. (Siegbahn, 1931). The most suitable terminology to describe the excitation of an inner electron to a bound state is probably the complete notation $1s^2 2s^2 \cdots 3p^6 \rightarrow 1s' 2s'^2 \cdots 3p'^6 4p'$, although the shorthand $1s \rightarrow 4p'$ is also used. The primes indicate excited-state orbitals. The energies of formation of all such ionization and excitation states from the ground state, being considered positive, are often indicated on a schematic energy-level diagram, which, when complete, is necessarily very complex (Parratt, 1959).

If a frozen orbital model is adapted (Koopmans'-theorem approximation), the state energies can be associated with the active electron, and the diagram is then less complex. K-state formation on the more complete diagram is represented by an arrow from the ground state up to the K state. In the frozen orbital case, the transition would be represented by an arrow going up from the $1s$ to zero level; *zero level* refers to the energy of an electron removed to rest at an infinite distance from the atom.

Emission lines which arise from one-electron jumps between singly-ionized states, with the inactive electrons retaining their original quantum numbers, are known as *diagram lines*; all others are *satellites*.[1] Only ionization thresholds and diagram lines can be consistently represented on the simpler type of energy-level diagram.

Diagrams, and the associated energy-level tables (Bearden and Burr, 1967;

---

[1] This definition is somewhat broader than the traditional one, although it is in accord with modern usage (Ekstig et al., 1970). During the 1920s and 1930s the term "satellite" referred ambiguously either to Wentzel-Druyvesteyn satellites or to any faint line accompanying a strong line. When such a line was identified satisfactorily, it became a diagram line, although from the more complex diagram.

Siegbahn et al., 1967), can be constructed whenever a known definite energy is associated with each state that is to be recognized in the diagram or table. The same situation is required for the Ritz combination principle to be used (Parratt, 1959; Deslattes, 1969). This principle is used to describe the relationship between x-ray emission energies and differences in binding energies deduced from x-ray photoemission data (Bearden and Burr, 1967; Siegbahn et al., 1967). One requirement for this principle to hold is that the states be at least quasi-stationary states; i.e., the time for hole formation must be very much shorter than the lifetime of the inner vacancy (Schnopper, 1967). This situation may not exist for K states of elements of high atomic number, and there is the possibility that the energies of K emission lines will depend on the excess kinetic energy carried away with the ejected electron, total energy being conserved (Schnopper, 1967). For such cases there is no reason that term value differences deduced from K x-ray emission should agree with those deduced from both L emission and photoemission measurements. Discrepancies of this type can be found for uranium and thorium, for example (from data compiled by Bearden, 1967, and by Siegbahn et al., 1967), although the data are not such that other possible causes of the discrepancy can be ruled out. Spectral features which depend on the time scale of valence-electron processes, relative to the inner-vacancy lifetime, have been discussed briefly in the literature (Parratt, 1959; Stott, 1969; Best, 1971).

Most x-ray absorption in crystalline solids probably can be understood in terms of the excitation of one or more electrons to quasi-stationary states. In the case where the ejected electron has sufficient energy, the wavefunction appropriate to it is that of a modified Bloch state, one that represents transmission through the lattice (Shiraiwa et al., 1958). If a situation occurred in which the inner hole decayed before the ejected electron could sample the lattice (Parratt, 1959), it is possible that the absorption spectra would exhibit maxima where local reflection occurred, rather than transmission as above. This difference is similar in effect to the difference predicted by Kronig and Hayasi-type absorption theories, respectively (Azároff, 1963).

V. TIME SCALES AND PROCESSES

We first discuss the models that are used to describe some processes which are involved in the emission and absorption of x rays. Only dipole transitions between quasi-stationary states will be considered. Finally we make some comments about x-ray spectra of more extended systems, namely molecules and solids, keeping in mind that the discussion to that point is applicable to all states of matter.

Two limiting cases of time scales are considered: those in which the change of potential brought about by the production of the initial state, or by its decay, is fast or slow, respectively, relative to the natural relaxation times of the inactive electrons. In each approximation we are interested first in the process by which only one electron changes quantum numbers, and then those by which more than one electron changes quantum numbers.

Processes involving x-ray absorption can be considered in two limiting approximations (Schiff, 1955). During the time of electron ejection the Hamiltonian of the system relaxes from H_0 to H_1. If the ejected electron travels slowly enough through the system such that $\partial H/\partial t$ between H_0 and H_1 is sufficiently small, the wavefunctions distort gradually (adiabatic approximation) (Schiff, 1955; Schnopper, 1967), the dipole-transition probability being proportional to the square of the dipole moment, which is

$$M = \int \Psi_f \sum r_i \Psi_0 \, d\tau \qquad (1\text{-}2)$$

This approximation is valid when $[h/(E_{KS} - E_K)^2] \, \partial H/\partial t \ll 1$, where E_{KS} and E_K are the energies of the KS and K states, and where $S = K, L, M$, etc. Expression (1-2) will be examined more closely later.

On the other hand, if the change in Hamiltonian is instantaneous, multiple-hole states may be formed by electron shake-off (sudden approximation) (Carlson and Krause, 1965; Aberg, 1967b, 1969). In the independent-particle model the probability that the inactive electrons retain the same quantum numbers during the excitation is proportional to

$$(\langle \phi'_{nl} \mid \phi_{nl} \rangle)^2$$

and the total probability for other electrons being simultaneously excited in the ejection process is just proportional to

$$1 - (\langle \phi'_{nl} \mid \phi_{nl} \rangle)^2$$

The overlap between the initial- and final-state wavefunction is never unity, since the radial parts of the wavefunctions contract slightly during the process. Thus two- and many-electron excitations occur in the sudden ejection of an inner electron.

The criterion that the sudden approximation be reasonably valid is that

$$\frac{E_{KS} - E_K}{h} \tau_n \ll 1$$

where τ_n is the time of transit of the ejected electron past the n shell, $n = 1, 2$, etc. This expression brings out the point that a particular process may be sudden for one electron shell and approximately adiabatic for another more tightly bound shell. It has been shown experimentally and theoretically that this criterion is too severe in some cases (Carlson and Krause, 1965). Note that the application of the sudden approximation to x-ray states is similar to its application to vibronic states, which is described by the Franck-Condon principle (Manne and Aberg, 1970). In that case the change in potential due to an electron transition occurs rapidly compared with the time for nuclear motion.

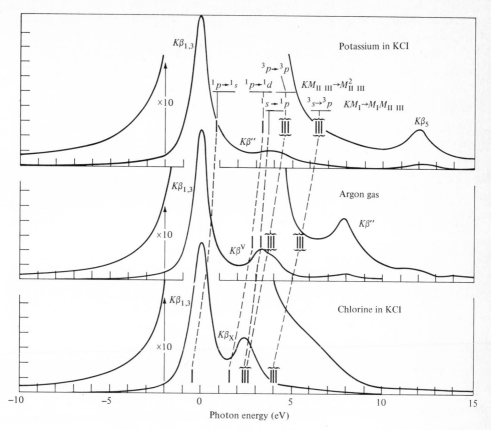

FIGURE 1-2
Corrected $K\beta$ spectra of potassium and chlorine in KCl, and of argon, together with the calculated relative energy positions of the $KM \to M^2$ satellites. (*After Deslattes*, 1964.)

Much of the evidence for the sudden approximation has come from the analysis of the distribution of charged states formed by x-ray photoemission, and in the analysis of the kinetic energies of photo-ejected electrons (Carlson, 1967; Krause and Carlson, 1966). Also, relative intensities of Wentzel-Druyvesteyn satellites, which arise from a one-electron jump between double-vacancy states [Horak, 1961; Deslattes, 1964 (see Fig. 1-2)], have been used to monitor the rate of formation of double vacancies in x-ray excitation (Sachenko and Demekhin, 1966; Aberg, 1967*b*).

Next the case of an adiabatic transition will be considered. This is the approximate situation for absorption close to threshold, and for all emission. For absorption it has been shown how expression (1-2) for the adiabatic case joins with that for the sudden approximation (Aberg, 1967b, 1969). In each case, both single- and multiple-electron transitions can occur (Aberg, 1967b, 1971; Best, 1967).

Expression (1-2) has been considered in detail by a number of authors. We will expand it in the manner described by Knox (1958). Both initial- and final-state wavefunctions are described by an independent-particle model; the example being used is for the $1s \rightarrow 4p$ transitions in argon (Knox has discussed the $3p \rightarrow 4s$ transition)

$$M = \sqrt{2} \int \det (\phi'_{1s\beta}, \phi'_{2s\alpha}, \ldots, \phi'_{3p_z\beta}, \phi'_{4p_x\alpha})$$

$$\times \sum r_i(\phi_{1s\alpha}, \phi_{1s\beta}, \phi_{2s\alpha}, \ldots, \phi_{3p_z\beta}) \, dr_1 \cdots dr_{18} \qquad (1\text{-}3)$$

If the quantities $S = \int \phi'_\mu(1)\phi_\nu(1) \, d\tau_1 - \delta_{\mu\nu}$ are small, (1-3) may be expanded to obtain an expression valid up to order S^2.

$$M = DR - \sum \langle \phi'_{4p_x\alpha} | \phi_l \rangle \langle \phi'_l | r | \phi_{1s\alpha} \rangle$$

$$- \sum_l \langle \phi'_{4p_x\alpha} | r | \phi_l \rangle \langle \phi'_l | \phi_{1s\alpha} \rangle \qquad l \neq 1s\alpha, 4p_x\alpha \qquad (1\text{-}4)$$

where

$$R = \int \phi'_{4p_x\alpha} r \phi_{1s\alpha} \, dr \qquad (1\text{-}5)$$

and

$$D = \sum_{i=1}^{17}{}' \langle \phi'_i | \phi_i \rangle - \sum_i{}' \sum_{j \neq i} \langle \phi'_i | \phi_j \rangle \langle \phi'_j | \phi_i \rangle$$

In the first complete calculation of x-ray intensities using (1-2), Bagus (1964) found that the active-electron approximation (1-5), which is obtained by assuming $\langle \phi_\mu | \phi_\nu \rangle = \delta_{\mu\nu}$, underestimated intensities by as much as 25% when a valence electron was directly involved. For molecules the effect of the inner hole on emission intensities has been discussed semiquantitatively (Best, 1968). Friedel (1969) found that when expression (1-2) is applied to x-ray emission and absorption involving conduction electrons of certain metals, it leads to the Fermi edge depression or singularity which is observed in the spectra (Chap. 4). For other metals and solids, in general, the effects of the inner hole on the spectral shape is still a matter for debate (Stott, 1969; Brouers, Longe, and Bergersen, 1970; Allotey, 1971). Some of these effects which cannot be explained in terms of an independent-particle model are discussed in Chap. 5 of this book.

It is very common to assume that the spectral shape in valence emission is dependent only on the active-electron term (1-5), the other factors being effectively constant over the small energy region of interest. This is often a satisfactory approximation, and spectra that can be described by this model are discussed in detail in later chapters. Here, we will only mention some effects of the active-electron term on absorption spectra.

In the shorter wavelength region, a hydrogenic model gives an adequate description of the main features of the absorption spectra of many atoms (Barinskii, 1961; Watanabe, 1965). At longer wavelengths, this is no longer the case, and three effects due specifically to nonhydrogenic behavior are mentioned.

The final state of the x-ray absorption process is always autoionizing, and in this situation the direct transitions to a bound state interfere with the indirect transitions to the underlying continuum. In such cases the absorption peaks can be markedly asymmetric, the curves even exhibiting minima below the surrounding level of continuum absorption (Fano, 1961; Codling, Madden, and Ederer, 1967).

In addition, there are resonance effects which occur when the dipole moment of a transition to continuum states has positive and negative contributions at different energies. With increasing kinetic energy of the ejected electron, a node of its wavefunction may traverse the outer lobe of the discrete state, giving rise to a minimum in absorption (Cooper, 1962; Fano and Cooper, 1968).

Again, for low-energy continuum states of high angular momentum l there is a centrifugal barrier $l(l + 1)\hbar^2/(2mr^2)$, which the continuum orbital has difficulty traversing. Thus absorption to these states is strongly reduced at and near threshold, giving rise to delayed absorption thresholds (Manson and Cooper, 1968).

For the absorption spectra of some polar molecules there is marked nonhydrogenic behavior even at short wavelengths, due mainly to an effective barrier in the outer regions of the molecule (Nefedov, 1970; Dehmer, 1971). This barrier is due mainly to the Pauli exclusion principle in regions of high electron density about the negative species in the molecule. The barrier divides space into inner and outer regions. A state whose wavefunction is mainly localized in the inner valley may be discrete or quasi-discrete, depending on whether its energy is less or greater, respectively, than the ionization energy. Quasi-discrete states are very similar to virtual-bound states of alloy theory. In such situations there may be reduced x-ray absorption to bound (Rydberg) states lying outside the barrier, and enhanced absorption to nonbound states lying within the barrier (Dehmer, 1971). The sulfur K-absorption spectrum of SF_6 shows this type of behavior (Fig. 1-3).

We next discuss the situation in which more than one electron changes quantum numbers in an adiabatic transition. In the active-electron approximation there would be no such transitions. However, in general, such many-electron transitions (mostly two) occur for all systems in first order. This is because $\langle \phi'_\mu \mid \phi_\nu \rangle \neq \delta_{\mu\nu}$ and expression (1-2) will have finite amplitude for transitions to all states which have the correct symmetry. Multiple-electron transitions have been observed in absorption for atoms [Schnopper, 1963; Wuilleumier and Bonnelle, 1963 (Fig. 1-4)]. In large-band-gap solids quantitative interpretations of spectra based on one-electron-jump models are just becoming available (Kunz, 1970), and the assignment of absorption features to two-electron-jump processes is still a matter for debate (Vinograd and Zimkina, 1970; Brown et al., 1970; Sato et al., 1971).

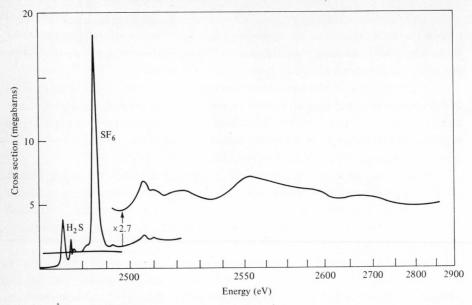

FIGURE 1-3

Absorption spectra of SF_6 and H_2S above the K threshold. The spectrum of H_2S exhibits structure which probably can be interpreted as arising from transitions involving a Rydberg series for the ejected electron. In SF_6, the Rydberg states lie outside the potential barrier which is in the region of the fluorines. The transition probability to these states is very small, while to nonbound states within the barrier, the transition probability is large.

In emission, two-electron transitions give rise to the so-called "radiative Auger effect." These transitions, with single-vacancy states as their initial states, give rise to low-energy satellites, often with a distinctive edge appearance [Aberg and Utriainen, 1969; Cooper and LaVilla, 1970; Siivola et al., 1970 (Figs. 1-5 and 1-6)].

In certain situations, larger effects can arise from transitions in which more than one electron changes quantum numbers. This is where there is a change in configuration interaction between the initial and final states (Best, 1967; Cooper and LaVilla, 1970; Aberg, 1971). For example, a series of ^1S states in Ar II can be written as a linear combination of simple configurations (Cooper and LaVilla, 1970)

$$\Psi^1S = a.1s^22s^22p^63s3p^6 + b.1s^22s^22p^63s^23p^43d + c.1s^22s^22p^63s^23p^44d + \cdots$$

rather than each of the ^1S terms appearing as a simple configuration. The L spectrum of argon then (the transitions $L_{2,3} \to M_1$) consists of a multiplet, since there is dipole intensity from the $3s^{-1}$ component in each of the ^1S states. Correlation between x-ray and photoemission data is again expected (Best, 1967; Wertheim and Rosencwaig, 1971).

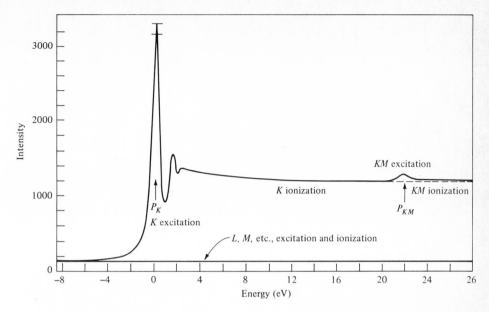

FIGURE 1-4
Corrected *K* absorption of argon gas, showing *K* and *KM* threshold regions.
(*After Schnopper*, 1963.)

Theories for many-electron jumps in atoms have been related to those for metals by Aberg (1971). In particular, the low-energy tail of soft x-ray emission bands of metals has been discussed using two models, each involving two electrons leaving the valence band in the emission process. The theory of Landsberg (1949) attributed the tailing to an internal Auger effect in the final state; this mechanism corresponding to configuration interaction between final single- and double-hole states, as discussed above for atoms. The theory of Pirenne and Longe (1964) for tailing is based on an independent-particle model, which leads to (1-4). These two contributions can interfere with each other (Longe and Glick, 1969).

That completes our main description of electronic processes as they apply to the x-ray spectrum, but there are some additional points we wish to make, even though they will probably be repeated in later, more detailed chapters. For molecules or ions in a solid, if the solid-state broadening is less than the inner-state lifetime width, the x-ray spectrum can be considered as perturbed ionic or molecular spectra. In general, the outer levels of an atomic ion will be affected more than those of a molecular ion, absorption being affected much more than emission in each case.

A general feature of absorption spectra of isolated systems, atoms, and molecules, is the presence of bound-ejected-electron (BEE) states, in which the excited electron is bound by virtue of the attraction to the inner hole (Parratt, 1959). In solids,

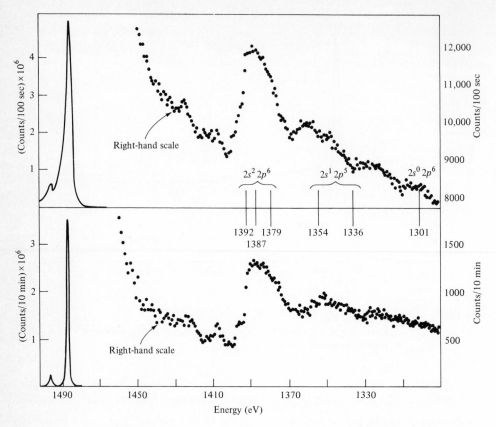

FIGURE 1-5
The low-energy side of the Al $K\alpha$ line in secondary (lower curve) and primary (upper curve) excitation. The vertical lines correspond to the KLL Auger-electron energies. These spectra support the $K \to L^2$ double-transition hypothesis. (*After Siivola, Utriainen, Linkoaho, Graeffe, and Aberg, 1970.*)

the valence electrons adjust to screen the interaction, and thus might effectively eliminate the states (Cauchois and Mott, 1949; Cauchois, 1967). The effectiveness of the screening depends on time factors, the radius of the unperturbed orbital, and the dielectric constant of the solid. It is noted that there is a direct relationship between the dielectric constant and the valence-band widths (Phillips and Van Vechten, 1969), and so BEE states are more likely for solids with narrow bands. Such states will occur more readily if the outer orbital involved has a radius that lies within a Wigner-Seitz cell, i.e., particularly for d and f orbitals.

Final states of the radiative Auger effect occur in which an electron appears to be bound in the presence of two inner holes, rather than one hole as in the case of

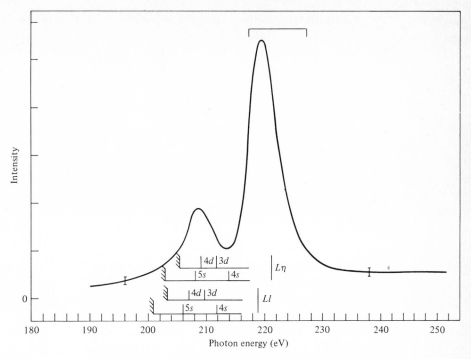

FIGURE 1-6
The emission spectrum in the $L_{2,3}$ region for argon gas. Beneath the spectrum are plotted the positions and limits of the ^2S levels. The low-energy structure is due to configuration interaction between ^2S states involving a $3s$ vacancy. (*After Cooper and LaVilla*, 1970.)

x-ray absorption (Aberg and Utriainen, 1969). The bound-ejected orbital will have a reduced radius due to the extra inner hole. Comparison between the two spectra— x-ray absorption and the radiative Auger emission—may make the study of BEE in solids more quantitative than it is now.

As mentioned above, Wentzel-Druyvesteyn (W-D) satellites result from one-electron jumps between double-vacancy states. They have been observed in atomic (Deslattes, 1964), molecular (LaVilla and Deslattes, 1966*b*; Deslattes and LaVilla, 1967), and solid systems [Deslattes, 1964 (fig. 1-2); Liefeld, 1968]. With adequate input energy, they will always occur for atoms and molecules. In solids this type of satellite will be observed for the case when both vacancies are inner ones, but not always when the spectator hole is in the valence band. For energy reasons, two holes formed at one atomic site tend to separate. If the time for dispersal of the spectator valence hole is less than that of the inner-hole lifetime, then the satellite intensity will be greatly reduced. The time of localization of a valence hole at one site is roughly inversely proportional to the bandwidth, and hence only solids with narrow valence bands will exhibit strong W-D satellites of the above kind. Such satellites involving a spectator valence hole have been observed for some (Deslattes, 1964), but not all (O'Bryan and Skinner, 1940), ionic crystals and for transition-metal compounds having narrow d or f bands (Liefeld, 1968), but not for covalently bonded semiconductors (Wiech, 1967; Zhukova et al., 1969) or "good" metals.

VI. SUMMARY

In this chapter we have presented a catalog of processes for which evidence has been cited in x-ray spectra. In most cases the evidence occurred in a spectral region more or less isolated from other features, and in such circumstances the interpretation of x-ray spectra can be relatively straightforward. However, quite often diagram lines of interest in emission overlap with other lines of more complicated origin. In absorption, there is overlap between features due to both one- and many-electron jumps. In these circumstances careful consideration of all mechanisms possible in the situation is necessary, and sometimes an unambiguous interpretation is not yet possible.

Overall, the last decade has seen a marked improvement in our understanding of x-ray processes in both isolated and condensed systems. Most progress has come in that part of the subject dealing with quasi-stationary states. In these cases, understanding of one- and many-electron-jump processes in both the sudden and adiabatic approximations has increased significantly. For spectra involving nonstationary states (those that are very time dependent) there have been very few studies, and relatively little progress has been made.

APPENDIX

There are a number of topics which are peripheral to the subject of x-ray spectra as discussed here, which are not treated in detail in this book. We mention them, with brief references, for the interested reader. X-ray astronomy is at present one of the most lively of the x-ray fields. In it, rapidly improving techniques have detected radiation from supernovas, from pulsars, and from quasars. Ideally, this involves precisely locating the position of the x-ray source, as well as studying its spectra (Giacconi et al., 1969). The understanding of all these spectra is incomplete. Attempts to explain the growing array of data are a part of the frontier of astrophysics.

Another is the study of x rays from man-made plasmas. Most of the x-ray processes occurring in plasmas are similar to those occurring in "normal" matter, and the x-ray probe can therefore give information about the plasma environment (Henke, 1969).

Although not discussed in this book, the topic of the index of refraction for x rays is an important part of x-ray physics. The x-ray interferometer has proved to be an admirable new instrument for measuring the refractive index, although only a small number of groups are making such measurements (Bonse and Hellköter, 1969; Creagh and Hart, 1970). Refractive indices are used in studies of the total reflection of x rays by matter. In these measurements the intensity of monochromatic radiation is recorded as a function of angle of incidence near the critical angle for total reflection. Structure in the plot contains information about the electron-density distribution normally into the plane surface (Parratt, 1954; Wainfan, Scott, and Parratt, 1959; Guentert, 1965).

Studies of the short-wavelength limit of the Bremsstrahlung spectra, and also of excitation curves (line-emission intensity vs. the input energy near threshold), offer different views of the unoccupied electron orbitals. In the former the high-energy electron is decelerated into vacant conduction levels. The short-wavelength limit of such spectra, for metals, has been associated with the impacting electron coming to rest at the Fermi level (Böhm and Ulmer, 1969). The intensity of structure at slightly longer wavelengths is, to lowest order, dependent on the corresponding density of vacant levels in the conduction band. The results from excitation curves, in addition to yielding information about atomic and solid-state electronic structures, have application as a tool in surface analysis (Park, Houston, and Schreiner, 1970).

ACKNOWLEDGMENTS

I am grateful to Prof. L. G. Parratt for many helpful discussions on this topic and for his critical reading of the manuscript.

REFERENCES

ABERG, T. (1967a): Correlation energy of K-shell electrons, *Phys. Rev.*, **162**: 5–6.

—— (1967b): Theory of x-ray satellites, *Phys. Rev.*, **156**: 35–41.

—— (1969): Multiple excitation of a many-electron system by photon and electron impact in the sudden approximation, *Ann. Acad. Sci. Fennicae*, **Ser. A VI. Physica**, no. 308, 7–46.

—— (1971): Theory of the radiative Auger effect, *Phys. Rev.*, **A4**: 1735–1740.

—— and UTRIAINEN, J. (1969): Evidence for a radiative Auger effect in x-ray photon emission, *Phys. Rev. Lett.*, **22**: 1346–1348.

ALLOTEY, F. K. (1971): Effect of threshold behavior on the calculations of the soft x-ray spectra of lithium, *Solid State Comm.*, **9**: 91–94.

AZÁROFF, L. V. (1963): Theory of extended fine structure of x-ray absorption edges, *Rev. Mod. Phys.*, **35**: 1012–1022.

BAGUS, P. S. (1964): SCF excited states and transition probabilities of some Ne-like and Ar-like ions, *Argonne Natl. Lab. Tech. Rept.*, ANL 6959.

—— (1965): Self-consistent-field wave functions for hole states of some Ne-like and Ar-like ions, *Phys. Rev.*, **A139**: 619–634.

BARINSKII, R. L. (1961): Experimental investigation and calculation of the L absorption spectra of noble gases, *Bull. Acad. Sci. USSR, Phys. Ser. (English Transl.)*, **25**: 958–964.

—— and NEFEDOV, V. I. (1969): *Röntgenspektroskopische bestimmung effectiver Atomladungen*, Akademische Verlagsges, Leipzig.

BEARDEN, J. A. (1967): X-ray wavelengths, *Rev. Mod. Phys.*, **39**: 78–124.

—— and BURR, A. F. (1967): Reevaluation of x-ray atomic energy levels, *Rev. Mod. Phys.*, **39**: 125–142.

BEST, P. E. (1967): X-ray absorption in molecules; SF_6, *J. Chem. Phys.*, **47**: 4002–4006.

—— (1968): Electronic structures from x-ray spectra. II. Mostly $ClO_3{}^-$ and $ClO_4{}^-$, *J. Chem. Phys.*, **49**: 2797–2805.

—— (1971): K x-ray emission from KCl, *Phys. Rev.*, **B3**: 4377–4382.

BLOKHIN, M. A. (1957): *The physics of x-rays*, 2d ed. (State Publishing House of Technical-Theoretical Literature, Moscow), English translation by U.S. Atomic Energy Commission, document AEC-tr-4502.

—— and NIKIFOROV, I. Y. (1964): Shape of the $K\alpha_{1,2}$ lines of iron-group elements, *Bull. Acad. Sci. USSR, Phys. Ser. (English Transl.)*, **27**: 689–694.

BÖHM, G., and ULMER, K. (1969): Energy dependence of isochromat structures of tungsten for energies of 0.15 to 6 keV, *Z. Physik.*, **228**: 473–488.

BONNELLE, C., and JØRGENSEN, C. K. (1964): Influence de l'operateur biélectronique sur les transitions $2p \rightarrow 3d$ dans l'oxyde de nickel II, *J. Chim. Phys.*, **61**: 826–829.

BONSE, U., and HELLKÖTER, H. (1969): Interferometrische Messung des Brechungsindex für Röntgenstrahlen, *Z. Physik.*, **223**: 365–352.

BROUERS, F., LONGE, P., and BERGERSEN, B. (1970): The effect of the core hole on the shape of the soft x-ray spectra, *Solid State Comm.*, **8**: 1423–1426.

BROWN, F. C., GÄHWILLER, C., FUJITO, H., KUNZ, A. B., SCHEIFLEY, W., and CARRERA, N. (1970): Extreme ultraviolet spectra of ionic crystals, *Phys. Rev.*, **B2**: 2126–2138.

BURCH, D., RICHARD, P., and BLAKE, R. C. (1971): Resolved structure in Fe $K\alpha$ x-rays produced by 30 MeV oxygen ions, *Phys. Rev. Lett.*, **26**: 1355–1359.

CARLSON, T. A. (1967): Double electron ejection resulting from photoionization in the outermost shell of He, Ne, and Ar, and its relationship to electron correlation, *Phys. Rev.*, **156**: 142–149.

—— and KRAUSE, M. O. (1965): Electron shake-off resulting from K-shell ionization in neon measured as a function of photoelectron velocity, *Phys. Rev.*, **A140**: 1057–1064.

CATZ, A. L. (1970): Some evidence for admixture of magnetic quadrupole radiation in atomic x-ray transitions from measurements of angular correlation of x-rays, *Phys. Rev. Lett.*, **24**: 127–130.

CAUCHOIS, Y. (1948): *Les spectres des rayons X et la structure électronique de la matière*, Gauthier-Villars, Paris.

—— (1967): Rappel de quelques données générales sur les spectres, *J. Phys. Radium*, **28**: Suppl. C3-59 to C3-64.

CAUCHOIS, Y., and MOTT, N. F. (1949): The interpretation of x-ray absorption spectra of solids, *Phil. Mag.*, **40**: 1260–1269.

CODLING, K., MADDEN, R. P., and EDERER, D. L. (1967): Resonances in the photo-ionization continuum of NeI (20–150 eV), *Phys. Rev.*, **155**: 26–37.

COMPTON, A. H., and ALLISON, S. K. (1935): *X-rays in theory and experiment*, 2d ed., D. Van Nostrand Company, Inc., Princeton, N.J.

COOPER, J. W. (1962): Photoionization from outer atomic subshells: A model study, *Phys. Rev.*, **128**: 681–693.

—— and LAVILLA, R. E. (1970): Semi-Auger processes in $L_{2,3}$ emission in Ar and KCl, *Phys. Rev. Lett.*, **25**: 1745–1748.

CREAGH, D. C., and HART, M. (1970): X-ray interferometric measurements of the forward scattering amplitude for lithium fluoride, *Phys. Stat. Sol.*, **37**: 753–758.

CUTHILL, J. R., MCALISTER, A. J., and WILLIAMS, M. L. (1967): Density of states of Ni: Soft x-ray spectrum and comparison with photoemission and ion neutralization studies, *Phys. Rev.*, **164**: 1006–1017.

DEHMER, J. L. (1971): Evidence of effective barriers in the x-ray absorption spectra of molecules, *J. Chem. Phys.*, **56**: 4496–4504.

DESLATTES, R. D. (1964): $K\beta$ spectra of argon and KCl. II. Satellite excitation, *Phys. Rev.*, **A133**: 399–407.

—— (1969): Relative energy measurements in the K series of argon, *Phys. Rev.*, **186**: 1–4.

—— and LAVILLA, R. E. (1967): Molecular emission spectra in the soft x-ray region, *Appl. Opt.*, **6**: 39–42.

EKSTIG, B., KÄLLNE, E., NORELAND, E., and MANNE, R. (1970): Electron interaction in transition metal x-ray emission spectra, *Physica Scripta*, **1**: 1–7.

FADLEY, C. S., and SHIRLEY, D. A. (1970): Multiplet splitting of metal-atom electron binding energies, *Phys. Rev.*, **A2**: 1109–1120.

FANO, U. (1961): Effects of configuration interaction on intensities and phase shifts, *Phys. Rev.*, **124**: 1866–1877.

—— and COOPER, J. W. (1968): Spectral distribution of atomic oscillator strengths, *Rev. Mod. Phys.*, **40**: 441–507.

FRIEDEL, J. (1969): X-ray absorption and emission edges in metals, *Comm. Solid State Phys.*, **2**: 21–29.

GIACCONI, R., REIDY, W. P., VAIANA, G. S., VAN SPEYBROECK, L. P., and ZEHNPFENNIG, T. F. (1969): Grazing-incidence telescopes for x-ray astronomy, *Space Sci. Rev.*, **9**: 3–57.

GILBERG, E. (1970): Das K-Röntgenemissionsspektrum des Chlors in freien Molekulen, *Z. Physik.*, **236**: 21–41.

GOULDING, F. S., and STONE, Y. (1970): Semiconductor radiation detectors, *Science*, **170**: 280–289.

GRAEFFE, G., SIIVOLA, J., UTRIAINEN, J., LINKOAHO, M., and ABERG, T. (1969): X-ray $K\alpha$ satellite spectra in primary and secondary excitation, *Phys. Lett.*, **A29**: 464–465.

GREGORY, T. K., and BEST, P. E. (1971): An automated two-crystal spectrometer employing direct angular positioning and readout, *Advan. X-Ray Anal.*, **15**: 90–101.

GUENTERT, O. J. (1965): Study of the anomalous surface reflection of x-rays, *J. Appl. Phys.*, **36**: 1361–1366.

HANSON, H. P., and HERRERA, J. (1957): Self-absorption in the x-ray spectroscopy of valence electrons, *Phys. Rev.*, **105**: 1483–1485.

HANZELY, S., and LIEFELD, R. J. (1969): An *L*-series x-ray spectroscopic study of the valence bands in iron, cobalt, nickel, copper and zinc, L. H. Bennett (ed.), in *Electronic density of states*, p. 319, Natl. Bur. Std. Spec. Publ. 323.

HENKE, B. L. (1969): An introduction to low energy x-ray and electron analysis, *Advan. X-Ray Anal.*, **13**: 1–25.

HORAK, Z. (1961): On the identification of the *K*α satellites. I. *LS* region, *Proc. Phys. Soc. (London)*, **77**: 980–986.

HUBBELL, J. H., MCMASTER, W. H., KERR DEL GRANDE, N., and MALLETT, J. H. (1973): Chap. 2 in Volume 4, *International tables for x-ray crystallography*, The International Union of Crystallography.

KNOX, R. S. (1958): Exciton states in solid argon, Ph.D. thesis, University of Rochester, Rochester, N.Y. (unpublished).

KNUDSON, A. R., NAGEL, D. J., BURKHALTER, P. G., and DUNNING, K. L. (1971): Aluminum x-ray satellite enhancement by ion impact excitation, *Phys. Rev. Lett.*, **26**: 1149–1152.

KOOPMANS, T. (1934): Uber die Zuordnung von Wellenfunktionen und Eigenwerten zu den einzelnen Elektronen eines Atoms, *Physica*, **1**: 104–113.

KOSTROUN, V. O., CHEN, M. H., and CRASEMAN, B. (1971): Atomic radiation transition probabilities to the 1*s* state and theoretical *K*-shell fluorescence fields, *Phys. Rev.*, **A3**: 533–545.

KRAUSE, M. O., and CARLSON, T. A. (1966): Charge distribution of krypton ions following photo-ionization in the *M* shell, *Phys. Rev.*, **149**: 52–61.

————, STEVIE, F. A., LEWIS, L. F., CARLSON, T. A., and MODDEMAN, W. E. (1970): Multiple excitation of neon by photon and electron impact, *Phys. Lett.*, **A31**: 81–82.

KUHN, H. G. (1969): *Atomic spectra*, chap. 5, Academic Press, Inc., New York.

KUNZ, A. B. (1970): Energy bands and the optical properties of LiCl, *Phys. Rev.*, **B2**: 5015–5024.

LANDSBERG, P. T. (1949): A contribution to the theory of soft x-ray emission bands of sodium, *Proc. Phys. Soc. (London)*, **A62**: 806–816.

LASSETTRE, E. N., and FRANCIS, S. A. (1964): Inelastic scattering of 390-V electrons by helium, hydrogen, methane, ethane, cyclohexane, ethylene and water, *J. Chem. Phys.*, **40**: 1208–1217.

LAVILLA, R. E., and DESLATTES, R. D. (1966a): *K*-absorption fine structures of sulfur in gaseous SF_6, *J. Chem. Phys.*, **44**: 4399–4400.

———— and ———— (1966b): Chlorine *K*β x-ray emission spectra from several chlorinated hydrocarbon and fluorocarbon molecular gases, *J. Chem. Phys.*, **45**: 3446–3448.

LEONHARDT, G., and MEISEL, A. (1970): Determination of effective atomic charges from the chemical shifts of x-ray emission lines, *J. Chem. Phys.*, **52**: 6189–6198.

LIEFELD, R. J. (1968): Soft x-ray emission spectra at threshold excitation, in D. J. Fabian (ed.), *Soft x-ray band spectra and the electronic structures of metals and materials*, Academic Press, Inc., London, pp. 133–149.

———— (1971): Private communication.

LONGE, P., and GLICK, A. J. (1969): Electron interaction effects on the soft x-ray emission spectrum of metals. I. Formalism and first-order theory, *Phys. Rev.*, **177**: 526–539.

MANNE, R., and ABERG, T. (1970): Koopmans' theorem for inner shell ionization, *Chem. Phys. Lett.*, **7**: 282–284.

MANSON, S. T., and COOPER, J. W. (1968): Photo-ionization in the soft x-ray range: Z-dependence in a central potential model, *Phys. Rev.*, **165**: 126–138.

MCGUIRE, E. J. (1968): Photo-ionization cross sections of the elements helium to xenon, *Phys. Rev.*, **175**: 20–30.

——— (1970): *K*-shell Auger transition rates—Fluorescence yields for elements Ar-Xe, *Phys. Rev.*, **A2**: 273–278.

NEFEDOV, V. I. (1970): Quasistationary states in x-ray absorption spectra of chemical compounds, *J. Struct. Chem. (USSR) (English Transl.)*, **11**: 272–276.

O'BRYAN, H. M., and SKINNER, H. W. B. (1940): The soft x-ray spectroscopy of solids. II. Emission spectra from simple chemical compounds, *Proc. Roy. Soc. (London)*, **176**: 229–262.

OVERHAUSER, A. W. (1959): Quoted by Parratt (1959).

PADALIA, B. D., and RAO, S. K. (1969): Classification of forbidden transitions in x-ray spectra, *J. Phys.*, **B2**: 134, 136.

PARK, R. L., HOUSTON, J. E., and SCHREINER, D. G. (1970): A soft x-ray appearance potential spectrometer for the analysis of solid surfaces, *Rev. Sci. Instr.*, **41**: 1810–1812.

PARRATT, L. G. (1954): Surface studies of solids by total reflection of x-rays, *Phys. Rev.*, **95**: 359–369.

——— (1959): Electronic band structure of solids by x-ray spectroscopy, *Rev. Mod. Phys.*, **31**: 616–645.

PHILLIPS, J. C., and VAN VECHTEN, J. A. (1969): Spectroscopic analysis of cohesive energies and heats of formation of tetrahedrally coordinated semiconductors, *Phys. Rev.*, **B2**: 2147–2156.

PIECH, K. R., and LEVINGER, J. S. (1964): Sum rules for neon photoeffect, *Phys. Rev.*, **A135**: 332–334.

PIRENNE, J., and LONGE, P. (1964): Contribution of the double electron transitions to the soft x-ray emission bands of metals, *Physica*, **30**: 277–291.

PORTEUS, J. O. (1962): Optimized method for correcting smearing aberrations: Complex x-ray spectra, *J. Appl. Phys.*, **33**: 700–707.

RAO, P. V., PALMS, J. M., and WOOD, R. E. (1971): High-*Z* *L*-subshell x-ray emission rates, *Phys. Rev.*, **A3**: 1568–1575.

ROSNER, H. R., and BHALLA, C. P. (1970): Relativistic calculation of atomic x-ray transition rates, *Z. Physik.*, **231**: 347–356.

RUDERMAN, I. W., NESS, K. J., and LINDSAY, J. C. (1965): Analyzer crystals for x-ray spectroscopy in the region 25–100 A, *Appl. Phys. Lett.*, **7**: 17–19.

SACHENKO, V. P., and DEMEKHIN, V. F. (1966): Satellites of x-ray spectra, *Soviet Phys. JETP (English Transl.)*, **22**: 532–535.

SANDSTRÖM, A. E. (1957): Experimental methods of x-ray spectroscopy: ordinary wavelengths, in S. Flügge (ed.), *Handbuch der physik*, vol. 30, pp. 78–245, Springer-Verlag, Berlin.

SATO, S., ISHIRI, T., NAGAKURA, I., AITA, O., NAKAI, S., YOKOTA, M., ICHIKAWA, K., MATSUOKA, G., KONO, S., and SAGAWA, T. (1971): Soft x-ray absorption spectra of some metal chlorides, *J. Phys. Soc. Japan*, **30**: 459–469.

SAWADA, T., TSUTSUMI, K., SHIRAIWA, T., ISHIMURA, T., and OBASHI, M. (1959): Some contributions to the x-ray spectroscopy of solid state, *Ann. Rept. Sci. Works, Fac. Sci., Osaka Univ.*, **7**: 1–87.

SCHIFF, L. I. (1955): *Quantum mechanics*, 2d ed., McGraw-Hill Book Company, New York.

SCHNOPPER, H. W. (1963): Multiple excitation and ionization of inner atomic shells by x-rays, *Phys. Rev.*, **131**: 2558–2560.

――― (1966): Chlorine K absorption edge in single crystal $KClO_3$ with polarized x-rays, in A. Meisel (ed.), *Röntgenspektren und chemische bindung*, pp. 303–313, Physikalisch-Chemisches Institut der Karl-Marx-Universitat, Leipzig.

――― (1967): Atomic readjustment to an inner-shell vacancy: Manganese K x-ray emission spectra from an Fe^{55} K-capture source and from the bulk metal, *Phys. Rev.*, **154**: 118–123.

SCOFIELD, J. H. (1969): Radiative decay rates of vacancies in the K and L shells, *Phys. Rev.*, **179**: 9–16.

SHAW, C. H. (1956): The x-ray spectroscopy of solids, in *Theory of alloy phases*, pp. 13–62, American Society for Metals, Cleveland.

SHIRAIWA, T., ISHIMURA, T., and SAWADA, M. (1958): The theory of the fine structure of the x-ray absorption spectrum, *J. Phys. Soc. Japan*, **13**: 847–859.

SIEGBAHN, K., et al. (1967): ESCA atomic molecular and solid-state structure studied by means of electron spectroscopy, *Nova Acta Regiae Soc. Sci. Upsalien.*, **Ser. IV**: 20.

SIEGBAHN, M. (1931): *Spectroskopie der Röntgenstrahlen, Zweite Auflage*, 2d ed., Julius Springer, Berlin.

SIIVOLA, J., UTRIAINEN, J., LINKOAHO, M., GRAEFFE, G., and ABERG, T. (1970): The low-energy structure of the K line in primary and secondary excitation, *Phys. Lett.*, **A32**: 438–439.

SMITHER, R. K., FREEDMAN, M. S., and PORTER, F. T. (1970): Intensity of K-L_1 x-ray transition, *Phys. Lett.*, **A32**: 405–407.

STOTT, M. J. (1969): The effect of localized states on the impurity soft x-ray spectrum of dilute alloys, *J. Phys.*, **C2**: 1474–1486.

SWANSON, N., and POWELL, C. J. (1968): Excitation of L-shell electrons in Al and Al_2O_3 by 20-keV electrons, *Phys. Rev.*, **167**: 592–600.

TOMBOULIAN, D. H. (1957): The experimental methods of soft x-ray spectroscopy and the valence band of light elements, in S. Flügge (ed.), *Handbuch der physik*, vol. 30, pp. 246–304, Springer-Verlag, Berlin.

TSUTSUMI, K., and NAKAMORI, H. (1968): X-ray K emission spectra of chromium in various chromium compounds, *J. Phys. Soc. Japan*, **25**: 1418–1423.

VINOGRAD, A. S., and ZIMKINA, T. M. (1970): Singularities of the x-ray absorption spectrum of lithium and fluorine in LiF crystals, *Soviet Phys.-Solid State* (*English Transl.*), **12**: 1147–1151.

WAINFAN, N., SCOTT, N. J., and PARRATT, L. G. (1959): Density measurements of some thin copper films, *J. Appl. Phys.*, **30**: 1604–1609.

WATANABE, M., YAMASHITA, H., NAKAI, Y., SATO, S., and ONARI, S. (1971): Polarization effect in the soft x-ray absorption of CdS, *Phys. Stat. Sol.*, **B43**: 631–636.

WATANABE, T. (1965): Theoretical fitting of the argon *K* absorption spectrum on a one-electron model, *Phys. Rev.*, **A139**: 1747–1751.

WEISS, A. W. (1969): Series perturbations and atomic oscillator strengths: The ^2D series of Al I, *Phys. Rev.*, **178**: 82–89.

WERTHEIM, G. K., and ROSENCWAIG, A. (1971): Configuration interaction in the x-ray photo-electron spectra of alkali halides, *Phys. Rev. Lett.*, **26**: 1179–1182.

WIECH, A. (1967): Röntgenspektroskopische Untersuchung der Struktur der valenz Bandes von Silicium, Silicium Carbid und Silicium Dioxid, *Z. Physik.*, **207**: 428–445.

WUILLEUMIER, F., and BONNELLE, C. (1963): Processes électroniques multiples dans le spectre *K* du neon, *Compt. Rend.*, **B270**: 1029–1032.

ZHUKOVA, I. I., FOMICHEV, V. A., VINOGRAD, A. S., and ZIMKINA, T. M. (1969): Investigation of the energy structure of silicon carbide and silicon nitride by ultrasoft x-ray spectroscopy, *Soviet Phys.-Solid State (English Transl.)*, **10**: 1097–1103.

2

HIGH-PRECISION X-RAY SPECTROSCOPY

John S. Thomsen

I. INTRODUCTION

A. Scope of field

High-precision x-ray spectroscopy implies a measurement of x-ray intensity as a function of wavelength with as great an accuracy as is experimentally feasible. The most common application is in the measurement of the profile of a characteristic x-ray emission line; the primary output of such a measurement is usually the wavelength. Line width and line asymmetries can also be obtained in measurements of this type. The investigation of the continuous x-ray spectrum, i.e., Bremsstrahlung, does not usually require such high precision; however, in certain applications, e.g., determination of the high-frequency limit, it is essential.

   Extensive treatments of this field have been given by Siegbahn (1925; 1931), Compton and Allison (1935), and Sandström (1957). In the present chapter we will review the subject with particular emphasis on corrections and sources of errors which must be considered in obtaining highest accuracy.

B.   Relative and absolute measurements

Most x-ray spectroscopic measurements are obtained with crystal spectrometers and are made relative to some length standard of atomic dimensions rather than directly in terms of SI (metric) units. The relative standard is usually of the same order of magnitude as the x-ray wavelength itself. This relative standard may be either the grating constant of a crystal (usually calcite) or a wavelength (e.g., the W $K\alpha_1$ or Mo $K\alpha_1$ line). Such measurements can be carried out on a given instrument with a reproducibility of the order of 1 ppm (part per million).

Direct absolute measurements of x-ray wavelengths have thus far been made only by means of ruled gratings. These have only been practicable near grazing incidence, where the grating surface is almost totally reflecting and the grating space need not be unduly small. Such measurements are more difficult and time-consuming than relative ones. (Indirect determinations may be carried out by a number of methods; these will be discussed in Section VIII.)

Of course, accuracy requires more than mere reproducibility. On this basis relative measurements are not always as reliable as might be inferred from reproducibility of the data; the true accuracy is likely to be substantially lower. Nevertheless, at the present time relative measurements are still substantially more accurate than absolute ones. The vast majority of x-ray spectroscopic results are relative measurements made with crystal spectrometers. In the present treatment primary emphasis will be given to such instruments.

C.   Types of crystal instruments

Crystal spectrometers are all based on the Bragg law

$$n\lambda = 2d_n \sin \theta_n \qquad (2\text{-}1)$$

where n is the order of diffraction, λ the wavelength, d_n the crystal grating constant for the nth order (the small dependence on n being due to index of refraction), and θ_n the angle between the incident ray and the diffracting plane of the crystal for the nth order.

The crystal spectrometer is designed to measure the angle θ_n. For high-precision work it is desirable to minimize the number of critical adjustments and the number of precision scales involved. Hence it is convenient to design an instrument in which the crystal is rotated and the angle of rotation measured directly on a carefully calibrated divided circle and for which no other motions need be known to high accuracy.

Three forms of such an instrument have been used. In the Siegbahn "tube spectrometer" the beam which is Bragg-diffracted from the crystal passes through a slit, followed by a long tube, and is recorded on a photographic plate. To complete a measurement, the x-ray source is moved from one position to another, but its exact position is not critical; the Bragg angle is derived primarily from the measured angle of rotation of the crystal, with a small correction determined by measurements on the photographic plate.

The Bragg spectrometer is essentially the inverse of the tube type. The slit system comes before the crystal, with the x-ray source being kept stationary. The detector (either an ionization chamber or a counter) is rotated, but its position is not critical. The only precision measurement required is that of the rotation of the crystal itself.

The two-crystal spectrometer is essentially a modification and extension of the Bragg type. In the usual two-crystal instrument the slit system is replaced by a stationary first crystal, adjusted so that a portion of the radiation from the x-ray source in the wavelength region under investigation is incident on it at the Bragg angle. All radiation appreciably reflected by the crystal then leaves at or near the appropriate Bragg angle; hence the radiation composing a particular x-ray line forms an almost parallel beam. This beam then strikes the second crystal, which is used to measure wavelength in the same way as in the Bragg instrument.

D.   Experimental results

Most of the high-precision x-ray spectroscopic work is directed toward obtaining one of two types of data:

1 Wavelengths corresponding to some characteristic feature (usually the peak) of emission lines.

2 Wavelengths corresponding to some characteristic feature (usually the half-intensity point or the inflection point) of absorption edges.

These wavelengths provide important working tools for crystallography and for low-energy nuclear spectroscopy. They are also valuable in computing atomic energy levels for the various excited states involved in x-ray transitions. In most cases such wavelength data (preferably supplemented by photoelectron measurements, which are roughly equivalent to absorption-edge determinations but usually more accurate) furnish an overdetermined set of equations for the values of atomic energy levels. These levels can then be obtained by a statistical adjustment, e.g., by a least-squares analysis.

X-ray spectroscopy is also needed in more specialized problems, e.g., the determination of the high-frequency cutoff of the continuous spectrum in the so-called *h/e* experiment. It is, of course, applicable to determining other parameters of the emission lines, such as width and asymmetry. In addition, the detailed structure of emission spectra has been studied in investigations of the band structure of solids. However, these applications will not be treated in the present chapter.

II.   SPECTROSCOPIC WAVELENGTH MEASUREMENT

A.   Wavelength profile

For high-precision work (or, in fact, for any wavelength measurement with a resolution substantially better than one part per thousand) an x-ray line cannot be considered as a monochromatic wavelength, but rather must be recognized as having a wavelength profile. The intensity in a small range $d\lambda$ may be described by the expression

$$\mathscr{J}(\lambda - \lambda_0)\, d\lambda \qquad (2\text{-}2)$$

where λ_0 is the wavelength at the peak of the profile and $\mathscr{J}(\lambda - \lambda_0)$ is the intensity distribution. In many cases this function is closely approximated by the classical Lorentzian expression

$$\mathscr{J}(\lambda - \lambda_0) = \frac{A}{1 + [2(\lambda - \lambda_0)/w]^2} \qquad (2\text{-}3)$$

where w represents the width, i.e., the full width of the line profile at half-maximum intensity. [For a classical derivation, see Compton and Allison (1935); for a quantum-mechanical treatment, see Weisskoff and Wigner (1930) or Brogren (1954).] However, this function is not universally valid; it is clearly inapplicable to the numerous x-ray line profiles which possess appreciable asymmetry. The exact profile may have some dependence on conditions of excitation and angle of observation, particularly in the soft x-ray region.

B.   Wavelength criterion

It should be noted that the ratio of line width to wavelength (i.e., w/λ) usually lies in the range of 10^{-4} to 10^{-3}. This is much larger than that for most optical lines. Hence in high-precision work, particularly in the case of asymmetric lines, it is essential to specify some particular feature of the line profile to define wavelength.

Among the possible choices for a wavelength criterion are (1) the *peak* of the line, (2) the *centroid* of a specified truncated portion of the line profile, (3) the *median* of a specified truncated portion of the line profile, or (4) the *midpoint of the chord* drawn at a specified fraction of the peak intensity. The peak has been the most commonly used criterion in x-ray spectroscopy; some crystallographers have found the centroid more useful for their particular applications. The various choices have been compared in a detailed study by Thomsen and Yap (1968). They found that some reduction in statistical error could be attained by using the centroid or median with a rather narrow truncation range ($\sim w$). Since such a definition would redefine wavelengths of all asymmetric lines and appeared lacking in any fundamental physical significance, they recommended retention of the generally accepted peak. Hence, in the remainder of the present treatment the term "wavelength" will always be understood to denote the *peak* wavelength of the line profile.

C. Instrumental window

The true line profile, discussed above in Subsection II A, is always modified to some extent by the properties of the spectrometer; in many cases the change may be quite significant. The result is usually approximated by the expression

$$\mathscr{F}(\tilde{\lambda}) = \int_0^\infty \mathscr{W}(\tilde{\lambda} - \lambda)\mathscr{J}(\lambda)\, d\lambda \qquad (2\text{-}4)$$

In this expression $\tilde{\lambda}$ denotes the "spectrometer setting" (to be defined more precisely in the discussion of various types of spectrometers), $\mathscr{F}(\tilde{\lambda})$ the corresponding intensity, and $\mathscr{W}(\tilde{\lambda} - \lambda)$ the *spectrometer window function*. Actually this window is a function of λ as well as $\tilde{\lambda} - \lambda$. Such dependence is usually negligible over a single wavelength profile $\mathscr{J}(\lambda)$; however, for a different spectral range a different window function must be used.

It is often more convenient to work in terms of angles rather than wavelengths. For this purpose the angular setting of the spectrometer will be denoted as β, where $\beta = 0$ corresponds to the spectrometer setting for the peak wavelength λ_0. We also define a new variable z by the relation

$$z = \left(\frac{d\beta}{d\lambda}\right)_{\lambda_0} (\lambda - \lambda_0) \approx \beta_\lambda \qquad (2\text{-}5)$$

where β_λ is the spectrometer setting corresponding to λ. Equation (2-4) may now be rewritten as

$$F(\beta) = \int_{-\infty}^\infty W(\beta - z)J(z)\, dz \qquad (2\text{-}6)$$

where the lower limit has been written as $-\infty$ for mathematical convenience. (The effect on the integral is negligible due to the asymptotic behavior of the integrand.)

The problem of obtaining the true line profile now consists of computing $J(z)$, given the observed curve $F(\beta)$ and the instrumental window $W(\beta - z)$. In its full generality this is a difficult task. The subject has recently been treated in extended analyses by Porteus (1962) and Sauder (1966); their papers include brief summaries of earlier work in the field.

D.   Peak-wavelength correction

If we assume that the line profile is substantially wider than the instrumental window (e.g., the ratio of full widths at half-maximum is roughly 3:1 or greater), it becomes relatively easy to obtain a first approximation for the peak-wavelength correction. As a first step we introduce the new variable

$$t \equiv \beta - z \qquad (2\text{-}7)$$

and transform (2-6) to

$$F(\beta) = \int_{-\infty}^{\infty} W(t)J(\beta - t)\, dt \qquad (2\text{-}8)$$

We now expand $J(\beta - t)$ as a Taylor series around the peak (i.e., $\beta - t = 0$), retaining terms through the second order. This yields

$$F(\beta) = \int_{-\infty}^{\infty} W(t)[J_p + \tfrac{1}{2}(\beta - t)^2 J_p'']\, dt \qquad (2\text{-}9)$$

where the subscript p denotes peak values of J and its derivatives, the J_p'' term vanishing at the peak.

To determine the observed peak β_p, this expression may be differentiated and equated to zero, i.e.,

$$F'(\beta_p) = \int_{-\infty}^{\infty} W(t)[(\beta_p - t)J_p'']\, dt = 0 \qquad (2\text{-}10)$$

Since J_p'' is a constant, this yields

$$\beta_p = \frac{\int_{-\infty}^{\infty} tW(t)\, dt}{\int_{-\infty}^{\infty} W(t)\, dt} \qquad (2\text{-}11)$$

i.e., the observed peak occurs at the centroid of $W(t)$. This is in agreement with the more general approach of Sauder (1966), who obtained the centroid as the lowest-order peak correction.

E.   Crystal window

For real crystals, Bragg reflection of a given wavelength is not restricted to a single angle, but is significant over a narrow angular range. The fraction reflected may be represented by a function $C_n(\theta - \theta_n)$, where θ is the actual angle of incidence and θ_n is the Bragg angle for the nth order, corrected for index of refraction.

The first theoretical derivation of this function was carried out by Darwin (1914), whose results for the nth-order window function give

$$C_n(\theta - \theta_n) = \begin{cases} 1 & |\theta - \theta_n| \leq k_n \\ \left[\left|\dfrac{\theta - \theta_n}{k_n}\right| + \sqrt{\left(\dfrac{\theta - \theta_n}{k_n}\right)^2 - 1}\right]^{-2} & |\theta - \theta_n| > k_n \end{cases} \quad (2\text{-}12)$$

where k_n is a parameter proportional to $\tilde{w}_n$, the width of the curve ($\tilde{w}_n \approx 2.12\, k_n$), and $\theta - \theta_n$ represents the angular deviation from the Bragg angle θ_n. (As used here, "width" denotes *full width at half-intensity*.)

The quantity k_n is given by

$$k_{n\perp} \equiv \frac{4|F_H|\,\delta}{F_0 \sin 2\theta_n} \qquad \begin{array}{l}(\sigma \text{ polarization, electric vector perpendicular}\\ \quad \text{to plane of incidence})\end{array} \qquad (2\text{-}13)$$

or

$$k_{n\parallel} \equiv \frac{4|F_H|\,\delta}{F_0|\tan 2\theta_n|} \qquad \begin{array}{l}(\pi \text{ polarization, electric vector parallel to}\\ \quad \text{plane of incidence})\end{array} \qquad (2\text{-}14)$$

F_H and F_0 denote, respectively, the structure factors of the cell for the reciprocal-lattice vector of the reflecting plane and for the zero reciprocal-lattice vector, i.e., the d.c. value. [For a detailed discussion, see Compton and Allison (1935) or Batterman and Cole (1964).] The quantity δ is the decrement of the index of refraction, i.e.,

$$\delta = 1 - \mu \qquad (2\text{-}15)$$

where μ is the index of refraction (usually $0 < \delta \ll 1$ for x rays).

If there is appreciable absorption in the crystal, the Darwin function given above no longer holds exactly. A modified form including absorption effects was given by Prins (1930). His result indicated that the function was then no longer symmetric, but exhibited lower reflectivity on the long-wavelength side. This result was experimentally verified, e.g., by Bearden, Marzolf, and Thomsen (1968). The theoretical analysis has been recently elaborated in the comprehensive treatment of Batterman and Cole (1964).

If absorption is negligible and the Darwin function is a satisfactory representation, then the centroid will be located at the origin of the crystal window and no correction will be needed (other than index of refraction, which is discussed in the following section). However, in the more general case, where the Darwin-Prins function applies, the centroid will usually occur on the short-wavelength side of the origin. Its position may be computed from the rather complicated theoretical formulas or from experimental measurements such as those of Bearden, Marzolf, and Thomsen (1968). Furthermore, if the crystal window is comparatively wide, the centroid correction may be insufficient and a more refined technique needed, such as that given by Sauder (1966).

F.   Crystal corrections: refractive index and temperature

If we assume that the simple Bragg equation holds in the interior of the crystal and that the usual optical principles of refraction apply, then the refractive-index correction is easily derived. Let the true grating constant be denoted by d_∞ (so designated since it will be shown that the effective nth-order constant $d_n \to d_\infty$ as $n \to \infty$) and let λ' and θ'_n denote wavelength and Bragg angle, respectively, *inside* the crystal (analogous to λ in air and θ_n at the surface). Then $\lambda' = \lambda/\mu$ and $\cos \theta'_n = (\cos \theta_n)/\mu$ (note that θ_n is the complement of the optical angle of incidence). Since equation (2-1) is assumed to hold precisely inside the crystal, i.e., for λ' and θ'_n, we then obtain

$$\frac{n\lambda}{\mu} = 2d_\infty \sin \theta'_n = 2d_\infty \left(1 - \frac{\cos^2 \theta_n}{\mu^2} \right)^{1/2} \qquad (2\text{-}16)$$

It follows that

$$n\lambda = 2d_\infty \left(\frac{\mu^2 - \cos^2 \theta_n}{\sin^2 \theta_n} \right)^{1/2} \sin \theta_n \qquad (2\text{-}17)$$

This reduces to equation (2-1), and thus takes the same form as the simple Bragg equation if we define

$$d_n = d_\infty \left(\frac{\mu^2 - \cos^2 \theta_n}{\sin^2 \theta_n} \right)^{1/2} \qquad (2\text{-}18)$$

With the help of (2-15) this reduces to

$$d_n = d_\infty \left(1 - \frac{2\delta - \delta^2}{\sin^2 \theta_n} \right)^{1/2} \approx d_\infty \left(1 - \frac{\delta}{\sin^2 \theta_n} \right) \qquad (2\text{-}19)$$

since $\delta \ll 1$. Substitution of (2-1) to eliminate $\sin \theta_n$ now yields

$$d_n = d_\infty \left(1 - \frac{4d_\infty^2}{n^2} \frac{\delta}{\lambda^2} \right) \qquad (2\text{-}20)$$

For most crystals the quantity δ/λ^2 is approximately constant over a range of wavelengths provided that no absorption edges occur. Under this condition the fractional correction to the grating constant, and hence to the wavelength, is independent of the wavelength, but inversely proportional to the square of the order.

In all high-precision work it is, of course, necessary to consider thermal expansion of the crystal. This may be accounted for by use of the usual expression

$$d_\infty = d_{0\infty}(1 + \tilde{\alpha}\Delta t) \qquad (2\text{-}21)$$

where $\tilde{\alpha}$ is the linear coefficient of expansion, $d_{0\infty}$ is the grating constant at a reference temperature (18 or 25°C for most published crystal data), and Δt is the temperature rise above the reference value. It should be noted that, if the crystal is anisotropic, a different value of $\tilde{\alpha}$ will be required for each reflecting plane. For example, in a biaxial crystal such as calcite with temperature coefficients $\tilde{\alpha}_{\parallel}$ and $\tilde{\alpha}_{\perp}$, respectively, parallel to and perpendicular to the optic axis, the effective coefficient is given by

$$\tilde{\alpha} = \tilde{\alpha}_{\parallel} \sin^2 \phi + \tilde{\alpha}_{\perp} \cos^2 \phi \qquad (2\text{-}22)$$

where ϕ denotes the angle between the plane of reflection and the optic axis (i.e., the normal to the plane of reflection forms an angle $\pi/2 - \phi$ with the optic axis).

III.   SINGLE-CRYSTAL SPECTROMETERS

A.   Elementary theory

There are two types of high-precision single-crystal spectrometers: the Bragg spectrometer and the tube spectrometer. In order to describe the basic principles of these instruments, idealized cases will first be considered. The axis of rotation of the crystal may always be assumed vertical, without loss of generality. In addition, we will *temporarily* make the following oversimplifying assumptions:

1   The x-ray source is monochromatic.
2   The radiation incident on the crystal consists of a geometrical ray in the horizontal plane.
3   The diffracting plane of the crystal is parallel to the axis of rotation.

Schematic diagrams of the two instruments are given in Fig. 2-1. Figure 2-1a shows the type introduced by Bragg and Bragg (1915); x rays produced by the source X are collimated by the slit system $S_1 S_2$ to form a narrow beam, idealized here as a single ray. The detector D, usually a solid-state detector (e.g., NaI-Th) or a proportional counter, has a relatively wide window and is placed so as to intercept the ray after it has been deviated through an angle $2\theta_n$ by crystal C, whose Bragg angle is θ_n. Crystal C is then slowly rotated through an appropriate range in the region of Bragg reflection, and the position for maximum count is located; its reflecting plane then makes an angle θ_n with the incident ray. The angular position of the crystal is then recorded on a high-precision divided circle attached to the crystal support.

The crystal is then rotated through an angle of approximately $180° - 2\theta_n$ and the detector moved through an angle of roughly $4\theta_n$ to give a mirror image of the original position. (For convenience, the first position described will be referred to as the A position and the second the B position.) The new crystal position for a

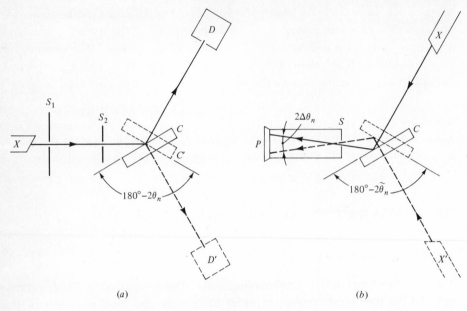

FIGURE 2-1
(*a*) Schematic diagram of Bragg spectrometer with x-ray source X and slit system S_1 and S_2. Crystal and detector are shown as C and D in A position and C' and D' in B position. (*b*) Schematic diagram of tube spectrometer. The "tube" portion extends from slit S to photographic plate P. Crystal and x-ray source are shown as C and X in A position and C' and X' in B position. [Positions of source and detector have been interchanged with respect to (*a*) to emphasize the analogy between the two instruments.]

maximum count is then located and recorded. The difference between these two positions is precisely $180° - 2\theta_n$, thus yielding the Bragg angle.

The tube spectrometer, which was developed by Siegbahn (1929) and Larsson (1927), is essentially the inverse of the Bragg type. It is shown schematically in Fig. 2-1*b*. A broad-focus x-ray source illuminates a crystal; in the A position, one ray strikes at the proper Bragg angle to be reflected through the slit S and recorded on the photographic plate P. In principle the x-ray source might then be moved through an angle of roughly $4\theta_n$ and the crystal rotated to a B position C' such that the reflected ray coincided with the first image on the plate P. The rotation of the crystal could then be measured on a high-precision divided circle; under these conditions the rotation angle would be exactly $180° - 2\theta_n$. In practice the rotation is intentionally chosen to give a small separation of the two photographic images; the divided circle records an angle $180° - 2\tilde{\theta}_n$. To obtain the Bragg angle, $\tilde{\theta}_n$ must be corrected by an amount $\Delta\theta_n = X/(2L)$, where X is the separation of the images on the plate and L

the distance between slit S and plate P, i.e., the length of the "tube." Also it is often awkward to move the x-ray source; the same result can be achieved by rotating the tube (i.e., slit and plate) through an angle of approximately $4\theta_n$ and then rotating the crystal through the proper angle to attain the B position. In either case the high-precision circle measures the rotation of the crystal with respect to the tube.

In both of these instruments the only high-precision measurement involved is the rotation of the crystal, which is determined by readings of the divided circle. The correction angle $\Delta\theta_n$ for the tube spectrograph is, of course, determined from geometrical measurements of X and L; however, this is of the order of 1% of θ_n so that the value of L is not unduly critical. It should be noted that, under the *ideal* conditions assumed above, it is immaterial whether or not the diffracting plane of the crystal coincides with the axis of rotation as long as they are parallel. (This conclusion depends on uniform crystal reflectivity.)

B.   Dispersion and resolution

The angular dispersion of the x-ray spectrum is readily calculated by differentiation of equation (2-1) with the result

$$D \equiv \frac{d\theta_n}{d\lambda} = \frac{\tan\theta_n}{\lambda} \qquad (2\text{-}23)$$

Clearly the dispersion increases with order; for small Bragg angles it is very nearly proportional to n, and for larger angles it increases somewhat faster. In the vicinity of 90° the dispersion becomes very large. In such cases it is possible to measure angles to moderate accuracy and still obtain relatively precise wavelength measurements; however, there may be other problems due to intensity, nonlinearity, etc.

The precise definition of resolution is somewhat arbitrary. In the present treatment we will follow that of Allison (1931) who assumed that two monochromatic lines can just be resolved if their separation is equal to the full width of the instrumental window. (While Allison assumed a Gaussian window, the same definition is a reasonable working criterion for most physically realistic window functions.) The expression for resolution then follows from equation (2-23) by setting $d\theta_n = \hat{w}_n$, where $\hat{w}_n$ is the effective width of the instrumental window in the nth order. The result is

$$\frac{\lambda}{d\lambda} = \tan\frac{\theta_n}{\hat{w}_n} \qquad (2\text{-}24)$$

The value of $\hat{w}_n$ will be primarily determined by the crystal window $\tilde{w}_n$ in the case of narrow slits and by the geometrical window $\hat{w}_g$ in the case of wide slits. In the intermediate case it will, of course, depend on both components.

An expression for the geometrical width is derived in Subsection III F and given in equation (2-54). As might be anticipated, it depends only on geometrical factors and is independent of order.

Equations (2-13) and (2-14) indicate that the larger crystal width $\tilde{w}_\perp$ varies *inversely* as $\sin 2\theta_n$, at least within the validity of the Darwin approximation (no absorption); there is, of course, also some variation in the structure factor F_H. Hence, when the crystal width is the controlling factor, the resolution includes a factor $\tan \theta_n \times \sin 2\theta_n$. For small glancing angles this factor is roughly proportional to n^2, while it increases even more rapidly with larger angles. Obviously there are three practical limitations to the improvement obtainable: (1) as the order increases, the geometrical window (due to slit system) ultimately becomes the controlling factor; (2) high orders may mean impractically low intensities; and (3) the order n is limited to values for which (2-1) yields $\sin \theta_n \leq 1$.

C. Geometry of a single ray

We will consider first the geometry of a single ray $\mathbf{r}$ reflected from a crystal. Let the axis of rotation of the crystal be taken as vertical and let the projection of the incident ray $\mathbf{r}$ make an angle η with the surface of the crystal in the horizontal plane, as shown in Fig. 2-2a. Let ψ denote the angle by which this ray lies *below* the horizontal plane, as illustrated in Fig. 2-2b. Finally let δ_1 be the crystal tilt, i.e., the angle by which the normal to the crystal surface lies *above* the horizontal plane, as shown in Fig. 2-2c.

We will now choose a coordinate system in which the z axis is vertical and the xy plane horizontal with the x axis along the surface of the crystal, as indicated in Fig. 2-2. The unit vector $\mathbf{r}$ along the direction of incident ray may then be represented as

$$\mathbf{r} = (\cos \psi \cos \eta, \ -\cos \psi \sin \eta, \ -\sin \psi) \qquad (2\text{-}25)$$

while the unit vector normal to the crystal is

$$\mathbf{n} = (0, \cos \delta_1, \sin \delta_1) \qquad (2\text{-}26)$$

The scalar product of $-\mathbf{r}$ and $\mathbf{n}$ will now give the cosine of the optical angle of incidence, i.e., the sine of the glancing angle θ (approximately equal to the Bragg angle). Thus we have

$$\sin \theta = -\mathbf{r} \cdot \mathbf{n} = \cos \psi \cos \delta_1 \sin \eta + \sin \psi \sin \delta_1 \qquad (2\text{-}27)$$

or

$$\sin \theta = \sin \eta + (\cos \psi \cos \delta_1 - 1) \sin \eta + \sin \psi \sin \delta_1 \qquad (2\text{-}28)$$

This represents a rigorous solution for the true glancing angle θ in terms of η, ψ, and δ_1.

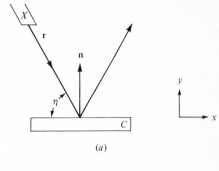

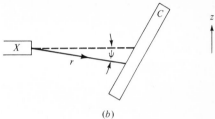

FIGURE 2-2
Angles and unit vectors used to describe geometry of single ray for Bragg spectrometer in Subsection III C. (x, y, and z show directions of coordinate axes.) (*a*) Horizontal projection showing incident ray **r**, crystal normal **n**, and horizontal angle of incidence η. X and C represent x-ray source and crystal. (*b*) Vertical plane containing incident ray, defining vertical angle ψ. (*c*) Vertical plane containing crystal normal **n** and defining tilt δ_1.

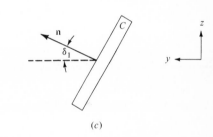

However, in practice we deal with the cases for which $\psi \ll 1$ and $\delta_1 \ll 1$. Hence we may set

$$\theta = \eta + \varepsilon \qquad (2\text{-}29)$$

with $\varepsilon \ll 1$ and thus obtain the Taylor-series approximation

$$\sin \theta \approx \sin \eta + \varepsilon \cos \eta \qquad (2\text{-}30)$$

Comparison of equations (2-28) and (2-30) immediately gives the value of ε as

$$\varepsilon \approx \frac{(\cos \psi \cos \delta_1 - 1) \sin \eta + \sin \psi \sin \delta_1}{\cos \eta} \qquad (2\text{-}31)$$

To a second-order approximation in ψ and δ_1, ε is given by

$$\varepsilon \approx -\tfrac{1}{2}(\psi^2 + \delta_1{}^2) \tan \eta + \frac{\psi \delta_1}{\cos \eta} \qquad (2\text{-}32)$$

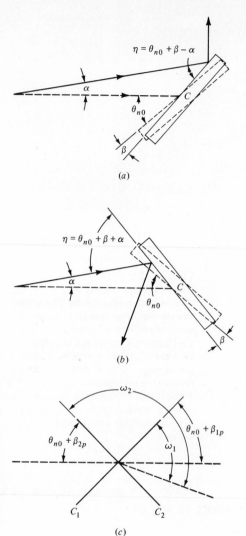

FIGURE 2-3
Angles used to describe Bragg spectrometer window in Subsection III D. (*a*) *A*
position. Reference ray makes angle θ_{n0}
with crystal in reference position (dashed
outline). Crystal has been rotated by
angle β and ray shown (solid line) forms
horizontal angle α with reference ray.
(*b*) *B* position, showing same ray. (*c*)
Schematic diagram showing angles ω_1
and ω_2 measured on spectrometer circle
for *A* and *B* positions; C_1 and C_2
indicate crystal planes for the respective
positions. Horizontal line represents
reference ray shown in (*a*) and (*b*).

D.   Bragg-spectrometer window

We will now apply the above result to the single-crystal Bragg spectrometer. Figure
2-3*a* shows schematically the horizontal projection with the crystal in the *A* position.
A reference ray (e.g., the central ray) makes an angle θ_{n0} (the Bragg angle for peak
wavelength λ_0) with the crystal surface when the crystal is in its reference position.
We assume that the crystal has been rotated from this position by a small angle β
(a positive β denotes a larger glancing angle and hence is in the direction of longer

wavelengths). We now consider a ray of wavelength λ making a small angle α with the reference direction, as shown. The glancing angle in the horizontal plane then becomes

$$\eta = \theta_{n0} + \beta - \alpha \qquad (2\text{-}33)$$

The Bragg angle for wavelength λ is approximately

$$\theta_{n\lambda} \approx \theta_{n0} + \left(\frac{d\theta}{d\lambda}\right)_{\lambda_0} (\lambda - \lambda_0) \qquad (2\text{-}34)$$

β may be considered as defining locally a wavelength scale on the spectrometer. Noting that the value of β is equal to the increment of θ, and employing equation (2-5), we may rewrite this expression as

$$\theta_{n\lambda} = \theta_{n0} + \left(\frac{d\beta}{d\lambda}\right)_{\lambda_0} (\lambda - \lambda_0) = \theta_{n0} + z \qquad (2\text{-}35)$$

Thus the Bragg angle is $\theta_{n0} + z$ while the actual glancing angle is obtained from equations (2-29) and (2-33). Combining these expressions yields

$$\theta - \theta_{n\lambda} = (\eta + \varepsilon) - (\theta_{n0} + z)$$
$$= \beta - z - \alpha + \varepsilon(\psi,\delta_1) \qquad (2\text{-}36)$$

where ε is a function of ψ and δ_1 given by (2-31). This now defines the argument in the crystal window [equation (2-12) when the simple Darwin function is applicable].

The intensity incident on the crystal will be given by an angular distribution function $G(\alpha,\psi)$, which will depend on the geometry of the slit system and the intensity distribution over that portion of the target subtended by the slits; any correlation between angular distribution and wavelength is usually negligible for the single-crystal case.

In light of the above discussion, the total intensity reflected from the crystal may be written as

$$F(\beta) = \iiint C_n[\beta - z - \alpha + \varepsilon(\psi,\delta_1)]G(\alpha,\psi)J(z) \, d\alpha \, d\psi \, dz \qquad (2\text{-}37)$$

where $G(\alpha,\psi)J(z) \, d\alpha \, d\psi \, dz$ describes the intensity in the angular ranges $d\alpha$ and $d\psi$ and the wavelength range dz, while C_n gives the fraction reflected by the crystal, used in nth order. For unpolarized incident radiation, C_n represents the average of the crystal window functions for the two polarizations; this point is further discussed in Subsection III I. If a detector D is employed with a sufficiently wide window to accept virtually all radiation from the line under investigation, then $F(\beta)$ will describe the recorded intensity as a function of the spectrometer setting β.

It will now be *assumed* that the distribution function $G(\alpha,\psi)$ may be represented as a product

$$G(\alpha,\psi) = h(\alpha)g(\psi) \qquad (2\text{-}38)$$

However, *this will not necessarily be true for every case.* The expression for $F(\beta)$ may now be rewritten in the form of relation (2-6) or (2-8) with the window function W given by

$$W(t) = W(\beta - z) = \iint C_n[\beta - z - \alpha + \varepsilon(\psi,\delta_1)]h(\alpha)g(\psi)\,d\alpha\,d\psi \qquad (2\text{-}39)$$

E. Bragg-spectrometer measurements

In wavelength measurements with the Bragg spectrometer the intensity in the A position (Fig. 2-3*a*) is recorded over a range of spectrometer settings and the peak position determined, i.e., the maximum of the function $F(\beta)$ defined by (2-37). As shown by equation (2-11), this occurs at the centroid of the window function (provided the natural line width is substantially wider than the window), i.e.,

$$\beta_{p1} = \int_{-\infty}^{\infty} t \iint \tilde{C}_n[t - \alpha + \varepsilon(\psi,\delta_1)]h(\alpha)g(\psi)\,d\alpha\,d\psi\,dt \qquad (2\text{-}40)$$

where $\tilde{C}_n$ represents the function C_n normalized to unity over the range of integration and $h(\alpha)$ and $g(\psi)$ are assumed normalized at the outset. [The denominator in (2-11) then becomes unity.] This may be immediately transformed to

$$\beta_{p1} = \int_{-\infty}^{\infty} \iint [u + \alpha - \varepsilon(\psi,\delta_1)]\tilde{C}_n(u)h(\alpha)g(\psi)\,d\alpha\,d\psi\,du \qquad (2\text{-}41)$$

or

$$\beta_{p1} = \bar{u}_n + \bar{\alpha} - \int \varepsilon(\psi,\delta_1)g(\psi)\,d\psi \qquad (2\text{-}42)$$

where $\bar{u}_n$ and $\bar{\alpha}$ represent the average values (centroids) over the distribution functions $\tilde{C}(u)$ and $h(\alpha)$, respectively.

The crystal is now rotated to the B position (Fig. 2-3*b*) and the process repeated. A positive β is again defined as being in the direction of increasing wavelength; with this sign convention $\eta = \theta_{n0} + \beta + \alpha$. An analysis similar to that above then yields for the peak value

$$\beta_{p2} = \bar{u}_n - \bar{\alpha} - \int \varepsilon(\psi,\delta_1)g(\psi)\,d\psi \qquad (2\text{-}43)$$

Figure 2-3*c* shows schematically the respective crystal positions as C_1 and C_2; ω_1 and ω_2 represent the two peak positions as recorded on the spectrometer with respect to some arbitrary zero. From the figure it is clear that

$$\omega_2 - \omega_1 = \pi - (\theta_{n0} + \beta_{2p}) - (\theta_{n0} + \beta_{1p}) \qquad (2\text{-}44)$$

Thus from the last three equations we find

$$\theta_{n0} = \tfrac{1}{2}[\pi - (\omega_2 - \omega_1)] - \bar{u}_n + \int \varepsilon(\psi,\delta_1)g(\psi) \, d\psi \qquad (2\text{-}45)$$

Hence in order to obtain the true Bragg angle we must correct the term in the brackets by subtracting the centroid of the crystal window (the centroid being referred to the Bragg angle *after correction for index of refraction*) and the integral on the right, which includes effects of vertical divergence, misalignment, and crystal tilt (as further discussed below). It should be noted that, if we are justified in assuming the product-function distribution (2-38), no correction is required for the horizontal-divergence distribution $h(\alpha)$, regardless of its form or asymmetry.

Recalling the expression for $\varepsilon(\psi,\delta_1)$ given by (2-32) and noting that $\eta \approx \theta_n$ for both positions, we find that

$$\int \varepsilon(\psi,\delta_1)g(\psi) \, d\psi \approx -(\delta_1{}^2 + \overline{\psi^2}) \frac{\tan \theta_n}{2} + \frac{\overline{\psi}\delta_1}{\cos \theta_n} \qquad (2\text{-}46)$$

Further simplification depends on the precise form of $g(\psi)$.

Let us first assume slits located between the x-ray source and the crystal. Consider the system shown in Fig. 2-4, with vertical-slit heights a and b separated by a distance L and with the central axis aligned precisely in the horizontal plane; let us further assume that the target intensity does not vary in the vertical direction. Under these conditions the figure shows that the intensity will be uniform for $|\psi| \leq \psi_c$ and then decrease linearly to zero over the range $\psi_c < |\psi| \leq \psi_m$. Thus the normalized distribution is trapezoidal, as indicated in Fig. 2-4c, and is described by

$$g(\psi) = \begin{cases} \dfrac{1}{\psi_m + \psi_c} & |\psi| \leq \psi_c \\[2ex] \dfrac{\psi_m - |\psi|}{\psi_m{}^2 - \psi_c{}^2} & \psi_c < |\psi| \leq \psi_m \end{cases} \qquad (2\text{-}47)$$

where

$$\psi_c = \frac{b - a}{2L} \qquad \psi_m = \frac{b + a}{2L} \qquad (2\text{-}48)$$

This distribution yields the result

$$\overline{\psi^2} = \int_{-\psi_m}^{\psi_m} \psi^2 g(\psi) \, d\psi = \tfrac{1}{6}(\psi_c{}^2 + \psi_m{}^2) \qquad (2\text{-}49)$$

or, by the above expressions for ψ_c and ψ_m,

$$\overline{\psi^2} = \frac{a^2 + b^2}{12L^2} \qquad (2\text{-}50)$$

Let us now consider the case in which the central ray makes an angle Ψ with the horizontal (Ψ will be referred to as *misalignment*), but the distribution about the

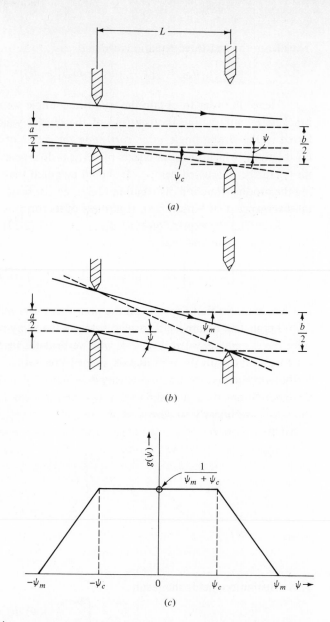

FIGURE 2-4
(*a*) Profile view of slits of unequal heights *a* and *b*, defining angle ψ_c and showing portion of the radiation transmitted with vertical angle ψ, where $|\psi| \leq \psi_c$. Center line is taken as horizontal. (*b*) Profile view of slits of unequal heights, defining angle ψ_m and showing portion of radiation transmitted with vertical angle ψ, where $\psi_c < |\psi| \leq \psi_m$. (*c*) Angular intensity distribution $g(\psi)$ produced by slit system of (*a*) and (*b*).

central ray has the same form as before. By a transformation of variables (i.e., with $\psi = \Psi + \phi$) the problem immediately reduces to

$$\overline{\psi^2} = \int_{-\psi_m}^{\psi_m} (\Psi + \phi)^2 g(\phi)\, d\phi = \Psi^2 + \frac{a^2 + b^2}{12L^2} \qquad (2\text{-}51)$$

since $\overline{\phi} = 0$. Combining equations (2-46) and (2-51) and noting that $\overline{\psi} = \Psi$, we now get

$$\int \varepsilon(\psi, \delta_1) g(\psi)\, d\psi = -\frac{1}{2}\left(\delta_1^2 + \Psi^2 + \frac{a^2 + b^2}{12L^2}\right) \tan \theta_n + \frac{\Psi \delta_1}{\cos \theta_n} \qquad (2\text{-}52)$$

Substitution in (2-45) yields finally

$$\theta_{n0} = \tfrac{1}{2}[\pi - (\omega_2 - \omega_1)] - \bar{u}_n - \frac{a^2 + b^2}{24L^2} \tan \theta_n$$

$$- \frac{\delta_1^2 + \Psi^2}{2} \tan \theta_n + \frac{\Psi \delta_1}{\cos \theta_n} \qquad (2\text{-}53)$$

The first term on the right is the uncorrected measured value of the Bragg angle. The next two represent corrections for crystal window and vertical divergence, respectively, and should be applied whenever of significant magnitude; note that the vertical-divergence correction is always negative.

The last two are error terms due to the crystal tilt δ_1 and the misalignment Ψ. Usually these parameters are unknown and no correction for them can be applied explicitly. Hence these last two terms are normally to be regarded as error estimates, which govern the allowable tolerances in the alignment procedure. For example, if $|\delta_1| < 10^{-3}$ rad, $|\Psi| < 10^{-3}$ rad, and $\theta_n < 45°$, then (2-53) shows that the resulting error is less than 2×10^{-6} rad.

It should be recalled that the above result was derived on the basis of two assumptions: (1) the product distribution, equation (2-38), and (2) uniform target intensity in the vertical direction, which led to equation (2-47). The latter is of crucial importance for the validity of the vertical-divergence correction. If the helical tube filament is vertical and the electron beam is properly focused on the target, the target intensity should be reasonably uniform in the vertical direction despite any variation in the horizontal plane. Under these conditions the above expression will be valid, and the final result becomes independent of the horizontal distribution function $h(\alpha)$.

F. Geometrical instrumental width

We are now in a position to discuss the instrumental width arising from purely geometrical effects. Consider first that due to the horizontal angular distribution, i.e., to $h(\alpha)$ alone. This may be found from equation (2-39). We set $\varepsilon = 0$ and replace the crystal window function by the δ function $\delta(t - \alpha)$. It follows that the instrumental window $W(t) = h(t)$ and is independent of order n, as might have been anticipated.

Now assume that the horizontal distribution is governed by two slits of widths a' and b' with separation L'. (If the x-ray source has a narrow-line focus, this itself may serve as one of the slits.) The geometry is thus identical to the vertical-slit system described by equation (2-47) and shown in Fig. 2-4c. It is obvious from the figure that the width (i.e., full width at half-intensity) is $\psi'_m + \psi'_c$, or with the help of (2-48),

$$\hat{w}_g = \frac{b'}{L'} \qquad (2\text{-}54)$$

where b' denotes the larger of the two slit widths. (This result may alternatively be deduced from a study of Fig. 2-4b.)

There is also a geometrical width due to vertical divergence. Let us assume that the crystal window function C and the angular distribution $h(\alpha)$ are δ functions and that tilt and misalignment are both zero. Under these conditions equations (2-39) and (2-32) show that we have a nonvanishing $W(t)$ if and only if $t = -\varepsilon = (\psi^2/2) \tan \theta$ and $g(\psi) \neq 0$ for this value of ψ. Since $d\psi/dt \approx 1/\psi$, it follows that any $g(\psi)$ distribution, such that $g(0) \neq 0$, will become infinite at the origin when expressed in terms of t. (See the discussion of the two-crystal geometrical window in Subsection IV J and also Fig. 2-13 for a more detailed analysis.) Hence the concept of the full width at half-intensity is meaningless for this case.

However, the above discussion implies that $W(t)$ is nonzero only over the range $0 \leq t \leq (\psi_m^2/2) \tan \theta$. Thus we might take $(\psi_m^2/4) \tan \theta$ as a reasonable measure of the width of this distribution; an alternative choice could be $(\overline{\psi^2}/2) \tan \theta$. The latter is simply the vertical-divergence term appearing in (2-53) of the previous subsection.

In order to keep the vertical divergence reasonably small so that any error in this correction is negligible, it is generally desirable to restrict this term to the order of magnitude of 1 second (of arc). On the other hand, under the assumptions of our analysis, no correction is required for $h(\alpha)$, and this function may be permitted to have a somewhat greater width, perhaps 10 seconds (of arc) or more, to obtain reasonable intensities. Hence the vertical distribution will normally contribute a negligible amount to $\hat{w}_g$, and (2-54) may then be taken as a satisfactory expression for the geometrical width.

G. Alternative slit locations

The above equations are based on the assumption that the slit system defining vertical divergence is located between the x-ray tube and the crystal. However, for experimental reasons it is often preferable to place the second slit near the detector; this

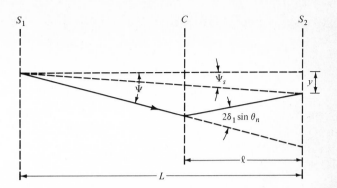

FIGURE 2-5
Schematic diagram for Bragg spectrometer, relating misalignment angle Ψ (for "central ray" incident on crystal) with slit misalignment angle Ψ_s and crystal tilt δ_1. S_1 and S_2 denote planes of slits and C a vertical plane through point of incidence on crystal. This figure is drawn schematically as if incident and reflected beam lay in the same vertical plane.

maximizes the length L and hence minimizes the vertical-divergence correction, as shown by (2-53).

Figure 2-5 shows schematically the vertical projection for this case; this is drawn as though the beam lay in a single vertical plane rather than changing horizontal direction at the crystal. The change in vertical angle at the crystal is approximately $2\delta \sin\theta$ since the component *normal* to the crystal face (i.e., the component in the yz plane) is rotated by twice the crystal tilt.

This may be verified as follows: The beam reflected by the crystal is obtained by reversing the component normal to the surface; i.e., the reflected ray $\mathbf{r}_{\text{refl}}$ is given by

$$\mathbf{r}_{\text{refl}} = \mathbf{r} - 2(\mathbf{r} \cdot \mathbf{n})\mathbf{n} \qquad (2\text{-}55)$$

Substitution of (2-25) and (2-26) yields

$$\mathbf{r}_{\text{refl}} = [\cos\psi \cos\eta, (2\cos^2\delta_1 - 1)\cos\psi \sin\eta + 2\sin\psi \sin\delta_1 \cos\delta_1,$$
$$2\cos\psi \sin\delta_1 \cos\delta_1 \sin\eta + (2\sin^2\delta_1 - 1)\sin\psi] \qquad (2\text{-}56)$$

or

$$\mathbf{r}_{\text{refl}} = (\cos\psi \cos\eta, \cos\psi \cos 2\delta_1 \sin\eta + \sin\psi \sin 2\delta_1,$$
$$\cos\psi \sin 2\delta_1 \sin\eta - \sin\psi \cos 2\delta_1) \qquad (2\text{-}57)$$

Comparing this with equation (2-25), we see that the vertical component (i.e., z component) has changed by an amount

$$\Delta r_z = (r_{\text{refl}})_z - r_z = \cos\psi \sin 2\delta_1 \sin\eta + (1 - \cos 2\delta_1)\sin\psi \qquad (2\text{-}58)$$

A second-order approximation in δ and ψ now yields (since $\eta \approx \theta_n$)

$$\Delta r_z = 2\delta_1 \sin \theta_n \qquad (2\text{-}59)$$

Since $\mathbf{r}$ denotes a unit vector and $\Delta r_z \ll 1$, it is clear that Δr_z is approximately the change in the vertical angle of the incident beam due to Bragg reflection. (If there is no crystal tilt, then there is no change in the vertical angle.)

From Fig. 2-5 it is clear that the net vertical displacement y between the two slits may be expressed in two ways:

$$L\Psi_s = y = L\Psi - l(2\delta_1 \sin \theta_n) \qquad (2\text{-}60)$$

In this expression Ψ_s is the slit misalignment angle, i.e., the vertical distance between the centers of the two slits divided by their separation L; l denotes the distance from the crystal to the second slit. It now follows that

$$\Psi = \Psi_s + 2\sigma\delta_1 \sin \theta_n \qquad (2\text{-}61)$$

where

$$\sigma = \frac{l}{L} \qquad (2\text{-}62)$$

Substitution of (2-61) into (2-53) and rearrangement now yield

$$\theta_{n0} = \tfrac{1}{2}[\pi - (\omega_2 - \omega_1)] - \bar{u}_n - \frac{a^2 + b^2}{24L^2} \tan \theta_n$$

$$- \left[\frac{\delta_1^2 + \Psi_s^2}{2} + 2\sigma\delta_1^2(\sigma \sin^2 \theta_n - 1) \right] \tan \theta_n$$

$$+ (1 - 2\sigma \sin^2 \theta_n)\left(\frac{\Psi_s \delta_1}{\cos \theta_n} \right) \qquad (2\text{-}63)$$

This expression now replaces (2-53) when the second slit comes after the crystal; it reduces to (2-53) for the case $\sigma = 0$.

The case in which both slits follow the crystal may be handled by a similar analysis. The result is equivalent to setting $\sigma = 1$ in equation (2-63). However, this arrangement does not appear to be of any great experimental interest.

H. Tube-spectrometer measurements

The principle of the tube spectrometer has been explained in Subsection III A. In most respects it is essentially the inverse of the Bragg type (i.e., with source and detector interchanged); however, there are two important differences:

1 There are *two* measurements involved in determining the Bragg angle: the rotation of the crystal as measured on spectrometer scale (as in the case of the

Bragg instrument) and a small additional term determined from the displacement of the two lines on the photographic plate.

2   The plate records a point-by-point response of the incident intensity rather than an integrated total; hence the theoretical analysis must consider intensity as a function of position on the plate.

A schematic diagram of the geometry of this instrument is shown in Fig. 2-6. A horizontal ray which is perpendicular to the plate and passes through Q (center of the slit system) forms an angle $\theta_{n0} + \gamma$ with the crystal, with $|\gamma| \ll \theta_{n0}$. In practice, γ is usually a few minutes of arc; it is deliberately chosen to be nonzero. An arbitrary horizontal ray through the center of the slit system forms an angle $\theta_{n0} + \beta + \gamma$, where $\beta \approx x/L$ and x and L represent, respectively, displacement along the plate and plate-to-slit distance (tube length). Finally, for a ray reaching this same point with horizontal divergence α we have a glancing angle $\eta = \theta_{n0} + \beta + \gamma - \alpha$, which is similar to (2-33) except for the addition of γ. (In comparing Figs. 2-3 and 2-6 we note that there is a close analogy provided that α and ψ are defined for the *reversed* ray in the case of the tube spectrometer; this is to be expected since the tube instrument is essentially the inverse of the Bragg type.)

Thus, in analogy to (2-42), we have for the A position

$$\frac{x_1}{L} \approx \beta_{1p} = \bar{u}_n + \bar{\alpha}_1 - \int \varepsilon(\psi,\delta_1)g(\psi)\,d\psi - \gamma_1 \qquad (2\text{-}64)$$

Equation (2-46) again applies to the evaluation of the integral in the above expression. However, $g(\psi)$ assumes a simpler form than before since we are now concerned with only one point on the plate at a time. Figure 2-6b gives a schematic diagram of the geometry in the vertical plane. Let the origin be taken as the point on the plate in the same horizontal plane as the center of the slit, not necessarily at the center of the plate. If we again assume uniform target intensity in the vertical direction, the distribution $g(\psi)$ at the origin will be given by

$$g(\psi) = \frac{1}{2\psi_m} \qquad -\psi_m \leq \psi \leq \psi_m \qquad (2\text{-}65)$$

where

$$\psi_m = \frac{b}{2L} \qquad (2\text{-}66)$$

(If the vertical divergence were limited by the height of the x-ray target, then b and L would be, respectively, target height and plate-to-target distance.) It follows that

$$\overline{\psi^2} = \frac{\psi_m{}^2}{3} = \frac{b^2}{12L^2} \qquad (2\text{-}67)$$

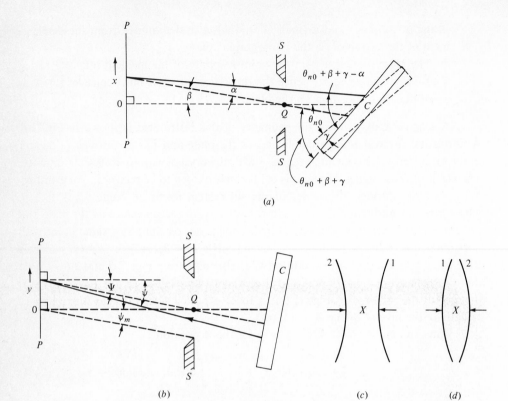

FIGURE 2-6
Tube-spectrometer geometry, showing angles used in Subsection III H. (*a*) Horizontal projection. *SS* represents slit with midpoint *Q*. Ray through this point perpendicular to plate has been reflected at angle θ_{no} when crystal is in reference position (dashed outline). Crystal has been rotated from reference position by angle γ. Ray shown (solid line) strikes plate at distance *x* from reference ray (angular separation β as measured at *Q*) and has horizontal divergence α. (*b*) Vertical projection. Point at *y* has an angular height Ψ as measured at *Q*. Incident ray has vertical angle ψ. ψ_m represents half the angular aperture of slit as seen from plate. (*c*) Observed photographic images when *X* is minimum separation and crystal has been rotated by slightly more than $\pi - 2\theta_n$. 1 and 2 denote images for *A* and *B* positions, respectively. (*d*) Photographic images when crystal rotation is slightly less than $\pi - 2\theta_n$ and *X* is maximum separation.

For any other point on the plate the total angle subtended $2\psi_m$ is approximately the same (provided all angles involved are small), but the central ray through the slit makes an angle with the horizontal $\Psi = y/L$. This angle plays the same role as misalignment in the Bragg case, but is now a function of position.

Then, in analogy to (2-52), we have

$$\int \varepsilon(\psi,\delta_1)g(\psi)\, d\psi = -\frac{1}{2}\left(\delta_1{}^2 + \frac{y^2}{L^2} + \frac{b^2}{12L^2}\right)\tan\theta_n + \frac{y\delta_1}{L\cos\theta_n} \qquad (2\text{-}68)$$

Substitution in (2-64) then yields

$$\frac{x_1}{L} = \bar{u}_n + \bar{\alpha}_1 + \frac{1}{2}\left(\delta_1{}^2 + \frac{y^2}{L^2} + \frac{b^2}{12L^2}\right)\tan\theta_n - \frac{y\delta_1\tan\theta_n}{L\sin\theta_n} - \gamma_1$$

$$= \bar{u}_n + \bar{\alpha}_1 + \frac{1}{2}\left[\frac{b^2}{12L^2} + \delta_1{}^2 - \frac{\delta_1{}^2}{\sin^2\theta_n} + \left(\frac{y}{L} - \frac{\delta_1}{\sin\theta_n}\right)^2\right]\tan\theta_n - \gamma_1 \qquad (2\text{-}69)$$

This describes a parabola (x vs. y) on the photographic plate with an extreme abscissa value at the ordinate

$$y_m = \frac{L\delta_1}{\sin\theta_n} \qquad (2\text{-}70)$$

at which point

$$\frac{x_{1m}}{L} = \bar{u}_n + \bar{\alpha}_1 - \frac{1}{2}\left(\delta_1{}^2\cot^2\theta_n - \frac{b^2}{12L^2}\right)\tan\theta_n - \gamma_1 \qquad (2\text{-}71)$$

The B position may be treated in a similar way. As before, we define β (and also γ) to be in the direction of increasing wavelength while the sign convention for α remains unchanged. It follows that $\beta_2 = -x_2/L$. Following the same approach as above, we find for the extreme value

$$-\frac{x_{2m}}{L} = \bar{u}_n - \bar{\alpha}_2 - \frac{1}{2}\left(\delta_1{}^2\cot^2\theta_n - \frac{b^2}{12L^2}\right)\tan\theta_n - \gamma_2 \qquad (2\text{-}72)$$

The measured angle of rotation of the crystal is, in analogy to equation (2-44),

$$\omega_2 - \omega_1 = \pi - (\theta_{n0} + \gamma_2) - (\theta_{n0} + \gamma_1) \qquad (2\text{-}73)$$

Solving (2-71) and (2-72) for γ_1 and γ_2 and substituting in the above, we now get

$$\theta_{n0} = \tfrac{1}{2}[\pi - (\omega_2 - \omega_1)] + \frac{X}{2L} - \bar{u}_n - \frac{b^2}{24L^2}\tan\theta_n$$

$$+ \frac{\delta_1{}^2}{2}\cot\theta_n - \tfrac{1}{2}(\bar{\alpha}_1 - \bar{\alpha}_2) \qquad (2\text{-}74)$$

where

$$X = x_{1m} - x_{2m} \qquad (2\text{-}75)$$

X is thus the maximum or minimum distance of separation of the two lines on the plate, which is the criterion employed by Siegbahn (1929) and Larsson (1927).

Figure 2-6c and d shows schematically the distance X as it appears on the plate. In Fig. 2-6c, where X is the minimum separation, the rotation of the crystal has been slightly greater than $\pi - 2\theta_n$, while in Fig. 2-6d it has been slightly less.

The above result is quite similar to that for the Bragg spectrometer, as given by equation (2-53). The main contribution comes from the term in brackets, which is obtained directly from the measured angular rotation of the crystal and is identical for the two cases. The next term represents the small angle $\Delta\theta$, discussed in Subsection III A, which is determined by measurement of the plate with a precision microscope. The two following terms give the crystal window and vertical-divergence corrections (the latter may be obtained from the Bragg case by setting $a = 0$, which is appropriate when we are dealing with a single point on the detector).

The last two terms, due to crystal tilt and horizontal divergence, will usually serve as error estimates rather than explicit corrections. It is clear that, when the distance X is measured at the extremum separation, the crystal-tilt effect may be considerably more serious than for the case of the Bragg spectrometer, particularly with small values of θ. The danger of such an error has been noted by Siegbahn (1929). The horizontal-divergence term will vanish provided $\bar{\alpha}_1 = \bar{\alpha}_2$, but this equality need not necessarily hold. The geometrical window will, of course, be identical in both positions. However, between the two exposures, the crystal, slit, and plate have been rotated with respect to the x-ray source. Thus, in general, a different portion of the x-ray target is viewed. Hence we cannot assume $h(\alpha)$ as identical for the two positions unless we know that the target intensity is uniform horizontally. For a nonuniform target there will be an error of the order of the displacement of the centroid from the geometrical center.

I.   Wavelength corrections and errors

For some purposes it is convenient to express the various corrections and errors in terms of wavelength rather than angle. This may readily be done by dividing the respective angles by the dispersion D, which is given by equation (2-23).

The Bragg-spectrometer result is given by (2-53); all terms on the right except for the first bracket represent corrections and errors. Division by the dispersion D now yields

$$\frac{\Delta\lambda}{\lambda} = -\frac{\bar{u}_n}{\tan\theta_n} - \frac{a^2 + b^2}{24L^2} - \frac{\delta_1^2 + \Psi^2}{2} + \frac{\Psi\delta_1}{\sin\theta_n} \qquad (2\text{-}76)$$

where $\Delta\lambda$ represents the correction to the apparent wavelength (i.e., *a negative $\Delta\lambda$ implies that the true value is smaller than the uncorrected one*).

The first two terms in the above expression represent explicit corrections, while the last two are essentially error terms. If we can limit tilt and misalignment so that $|\delta_1| < 10^{-4}$ and $|\Psi| < 10^{-3}$, then such errors will be of the order of 1 ppm. Under these circumstances it is also desirable to limit the vertical divergence so that $(a^2 + b^2)/(24L^2) \approx 10$ ppm so that the result will not be unduly sensitive to the values of the slit heights.

Two further points should be noted. First, in the absence of any crystal-tilt or centroid correction, $(\Delta\lambda)/\lambda$ will be independent of wavelength; hence under such conditions *relative* wavelength values will be unaffected by misalignment. Second, the centroid correction will decrease with increasing order since $\tan\theta_n$ will increase while the crystal width usually tends to decrease, as indicated in Subsection II D.

The tube-spectrometer result is given by (2-74); the first two terms on the right now comprise the uncorrected value. Following the same procedure as above, we find for this case

$$\frac{\Delta\lambda}{\lambda} = -\frac{\bar{u}_n}{\tan\theta_n} - \frac{b^2}{24L^2} + \frac{\delta_1{}^2}{2\tan^2\theta_n} - \frac{\bar{\alpha}_1 - \bar{\alpha}_2}{2\tan\theta_n} \qquad (2\text{-}77)$$

The last two terms represent error terms; it will be noted that these both increase rapidly for small Bragg angles.

It should be reemphasized here that the centroid terms computed in this section are only a first approximation to the instrumental-window correction. Sauder's (1966) more comprehensive analysis indicates that the next two terms are functions of the third moments of the observed line profile and the crystal window, respectively. For symmetric line profiles and crystal windows both these terms will vanish; this condition holds closely for most short-wavelength measurements. With moderate asymmetries these additional corrections should not exceed a few parts per million.

One note of caution regarding the centroid should be added here. It was indicated in Subsection III D that the effective window for unpolarized incident radiation is the average of those for the two polarizations, i.e., $\frac{1}{2}(C_n{}^\perp + C_n{}^\|)$. It does *not* follow that the effective centroid may be similarly found by averaging the centroids computed for the two polarizations; the correct expression is clearly

$$\bar{u}_n = \frac{\int_{-\infty}^{\infty} u C_n{}^\perp(u)\,du + \int_{-\infty}^{\infty} u C_n{}^\|(u)\,du}{\int_{-\infty}^{\infty} C_n{}^\perp(u)\,du + \int_{-\infty}^{\infty} C_n{}^\|(u)\,du} \qquad (2\text{-}78)$$

While an explicit result would involve use of the Darwin-Prins solution, the general features of the above expression may easily be deduced. For $\theta \ll 1$, or $\pi/2 - \theta \ll 1$, equations (2-12) through (2-14) show that the windows are substantially identical for both polarizations; hence we may say $\bar{u}_n \approx \bar{u}_n{}^\perp$. For $\theta = \pi/4$ the width of the $C_n{}^\|$ function goes to zero, and hence the corresponding integrals vanish; thus $\bar{u}_n = \bar{u}_n{}^\perp$ for this point also.

For intermediate angles we may assume that $C_n^{\parallel}$ is similar to $C_n^{\perp}$, but reduced in width by the scale factor $\cos 2\theta_n$. Thus the $C_n^{\parallel}$ integral in the denominator varies as $\cos 2\theta_n$ while that in the numerator is proportional to $\cos^2 2\theta_n$ and decreases more rapidly. It follows that $\bar{u}_n < \bar{u}_n^{\perp}$ for all intermediate points. However, when $C_n^{\parallel}$ differs most from $C_n^{\perp}$, both integrals containing $C_n^{\parallel}$ become too small to have much effect on (2-78); hence $\bar{u}_n^{\perp}$ will never greatly exceed $\bar{u}_n$ and may therefore be used as a rough approximation for effective centroid; this estimate will generally be on the high side.

IV.   TWO-CRYSTAL SPECTROMETERS

A.   Elementary theory

As in the single-crystal case in Subsection III A, we will first outline the basic theory of the two-crystal spectrometer[1] under idealized conditions. The axis of rotation of the second crystal may again be assumed as vertical, without loss of generality. We will also *temporarily* make the following oversimplifying assumptions:

1   The x-ray source is monochromatic.
2   The x rays incident on the first crystal lie entirely in the horizontal plane.
3   The diffracting planes of both crystals are vertical (and hence parallel to the axis of rotation of the second crystal).

The principle of operation of the two-crystal spectrometer is shown schematically in Fig. 2-7. This form of the instrument was introduced by Compton (1917) and extensively used for precision-wavelength measurements (see Compton, 1931; Bearden and Shaw, 1935; Merrill and DuMond, 1961; Bearden et al., 1964). The theory of the instrument has been treated in detail by Compton and Allison (1935). The alignment procedure, corrections, and errors involved have recently been treated by Bearden and Thomsen (1971).

As shown by the figure, radiation from the x-ray source strikes a relatively broad portion of the first-crystal surface. However, only those rays which are incident very nearly at the mth-order Bragg angle θ_m are appreciably reflected. Thus there is a collimated beam incident on the second crystal, which may then be employed to measure wavelength in the same manner as in the single-crystal instrument. The beam is Bragg-reflected in the nth order first in one direction and then in the other, the angular difference again being $\pi - 2\theta_n$.

In accord with the terminology introduced by Allison (1930), the position in

---

[1] The term *two-crystal spectrometer* is synonymous with *double-crystal spectrometer*, frequently used in the past.

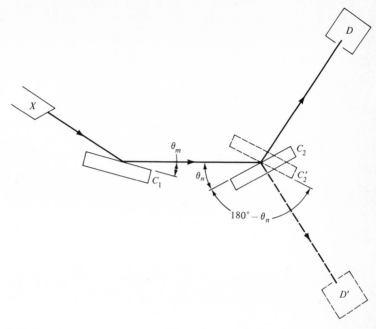

FIGURE 2-7
Schematic diagram of two-crystal spectrometer with x-ray source X and first crystal C_1. Second crystal and detector are shown as C_2 and D in the plus position and C_2' and D' in minus position.

which the beam is deviated in the same sense by both crystals (solid lines in Fig. 2-7) is called the *plus position* and denoted as (m,n). The opposite situation (dashed lines in Fig. 2-7) is called the *minus position* and denoted as $(m,-n)$. The $(n,-n)$ or *parallel position* represents a special case; there is no net angular deviation, and the beam emerges parallel to itself.

When the simplifying assumptions are dropped, we find two important differences between the single-crystal and two-crystal instruments. In the latter case the first crystal reflects a given wavelength only at or near the Bragg angle; hence when the incident beam covers a range of wavelengths, there will be a correlation between wavelength and angle (and also position) for the radiation incident on the second crystal. Also, since the angular spread of the beam incident on the first crystal will be limited by the crystal width (and perhaps also by a relatively wide horizontal-slit system, not shown in Fig. 2-7), the wavelength range incident on the second crystal is correspondingly restricted; i.e., the first crystal acts as a wavelength filter.

In practice the first crystal is almost always used in the first order to minimize any loss in intensity. The second crystal is frequently used in a higher order so that

$2\theta_n$ becomes a rather large angle, which can be measured to high relative accuracy. In some cases a fifth or sixth order has been used, but this also is limited by available intensity.

Since the two-crystal spectrometer is somewhat more involved than the single-crystal one, it is pertinent to ask what advantages justify the additional complication. There are at least three points in its favor, viz.: (1) because the first crystal acts as a wavelength filter, scattered radiation is reduced and the signal-to-noise ratio improved; (2) the resolving power is somewhat superior for the two-crystal case; (3) for short wavelengths ($\lambda < 0.5$ A*) the first crystal can achieve better angular collimation than is feasible with a geometrical slit system. On the other hand the two-crystal instrument gives a lower total intensity and requires repositioning of the first crystal each time the wavelength range is changed.

An alternate form of the two-crystal spectrometer, in which both crystals rotate, was introduced by DuMond and Hoyt (1930). Their application is restricted to the (n,n) position; for this case they point out that the wavelength is given directly by the dihedral angle between the planes of the two crystals. The symmetry of their arrangement eliminates certain sources of error, particularly those due to variations of reflectivity or grating constant along the crystal surfaces. However, the mechanical requirements and the problem of precision angular measurement are more severe; thus far it has not been extensively employed for wavelength measurement. Sauder (1971) has recently reported the construction of a γ-ray spectrometer of somewhat similar design, with the dihedral angle to be measured by means of an angular interferometer; his crystals are to be used in transmission (see Section VI).

B.   Dispersion and resolution

Let us compute the dispersion for the two-crystal spectrometer, first considering the design in which only the second crystal is rotated. Figure 2-8a and b shows the angles involved for the plus and minus positions, respectively. We will adopt the notation of Subsection III E and Fig. 2-3c; the angle ω, which describes the rotation of the second crystal C_2, will be measured from a line RR through its axis and parallel to the face of the first crystal C_1.

For the plus position, it is clear from Fig. 2-8a that

$$\omega_+ = \theta_m + \theta_n \qquad (2\text{-}79)$$

The dispersion is thus

$$\frac{d\omega_+}{d\lambda} = \frac{d\theta_m}{d\lambda} + \frac{d\theta_n}{d\lambda} \qquad (2\text{-}80)$$

Applying equation (2-1) for Bragg reflection at the two-crystal surfaces, we find that

$$\frac{d\omega_+}{d\lambda} = \frac{\tan \theta_m + \tan \theta_n}{\lambda} \qquad (2\text{-}81)$$

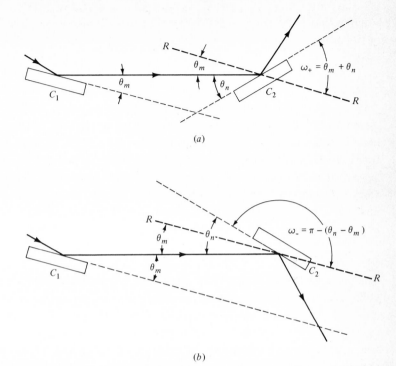

(a)

(b)

FIGURE 2-8
Schematic diagram of two-crystal spectrometer, showing angles ω_+ and ω_-, measured from reference line RR, which is drawn through axis of second crystal C_2 parallel to face of first crystal C_1. (a) Plus position. (b) Minus position.

Similarly, for the minus position, Fig. 2-8b shows

$$\omega_- = \pi + \theta_m - \theta_n \qquad (2\text{-}82)$$

The dispersion is identical to the previous case except for a change of sign of $\tan \theta_n$. Thus the general expression becomes

$$D \equiv \frac{d\omega}{d\lambda} = \frac{\tan \theta_m \pm \tan \theta_n}{\lambda} \qquad (2\text{-}83)$$

with the plus and minus signs applying to the corresponding positions. As shown by Allison (1931), an alternate form can be derived through the use of (2-1):

$$D \equiv \frac{d\omega}{d\lambda} = \frac{m}{2d_m \cos \theta_m} + \frac{n}{2d_n \cos \theta_n} \qquad (2\text{-}84)$$

where n is taken as an algebraic quantity which is negative for the minus position.

To compute resolution we again employ the Allison (1931) criterion, which assumes that two lines may just be resolved if their separation is equal to the width of the instrumental window. Equation (2-83) then immediately yields the result

$$\frac{\lambda}{d\lambda} = \frac{\tan \theta_m \pm \tan \theta_n}{\hat{w}_{m,n}} \qquad (2\text{-}85)$$

The factors governing the instrumental width $\hat{w}_{m,n}$ are treated in Subsection IV J. As might be expected from the elementary theory outlined in the previous section, horizontal divergence is not a significant factor. The crystal width $\tilde{w}_{m,\pm n}$ will, of course, be somewhat greater than the width of either crystal, but it is roughly equal to the larger value. The geometrical width $\hat{w}_g$ will be determined by vertical-divergence effects. In most instances $\hat{w}_g$ will be less than $\tilde{w}_{m,\pm n}$; however, for short wavelengths and higher orders it may become a significant or even dominant factor.

C.   Geometry of a single ray

The geometry of the two-crystal spectrometer has been analyzed by Schwarzschild (1928). Schnopper (1965) further developed this theory and applied it to obtain an alignment procedure. Unfortunately a typographical error in the Schwarzschild work led to erroneous results in Schnopper's papers, which he subsequently corrected. Using the corrected Schwarzschild equation, Bearden and Thomsen (1971) treated the same problem and gave a somewhat simplified alignment procedure. An earlier analysis was made by Merrill and DuMond (1961); their result is restricted to the $(n, \pm n)$ positions, but includes quadratic terms in the tilt which were neglected in the other treatments.

In the present analysis the result will be developed from fundamentals without explicit use of the previous work. It will be applicable for all $(m, \pm n)$ positions and will include quadratic terms in both misalignment and tilt. As in the single-crystal case, we will begin by treating the geometry of a single ray.

The notation used for the single-crystal geometry is easily extended to the present case, as shown in Fig. 2-9. The unit vectors representing the direction of the ray as

FIGURE 2-9
Angles and unit vectors used to describe geometry of two-crystal spectrometer (x, y, and z indicate coordinate axes). (a) Horizontal projection for plus position, showing x-ray source X, rays r_1 and r_2 incident on crystals C_1 and C_2, crystal normals n_1 and n_2, and spectrometer angle ω_+ (measured with respect to line RR which is parallel to the face of C_1). In this plane the projected incident ray forms an angle η with first crystal C_1 and $\omega_+ - \eta$ with second crystal C_2. (b) Horizontal projection for minus position, showing spectrometer angle ω_-. In this plane the projected ray now makes an angle $\eta + (\pi - \omega_-)$ with C_2. (c) Vertical plane containing ray r_1 incident on C_1, defining vertical angle ψ. (d) and (e) Vertical planes containing normals n_1 and n_2 to respective crystals C_1 and C_2, defining corresponding tilts δ_1 and δ_2.

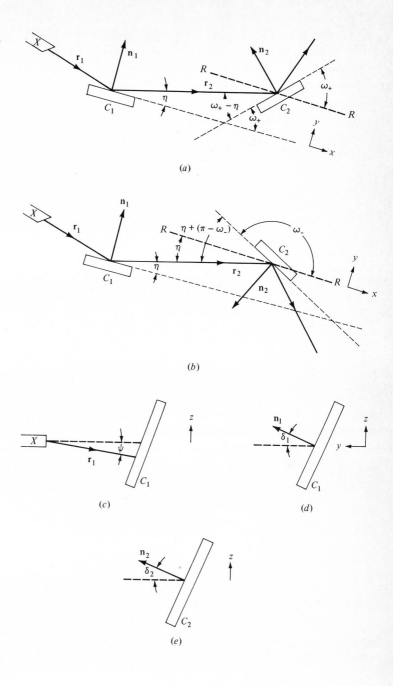

(a)

(b)

(c)

(d)

(e)

incident on crystals C_1 and C_2, respectively, are denoted as $\mathbf{r}_1$ and $\mathbf{r}_2$; in the horizontal plane the projection of $\mathbf{r}_1$ strikes crystal C_1 at glancing angle η. The angle which the surface of the second crystal forms with that of the first is denoted by ω, as in the previous section. The incident ray $\mathbf{r}_1$ lies below the horizontal plane by an angle $\psi(\psi \ll 1)$. The normals to the two-crystal surfaces are unit vectors $\mathbf{n}_1$ and $\mathbf{n}_2$, which lie above the horizontal plane by angles δ_1 and δ_2 ($\delta_1 \ll 1$ and $\delta_2 \ll 1$). The x axis is taken along the surface of C_1, and the z axis is defined to be vertical, as shown.

With these definitions the vectors $\mathbf{r}_1$, $\mathbf{n}_1$, and $\mathbf{n}_2$ may be represented as

$$\mathbf{r}_1 = (\cos \psi \cos \eta, -\cos \psi \sin \eta, -\sin \psi) \qquad (2\text{-}86)$$

$$\mathbf{n}_1 = (0, \cos \delta_1, \sin \delta_1) \qquad (2\text{-}87)$$

$$\mathbf{n}_2 = (-\cos \delta_2 \sin \omega, \cos \delta_2 \cos \omega, \sin \delta_2) \qquad (2\text{-}88)$$

These expressions are valid for both plus and minus positions. [Note that the first two expressions, (2-86) and (2-87), are analogous to equations (2-25) and (2-26).]

As in the single-crystal analysis, the sine of the glancing angle at the surface of the first crystal, which will be denoted by θ', is obtained from the relation $\sin \theta' = -\mathbf{r}_1 \cdot \mathbf{n}_1$.

Setting

$$\theta' = \eta + \varepsilon_1 \qquad (2\text{-}89)$$

we follow the procedure of Subsection III C and obtain a second-order approximation similar to (2-32):

$$\varepsilon_1(\psi, \delta_1) \approx -\tfrac{1}{2}(\psi^2 + \delta_1{}^2) \tan \eta + \frac{\psi \delta_1}{\cos \eta} \qquad (2\text{-}90)$$

The ray $\mathbf{r}_2$ is simply the ray reflected from crystal C_1, i.e., the unit vector $\mathbf{r}_{\text{refl}}$ calculated in Subsection III G. Hence we obtain from (2-55) and (2-57)

$$\mathbf{r}_2 = \mathbf{r}_1 - 2(\mathbf{n}_1 \cdot \mathbf{r}_1)\mathbf{n}_1 = (\cos \psi \cos \eta, \cos \psi \cos 2\delta_1 \sin \eta + \sin \psi \sin 2\delta_1,$$
$$\cos \psi \sin 2\delta_1 \sin \eta - \sin \psi \cos 2\delta_1) \qquad (2\text{-}91)$$

The sine of the glancing angle at the second crystal, denoted by θ'', is now obtained from (2-88) and (2-91) as follows:

$$\sin \theta'' = -\mathbf{r}_2 \cdot \mathbf{n}_2 = \cos \psi \cos \delta_2 \sin \omega \cos \eta$$
$$-\cos \psi \cos 2\delta_1 \cos \delta_2 \cos \omega \sin \eta - \sin \psi \sin 2\delta_1 \cos \delta_2 \cos \omega$$
$$-\cos \psi \sin 2\delta_1 \sin \delta_2 \sin \eta + \sin \psi \cos 2\delta_1 \sin \delta_2 \qquad (2\text{-}92)$$

This may be rewritten in the form

$$\sin \theta'' = [1 + (\cos \psi \cos \delta_2 - 1)] (\sin \omega \cos \eta - \cos \omega \sin \eta)$$
$$- (\cos 2\delta_1 - 1) \cos \psi \cos \delta_2 \cos \omega \sin \eta$$
$$- \sin \psi \sin 2\delta_1 \cos \delta_2 \cos \omega$$
$$- \cos \psi \sin 2\delta_1 \sin \delta_2 \sin \eta + \sin \psi \cos 2\delta_1 \sin \delta_2 \qquad (2\text{-}93)$$

or finally

$$\sin \theta'' = \sin (\omega - \eta) + (\cos \psi \cos \delta_2 - 1) \sin (\omega - \eta)$$
$$- (\cos 2\delta_1 - 1) \cos \psi \cos \delta_2 \cos \omega \sin \eta$$
$$- \sin \psi \sin 2\delta_1 \cos \delta_2 \cos \omega$$
$$- \cos \psi \sin 2\delta_1 \sin \delta_2 \sin \eta + \sin \psi \cos 2\delta_1 \sin \delta_2 \qquad (2\text{-}94)$$

A second-order approximation in the small angles ψ, δ_1, and δ_2 reduces the above expression to

$$\sin \theta'' \approx \sin (\omega - \eta) - \tfrac{1}{2}(\psi^2 + \delta_2{}^2) \sin (\omega - \eta) + 2\delta_1{}^2 \cos \omega \sin \eta$$
$$- 2\psi\delta_1 \cos \omega - 2\delta_1\delta_2 \sin \eta + \psi\delta_2 \qquad (2\text{-}95)$$

For the plus position Fig. 2-9a shows that the horizontal projection of the glancing angle on crystal C_2 is $\omega_+ - \eta$. Since this is very nearly equal to the glancing angle θ'', we may set

$$\theta''_+ = \omega_+ - \eta + \varepsilon_{2+} \qquad (2\text{-}96)$$

where $\varepsilon_{2+} \ll 1$. Inserting this into (2-95) and following the same procedure as in Subsection III C yield the approximate expression

$$\varepsilon_{2+} \approx - \frac{1}{\cos (\omega_+ - \eta)} \left[\tfrac{1}{2}(\psi^2 + \delta_2{}^2) \sin (\omega_+ - \eta) - 2\delta_1{}^2 \cos \omega_+ \sin \eta \right.$$
$$\left. + 2\psi\delta_1 \cos \omega_+ + 2\delta_1\delta_2 \sin \eta - \psi\delta_2 \right] \qquad (2\text{-}97)$$

Since ε_{2+} represents a small correction, approximate values will suffice here for η and ω_+. As in the single-crystal case, η must be very close to θ_m, the Bragg angle for the first crystal. Similarly $\theta''_+ \approx \theta_n$ throughout the rocking curve; hence, by equation (2-96), $\omega_+ - \eta \approx \theta_n$ and $\omega_+ \approx \theta_n + \theta_m$. Thus ε_{2+} becomes

$$\varepsilon_{2+} \approx -\tfrac{1}{2}(\psi^2 + \delta_2{}^2) \tan \theta_n$$
$$+ \frac{1}{\cos \theta_n} \left[2\delta_1{}^2 \cos (\theta_n + \theta_m) \sin \theta_m - 2\psi\delta_1 \cos (\theta_n + \theta_m) \right.$$
$$\left. - 2\delta_1\delta_2 \sin \theta_m + \psi\delta_2 \right] \qquad (2\text{-}98)$$

The glancing angle is by definition never greater than a right angle; hence for the minus position its horizontal projection becomes $\pi - (\omega_- - \eta)$, as indicated by Fig. 2-9b. Equation (2-96) is thus replaced by

$$\theta''_- = \pi - (\omega_- - \eta) + \varepsilon_{2-} \qquad (2\text{-}99)$$

where $\varepsilon_{2-} \ll 1$. Substitution of this into (2-95) then gives the approximate result

$$\varepsilon_{2-} \approx - \frac{1}{\cos (\pi - \omega_- + \eta)}$$
$$\times \left[\tfrac{1}{2}(\psi^2 + \delta_2{}^2) \sin (\omega_- - \eta) - 2\delta_1{}^2 \cos \omega_- \sin \eta \right.$$
$$\left. + 2\psi\delta_1 \cos \omega_- + 2\delta_1\delta_2 \sin \eta - \psi\delta_2 \right] \qquad (2\text{-}100)$$

Since θ''_- must be close to the Bragg angle for crystal C_2, equation (2-99) shows that $\theta_n \approx \pi - (\omega_- - \eta)$. As indicated above, we have $\eta \approx \theta_m$. It follows that $\omega_- - \eta \approx \pi - \theta_n$, $\omega_- \approx \pi - (\theta_n - \theta_m)$, $\cos(\pi - \omega_- + \eta) \approx \cos\theta_n$, $\sin(\omega_- - \eta) \approx \sin\theta_n$, and $\cos\omega_- \approx -\cos(\theta_n - \theta_m)$. Thus ε_{2-} becomes

$$\varepsilon_{2-} \approx -\tfrac{1}{2}(\psi^2 + \delta_2{}^2)\tan\theta_n$$

$$+ \frac{1}{\cos\theta_n}\left[-2\delta_1{}^2\cos(\theta_n - \theta_m)\sin\theta_m + 2\psi\delta_1\cos(\theta_n - \theta_m)\right.$$

$$\left. - 2\delta_1\delta_2\sin\theta_m + \psi\delta_2\right] \qquad (2\text{-}101)$$

The above results may be compactly reexpressed in terms of one set of equations valid for both positions. Equations (2-96) and (2-99) may be written

$$\theta'' = \frac{\pi}{2} \pm \left(\omega - \eta - \frac{\pi}{2}\right) + \varepsilon_2(\psi,\delta_1,\delta_2) \qquad (2\text{-}102)$$

where the arguments of the function ε are indicated explicitly and the upper and lower signs apply to the plus and minus positions, respectively.

Similarly (2-98) and (2-101) become

$$\varepsilon_2(\psi,\delta_1,\delta_2) \approx -\tfrac{1}{2}(\psi^2 + \delta_2{}^2)\tan\theta_n$$

$$+ \frac{1}{\cos\theta_n}\left[\pm 2\delta_1{}^2\cos(\theta_n \pm \theta_m)\sin\theta_m \mp 2\psi\delta_1\cos(\theta_n \pm \theta_m)\right.$$

$$\left. - 2\delta_1\delta_2\sin\theta_m + \psi\delta_2\right] \qquad (2\text{-}103)$$

with the same sign convention.

D.   Two-crystal-spectrometer window

To compute the spectrometer window, we must obtain the arguments for the two-crystal window functions (see Subsection II D), i.e., the differences between the glancing angles for a given ray and the respective Bragg angles. For the first crystal this argument is found by setting β equal to zero in (2-36) (since the first crystal is stationary in the type of spectrometer under consideration). This yields the expression

$$\theta' - \theta_{m\lambda} = -z_m - \alpha + \varepsilon_1(\psi,\delta_1) \qquad (2\text{-}104)$$

where θ' denotes the glancing angle and

$$z_m \equiv \left(\frac{d\theta_m}{d\lambda}\right)_{\lambda_0}(\lambda - \lambda_0) = \frac{\tan\theta_{m0}}{\lambda_0}(\lambda - \lambda_0) \qquad (2\text{-}105)$$

with the last equality resulting from equation (2-1). Similarly, in analogy to (2-33), the angle η may be expressed as

$$\eta = \theta_{m0} - \alpha \qquad (2\text{-}106)$$

As shown in the previous section, the values of ω_+ and ω_- are approximately $\theta_n + \theta_m$ and $\pi - (\theta_n - \theta_m)$, respectively. (This is also clear from the simplified description of Subsection IV A, illustrated in Fig. 2-8.) Hence we will now take the spectrometer setting β with respect to these reference positions, i.e.,

$$\omega = \frac{\pi}{2} \pm \left(\theta_{n0} + \beta - \frac{\pi}{2} \right) + \theta_{m0} \qquad (2\text{-}107)$$

with the upper sign holding for the plus position and the lower for the minus. This expression has been defined so that β increases with increasing wavelength in either position.

Combining equations (2-106) and (2-107) gives

$$\omega - \eta - \frac{\pi}{2} = \pm \left(\theta_{n0} + \beta - \frac{\pi}{2} \right) + \alpha \qquad (2\text{-}108)$$

Inserting this into (2-102) yields

$$\theta'' = \theta_{n0} + \beta \pm \alpha + \varepsilon_2 \qquad (2\text{-}109)$$

The Bragg angle for the second crystal may be taken as

$$\theta_{n\lambda} = \theta_{n0} + z_n \qquad (2\text{-}110)$$

where

$$z_n \equiv \left(\frac{d\theta_n}{d\lambda} \right)_{\lambda_0} (\lambda - \lambda_0) = \frac{\tan \theta_{n0}}{\lambda_0} (\lambda - \lambda_0) \qquad (2\text{-}111)$$

Finally from (2-109) and (2-110) we obtain the result

$$\theta'' - \theta_{n\lambda} = \beta - z_n \pm \alpha + \varepsilon_2 \qquad (2\text{-}112)$$

This indicates that the angle of incidence of a given ray of wavelength λ will differ from the Bragg angle by four correction terms: (1) β, the rotation of the second crystal from its reference position; (2) z_n, arising from the difference in Bragg angle for wavelengths λ and λ_0; (3) α, the horizontal divergence of the selected ray at the first crystal (referred to the reference angle θ_{m0}); (4) $\varepsilon_2(\psi, \delta_1, \delta_2)$, a correction due to vertical-divergence and crystal-tilt effects.

The general expression for the intensity reflected by the instrument may now be written in a form very similar to (2-37), with the arguments for the crystal functions given by (2-104) and (2-112). Assuming unpolarized incident radiation, we obtain the result

$$F(\beta) = \tfrac{1}{2} \iiint \sum_P C_m{}^P(\varepsilon_1 - z_m - \alpha) C_n{}^P(\beta - z_n \pm \alpha + \varepsilon_2)$$
$$\times \, G(\alpha, \psi) \mathscr{I}(\lambda) \, d\alpha \, d\psi \, d\lambda \qquad (2\text{-}113)$$

where summation over P denotes the sum over the two polarization states, with corresponding window functions given by equations (2-12) through (2-14). [Note that z_m and z_n are both functions of λ as shown by equations (2-105) and (2-111).]

In the general case, (2-113) cannot be reduced to the form of (2-6) and is difficult to integrate except by numerical methods. Hence we will now introduce some simplifying physical assumptions; possible departures from these conditions will be considered in a subsequent section. It will be assumed that there are no physical slits limiting horizontal divergence and that the first crystal is sufficiently wide so that all rays from the target incident near the angle θ_{m0} (the Bragg angle for the peak wavelength λ_0) are far removed from the edge of the crystal. Furthermore the crystal window function C_m will be assumed uniform over its entire surface; i.e., crystal imperfections will be ignored.

The x-ray line profile is relatively narrow, and the first crystal reflects appreciably only for angles of incidence near θ_m; hence only a very limited range of values of α will be significant. Within this narrow angular range, rays from each portion of the target strike the crystal according to our assumptions (subject to limitation by the vertical-slit system); thus the incident intensity will be substantially independent of α, and we may set

$$G(\alpha,\psi) = g(\psi) \qquad (2\text{-}114)$$

Furthermore, since the integrand is essentially zero outside of a narrow range of α, we may extend integration limits to infinity without appreciable error. Thus equation (2-113) becomes

$$F(\beta) = \tfrac{1}{2} \iiint_{-\infty}^{\infty} \sum_P C_m{}^P(\varepsilon_1 - z_m - \alpha) C_n{}^P(\beta - z_n \pm \alpha + \varepsilon_2)\, g(\psi) \mathscr{J}(\lambda)\, d\alpha\, d\psi\, d\lambda$$

$$(2\text{-}115)$$

A transformation of variables now yields

$$F(\beta) = \tfrac{1}{2} \int \mathscr{J}(\lambda) \int g(\psi) \int_{-\infty}^{\infty} \sum_P C_m{}^P(v) C_n{}^P(\beta - z_n \mp z_m \mp v \pm \varepsilon_1 + \varepsilon_2)\, dv\, d\psi\, d\lambda$$

$$(2\text{-}116)$$

Let us define, with the aid of (2-105) and (2-111),

$$z \equiv z_n \pm z_m = \frac{\tan \theta_{n0} \pm \tan \theta_{m0}}{\lambda_0} (\lambda - \lambda_0) \qquad (2\text{-}117)$$

If $\mathscr{J}(\lambda)$ now is reexpressed as $J(z)$, as in the single-crystal case, the above expression for $F(\beta)$ becomes

$$F(\beta) = \tfrac{1}{2} \sum_P \int J(z) \int g(\psi) \int_{-\infty}^{\infty} C_m{}^P(v)\, C_n{}^P(\beta - z \mp v \pm \varepsilon_1 + \varepsilon_2)\, dv\, d\psi\, dz$$

$$(2\text{-}118)$$

Comparing this with (2-6) and (2-8), we see that the window function is given by

$$W(t) = W(\beta - z) = \tfrac{1}{2} \sum_P \int g(\psi) \int_{-\infty}^{\infty} C_m{}^P(v) C_n{}^P(\beta - z \mp v \pm \varepsilon_1 + \varepsilon_2) \, dv \, d\psi$$

(2-119)

where it is understood that $\varepsilon_1 = \varepsilon_1(\psi,\delta_1)$ and $\varepsilon_2 = \varepsilon_2(\psi,\delta_1,\delta_2)$.

Equation (2-11) shows that the value β_p for the peak of the rocking curve is given by the centroid of the window function. Substitution of (2-119) in (2-11) then gives

$$\beta_p = \bar{t} = \frac{\sum_P \int_{-\infty}^{\infty} t \int g(\psi) \int_{-\infty}^{\infty} C_m{}^P(v) C_n{}^P(t \mp v \pm \varepsilon_1 + \varepsilon_2) \, dv \, d\psi \, dt}{\sum_P \int_{-\infty}^{\infty} \int g(\psi) \int_{-\infty}^{\infty} C_m{}^P(v) C_n{}^P(t \mp v \pm \varepsilon_1 + \varepsilon_2) \, dv \, d\psi \, dt}$$

(2-120)

This may be transformed to the form

$$\beta_p = \frac{\sum_P \int g(\psi) \int_{-\infty}^{\infty} C_m{}^P(v) \int_{-\infty}^{\infty} (u \pm v \mp \varepsilon_1 - \varepsilon_2) C_n{}^P(u) \, du \, dv \, d\psi}{\sum_P \int g(\psi) \int_{-\infty}^{\infty} C_m{}^P(v) \int_{-\infty}^{\infty} C_n{}^P(u) \, du \, dv \, d\psi}$$

(2-121)

As noted above, ε_1 and ε_2 depend only on ψ, δ_1, and δ_2, the last two parameters not being involved in the integration. Hence, since $g(\psi)$ is normalized, we obtain at once

$$\beta_p = \bar{u}_n{}^e \pm \bar{v}_m{}^e - \int g(\psi)[\varepsilon_2(\psi,\delta_1,\delta_2) \pm \varepsilon_1(\psi,\delta_1)] \, d\psi \qquad (2\text{-}122)$$

where the "effective crystal window centroids" $\bar{u}_n{}^e$ and $\bar{v}_m{}^e$ are given by

$$\bar{u}_n{}^e = \frac{\sum_P \int_{-\infty}^{\infty} \int_{-\infty}^{\infty} u C_m{}^P(v) C_n{}^P(u) \, du \, dv}{\sum_P \int_{-\infty}^{\infty} \int_{-\infty}^{\infty} C_m{}^P(v) C_n{}^P(u) \, du \, dv}$$

(2-123)

$$\bar{v}_m{}^e = \frac{\sum_P \int_{-\infty}^{\infty} \int_{-\infty}^{\infty} v C_m{}^P(v) C_n{}^P(u) \, du \, dv}{\sum_P \int_{-\infty}^{\infty} \int_{-\infty}^{\infty} C_m{}^P(v) C_n{}^P(u) \, du \, dv}$$

The significance of these "effective values" is further considered in Subsection IV K.

E.   Two-crystal-spectrometer measurements

As in the single-crystal case, the Bragg angle is obtained from the difference of the two measured angles. In computing the desired expression from the above results we must recall that various signs in the expressions for ω, β, and ε_2 (but not ε_1)

differ for the plus and minus positions. Thus equation (2-107) for ω may be rewritten for the two cases with appropriate subscripts as follows:

$$\omega_+ = \theta_{n0} + \beta_+ + \theta_{m0} \qquad \omega_- = \pi - \theta_{n0} - \beta_- + \theta_{m0} \qquad (2\text{-}124)$$

Subtraction of these two expressions yields

$$\theta_{n0} = \tfrac{1}{2}[\pi - (\omega_- - \omega_+)] - \tfrac{1}{2}(\beta_+ + \beta_-) \qquad (2\text{-}125)$$

For the peak positions the values of β are given by (2-122). Writing the appropriate forms for β_{p+} and β_{p-} gives

$$\beta_{p+} + \beta_{p-} = 2\bar{u}_n^{\,e} - \int g(\psi)(\varepsilon_{2+} + \varepsilon_{2-})\, d\psi \qquad (2\text{-}126)$$

The functions ε_{2+} and ε_{2-} may be obtained from (2-103) or more explicitly from (2-98) and (2-101). When these are inserted into the above equation, it becomes

$$\beta_{p+} + \beta_{p-} = 2\bar{u}_n^{\,e} - \int g(\psi)\left\{-(\psi^2 + \delta_2^{\,2})\tan\theta_n + 2\frac{\delta_1^{\,2}\sin\theta_m - \psi\delta_1}{\cos\theta_n}\right.$$
$$\times \left[\cos(\theta_n + \theta_m) - \cos(\theta_n - \theta_m)\right]$$
$$\left. - 4\delta_1\delta_2\frac{\sin\theta_m}{\cos\theta_n} + \frac{2\psi\delta_2}{\cos\theta_n}\right\} d\psi \qquad (2\text{-}127)$$

The integration over ψ consists merely of finding the mean values $\bar{\psi}$ and $\overline{\psi^2}$, as in the single-crystal case; $g(\psi)$ is, of course, normalized over the range of integration. Denoting $\bar{\psi}$ by Ψ and employing a trigonometric identity, we now simplify the above expression to obtain

$$\beta_{p+} + \beta_{p-} = 2\bar{u}_n^{\,e} + (\overline{\psi^2} + \delta_2^{\,2})\tan\theta_n + 4\frac{\delta_1^{\,2}\sin\theta_m - \Psi\delta_1}{\cos\theta_n}$$
$$\times \sin\theta_n\sin\theta_m + 4\delta_1\delta_2\frac{\sin\theta_m}{\cos\theta_n} - \frac{2\Psi\delta_2}{\cos\theta_n} \qquad (2\text{-}128)$$

Finally substitution into (2-125) yields

$$\theta_{n0} = \tfrac{1}{2}[\pi - (\omega_- - \omega_+)] - \bar{u}_n^{\,e} - \frac{\overline{\psi^2}}{2}\tan\theta_n$$
$$+ \left(2\delta_1\sin\theta_m\tan\theta_n + \frac{\delta_2}{\cos\theta_n}\right)\Psi$$
$$- 2\delta_1^{\,2}\sin^2\theta_m\tan\theta_n - 2\delta_1\delta_2\frac{\sin\theta_m}{\cos\theta_n} - \frac{\delta_2^{\,2}}{2}\tan\theta_n \qquad (2\text{-}129)$$

If we again assume the vertical-slit system of Fig. 2-4, described by (2-47), the value of $\overline{\psi^2}$ may be taken from (2-51) with the result

$$\theta_{n0} = \tfrac{1}{2}[\pi - (\omega_- - \omega_+)] - \bar{u}_n^{\,e} - \frac{a^2 + b^2}{24L^2}\tan\theta_n - \frac{\Psi^2}{2}\tan\theta_n$$

$$+ \left(2\delta_1\sin\theta_m\tan\theta_n + \frac{\delta_2}{\cos\theta_n}\right)\Psi - 2\delta_1^{\,2}\sin^2\theta_m\tan\theta_n$$

$$- \frac{2\delta_1\delta_2\sin\theta_m}{\cos\theta_n} - \frac{\delta_2^{\,2}}{2}\tan\theta_n \qquad (2\text{-}130)$$

This may be written in the alternative form

$$\theta_{n0} = \tfrac{1}{2}[\pi - (\omega_- - \omega_+)] - \bar{u}_n^{\,e} - \frac{a^2 + b^2}{24L^2}\tan\theta_n$$

$$- \frac{\delta_2^{\,2} + \tilde{\Psi}^2}{2}\tan\theta_n + \frac{\tilde{\Psi}\delta_2}{\cos\theta_n} \qquad (2\text{-}131)$$

where

$$\tilde{\Psi} = \Psi - 2\delta_1\sin\theta_m \qquad (2\text{-}132)$$

This now has exactly the same form as (2-53) for the single-crystal case. $\tilde{\Psi}$ now represents the effective misalignment of the beam incident on the second crystal and is due to two causes, viz.: (1) the misalignment of the slit system, as described by Ψ, and (2) the departure of the beam from horizontal due to the tilt of the first crystal, i.e., δ_1.

F. Parallel-position window

When the two crystals have the same grating constant and are used in the $(n, -n)$, or parallel, position, (2-83) shows that there is no dispersion. Thus the position β_{p-} will occur at substantially the same position for all wavelengths. In this case (2-116) cannot be reduced to the form of (2-6), and a special analysis is needed.

Since $m = n$ and $z_n = z_m$ for the $(n, -n)$ case, the argument of $C_n^{\,P}$ in (2-116) is independent of wavelength, and the equation may be rewritten

$$F(\beta) = \tfrac{1}{2}A \sum_P \int g(\psi) \int_{-\infty}^{\infty} C_n^{\,P}(v)C_n^{\,P}(\beta + v - \varepsilon_1 + \varepsilon_{2-})\,dv\,d\psi \qquad (2\text{-}133)$$

where $A = \int \mathscr{I}(\lambda)\,d\lambda$.

At this point we note from (2-90), with $\eta \approx \theta_n$, and (2-101) that the ψ^2 terms in $-\varepsilon_1 + \varepsilon_{2-}$ cancel for the $(n, -n)$ position. Thus there is *no vertical-divergence correction for this case;* even the error terms will vanish in the absence of tilts. It also follows that $-\varepsilon_1 + \varepsilon_2$ is a linear function of ψ.

The peak value of β is obtained by differentiation of (2-133). (In the rest of this particular derivation the P superscripts will be omitted to avoid confusion with exponents and differentiation.) Thus we obtain

$$F'(\beta_P) = \tfrac{1}{2}A \sum_P \int g(\psi) \int_{-\infty}^{\infty} C_n(v)C_n'(\beta_p + v - \varepsilon_1 + \varepsilon_2)\,dv\,d\psi = 0 \qquad (2\text{-}134)$$

At this point we will assume that $-\varepsilon_1 + \varepsilon_2$ is relatively small compared to the width of the crystal window C_n. [If $|\delta_1 + \delta_2| < 10^{-4}$ and the maximum value of ψ is less than 10^{-2}, then $|-\varepsilon_1 + \varepsilon_2| \sim 10^{-6} \approx 0.2$ sec. This angle is sufficiently small to justify the assumption in most cases, particularly since the only $(n, -n)$ case frequently arising in practice is the $(1, -1)$ position.] With this condition the above expression may be approximated

$$\tfrac{1}{2}A \sum_P \int g(\psi) \int_{-\infty}^{\infty} C_n(v)[C_n'(v) + (\beta_p - \varepsilon_1 + \varepsilon_2)C_n''(v)]\,dv\,d\psi = 0 \qquad (2\text{-}135)$$

After integration of the first term, the above equation may be rewritten

$$\sum_P [C_n^2(v)]_{-\infty}^{\infty} + \sum_P \int_{-\infty}^{\infty} C_n(v)C_n''(v)\,dv \int (\beta_p - \varepsilon_1 + \varepsilon_2)g(\psi)\,d\psi = 0 \qquad (2\text{-}136)$$

Since the window function must always tend to zero in the tails, the first term clearly vanishes; however, the integral of $C_n(v)C_n''(v)$ will be nonzero. It follows that

$$\beta_{p-} = -\int g(\psi)[\varepsilon_2(\psi,\delta_1,\delta_2) - \varepsilon_1(\psi,\delta_1)]\,d\psi \qquad (2\text{-}137)$$

This is precisely the same result as given by (2-122) for the case of identical crystals in the minus position with $m = n$. Hence all the previous results in Subsection IV E remain valid even when the minus position is the nondispersive parallel one.

One additional feature of the parallel position may be noted at this point. If we assume that vertical divergence, misalignment, and tilts are all negligible (i.e., $\psi = 0$ and $\delta_1 = 0 = \delta_2$) and write (2-133) for this case, the result is

$$F(\beta) = \tfrac{1}{2}A \sum_P \int_{-\infty}^{\infty} C_n^P(v)C_n^P(\beta + v)\,dv$$

$$= \tfrac{1}{2}A \sum_P \int_{-\infty}^{\infty} C_n^P(-v)C_n^P(\beta - v)\,dv \qquad (2\text{-}138)$$

In the case of *symmetric-crystal window functions*, this may be transformed to give

$$F(\beta) = \tfrac{1}{2}A \sum_P \int_{-\infty}^{\infty} C_n^P(v)C_n^P(\beta - v)\,dv \qquad (2\text{-}139)$$

For the $(n, +n)$ position the window function $W(t)$ is obtained from (2-119); under the same simplifying conditions just listed this becomes

$$W(t) = \tfrac{1}{2} \sum_P \int_{-\infty}^{\infty} C_n^P(v)C_n^P(t - v)\,dv \qquad (2\text{-}140)$$

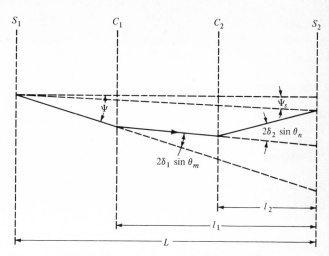

FIGURE 2-10
Schematic diagram for two-crystal spectrometer, relating misalignment angle Ψ (for "central ray" incident on first crystal) with slit misalignment angle Ψ_s and crystal tilts δ_1 and δ_2. S_1 and S_2 denote planes of slits and C_1 and C_2 vertical planes through points of incidence on the respective crystals. This figure is drawn schematically as if incident and reflected beams lay in the same vertical plane.

It is evident that $F(\beta)$ is proportional to the $(n, +n)$ window function. This property is useful in obtaining an approximate representation of the window to correct measured line widths. However, it is of little value in correcting wavelengths, since the assumption of symmetric crystal window functions implies that $\bar{u}_n^e = 0$ and hence eliminates the crystal-correction term in (2-130).

G. Alternative slit locations

Unless both slits are located between the target and the first crystal, the angle Ψ will depend on crystal tilts as well as on slit misalignment. As in the single-crystal case, various alternative slit locations are possible. We will consider here only the case in which one slit is placed between the target and first crystal (or perhaps formed by the line focus on the x-ray tube itself) and the other is located at or near the detector. This choice minimizes the vertical-divergence correction; it is shown schematically in Fig. 2-10. (Compare with Fig. 2-5.)

The analogous single-crystal case was treated in Subsection III F. By an obvious extension of (2-61) we now obtain

$$\Psi = \Psi_s + 2\sigma_1\delta_1 \sin \theta_m + 2\sigma_2\delta_2 \sin \theta_n \qquad (2\text{-}141)$$

where

$$\sigma_1 = \frac{l_1}{L} \qquad \sigma_2 = \frac{l_2}{L} \qquad (2\text{-}142)$$

and l_1 and l_2 represent distances from the respective crystals to the slits as shown in Fig. 2-10.

It was noted in Subsection IV E that the two-crystal result could be reduced to the single-crystal form by use of an equivalent angle $\bar{\Psi}$. Substituting (2-141) in (2-132) and collecting the terms in Ψ_s and δ_1 give

$$\bar{\Psi} = \hat{\Psi} + 2\sigma_2\delta_2 \sin \theta_n \qquad (2\text{-}143)$$

where

$$\hat{\Psi} = \Psi_s + 2(\sigma_1 - 1)\delta_1 \sin \theta_m \qquad (2\text{-}144)$$

Inserting (2-143) into the error terms contained in (2-131) and rearranging yield

$$-\frac{\delta_2{}^2 + \bar{\Psi}^2}{2} \tan \theta_n + \frac{\bar{\Psi}\delta_2}{\cos \theta_n}$$

$$= -\left\{ \frac{\hat{\Psi}^2 - \delta_2{}^2}{2} + (2\sigma_2 \sin^2 \theta_n - 1)\frac{\hat{\Psi}\,\delta_2}{\sin \theta_n} \right.$$

$$\left. + [1 - 2\sigma_2(1 - \sigma_2 \sin^2 \theta_n)]\delta_2{}^2 \right\} \tan \theta_n \qquad (2\text{-}145)$$

From the definitions given by (2-142) and Fig. 2-10 it is clear that $0 < \sigma_1 < 1$ and $0 < \sigma_2 < 1$. It follows that $|2\sigma_2 \sin^2 \theta_n - 1| < 1$ and $|1 - 2\sigma_2(1 - \sigma_2 \sin^2 \theta_n)| < 1$, and hence

$$\left| -\frac{\delta_2{}^2 + \bar{\Psi}^2}{2} \tan \theta_n + \frac{\bar{\Psi}\delta_2}{\cos \theta_n} \right| < \left(\left| \frac{\hat{\Psi}^2 - \delta_2{}^2}{2} \right| + \frac{|\hat{\Psi}\delta_2|}{\sin \theta_n} + \delta_2{}^2 \right) \tan \theta_n \qquad (2\text{-}146)$$

The upper limit for $|\hat{\Psi}|$ is readily obtained from (2-144) as

$$|\hat{\Psi}| \le |\Psi_s| + 2(1 - \sigma_1)|\delta_1| \sin \theta_m \qquad (2\text{-}147)$$

If we know the maximum absolute values for Ψ_s, δ_1, and δ_2, these two equations now permit calculation of an upper limit to the resulting error in θ_n.

It should be noted that, if δ_1 is set equal to zero, $\hat{\Psi} = \Psi_s$ by (2-144), and the analysis reduces to the single-crystal case. Hence the last two equations may also be employed to provide an error estimate for the single-crystal instrument.

H.   Horizontal-divergence slit system errors

It was assumed in Subsection IV D that there were no slits significantly limiting horizontal divergence. Such slits will usually cause an error in measured wavelength. This may be readily seen by consideration of Fig. 2-11. Let P and Q be two

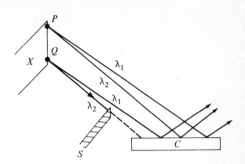

FIGURE 2-11
Illustration of stop partially blocking incident x-ray beam. P and Q denote two points on x-ray source X. Rays originating from each point and satisfying the Bragg law at crystal C are shown for wavelengths λ_1 and λ_2.

points on x-ray target X. For simplicity we will assume an ideal crystal with zero width; i.e., each wavelength is reflected only at its own Bragg angle. Now consider two wavelengths λ_1 and λ_2 (with $\lambda_2 > \lambda_1$), which typify the short- and long-wavelength sides of the line profile under investigation. In the absence of any slits each wavelength would be reflected by crystal C_1 at its appropriate angle. However, if a knife edge is inserted from the lower side as shown, it will block the λ_2 ray originating from point Q, but not interfere with either of the λ_1 rays. Hence the wavelength reflected from C_1 will consist predominantly of the λ_1 component, and the measured peak will be shifted in that direction.

This conclusion may be verified by a mathematical analysis. For simplicity we still restrict ourselves to crystals of zero width, which may be represented by δ functions; furthermore we will assume that tilts, misalignment, and vertical divergence are all negligible, so that the functions $\varepsilon_1(\psi, \delta_1)$ and $\varepsilon_2(\psi, \delta_1, \delta_2)$ may be ignored. Under these circumstances (2-113) reduces to

$$F(\beta) = \iiint \delta(-z_m - \alpha)\delta(\beta \pm \alpha - z_n)G(\alpha,\psi)J(z)\,d\alpha\,d\psi\,dz \qquad (2\text{-}148)$$

where $J(z)\,dz$ has replaced $\mathscr{J}(\lambda)\,d\lambda$.

We continue to assume for $G(\alpha,\psi)$ the product function given by (2-38); since ψ does not appear elsewhere in the integrand, and $g(\psi)$ is normalized, we have immediately

$$F(\beta) = \int J(z) \int h(\alpha)\delta(\beta \pm \alpha - z_n)\delta(-z_m - \alpha)\,d\alpha\,dz \qquad (2\text{-}149)$$

or, with the aid of (2-117),

$$F(\beta) = \int J(z)h(-z_m)\delta(\beta - z)\,dz \qquad (2\text{-}150)$$

To complete this integration we must compute the value of z_m when $z = \beta$; equations (2-105) and (2-117) yield the relation

$$z_m = \frac{z \tan \theta_{m0}}{\tan \theta_{n0} \pm \tan \theta_{m0}} \qquad (2\text{-}151)$$

and hence

$$F(\beta) \approx J(\beta)h\left(-\frac{\beta \tan \theta_{m0}}{\tan \theta_{n0} \pm \tan \theta_{m0}}\right) \qquad (2\text{-}152)$$

This result may also be expressed in terms of wavelength by converting β to a nominal wavelength scale through use of the dispersion, as given by (2-83). Noting that the variation in ω is equal to $\pm \beta$ (the signs referring to the plus or minus positions), we can rewrite (2-152)

$$\mathscr{F}(\lambda) \approx \mathscr{J}(\lambda)h\left[-\frac{(\lambda - \lambda_0) \tan \theta_{m0}}{\lambda_0}\right] \qquad (2\text{-}153)$$

The above expression indicates that the observed curve will reproduce the true wavelength profile if $h(\alpha)$ is constant over the region of observation. [In terms of α this region will have a width of the order of $(w \tan \theta_{m0})/\lambda_0$, where w is the width of the line profile.] Such a result is to be expected here since all other perturbing factors have been idealized in the present calculation.

However, if $h(\alpha)$ varies in the region of observation, the wavelength profile will be correspondingly distorted. In particular, if it has a nonzero slope at the peak, the measured wavelength will be shifted. If the oversimplified assumption of zero-width crystal windows is dropped, we would expect that the observed peak might be influenced by a slope of $h(\alpha)$ at any point close to the origin (i.e., within the order of width of the crystal windows).

Such an asymmetric distribution function $h(\alpha)$ can arise from at least three different causes:

1   It may be caused by a relatively narrow asymmetrically located slit which limits horizontal divergence. This may be located before the first crystal as in Fig. 2-11; however, a slit system following the crystal can produce the same effect.

2   If the slit system is narrow and the target has a nonuniform intensity, this can produce a similar effect even if the slit is symmetrically located about the "central ray" (i.e., the ray leaving the center of the target and incident on the first crystal at angle θ_{m0}).

3   An effect of the same type can also be produced by nonuniform reflectivity of either crystal. This is readily seen by considering a limiting case; if we move the slit in Fig. 2-11 up to the surface of the crystal, this is equivalent to a crystal with zero reflectivity over a portion of its surface.

The combination of source, slits, and *stationary* first crystal effectively forms a filter, which will selectively pass certain components of the incident wavelength under the above conditions. If this occurs, then clearly there is no way for the second crystal to recover the true wavelength profile.

These drawbacks to the conventional two-crystal spectrometer were noted by DuMond and Hoyt (1930). The best experimental precaution for avoiding such effects is to verify that the observed wavelength is insensitive to small variations in the portion of crystals and x-ray target used and to small displacements in the positions of any slits employed.

I. Two-crystal rotation

When the conventional two-crystal spectrometer is used to survey a narrow wavelength region (e.g., in determining wavelengths of a doublet or multiplet or in measuring line widths), the first crystal remains stationary while the second is rotated. Subsection IV H indicates some of the problems which this entails. For determination of a single wavelength peak, it is essential only that the function $h(\alpha)$ be constant over a small region of observation of the order of the width of the instrumental window. However, when a wavelength region $\Delta\lambda$ is to be surveyed, $h(\alpha)$ must be essentially constant over a range $D\,\Delta\lambda$, where D is the dispersion. The latter requirement is obviously a much more restrictive one. This is particularly true if the x-ray source is a narrow-line-focus x-ray tube. In this case the effective reflecting region of the first crystal for a single wavelength is primarily determined by the crystal window and is quite narrow; when even a small wavelength range is involved, the region becomes much more extensive.

The problem can be avoided by an ingenious scheme which involves rotation of both crystals in the $(n, +n)$ position; this is shown schematically in Fig. 2-12. In the $(n, +n)$ position a ray of wavelength λ_0 which satisfies the Bragg condition will make the same angle θ_{n0} with both crystals. Let C_1 and C_2 represent the positions of the respective crystals under these conditions and X and D the corresponding positions of the x-ray tube and the detector.

Now consider another ray of wavelength $\lambda > \lambda_0$. In order to satisfy the Bragg condition for this ray, the angle between the two crystals must be increased by an angle β, which is given by $2(\theta_n - \theta_{n0})$, as may be deduced from equation (2-80). This may be accomplished by rotating each crystal through an angle $\beta/2$ to positions C_1' and C_2' (see Fig. 2-12) rather than by turning only the second crystal. If we also rotate the x-ray tube and the detector each by an angle β, as shown, then we see that the ray can satisfy the Bragg condition exactly and still follow the *same path between crystals* as the initial ray of wavelength λ_0. Under these conditions the portion of each crystal which plays a significant part for wavelength λ_0 will play the same role for all other wavelengths. It follows that, if the requirement on uniform $h(\alpha)$ is satisfied for a single wavelength, it holds for all.

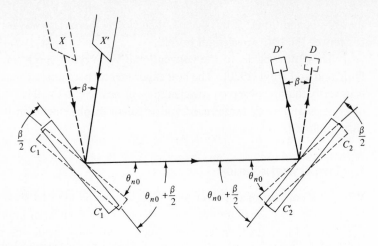

FIGURE 2-12
Schematic diagram of two-crystal spectrometer in which both crystals rotate. Crystals C_1 and C_2 (dashed outlines) each rotate, in opposite senses, by an angle $\beta/2$ to positions C_1' and C_2'. Similarly x-ray source X and detector D rotate by angle β to positions X' and D'. In this way rays satisfying the Bragg law always follow same path between crystals.

To describe this arrangement mathematically, it is only necessary to consider (2-113) for the plus position and modify it to describe the two-crystal rotation with $m = n$; i.e.,

$$F(\beta) = \tfrac{1}{2} \iiint \sum_P C_n^P \left(\frac{\beta}{2} + \varepsilon_1 - z_n - \alpha \right) C_n^P \left(\frac{\beta}{2} - z_n + \alpha + \varepsilon_2 \right)$$
$$\times\ G(\alpha,\psi) \mathscr{I}(\lambda)\ d\alpha\ d\psi\ d\lambda \qquad (2\text{-}154)$$

It is clear from (2-111) and (2-117) that $z_n = z/2$ in the $(n, +n)$ position. Hence, if $J(\lambda)$ is replaced by $J(z)$, the above expression now becomes

$$F(\beta) = \tfrac{1}{2} \sum_P \int J(z) \iint C_n^P \left(\frac{\beta - z}{2} - \alpha + \varepsilon_1 \right) C_n^P \left(\frac{\beta - z}{2} + \alpha + \varepsilon_2 \right)$$
$$\times\ G(\alpha,\psi)\ d\alpha\ d\psi\ dz \qquad (2\text{-}155)$$

Unlike the result in Subsection IV D, equation (2-113), this expression is in the form of (2-6) without further assumptions, and the window function may be written

$$W(t) = W(\beta - z) = \tfrac{1}{2} \sum_P \iint C_n^P \left(\frac{\beta - z}{2} - \alpha + \varepsilon_1 \right)$$
$$\times\ C_n^P \left(\frac{\beta - z}{2} + \alpha + \varepsilon_2 \right) G(\alpha,\psi)\ d\alpha\ d\psi \qquad (2\text{-}156)$$

At this point we introduce the same mathematical simplifications as employed in Subsection IV D; i.e., we replace $G(\alpha,\psi)$ by $g(\psi)$ and extend the α integral over an infinite range. Note, however, that this restriction on $h(\alpha)$ is far less stringent physically, as discussed above. A transformation of variables then yields

$$W(t) = W(\beta - z)$$

$$= \tfrac{1}{2} \sum_P \int g(\psi) \int_{-\infty}^{\infty} C_n^P(v) C_n^P(\beta - z - v + \varepsilon_1 + \varepsilon_2) \, dv \, d\psi \qquad (2\text{-}157)$$

which is identical with (2-119) for the $(n,+n)$ position. Thus the ensuing results of Subsections IV D and E hold also for this type of two-crystal instrument.

While the description given above assumes that the x-ray target, both crystals, and the detector all rotate, obviously only the relative motion is important. For example, in the DuMond and Hoyt (1930) design the detector was stationary. As the second crystal was rotated by an angle $\beta/2$, the spectrometer itself (i.e., the line joining the axes of rotation of the two crystals) was rotated by β about the axis of the second crystal. At the same time the x-ray source was rotated by β about the axis of the first crystal. On the other hand, Allison (1933) constructed an instrument in which the x-ray tube remained stationary and all other components described the proper relative motions.

Despite the advantages of this instrument, severe problems would arise in applying it to a direct high-precision measurement of the Bragg angle as described in the previous sections. Mechanical requirements for maintaining proper relative motion within the desired accuracy would be formidable. In most cases it would probably be necessary to have a divided circle of high precision associated with each crystal rather than relying solely on the 2:1 gear ratio.

However, these problems are not too serious when only small angular rotations are involved. Thus this design may be used to good advantage to measure a doublet separation when the wavelength of one component is assumed known. The design is equally advantageous in measuring line widths; Allison (1933) employed his instrument for this purpose. Sauder (1971) has recently designed a spectrometer with two rotating crystals for a precision measurement of the wavelength in electron-positron annihilation; his angles are to be measured by angular interferometers.

J. Instrumental width

We are now in a position to analyze the factors which determine the instrumental width of the two-crystal spectrometer. Let us first consider the width $\tilde{w}_{m,n}$ due to the crystals alone. If vertical divergence, misalignment, and tilts are all negligible, (2-119) reduces to

$$W(t) = \tfrac{1}{2} \sum_P \int_{-\infty}^{\infty} C_m^P(v) C_n^P(t \mp v) \, dv \qquad (2\text{-}158)$$

[This is a slightly generalized form of (2-140).]

If only one polarization component were present, this expression would be simply the convolution of two crystal-window functions. Compton and Allison (1935) implicitly assumed this restriction and computed the width of the $(n, \pm n)$ window when both window functions were given by the Darwin solution, (2-12) (i.e., with zero absorption). They obtained a width $\tilde{w}_{n, \pm n} = 1.32 \, \tilde{w}_n$. Thus the effective crystal width of the $(n, \pm n)$ two-crystal instrument is only slightly greater than for a single-crystal one in the nth order. For the case $m \neq n$, say $n > m$, $\tilde{w}_m$ will be considerably greater than $\tilde{w}_n$ and substantially fix the width of the convolution.

For unpolarized incident radiation, equation (2-158) shows that we must find the width of the average of the convolutions for the two polarizations. Qualitatively this problem is somewhat similar to that of the single-crystal centroid discussed in Subsection III I. If θ_m and θ_n are both small (i.e., $\theta_m \ll 1$ and $\theta_n \ll 1$), the window functions are substantially independent of polarization, and the width for the perpendicular component $\tilde{w}^{\perp}$ may be used. As the angles increase slightly, the effective width is approximately the average $\frac{1}{2}(\tilde{w}^{\perp} + \tilde{w}^{\parallel})$ and hence somewhat less than $\tilde{w}^{\perp}$. However, when either θ_m or θ_n approaches $\pi/4$, the width $\tilde{w}^{\parallel}$ for the corresponding crystal goes to zero. Thus, no matter what the width of the other crystal may be, the amplitude of the window for the parallel component becomes too small to influence the result appreciably. Hence the window width is again equal to $\tilde{w}^{\perp}$. A similar behavior occurs for higher angles, with the width again approaching $\tilde{w}^{\perp}$ as θ_m and θ_n approach $\pi/2$.

Thus it may be shown that the effective window width is generally less than $\tilde{w}^{\perp}$, but never less than about two-thirds of this value. From the above discussion together with Compton and Allison's computation of the convolution width, we see that the effective width of the window may be roughly estimated by taking $\tilde{w}_m^{\perp}$ or $\tilde{w}_n^{\perp}$, whichever is greater. Thus, if $n > m$, it follows that $\tilde{w}_{m, \pm n} \approx \tilde{w}_m^{\perp}$.

As indicated in the last two subsections, it is a requirement of the two-crystal spectrometer that $h(\alpha)$ be essentially constant over an appropriate range. When this condition is satisfied, $h(\alpha)$ drops out of the expression for the window function, as shown in (2-115), and thus has no effect on the observed window.

However, there will still be a geometrical window due to vertical divergence. Let us eliminate nongeometrical effects from (2-119) by assuming that the crystal windows are δ functions (for both polarizations); this expression then becomes

$$W(t) = B \int g(\psi) \int_{-\infty}^{\infty} \delta(v)\delta(t \mp v \pm \varepsilon_1 + \varepsilon_2) \, dv \, d\psi$$

$$= B \int g(\psi)\delta(t \pm \varepsilon_1 + \varepsilon_2) \, d\psi \tag{2-159}$$

where B is a constant.

If tilts are assumed as negligible ($\delta_1 = 0 = \delta_2$), then equation (2-90), with $\eta \approx \theta_m$, and (2-103) give

$$\pm \varepsilon_1 + \varepsilon_2 = -\rho\psi^2 \qquad (2\text{-}160)$$

where

$$\rho = \tfrac{1}{2}(\tan \theta_n \pm \tan \theta_m) \qquad (2\text{-}161)$$

Hence the geometrical window becomes

$$W(t) = B \int g(\psi)\delta(t - \rho\psi^2)\, d\psi \qquad (2\text{-}162)$$

A transformation of variables now yields the result

$$W(t) = B \int g\left(\sqrt{\frac{u}{\rho}}\right) \delta(t - u)\, \frac{du}{2\sqrt{\rho u}} \qquad (2\text{-}163)$$

or

$$W(t) = \frac{Bg(\sqrt{t/\rho})}{2\sqrt{\rho t}} \qquad (2\text{-}164)$$

If $g(\psi)$ is given by (2-47), this becomes

$$W(t) = \begin{cases} B_1(\psi_m - \psi_c)\left(\sqrt{\dfrac{\rho}{t}}\right) & 0 \le t \le \rho\psi_c^2 \\[4mm] B_1\left[\psi_m\left(\sqrt{\dfrac{\rho}{t}}\right) - 1\right] & \rho\psi_c^2 < t < \rho\psi_m^2 \end{cases} \qquad (2\text{-}165)$$

where $B_1 \equiv B/[2\rho(\psi_m^2 - \psi_c^2)]$ is a multiplicative constant. $W(t)$, of course, vanishes outside of this range. Shacklett and DuMond (1957) obtained a similar result for the case of equal slit heights, i.e., with $\psi_c = 0$.

Figure 2-13 shows a plot of the above equation with $\psi_c = \psi_m/2$. Recalling that $t \equiv \beta - z$ and $z \sim \lambda - \lambda_0$, we can interpret the physical significance of this window function. It attains an (infinite) maximum value at the wavelength for which the spectrometer is set and passes a narrow band of radiation on the *short-wavelength side* of this.

Since $W(t)$ becomes infinite at the origin, it is obvious that the concept of full width at half-intensity is meaningless here. It is equally clear that the effective width must be a relatively small fraction of $\rho\psi_m^2$, since $W(t)$ vanishes outside the range $0 \le t < \rho\psi_m^2$ and approaches zero rapidly near the upper limit. As suggested in Subsection III F, it seems reasonable to use $\rho\overline{\psi^2}$ as a measure of the window width $\hat{w}_g$. Equations (2-50) and (2-161) then yield

$$\hat{w}_g = \rho\overline{\psi^2} = \frac{a^2 + b^2}{24L^2}(\tan \theta_n \pm \tan \theta_m) \qquad (2\text{-}166)$$

This is similar to the vertical-divergence correction except that the factor $\tan \theta_n \pm \tan \theta_m$ replaces $\tan \theta_n$.

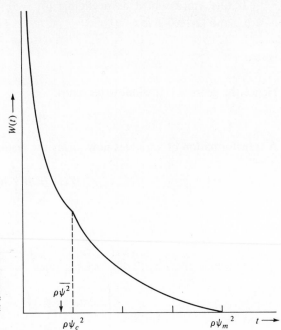

FIGURE 2-13
Geometrical window for two-crystal spectrometer due to vertical divergence alone. This window is computed from equation (2-165) for $\psi_c = \psi_m/2$. The abscissa value $\rho\overline{\psi^2}$ is indicated, where $\overline{\psi^2}$ is mean-square value of ψ^2 over $W(t)$.

For measurements of the Cu $K\alpha_1$ ($\lambda \approx 1.54$ A) in the $(1,+1)$ position, the crystal window $\tilde{w}_{1,+1}$ will be of the order of 10 seconds (of arc) and only slightly smaller in the $(1,\pm n)$ positions. If the vertical-divergence correction is held to within reasonable limits (~ 10 ppm), the geometric width $\hat{w}_g$ will be of the order of 1 second (of arc) and thus much less than $\tilde{w}_{1,+1}$. However, if we consider $(n,\pm n)$ positions, the crystal width decreases with n while the geometric width increases; hence the two may become comparable. Both widths are roughly proportional to wavelength so that the same *relative* situation holds for other x-ray lines.

K. Wavelength corrections and errors

As in the case of the single-crystal spectrometer, it is now convenient to reexpress the various errors and corrections in terms of wavelength rather than angle. These are contained in equation (2-130) and include all terms beyond the first bracket. It must be noted that the two-crystal-measurement process determines the wavelength essentially through the angle θ_n and the corresponding grating constant d_n. Hence, in converting angular error to wavelength error we may simply divide by the single-crystal dispersion D, given by (2-23). Alternatively the required factor may be computed as

$$\frac{d\theta_n}{d\lambda} = \frac{1}{2}\left[\frac{d}{d\lambda}\left(\omega_+ - \omega_-\right)\right] = \frac{\tan \theta_n}{\lambda} \qquad (2\text{-}167)$$

which follows from (2-83).

With this factor we obtain from (2-130)

$$\frac{\Delta\lambda}{\lambda} = -\frac{\bar{u}_n{}^e}{\tan \theta_n} - \frac{a^2 + b^2}{24L^2} - \frac{\delta_2{}^2 + \Psi^2}{2} + \left(2\delta_1 \sin \theta_m + \frac{\delta_2}{\sin \theta_n}\right)\Psi$$

$$- 2\delta_1{}^2 \sin^2 \theta_m - \frac{2\delta_1\delta_2 \sin \theta_m}{\sin \theta_n} \qquad (2\text{-}168)$$

(*A negative $\Delta\lambda$ implies a true value smaller than the uncorrected one.*) This equation is similar to that previously given by Bearden and Thomsen (1971), but includes the crystal centroid correction as well as second-order terms in δ_1 and δ_2. It is also consistent with the *corrected* forms of the results obtained for the $(n, \pm n)$ case by Merrill and DuMond (1961) and by Schnopper (1965). The vertical-divergence term is identical with that obtained by Williams (1932) and subsequently by Merrill and DuMond, but in clear disagreement with that given by Parratt (1935).

Equation (2-168) is analogous to (2-76) for the single-crystal instrument. The first two terms represent explicit corrections while the remainder are useful chiefly for error estimates.

If tilt and misalignment are limited such that $|\delta_1| < 10^{-4}$, $|\delta_2| < 10^{-4}$, and $|\Psi| < 10^{-3}$, wavelength errors will be of the order of 1 ppm. Under such conditions the vertical-divergence term $(a^2 + b^2)/(24L^2)$, which is identical with that for the single-crystal case, should be limited to roughly 10 ppm in order that this correction be made without appreciable error. Once again, in the absence of centroid correction or tilts, *relative* wavelength values will be unaffected by vertical divergence or misalignment.

As previously noted, the theory of the two-crystal spectrometer depends crucially on a constant horizontal angular distribution $h(\alpha)$ over an appropriate range. It is important to verify experimentally that this condition is satisfied reasonably well. This point has been discussed at length in Subsection IV H.

It should be reemphasized that the above result for the two-crystal spectrometer is based on a first-order (i.e., centroid) calculation (see Subsection II D). Sauder's (1966) analysis should be consulted for higher order corrections; these may be important when either the line or the instrumental window is highly asymmetric.

A final point should be noted here in regard to the "effective centroid" $\bar{u}_n^e$. Equations (2-123) for $\bar{u}_n^e$ may be rewritten more explicitly:

$$\bar{u}_n^e = \frac{\int_{-\infty}^{\infty} C_m^{\perp}(v)\, dv \int_{-\infty}^{\infty} u C_n^{\perp}(u)\, du + \int_{-\infty}^{\infty} C_m^{\parallel}(v)\, dv \int_{-\infty}^{\infty} u C_n^{\parallel}(u)\, du}{\int_{-\infty}^{\infty} C_m^{\perp}(v)\, dv \int_{-\infty}^{\infty} C_n^{\perp}(u)\, du + \int_{-\infty}^{\infty} C_m^{\parallel}(v)\, dv \int_{-\infty}^{\infty} C_n^{\parallel}(u)\, du} \tag{2-169}$$

or

$$\bar{u}_n^e = \frac{[\int_{-\infty}^{\infty} C_m^{\perp}(v)\, dv \int_{-\infty}^{\infty} C_n^{\perp}(u)\, du]\bar{u}_n^{\perp} + [\int_{-\infty}^{\infty} C_m^{\parallel}(v)\, dv \int_{-\infty}^{\infty} C_n^{\parallel}(u)\, du]\bar{u}_n^{\parallel}}{\int_{-\infty}^{\infty} C_m^{\perp}(v)\, dv \int_{-\infty}^{\infty} C_n^{\perp}(u)\, du + \int_{-\infty}^{\infty} C_m^{\parallel}(v)\, dv \int_{-\infty}^{\infty} C_n^{\parallel}(u)\, du} \tag{2-170}$$

where $\bar{u}_n^{\perp}$ and $\bar{u}_n^{\parallel}$ are the centroids of the second crystal for the respective polarization components. Equation (2-170) shows clearly that $\bar{u}_n^e$ is a *weighted average* of $\bar{u}_n^{\perp}$ and $\bar{u}_n^{\parallel}$, but that *both* crystals play a role in determining the weights. For example, if it happens that $\theta_m = \pi/4$, the width of the window $C_m^{\parallel}$ vanishes, as indicated by (2-14); physically this means that none of the parallel components ever reaches the second crystal. In this case $\int_{-\infty}^{\infty} C_m^{\parallel}(v)\, dv$ vanishes and (2-170) gives $\bar{u}_n^e = \bar{u}_n^{\perp}$, as might be expected.

V.   EXPERIMENTAL PROCEDURES

A.   General considerations

X-ray spectroscopy consists basically of the measurement of an angle, i.e., the Bragg angle. Through the use of (2-1) the desired wavelength is then obtained in terms of a known grating constant, or *vice versa*; by two such measurements a wavelength ratio is obtained in terms of the ratio of sines of the corresponding angles.

In the cases of the Bragg or the two-crystal spectrometers there are two aspects to the precision measurement of the Bragg angle: (1) the peak-wavelength positions between which the angle is to be measured must be accurately determined by x-ray measurements of the rocking curves, and (2) the angle must then be measured on a high-precision divided circle or its equivalent. (For the tube spectrometer the first step is replaced by the measurement of the photographic plate with a precision microscope.)

Location of the peak involves questions of optimum intensity and resolution. If the statistical error in determining the peak of the observed curve is to be held to the order of 1 ppm, the total number of counts recorded over the entire curve must be about 10^6, as shown by the analysis of Thomsen and Yap (1968).

It is also important that the observed peak be close to the true peak so that the necessary corrections may be minimized; this condition requires good resolution. If the line profile approximates a Lorentzian curve, (2-3), an instrumental-window width about one-third that of the line will be satisfactory provided that the window is reasonably symmetric. As the resolution is further increased, a narrower band of radiation is included at each point, and the number of counts in a given time interval is correspondingly reduced. Consequently the statistical error increases without any appreciable improvement in determining the true line profile; thus very high resolution may actually sacrifice accuracy. Of course, if the line contains significant fine structure, high resolution can indeed be helpful and even essential in locating the peak.

One vital note of caution should be emphasized here. The preceding theoretical analyses do not cover exhaustively all possible systematic errors in the various types of spectrometers. Other systematic errors, often not well understood, can and do occur in practice; these are to be expected almost routinely. *Hence the mere reproducibility of the values for a particular Bragg angle as measured with a specific crystal must never be taken as an indication of the true accuracy of the result.*

In order to assess this accuracy properly, as many factors as possible should be changed during the course of the experiment; variations which should be considered include:

1 Use of several crystals, including both different species of crystals and different samples of the same species

2 Measurement of two or more orders of diffraction with each crystal

3 Use of different portions of the surface of each crystal, including rotation of the crystal about its surface normal and interchange of crystal faces

4 Small displacements of the detector to ensure that the results are not significantly influenced by the detector window or by inhomogeneities in the detector surface

5 Small displacements of any slits or stops employed and small variations in slit widths

6 If feasible, measurements with both single-crystal and two-crystal spectrometers

7 Where possible, measurements both in reflection, as discussed in the preceding sections, and in transmission, as described in Section VI

Only through the consistency of a series of measurements which include most of the precautions suggested above can one obtain a reasonable estimate of the true accuracy of a spectroscopic measurement.

B. Angular measurement

The basic purpose of the x-ray spectrometer is to determine the Bragg angle to high accuracy. The primary angular measurement is usually carried out on a high-precision divided circle, which is thus the most essential component of the instrument. The circle should be read by one or more pairs of microscopes with vernier scales, located diametrically opposite one another. When the readings of such a pair are averaged, any error due to slight translation of the shaft in its bearings will be effectively removed.

When spectroscopic measurements are taken, repeated readings should be made with each microscope so that the resulting statistical error is consistent with the accuracy of the rest of the experiment. These may be taken in the usual way by a number of visual settings on the circle ruling with the vernier scale on the microscope. Much drudgery and occasional reading blunders can be avoided by equipping each microscope with a photocell and scanning the rulings photoelectrically. The effective center of the ruling may then be found by passing a smooth curve through the recorded intensity points and determining the minimum; this procedure is very similar to finding the peak of an x-ray curve, as described in Subsection V I.

It is desirable that the circle be calibrated and rechecked from time to time. If there are two pairs of microscopes spanning a suitable submultiple of the full circle, it may be convenient to use a symmetric method outlined by Thomsen (1966), in which each line to be calibrated is read once by each microscope. Digital angular generators may also be employed, as was done by Deslattes (1967); the accuracy is probably somewhat lower.

An angular interferometer, such as that described by Marzolf (1964), may also be useful. Normally the interferometer itself must be calibrated in terms of a known angle; this may be done by closing a full circle or a half-circle with a series of interferometer measurements. (If the circle is equipped with diametrically opposite microscope pairs, a 180° interval may be calibrated directly, unaffected by any errors in the circle rulings.) Since the angular interferometer can cover only a restricted range at one step, it is not usually a convenient device for measuring the angle $\pi - 2\theta_n$. However, in the transmission method described in Section VI, the angle to be measured is $2\theta_n$; for a small Bragg angle it is then entirely practicable to measure this quantity directly with the interferometer.

C. Alignment requirements

In order that the apparent precision of the angular measurement not be degraded by serious systematic error, it is essential that the spectrometer components be properly aligned. The necessary tolerances have been discussed at some length in Subsections III I and IV K. It was shown there that in order to reduce such systematic errors to the order of 1 ppm, it is usually sufficient to hold the misalignment angle Ψ to within 10^{-3} rad and the crystal tilt(s) to within 10^{-4} rad, i.e., approximately 3 minutes and 20 seconds (of arc), respectively. These conditions can be satisfied relatively easily if the method given below is properly applied.

The procedure will be described first in detail for the two-crystal spectrometer

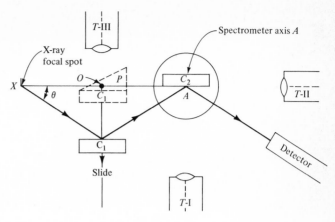

FIGURE 2-14

Plan view of the two-crystal spectrometer for which the alignment procedure is described. The first crystal C_1 is mounted on a slide; the second crystal C_2 is mounted on the axis A of the spectrometer; the distances are fixed such that $XO = OA$. T-I, T-II, and T-III represent the three positions used during the alignment procedure for the autocollimating telescope of the engineer's level. The 90° prism P is inserted to align the telescope on the line XOA; this then determines the focal spot position X, the zero position of the slide, and the angular position of crystal C_2 for the $(1, -1)$ position.

used at Johns Hopkins University. Modifications required for the single-crystal instrument will then be discussed.

The general plan of the two-crystal spectrometer for which the adjustment is to be described is shown in Fig. 2-14. In this arrangement, the first crystal C_1 is mounted on a slide, which is perpendicular to a line joining the axis A of the spectrometer to the focal spot of the x-ray tube. C_1 is equidistant between the focal spot and the second crystal C_2, i.e., $OX = OA$. Only the second crystal C_2 rotates, with its angular position determined by the high-precision divided circle, which is read by four microscopes in four different positions (two diametrically opposite pairs). The wavelength band that can be studied without changing the position of C_1 is limited by the lengths of the crystals; this restriction usually presents no problem since most high-resolving-power measurements involve relatively small wavelength ranges.

The selected surfaces of crystals C_1 and C_2, which have been cut and ground parallel to the desired atomic planes (by the procedure to be described later in Subsection V G), are held against three-point supports by light springs. The optical alignment of these three-point supports parallel to the axis of rotation A (for the most precise work the spectrometer axis A should lie in the plane of the face of C_2) constitutes the major portion of the alignment. This part of the procedure is identical for both two-crystal and single-crystal instruments.

The method of adjustment described below is based on the use of an *engineer's*

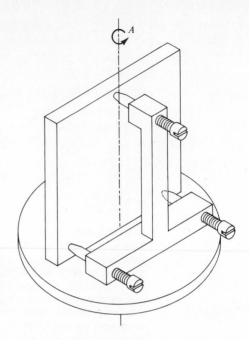

FIGURE 2-15
Schematic diagram of one of the three-point crystal supports for C_1 and C_2. Light springs (not shown in figure) hold the crystal against the three screws. The support for the top screw is designed to permit the x-ray beam to pass between it and the crystal for all Bragg angles employed.

level equipped with an autocollimating telescope whose optical axis can be made horizontal to within a few seconds of arc. These telescopes are readily available with accuracies as great as a fraction of a second.[1] A microscope with a magnification of approximately $100\times$, equipped with cross hairs and a divided scale, is required for locating the crystal surface C_2 on the axis of rotation A. This step is essential for wavelength measurements with imperfect crystals; otherwise it is unimportant.

A suitable three-point crystal support is illustrated in Fig. 2-15. It is desirable that the screw threads (of the order of 20 per cm) be sufficiently tight so that the screws can be turned only with difficulty. Three light springs (not shown) just sufficient to hold the crystal in place are located immediately back of the crystal. These are directly in line with the axes of the three screws in order not to produce a bending strain in the crystal.

D. Adjustment of the rotating crystal

In adjusting the holder for the rotating crystal (C_2 in the case of the two-crystal spectrometer), the primary requirement is to place the plane formed by the rounded ends of the three crystal-support screws parallel to the axis of rotation A. At the same

---

[1] For example, a Keuffel and Esser Company (Hoboken, N.J.) 9092-3A Tilting Level has been found satisfactory in our laboratory.

time the end of the top screw must be located approximately on the axis A; this requirement is less critical. To facilitate this last adjustment, the holder is constructed and mounted so that the axis of this screw passes through A within 0.1 mm.

The microscope is mounted near the axis of rotation and focused on the end of the top screw. Rotation of the spectrometer circle by 180° and adjustment of the screw now fix the position of its end very nearly on the spectrometer axis A.

Then a plane mirror, an optical flat aluminum-coated on one side, is mounted with the reflecting coating in contact with the three screws and adjusted parallel to the axis A as follows. The level with its autocollimating telescope is placed in a convenient location, T-I in Fig. 2-14, with its axis roughly parallel to a line joining two of the base-leveling screws of the spectrometer. The position is chosen so that the observer can reach the crystal-support adjusting screws while looking at the image of the reflected cross hairs; the precision of the telescope is unaffected by the object distance. The top screw support does not interfere with the telescope image. The two lower crystal-support screws and the telescope tilt are adjusted until the horizontal cross hairs coincide after repeated 180° rotations of the mirror. In this way the surface of the mirror can be located within 0.01 mm of the axis of rotation and parallel to it within 1 second of arc.

The next step is to make the spectrometer axis vertical. For this purpose the telescope axis is adjusted to be horizontal by means of the level attachment. The base-leveling screws of the spectrometer are then adjusted so that the mirror is perpendicular to the horizontal axis of the telescope. The level is then moved approximately 90° to position T-II (see Fig. 2-14) and the appropriate spectrometer leveling screw adjusted. If the initial location of the level T-I was properly chosen with respect to the spectrometer leveling screws, a recheck in position T-I will indicate a precisely vertical axis. This adjustment should be made to within a few seconds of arc, as it is of prime importance in minimizing the misalignment angle Ψ and the tilt angle δ.

E.   Other adjustments: two-crystal case

The plane of the three-point support of the first crystal C_1 (similar to the C_2 support shown in Fig. 2-15) must now be adjusted perpendicular to the axis of the slide ($\pm 1°$) and parallel to the spectrometer axis [± 5 seconds (of arc)]. The slide (with scale) which supports the first crystal C_1 should be about 20 to 25 cm in length and rigidly mounted to the spectrometer. It is placed so that one extreme position of the three-point support lies on the line joining the proposed location of the focal spot and the axis A. The telescope is put in position T-III and its axis adjusted to coincide with the slide. This is accomplished by moving the slide as close to the telescope as possible, focusing on a suitable mark on the slide, and then moving the slide as far away from the telescope as possible and refocusing. When the vertical cross hair and the mark remain in coincidence, the adjustment is complete. The plane mirror used for aligning the second-crystal support is then placed in contact with the three screws of the first-crystal holder, and, with the telescope level, the three-point support is adjusted normal to the telescope axis and hence parallel to the spectrometer axis A.

The final steps required are the location of the x-ray focal spot and the determination of the angular position of the second crystal C_2 for x-ray diffraction in the $(1,-1)$ position. For this purpose the axis of the telescope must pass horizontally through the axis of the spectrometer A and be parallel to the three-point plane of the first crystal C_1. The first-crystal support is adjusted to a position near the inner extreme of the slide and the mirror replaced by a 90° prism (as shown by dotted lines in Fig. 2-14) with an aluminized surface facing the spectrometer axis. The telescope is placed in position T-II and adjusted until level. It is then repositioned slightly until it is perpendicular to the prism, with its axis through the spectrometer axis A. This condition is verified by alternately focusing on the end of the top screw of the second-crystal support C_2 and on the reflected image from the normal prism face. The image of the horizontal cross hairs can be adjusted by sliding one end of the prism slightly up or down on the three-point support. When this adjustment has been completed, the prism is removed.

The slide is then moved until the end of the top screw on the first-crystal support is in line with the telescope axis; this fixes the zero reading from which the position of the slide is calculated for a particular wavelength λ and crystal grating constant d. The exact position of the focal spot can be determined by measuring along the telescope axis a distance $OX = OA$, where O denotes the center of the first-crystal support (see Fig. 2-14). For most x-ray tubes external measurements suffice to locate the focal spot in the correct position.

The mirror is now placed in the holder for the second crystal C_2 and rotated to a position normal to the telescope axis. An additional rotation of the mirror by 90° (determined by the divided circle) thus puts it in the $(1,-1)$ position. The prepared crystals are then placed in the respective supports for C_1 and C_2. The position of the slide is calculated from the Bragg angle θ and the distance OA. A detector without slits is set at the approximate position for $(1,-1)$ diffraction, and the x-ray source is turned on; a slight angular rotation of C_2 will locate the $(1,-1)$ peak position. Appropriate slits and lead stops should now be inserted to prevent direct or scattered x rays from entering the detector; in this process it is most important not to block part of the wavelength region under investigation, as emphasized in Subsection IV H.

The final step is to add the slits defining vertical divergence. A horizontal slit of height a near the x-ray tube and one of height b at the detector are centered on the horizontal plane containing the crystal centers, with the aid of the telescope and level. The slits can easily be positioned to within 0.1 mm; thus for a separation of 50 cm between the two horizontal slits $|\Psi| \leq 2 \times 10^{-4}$.

With the above procedure the magnitudes of δ_1 and δ_2 should not exceed 10^{-5}, provided that the atomic planes are precisely parallel to the crystal surfaces; hence for this case (2-168) indicates that $\Delta\lambda/\lambda < 0.03$ ppm ($\theta_n = 30°$), a completely negligible correction. If we allow a factor of 10 in δ_1 and δ_2 for the angle between the crystal surfaces and the atomic planes, we still have $\Delta\lambda/\lambda < 0.1$ ppm.

F.   Other adjustments: single-crystal case

The single-crystal Bragg spectrometer requires a beam of radiation with a narrow horizontal angular spread. Two slits, each 0.1 mm in width, separated by 1000 mm will give an angular width of about 20 seconds; this provides adequate resolution for long wavelengths ($\lambda \sim 1.5$ A) and higher order ($n > 1$). Intensity considerations will govern the minimum angular width; by using a rotating-target x-ray tube with a narrow-line focus, Bearden et al. (1964) achieved a width of only 6 seconds (of arc).

Recently Bearden[1] has employed a single-crystal instrument for the short-wavelength W $K\alpha_1$ line ($\lambda \sim 0.2$ A). With slit widths of 0.2-mm and 1500-mm separation he was able to obtain adequate intensity from a commercial tube to observe the line in orders as high as the fourth. Resolution was less than optimum, but still adequate for wavelength measurement.

Under ideal conditions it is sufficient that the narrow beam of incident radiation, which is formed by the slit system, lie in a vertical plane, i.e., parallel to the spectrometer axis. However, in practice it is also highly desirable that the beam be centered on the axis so that the same portion of the crystal is used in both the A and B positions. Hence the slits must be made vertical and the centers placed on a straight line passing through the spectrometer axis. (Prior to this adjustment, slits defining vertical divergence should be inserted, as described in the preceding section.)

In order to block effectively the unwanted portion of the beam, the slit jaws must be relatively broad rather than knife edges; thus the edges may be milled to give plane surfaces. These can be employed in the vertical adjustment as follows: After the slits have been set up in close approximation to the required alignment, one jaw is temporarily removed from the first slit and a small plane mirror placed against the other jaw. The telescope level, discussed in the last two subsections, is next employed to set the jaw vertical. This can be done within a few seconds (of arc) although a somewhat cruder adjustment should be adequate. Appropriate spacers are then inserted at the top and bottom and the second jaw clamped in place to form the slit. The process is repeated for the second slit.

The final operation is to place the second slit on a line joining the first slit and the spectrometer axis; for this purpose the second slit should be provided with a slide adjustable by a micrometer screw. This step may be carried out with the aid of the

[1] Private communication.

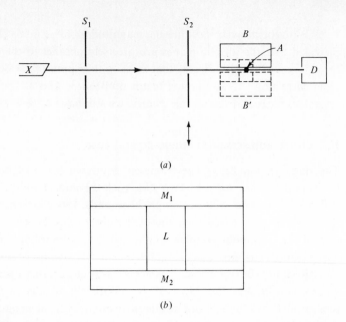

(a)

(b)

FIGURE 2-16
Method of aligning slit system for Bragg spectrometer. (a) Stop B in crystal
holder with face on spectrometer axis A. X-ray beam from source X passes
through slit system S_1S_2 and enters detector D. Slit S_2 is adjusted until equal
intensities are obtained with stop in positions B and B'. (b) Vertical face of
stop B. L denotes a lead strip which acts as the stop. M_1 and M_2 are metal
strips milled flat with the surface of L and used for mounting the assembly in
crystal holder. Remainder of block is recessed.

lead stop, as indicated in Fig. 2-16a. The stop, which is shown in detail in Fig. 2-16b,
is constructed so that the outer edge of the lead is on the spectrometer axis when the
block is mounted in the crystal holder. Thus, when the x-ray beam is properly centered
and the surface of the lead is parallel to it, the stop will pass just half of the incident
radiation.

The block is first placed in position B, as shown in Fig. 2-16a; the angle is varied
slightly to obtain a maximum detector reading, which indicates parallelism with the
beam. It is then rotated 180° to position B' and the process repeated. If the readings
are unequal, the slit S_2 is shifted slightly in the proper direction and a new set of readings
taken. These can be made to agree to within about 1%, which will usually be sufficient.

An alternative method of making the slits vertical is available at this point.
The stop is left in position to pass half the beam as described above. The detector is
then provided with an auxiliary slit, which limits the input to a small part of the upper
half of the beam and the intensity noted. The slit is then moved to select the corre-

sponding portion of the lower half of the beam and the process repeated. If the readings are unequal, the slits are rotated slightly in the appropriate directions to attain equality; this adjustment will be more sensitive to the second slit.

The procedure just discussed is only partially applicable to the case of the tube spectrometer. However, the necessary modifications will not be discussed here. Larsson (1927) has described a method of aligning this instrument for high-precision work.

G.   Spectroscopic crystals

The crystals most commonly utilized in x-ray spectroscopy are calcite, quartz, and silicon. Calcite and quartz occur naturally and have been used almost from the start; the (211) cleavage plane of calcite is usually employed, while several different possible choices have been extensively used for quartz. In recent years excellent low-impurity silicon crystals have been grown artificially and are highly suited for x-ray work. Indeed it is quite likely that silicon may largely displace the others for high-precision measurements. Good artificially grown germanium crystals provide another possibility. However, due to absorption, the germanium crystal window is quite asymmetric for wavelengths of 1 A* or longer; hence it is hardly competitive with silicon except for short wavelengths.

Table 2-1 shows some of the important characteristics of the three common spectroscopic crystals. The first group—temperature coefficient(s), index of refraction, and shortest-wavelength absorption edge—are general properties. (This absorption edge is listed to give a rough idea of crystal-window asymmetry; the more this edge exceeds the measured wavelength, the less the crystal-window asymmetry and hence the smaller the crystal-centroid correction.) The second group of properties, such as grating constants, applies only to the specific crystal plane indicated. This includes the maximum wavelength measurable; since $n \geq 1$ and $\sin \theta_n \leq 1$, relation (2-1) shows that $\lambda \leq 2d$.

It should be emphasized that these constants are given as typical values rather than being precisely applicable to any specific crystal. Differences of the order of 10 ppm have been reported between good calcite crystals (Bearden, 1965a) and between good quartz crystals (Bearden et al., 1964). Silicon values are more consistent, possibly an order of magnitude better. However, for the present it is best to consider that *each sample of a given crystal must be individually calibrated*, either in terms of reference wavelengths or against a known crystal. Hart (1969) has given a high-precision technique for comparing the nearly equal grating constants of two similar crystals.

It will be noted that the index-of-refraction term is of the same order of magnitude for all crystals listed in Table 2-1. This correction is roughly 100 ppm in the first order and decreases inversely as n^2 as the order is varied. Of course, the assumption of a constant value for δ/λ^2 breaks down as an absorption edge is approached; this effect is called *anomalous dispersion*. Of those crystals listed in Table 2-1 calcite has the lowest absorption edge; the crystal window has significant asymmetry and requires correction for anomalous dispersion for wavelengths of the order of 2 A* or higher.

Table 2-1 APPROXIMATE DATA ON SELECTED CRYSTALS USED IN X-RAY SPECTROSCOPY

Property	Crystal		
	CaCO₃ (Calcite)	SiO₂ (Quartz)	Si (Silicon)
Temperature coefficient(s) of expansion $\tilde{\alpha}$ (°C)⁻¹	$\tilde{\alpha}_{\parallel} = 25.14 \times 10^{-6}$† $\tilde{\alpha}_{\perp} = -5.58 \times 10^{-6}$†	$\tilde{\alpha}_{\parallel} = 7.97 \times 10^{-6}$† $\tilde{\alpha}_{\perp} = 13.37 \times 10^{-6}$†	2.32×10^{-6}‡
Index of refraction δ/λ^2 (A*)⁻²	3.69×10^{-6}§	3.60×10^{-6}§	3.22×10^{-6}††
Shortest-wavelength absorption edge λ_{abs} (A*)	3.07	6.74	6.74

Property	Plane		
	(211)	(10$\bar{1}$1)	(220)
Angle ϕ (2-22)	45.5°	38°	
$\tilde{\alpha}_{effective}$ (°C)⁻¹	10.2×10^{-6}	11.4×10^{-6}	2.32×10^{-6}
$4d^2\delta/\lambda^2$	136×10^{-6}	160×10^{-6}	47×10^{-6}
d_1 (A*)	3.035 53	3.342 93	1.920 04
d_2 (A*)	3.035 84	3.343 33	1.920 11
d_∞ (A*)	3.035 94‡‡	3.343 46‡‡	1.920 13‡‡
λ_m (A*)	6.07	6.69	3.84

Note: The A* unit, which is very nearly equal to the angstrom (A), is defined and discussed in Section VII. Grating constants for other orders may be computed from (2-20), viz.:

$$d_n = d_\infty \left(1 - \frac{4d^2}{n^2}\frac{\delta}{\lambda^2}\right) \qquad (2\text{-}20)$$

These grating constants may be considered as typical, but should *not* be assumed to apply precisely to every sample of a given species. *For precision work each crystal sample should be individually calibrated.*

† *Handbook of chemistry and physics*, 44th ed., pp. 2326–2329, The Chemical Rubber Publishing Co., Cleveland, 1962. There is a sign error in $\tilde{\alpha}_{\perp}$ for calcite.

‡ GIBBONS, D. F. (1958): *Phys. Rev.*, **112:** 137.

§ SANDSTRÖM, A. E. (1957): *Handbuch der physik*, vol. 30, p. 143, edited by S. Flugge, Springer-Verlag, Berlin.

†† HENINS, I., and BEARDEN, J. A. (1964): *Phys. Rev.*, **A135:** 890–898.

‡‡ BEARDEN, J. A., private communication.

While the planes listed in Table 2-1 do not permit wavelength measurement above 6.7 A*, other choices are available. The $(10\bar{1}0)$ plane of quartz permits measurements up to about 8.5 A*. Further extension of the wavelength range may be attained through the use of organic crystals, such as ammonium dihydrogen phosphate (ADP). Such crystals tend to deteriorate with age and are not competitive in the shorter wavelength region where more conventional crystals are practicable.

H. Preparation of crystals

Bearden and Henins (1965) recently employed a method of testing crystals which made use of the highly penetrating W $K\alpha_1$ x-ray line, used in either reflection or transmission. While this choice has certain advantages, it is not essential except for the most precise measurements involving short wavelengths. The Mo $K\alpha_1$ line is quite satisfactory both for checking the angle between the crystal surface and the atomic plane in the grinding procedure and for testing the geometrical perfection of the finished crystal.

Initially an optically adjusted two-crystal spectrometer can be equipped with cleaved calcite surfaces or other crystals which have been cut parallel to the desired planes within commercial tolerances. The rotating crystal is adjusted to the peak of the $(1, -1)$ position; it is then removed and ground with a medium grinding material on a cast-iron lap until a flat surface has been obtained. The crystal is returned to the spectrometer and the angle of the $(1, -1)$ peak established. The crystal is then rotated 180° about the normal to the ground surface, and the new $(1, -1)$ peak angle is established. The difference in the two $(1, -1)$ angles, as read on the divided circle or tangent screw, represents twice the angle between the ground surface and the true atomic plane for this orientation of the crystal. The crystal is then rotated 90 and 270° about the same axis; a new pair of $(1, -1)$ measurements establishes the error of the ground surface in this second orientation. From these observations one can determine the part of the crystal surface that requires the most grinding and proceed accordingly.

When the error has been reduced to about 10 minutes (of arc), $5-\mu$ powder is substituted for the final grinding. After the grinding has been completed, the crystal should be etched for approximately 20 seconds in 0.1 normal HCl for calcite or in 50% HF for quartz; for silicon an etchant consisting of 20 parts (by volume) glacial acetic acid, 12 parts concentrated nitric acid, and 3 parts concentrated hydrofluoric acid may be used. The finished crystal is now substituted for the first crystal and the process repeated. After some experience, one can complete the procedure in 2 to 4 hours and obtain a pair of crystals whose faces are within approximately 10 seconds (of arc) of the atomic planes. This technique has been used satisfactorily with crystals of calcite, quartz, and silicon, and should be applicable to most crystals used in x-ray spectroscopy.

In the single-crystal case there is, of course, only one crystal involved. The procedure is generally similar to that just described except that it is obviously impossible to employ the $(1, -1)$ position, where the observed width is due solely to the instrumental window. The alternative is to use as low dispersion as possible, i.e., a first-order position.

I.   Peak-wavelength location

To determine the peak wavelength, it is essential to obtain the profile of the line in both positions of the spectrometer. For this purpose the instrument should be provided with a reduction gear and vernier dial or some other means of stepping off small equal angular increments of the order of a fraction of a second. The intensity is recorded over a selected range symmetric about the peak; about 20 points will suffice to approximate a continuous curve. The range is usually of the order of w, the full width at half-intensity; it may be as low as $w/2$, i.e., between the 80% intensity points.

In earlier work the resulting data were generally plotted and the peak obtained graphically, as shown in Fig. 2-17. A number of chords were drawn across the curve and their midpoints fitted to a smooth bisector curve, which was extended to intersect the line profile at the "extrapolated peak." This process permits a rather precise peak location, which is insensitive to statistical errors in the data points near the maximum.

While such a procedure is very useful in graphical work, Thomsen and Yap (1968) showed that the extrapolated peak has little or no advantage when a more mathematical approach is used. With the wide availability of computers the latter method seems preferable. A least-squares fit of the experimental data to a suitable polynomial is obviously suggested. With higher order polynomials a direct application of this method may produce serious roundoff errors, even when a computer is used. The orthogonal-polynomial technique described by Forsythe (1957) yields the same result without any roundoff problem. This approach was treated in considerable detail by Thomsen and Yap (1968); the remainder of this subsection summarizes their analysis.

In principle each curve should be fitted with a set of orthogonal polynomials appropriate to the statistical weights of the various data points. However, when the range does not go beyond the half-intensity points, no serious error will result from assuming equal weights; in this case the required set is simply the well-known Legendre polynomials.

The statistical error in locating the peak of an x-ray profile may be represented by a semiempirical expression, provided that the range R is symmetric about the peak and $w/2 \leq R \leq w$. This result, which is valid to within a few percent for fourth-, sixth-, and eighth-degree polynomials, is

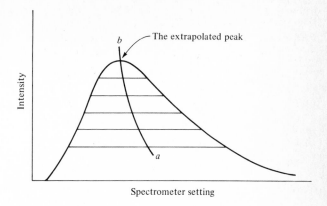

FIGURE 2-17
Location of the "extrapolated peak" by means of a smooth bisector curve through the chord midpoints.

$$\frac{\sigma_p}{w} \approx \frac{0.13}{(I_p T)^{1/2}} \left(\frac{Mw}{R}\right)^{3/2} \left(\tan^{-1}\frac{R}{w}\right)^{1/2} \qquad (2\text{-}171)$$

where σ_p is the standard deviation of the peak, M the degree of the polynomial, I_p the intensity (counts per unit time) at the peak, and T the *total counting time* for all points. From these definitions it is clear that $I_p T$ is of the same order as the total number of counts recorded, but somewhat greater; it is independent of the number of points involved. It should be noted that this result is based solely on statistical counting errors; in particular, it is assumed that there is no error in the abscissa steps.

It is clear from the expression (2-171) that the statistical error is minimized by using as low a degree of polynomial as possible. On the other hand, such a choice may increase the systematic error, particularly if the line has an appreciable asymmetry. The latter danger is apt to be a more serious one since it will not be evident from an inspection of the data. The adequacy of the fit may be verified by a χ^2 test or by comparison of the coefficients of the orthogonal polynomials with their respective statistical errors.

The statistical error σ_p depends explicitly on the range R, as shown by equation (2-171). There is also an implicit dependence on range through M since the degree of the polynomial must usually be increased with range to obtain an adequate fit. These two effects tend to cancel one another so that the error is not particularly sensitive to range.

We can now consider the requirements for holding the wavelength error to the order of 1 ppm. The ratio of w to λ is usually of the order 10^{-3}, and somewhat smaller for the prominent K lines. Hence it is necessary to locate the peak of each curve within a few parts per thousand of its width and to repeat this process for a number of curves. For this purpose a value of I_pT between 10^5 and 10^6 is usually required. This quantity may alternatively be regarded as the total count at the peak times the number of points on the curve; thus 20 points with a peak count of 2×10^4 will be a reasonable choice.

If the profile is a rough approximation to a Lorentzian curve, then for a range $R = w$ a sixth-degree polynomial will limit the systematic error to a value consistent with the statistical accuracy. In this special case (2-171) is replaced by the more exact expression

$$\frac{\sigma_p}{w} = \frac{1.67}{(I_pT)^{1/2}} \qquad M = 6, R = w \qquad (2\text{-}172)$$

For a profile of relatively high asymmetry (e.g., the Fe $K\alpha_1$ line), an eighth-degree polynomial may be preferable. As the range is decreased to roughly $2w/3$, a fourth-degree polynomial may suffice for a reasonably symmetric case.

If the line profile can be assumed as an exact Lorentzian, the parameters for this function may be found by a least-squares fit and the error of (2-172) halved. Alternatively it is possible to restrict the range to approximately $w/3$ (i.e., between the 90% intensity points) and fit the data with a parabola. For a symmetric line this approach yields an error slightly lower than that of (2-172) and is obviously simpler mathematically. However, in either case a small asymmetry can produce a significant systematic error; hence these methods should be used only with great caution.

J.   Resolving power

Approximate numerical values for the factors determining resolving power are listed in Table 2-2 for wavelengths of 1.54 and 0.71 A* (Cu $K\alpha_1$ and Mo $K\alpha_1$ lines). The table includes natural line widths, geometrical window widths for typical slit systems, and crystal window widths for the three selected crystal planes previously considered. Substantially narrower crystal windows may be obtained by asymmetric reflection (i.e., with crystal surface oblique to the atomic plane), as described by Kohra (1962), Renninger (1967), and Kikuta and Kohra (1970).

A number of features, some previously discussed, are evident from an inspection of this table, viz.:

1   The single-crystal geometrical windows are independent of wavelength, while crystal windows, two-crystal geometrical windows (due to vertical divergence), and natural line widths are all roughly proportional to wavelength. Thus the greater the wavelength, the more readily a single-crystal instrument can be applied. However, the narrower geometrical window, with $b = 0.015$ mm, is only practicable with a narrow-focus high-intensity x-ray source, such as the rotating-target tube used by Bearden et al. (1964).

2   As the order is raised, the natural line width increases while the crystal window decreases. Thus resolution can be increased by going to higher orders, provided that such orders are permitted by the Bragg law and that sufficient intensity is available.

Table 2-2   APPROXIMATE WINDOW WIDTHS AND NATURAL LINE WIDTHS (SECONDS OF ARC) FOR Cu $K\alpha_1$ AND Mo $K\alpha_1$ RADIATION

	Order	Cu $K\alpha_1$ ($\lambda \approx 1.54$ A*)	Mo $K\alpha_1$ ($\lambda \approx 0.71$ A*)
Single-crystal geometrical window $\hat{w}_g$ (with $b' > a'$)			
$b' = 0.015$ mm, $L = 500$ mm	all	6	6
$b' = 0.2$ mm, $L = 1500$ mm	all	28	28
Single-crystal window $\tilde{w}_n$			
$CaCO_3$ (211)	1	6.5	3.0
	2	1.3	0.6
SiO_2 (10$\bar{1}$1)	1	4.2	2.0
	2	0.9	0.4
Si (111)	1	6.5	3.0
	3	1.9	0.5
Two-crystal window $\tilde{w}_{m,n}$			
$CaCO_3$ (211)	(1,±1)	9.0	4.1
	(1,±2)	7.4	3.3
	(2,±2)	1.8	0.8
SiO_2 (10$\bar{1}$1)	(1,±1)	5.8	2.7
	(1,±2)	4.7	2.2
	(2,±2)	1.2	0.5
Si (111)	(1,±1)	9.0	·4.1
	(3,±3)	2.5	0.7
Two-crystal geometrical window $\hat{w}_g$			
$a = 5$ mm $= b$, $L = 500$ mm	(1,+1)	≈1	≈0.5
	(1,−1)	0	0
	(2,+2)	≈2	≈1
Natural line width w			
	1	≈15	≈10
	2	≈30	≈20
	(1,+1)	≈30	≈20

Note: The above data on crystal windows are based on unpublished calculations by Marzolf (private communication) on the basis of the Darwin-Prins theory. The figures quoted are only approximate and in any event will be somewhat modified for real crystals.

The single-crystal geometrical window is calculated from equation (2-54). The two-crystal geometrical windows are given by (2-166); the quoted values represent *only rough estimates* since these equations involve the Bragg angles for the specific crystals used.

The natural line widths (in seconds of arc) will depend on the Bragg angle; again the quoted values give only the order of magnitude.

3 It will be noted that the $(n,+n)$ two-crystal windows are roughly one-third greater than the single-crystal ones, as computed by Compton and Allison (1935). Equations (2-24) and (2-85) show that the resolutions for the respective cases are $(2 \tan \theta_n)/\hat{w}_{n,n}$ and $(\tan \theta_n)/\hat{w}_n$. Thus the resolution in the $(n,+n)$ two-crystal position exceeds the nth-order single-crystal value by about 50%, even if the geometrical window is negligible in the latter case. This is in general agreement with the conclusion of Allison (1931).

4 However, at longer wavelengths the single-crystal spectrometer, if used in the nth order with a narrow-slit system, may well be competitive in resolution with the two-crystal instrument employed in the $(1,\pm n)$ positions. At shorter wavelengths the two-crystal spectrometer has a clear advantage in resolution.

It should be recalled that, if the line profile is roughly resembling a Lorentzian in shape, a resolution of about one-third of the natural line width is desirable. This will avoid any serious distortion of the profile and ensure the validity of the peak correction as derived in Subsection II D. Further increase in resolution may unduly diminish intensity, as discussed in the following subsection.

K. Intensity

Up to this point we have been concerned with only the *shape* of the function $F(\beta)$ and the location of the peak. It is now useful to consider how some of the instrumental parameters affect intensity, i.e., the *amplitude* of this function. In this subsection we will derive approximate expressions for the *peak intensity* I_p under the restriction that *both crystal and geometric windows are small compared to natural line width*.

Consider first the single-crystal case. If the above restriction holds, $J(z)$ may be approximated by $J(0)$ in (2-37) with the result

$$I_p \approx F(0) = J(0) \int_{-\infty}^{\infty} \iint C_n[-z - \alpha + \varepsilon(\psi,\delta_1)]G(\alpha,\psi)\, d\alpha\, d\psi\, dz \qquad (2\text{-}173)$$

where the wavelength range has been extended from $-\infty$ to $+\infty$. (This change produces a completely negligible error.) The above result may now be transformed to give

$$I_p = J(0) \iint G(\alpha,\psi) \int_{-\infty}^{\infty} C_n(u)\, du\, d\alpha\, d\psi \qquad (2\text{-}174)$$

or, since $G(\alpha,\psi)$ is normalized,

$$I_p = J(0) \int_{-\infty}^{\infty} C_n(u)\, du \qquad (2\text{-}175)$$

The integral in this expression has been called the "coefficient of reflection of a single

crystal" [e.g., Compton and Allison (1935)]. This is perhaps an unfortunate terminology since it is expressed in angular units and is, in fact, roughly equal to the crystal
window $\tilde{w}_n$.

It remains to evaluate $J(0)$ in terms of slit dimensions. Equation (2-173) indicates that the radiation per unit wavelength per unit solid angle *incident* on the crystal
along the direction of the central axis is $J(0)G(0,0)$. This quantity should be proportional to the minimum cross-sectional area defining the beam. If the slits are such
that the width $a' < b'$ and the height $a < b$, then this area is aa', and we have with
the aid of (2-38)

$$J(0)G(0,0) = J(0)h(0)g(0) = J_s aa' \qquad (2\text{-}176)$$

where J_s is a constant characteristic of the x-ray source. It follows from (2-47) and
(2-48) that $g(0) = L/b$; similarly we will have $h(0) = L'/b'$, where L' is the separation
of the slits defining horizontal angular distribution. With these values equation
(2-176) becomes

$$J(0) = \frac{J_s aa'bb'}{LL'} \qquad (2\text{-}177)$$

Substitution in (2-175) now yields

$$I_p = \frac{J_s aa'bb'}{LL'} \int_{-\infty}^{\infty} C_n(u) \, du \qquad (2\text{-}178)$$

We note that, according to (2-54), $b'/L' = \hat{w}_g$, the geometrical width. Furthermore
the integral in expression (2-178) is roughly equal to the crystal width [actually about
25% greater if the Darwin function (2-12) holds precisely]. Thus (2-178) may be
rather crudely approximated by

$$I_p \approx \frac{J_s aa'b}{L} \hat{w}_g \tilde{w}_n \qquad (2\text{-}179)$$

The physical basis of this result is as follows: since $J(z)$ is assumed to be constant,
the crystal reflects the same fraction of the incident intensity for all angles of incidence.
The wavelength band reflected is a function of angle, but the fraction itself remains
constant and is proportional to the crystal width $\tilde{w}_n$. The total energy incident on the
crystal is proportional to both the minimum cross-sectional area, i.e., aa', and the
solid angle subtended, i.e., $bb'/(LL')$. Since $\hat{w}_g = b'/L'$, equation (2-179) follows. It
should be reemphasized that this result is valid *only if $\tilde{w}_n$ and $\hat{w}_g$ are small compared
to natural line width*.

A similar approach can be used for the two-crystal case. With the same restriction as to narrow instrumental window, (2-118) becomes

$$I_p \approx F(0)$$

$$= \tfrac{1}{2} J(0) \sum_P \int_{-\infty}^{\infty} \int g(\psi) \int_{-\infty}^{\infty} C_m^P(v) C_n^P(-z \mp v \pm \varepsilon_1 + \varepsilon_2) \, dv \, d\psi \, dz \qquad (2\text{-}180)$$

After a variable transformation, this yields

$$I_p = J(0)(\tfrac{1}{2}) \sum_P \int_{-\infty}^{\infty} C_m{}^P(v)\, dv \int_{-\infty}^{\infty} C_n{}^P(u)\, du \qquad (2\text{-}181)$$

In this case (2-180) implies that the radiation per unit wavelength per unit solid angle *incident* on the first crystal along the direction of the central axis (reference ray) is $J(0)g(0)$. [In Subsection IV D, $h(\alpha)$ was assumed constant over the region of interest and no longer appears explicitly.] Hence (2-177) is replaced by

$$J(0) = \frac{J_s a a_x' b}{L} \qquad (2\text{-}182)$$

In this expression a_x' denotes the width of the focal spot of the x-ray source since this parameter rather than a physical slit limits the horizontal cross section.

Thus in place of (2-178) and (2-179), we obtain

$$I_p = \frac{J_s a a_x' b}{2L} \sum_P \int_{-\infty}^{\infty} C_m{}^P(v)\, dv \int_{-\infty}^{\infty} C_n{}^P(u)\, du \qquad (2\text{-}183)$$

and

$$I_p \approx \frac{J_s a a_x' b}{2L} (\tilde{w}_m{}^{\perp}\tilde{w}_n{}^{\perp} + \tilde{w}_m{}^{\parallel}\tilde{w}_n{}^{\parallel}) \qquad (2\text{-}184)$$

These results are *valid only if both crystal widths are small compared to natural line width.*

It is also of interest to compare the two-crystal and single-crystal peak intensities. The single-crystal width w_n may be written with somewhat greater accuracy as $\tfrac{1}{2}(\tilde{w}_n{}^{\perp} + \tilde{w}_n{}^{\parallel})$. Thus from (2-179) and (2-184) we obtain the approximate forms

$$\frac{(I_p)_{TC}}{(I_p)_{SC}} = \frac{a_x'}{a'\hat{w}_g} \frac{\tilde{w}_m{}^{\perp}\tilde{w}_n{}^{\perp} + \tilde{w}_m{}^{\parallel}\tilde{w}_n{}^{\parallel}}{\tilde{w}_n{}^{\perp} + \tilde{w}_n{}^{\parallel}} \approx \frac{a_x'\tilde{w}_m}{a'\hat{w}_g} \qquad (2\text{-}185)$$

The focal-spot width a_x' in the two-crystal case may frequently be much greater than the slit width a' for the single-crystal instrument, thus producing substantially higher observed intensity. This factor may be partially counterbalanced if the geometrical window width $\hat{w}_g$ is considerably larger than that of the first-crystal window $\tilde{w}_m$.

It is evident that for either instrument the observed intensity is roughly proportional to the product of two window widths. Thus a decrease in the smaller of these factors will diminish intensity without any significant improvement in resolution, which will always be dominated by the larger window. Hence there is little point in ever attempting to make the single-crystal geometrical window smaller than the crystal one. On the other hand, it may often be desirable to employ an $n > 1$ and make $\tilde{w}_n$ substantially less than $\hat{w}_g$ or $\tilde{w}_m$ since this increases the magnitude of the measured angle $2\theta_n$.

In any event the intensity employed should be high enough so that the line profile can be recorded without any serious problem of drift in source or spectrometer. The total time should preferably not exceed an hour; a shorter interval is desirable, particularly if any significant drift problems are noted.

L. Comparison of instruments

The two-crystal spectrometer has a number of advantages. It can make use of a broad, relatively diffuse focal spot without difficulty. It gives a high signal-to-noise ratio since the first crystal effectively filters out most of the background continuum. The resolving power will always be comparable to that of any other instrument and in most cases definitely superior, particularly if there is sufficient intensity to operate in the higher order $(n, \pm n)$ positions, i.e., with $n > 1$.

The single-crystal Bragg spectrometer is basically simpler in design. If a narrow-focus x-ray source is available, then at longer wavelengths the resolution can become competitive with the value for the two-crystal instrument. The Bragg spectrometer is not subject to the type of errors discussed in Subsection IV H since there is no significant correlation between wavelength and angle of incidence on the rotating crystal. While it is not ideal for an x-ray source with a broad focal spot, this problem may be alleviated with the aid of a form of Soller slits, as used by Hrdý, Henins, and Bearden (1970). At the expense of some additional complication, the signal-to-noise ratio can be much improved by use of a proportional counter and channel analyzer.

The tube spectrometer is somewhat similar to the Bragg instrument and is comparable in resolution. It has an intensity advantage since all parts of the profile are being recorded simultaneously. However, the photographic registration clearly creates potential complications due to linearity, grain size, and particularly the method of reading plates. If this is done visually with a microscope, there is no assurance of setting on the peak of an asymmetric line; in this case one might expect a reading somewhere between the peak and the centroid although there is no clear evidence that such an error occurred in practice. If the plates are read with a microphotometer, this introduces a new element into the problem. It does not appear that the tube spectrometer has been recently employed in any precision measurements, and its revival seems questionable.

In concluding the discussion of instruments, we will briefly note the modifications required for the longer wavelength "vacuum region" (above about 2.5 A*). In this range absorption is too high to permit any significant path length in air. Siegbahn (1931) constructed a single-crystal vacuum spectrometer, in which all the components are enclosed in a vacuum tank; Haglund (1941) built a precision tube spectrometer of this type. On the other hand, Henins (1971) carried out a single-crystal measurement at 8.3 A* by providing a hydrogen path at atmospheric pressure for the x rays; this

BELMONT COLLEGE LIBRARY

technique (with either hydrogen or helium) overcomes the absorption problem and avoids the mechanical complications of a vacuum instrument.

M.   Summary of procedure

At this point it is desirable to summarize the steps involved in measuring a wavelength in terms of a crystal of known grating constant, viz.:

1   The type and design of instrument and the crystal and plane employed must be selected to give adequate resolving power and intensity with the available source; these considerations have been discussed in Subsections V J through V L.

2   The instrument should be properly aligned and adjusted, as described in Subsections V D through V F.

3   Suitable samples of the desired crystal should be carefully examined and the selected crystal or crystals prepared and mounted according to the procedure of Subsections V G and V H.

4   The desired line profiles may then be recorded in the *A* and *B* positions (plus and minus positions in the two-crystal case). A suitable polynomial should then be fitted to each curve, as described in Subsection V I and the peak positions obtained.

5   These positions must then be referred to the spectrometer circle through fiducial points noted in the recording of each curve. After any required calibration correction for the spectrometer circle, the angular difference between peak positions, i.e., $\omega_2 - \omega_1$, is thus obtained.

6   The Bragg angle θ_{n0} may then be computed by (2-53) or (2-131) for the Bragg spectrometer or two-crystal spectrometer, respectively. In this process the *centroid* and vertical-divergence corrections are included, but the subsequent error terms are ignored. [Alternatively the uncorrected Bragg angle may be used to compute wavelength and the final value corrected for these two effects by (2-76) or (2-168) for the respective cases.]

7   Corrections for temperature and index of refraction should be made according to Subsection II F. This is done through adjusting the grating constant d_∞ for temperature by (2-21) and then computing d_n for the order employed by (2-20).

8   The desired wavelength may then be obtained by inserting the value of θ_{n0} in (2-1).

In some instances it may be desired to calibrate a crystal grating constant in terms of a known wavelength; in others the λ/d_∞ ratio may be measured where neither is assumed to be precisely known. In such cases it is more convenient to substitute (2-20) into (2-1) and use the Bragg law in the following form:

$$n\lambda = 2d_\infty \left(1 - \frac{4d_\infty^2}{n^2}\frac{\delta}{\lambda^2}\right)\sin\theta_n \qquad (2\text{-}186)$$

VI. TRANSMISSION MEASUREMENTS

A. Basic theory

Thus far this chapter has dealt only with spectrometers in which the x-ray radiation is Bragg-reflected from atomic planes parallel to the crystal surface. The Bragg law, with one modification, applies equally well to reflection from internal planes of a crystal. Hence it is also possible to construct a spectrometer with the crystal or crystals used in transmission, i.e., with the Laue case of radiation reflected from an internal plane (usually normal to the surface) and emerging from the opposite face.

Obviously the x rays must be transmitted through an appreciable thickness of crystal without excessive absorption if this method is to be feasible. Hence it has normally been limited to thin crystals with wavelengths of the order of a few tenths of an angstrom. However, Borrmann (1959) has shown that a perfect crystal will provide anomalously high transmission for a portion of the internally reflected beam, particularly the component polarized perpendicular to the plane of incidence; the theory is also given in detail in the review of Batterman and Cole (1964). Thus, with the availability of high-quality, artificially grown crystals the way has been opened to carry out transmission measurements at considerably longer wavelengths.

Figure 2-18a shows schematically a single-crystal Bragg spectrometer used in transmission (compare with Fig. 2-1a); the basic principle is clearly identical with the reflection case. However, the angle through which the crystal is rotated is now $2\theta_n$ rather than $180° - 2\theta_n$. It is this distinction that often gives a rotation angle sufficiently small for direct measurement by an angular interferometer, as discussed in Subsection V B.

The tilt angle δ_1 must now refer to the tilt of the internal plane, as illustrated in Fig. 2-18b. There is an important distinction here between the reflection and transmission cases. In reflection the same surface of the crystal is used in both A and B positions, while in transmission the incident ray strikes the internal plane from opposite sides, as may be seen by inspection of Fig. 2-18a. (This is a consequence of a rotation angle of only $2\theta_n$.) It follows that the outward normals will be oppositely directed in the A and B positions. Thus if the tilt is δ_1 in the A position, it will become

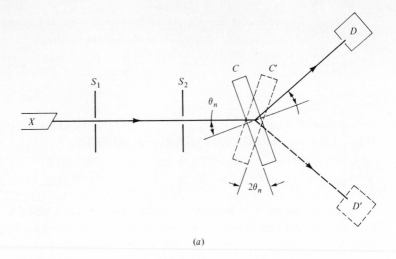

(a)

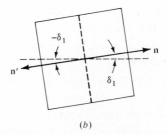

(b)

FIGURE 2-18
Single-crystal Bragg spectrometer used in transmission. (a) Schematic diagram, showing x-ray source X and slit system S_1 and S_2. Crystal and detector are denoted as C and D in the A position and C' and D' in the B position. The atomic plane employed is normal to crystal face; incident and reflected rays form Bragg angle θ_n with this plane. (b) Crystal face and atomic plane (heavy dashed line), defining normal and crystal tilt, $\mathbf{n}$ and δ_1 in the A position and $\mathbf{n}'$ and $-\delta_1$ in the B position.

$-\delta_1$ in the B position. The effect of this difference will be considered in Subsection VI C.

The two-crystal spectrometer may be modified in a similar way so that one or both of the crystals are used in transmission. Bearden et al. (1964) used this approach in part of their W $K\alpha_1$ measurements. Hart (1969) has employed an instrument of this type in the $(1, -1)$ position for comparing lattice constants of similar crystals. Sauder (1971) is constructing such a spectrometer for measurement of the electron-positron annihilation wavelength.

B.   Independence of refractive index

The index-of-refraction correction was derived in Subsection II F by assuming that the Bragg law held exactly inside the crystal, i.e.,

$$n\lambda' = 2d_\infty \sin \theta'_n \qquad (2\text{-}187)$$

where λ' and θ'_n denote the wavelength and the Bragg angle, respectively, *inside the crystal*. As in the reflection case, the internal wavelength is simply $\lambda' = \lambda/\mu$, where μ is the index of refraction. However, in transmission there is an important difference in regard to the angle. Since the atomic plane involved is normal to the surface, the glancing angle on this plane is identical with the optical angle of refraction in Snell's law (rather than the complement of this angle). Hence we now have $\sin \theta'_n = (\sin \theta_n)/\mu$. Substitution of these two values in the above equation yields

$$n\lambda = 2d_\infty \sin \theta_n \qquad \text{(transmission)} \qquad (2\text{-}188)$$

Thus, provided that the surface is normal to the atomic plane employed, *there is no index-of-refraction correction in transmission*; d_∞ is thus applicable for all orders in the Bragg law. While this result is strictly valid only for atomic planes normal to the surface, it is relatively insensitive to small departures from this condition. It can be shown that an angular error of $1°$ will produce an error of the order of 1 ppm in wavelength (see, for example, Inglestam, 1937).

C.   Errors and corrections

For one specific position (e.g., the A position) the reflection results may now be applied directly to the transmission case with the omission of any index-of-refraction correction. However, certain differences appear in combining the results for two positions to obtain the Bragg angle. These modifications arise from two causes: (1) rotation by $2\theta_n$, rather than $180° - 2\theta_n$, and (2) change in sign of the tilt between the two positions, as shown in Fig. 2-18b.

Figure 2-19a shows the two positions of the crystal in somewhat greater detail (compare with Fig. 2-3c). The incident ray is taken as horizontal; for simplicity it is also assumed as the reference line from which the spectrometer angle ω is measured. The crystal is rotated by an angle of magnitude $\omega_1 - \omega_2$ between the A and B positions. Note that the direction of rotation is opposite to that for the reflection case. This angle is, of course, equal to the angle of rotation of the surface normal and hence that of the atomic plane employed. An inspection of the figure now shows that this is $(\theta_{n0} + \beta_{2p}) + (\theta_{n0} + \beta_{1p})$. Thus we obtain instead of (2-44)

$$\omega_1 - \omega_2 = (\theta_{n0} + \beta_{2p}) + (\theta_{n0} + \beta_{1p}) \qquad (2\text{-}189)$$

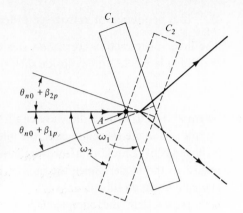

FIGURE 2-19
Schematic diagram for Bragg spectrometer in transmission, showing the measured spectrometer angles ω_1 and ω_2 for A and B positions (for simplicity angles are referred to horizontal ray). Crystal is designated as C_1 and C_2 and projected horizontal angles of incidence as $\theta_{n0} + \beta_{p1}$ and $\theta_{n0} + \beta_{p2}$ for the respective positions.

The *form* of the expression for the β's will be the same as in Subsection III E; however, due to the change in sign of the tilt between the A and B positions, the term $\varepsilon(\psi,\delta_1)$ in (2-43) must be replaced by $\varepsilon(\psi,-\delta_1)$. As a result, (2-45) is replaced by

$$\theta_{n0} = \tfrac{1}{2}(\omega_1 - \omega_2) - \bar{u}_n + \tfrac{1}{2} \int [\varepsilon(\psi,\delta_1) + \varepsilon(\psi,-\delta_1)]g(\psi)\,d\psi \qquad (2\text{-}190)$$

The consequence of this change is the cancellation of the terms in $\Psi\delta_1$, so that we have in place of (2-53)

$$\theta_{n0} = \tfrac{1}{2}(\omega_1 - \omega_2) - \bar{u}_n - \frac{a^2 + b^2}{24L^2} \tan \theta_n - \tfrac{1}{2}(\delta_1{}^2 + \Psi^2) \tan \theta_n \qquad (2\text{-}191)$$

It follows immediately that the $\Psi\delta_1$ term is also dropped from (2-76).

The treatment of the alternative slit location given in Subsection III G may be modified in a similar way. In this analysis we must note that Ψ (i.e., $\bar{\psi}$) differs for the A and B positions, due to the change of sign of δ_1 in (2-61). The resulting calculation shows that the $\Psi_s\delta_1$ term disappears from (2-63) while the other terms are unchanged.

These results could have been predicted from symmetry considerations. It is essentially arbitrary which position is denoted as A and which as B. Yet, if we reverse this convention, the sign of δ_1 would be changed in the transmission case. Since this reversal can have no physical significance in the present instance, we would expect that all terms linear in δ_1 must disappear; the calculations outlined above confirm this conclusion.

The two-crystal instrument may be treated in a similar way. In this case the sign reversal involves δ_2, the tilt of the second (i.e., the rotating) crystal. With the necessary modifications, equations (2-125) and (2-126) are replaced by

$$\theta_{n0} = \tfrac{1}{2}(\omega_+ - \omega_-) - \bar{u}_n{}^e + \tfrac{1}{2} \int g(\psi)[\varepsilon_{2+}(\psi,\delta_1,\delta_2) + \varepsilon_{2-}(\psi,\delta_1,-\delta_2)]\,d\psi \qquad (2\text{-}192)$$

Again the result is the cancellation of the terms which are linear in δ_2; thus we obtain in place of (2-130)

$$\theta_{n0} = \tfrac{1}{2}(\omega_+ - \omega_-) - \bar{u}_n^e - \frac{a^2 + b^2}{24L^2} \tan \theta_n - \frac{\Psi^2}{2} \tan \theta_n$$

$$+ 2\delta_1 \Psi \sin \theta_m \tan \theta_n - 2\delta_1^2 \sin^2 \theta_m \tan \theta_n - \frac{\delta_2^2}{2} \tan \theta_n \qquad (2\text{-}193)$$

It follows that the $\Psi \delta_2$ term disappears in (2-131) and the $\delta_2 \Psi$ and $\delta_1 \delta_2$ terms vanish in (2-168).

The results can again be understood in terms of symmetry. Basically the first crystal modifies the angular and wavelength distributions of the incident radiation; however, this modified distribution, which is incident on the second crystal, is identical for both plus and minus positions. Hence we may now apply the same symmetry argument as before to the rotating crystal and conclude that there can be no terms linear in δ_2.

This argument breaks down if the crystals are located between the slits, i.e., in the case treated in Subsection IV G. In this event we must modify (2-141) to read

$$\bar{\psi}_\pm \equiv \Psi_\pm = \Psi_s + 2\sigma_1 \delta_1 \sin \theta_m \pm 2\sigma_2 \delta_2 \sin \theta_n \qquad (2\text{-}194)$$

with the upper and lower signs applying to the plus and minus positions, respectively; this means that the $g(\psi)$ distribution differs for the two positions. Hence we cannot conclude that ε_1 disappears as in (2-126). On the contrary, we must go back to (2-122) and evaluate β_p separately for each position, taking into account both the explicit change of sign of δ_2 and its implicit effect on $\bar{\psi}$.

The ensuing calculations are generally similar to those of Subsection IV E, with the result

$$\theta_{n0} = \tfrac{1}{2}(\omega_+ - \omega_-) - \bar{u}_n^e - \frac{a^2 + b^2}{24L^2} \tan \theta_n$$

$$- \tfrac{1}{2}[\hat{\Psi}^2 + \delta_2^2(1 - 4\sigma_2 + 4\sigma_2^2 \sin^2 \theta_n)] \tan \theta_n$$

$$- \frac{2\sigma_2 \delta_2 \sin \theta_n}{\cos \theta_m} (\hat{\Psi} \sin \theta_m + \delta_1) \qquad (2\text{-}195)$$

with $\hat{\Psi}$ defined by (2-144). If $\sigma_1 = 0 = \sigma_2$, then we have $\hat{\Psi} = \Psi - 2\delta_1 \sin \theta_m$, and the above expression reduces to (2-193). If only $\sigma_2 = 0$, then the terms linear in δ_2 will vanish.

The physical basis for this result lies in the loss of symmetry. If we have one slit behind the second crystal, i.e., $\sigma_2 > 0$, then the tilt δ_2 will determine the effective misalignment of radiation incident on the first crystal, as can be seen from Fig. 2-10. Thus there is an interaction between the tilt δ_2 and the angular deviation at the first crystal; this destroys the symmetry property and produces the cross product in the above result.

Where the symmetry property does hold and the cross product disappears, the error estimate will obviously be reduced. This is particularly true when small Bragg angles are involved and $\tan \theta_n \ll (\cos \theta_n)^{-1}$; e.g., compare equations (2-53) and (2-191). Even in the case described by (2-195) we usually have $\sigma_2 \ll 1$ and therefore a significant reduction in error. Hence the alignment requirements are generally less restrictive for transmission measurements than for reflection ones.

D.   Alignment procedure: single-crystal case

Most of the steps in the alignment of the Bragg spectrometer are the same as in the case of reflection. The three-point support of the rotating crystal is located on the axis of rotation, and the axis is made vertical by the procedure described in Subsection V D. The method of Subsection V F is next employed to define the x-ray beam and center it on the axis of rotation. The slits limiting vertical divergence are then centered in a horizontal plane, as described in Subsection V E.

It remains only to eliminate the crystal tilt δ_1. Since the plane used in transmission is an internal one, it is not feasible to employ an optical method for this step. Hence it is necessary to introduce an x-ray alignment technique.

The method to be described here makes use of a plate which covers the second vertical-divergence slit and allows only two narrow beams to pass at the top and bottom. This plate, shown in Fig. 2-20, must be carefully centered with the telescope level. If there is a misalignment Ψ at the center line, then the effective misalignments of the two beams are approximated by

$$\Psi_{\text{up,low}} \approx \Psi \mp \frac{b}{2L} \qquad (2\text{-}196)$$

where b is the height of the slit and the upper and lower signs refer to the respective beams.

By use of an additional stop at the top or bottom it is possible to examine the line profile for each beam separately and determine the corresponding peak; these positions will usually differ due to the crystal tilt δ_1. If this check is made in the A position, equations (2-42) and (2-52) indicate that the change in peak angle between lower and upper positions will be

$$\Delta \beta_{1p} = \tfrac{1}{2}(\Psi_{\text{up}}^2 - \Psi_{\text{low}}^2) \tan \theta_n - \frac{\delta_1}{\cos \theta_n} (\Psi_{\text{up}} - \Psi_{\text{low}}) \qquad (2\text{-}197)$$

Combining the last two equations now gives

$$\Delta \beta_{1p} = \frac{b}{L \cos \theta_n} (\delta_1 - \Psi \sin \theta_n) \qquad (2\text{-}198)$$

Thus the change in peak angle is a linear function of the tilt δ_1.

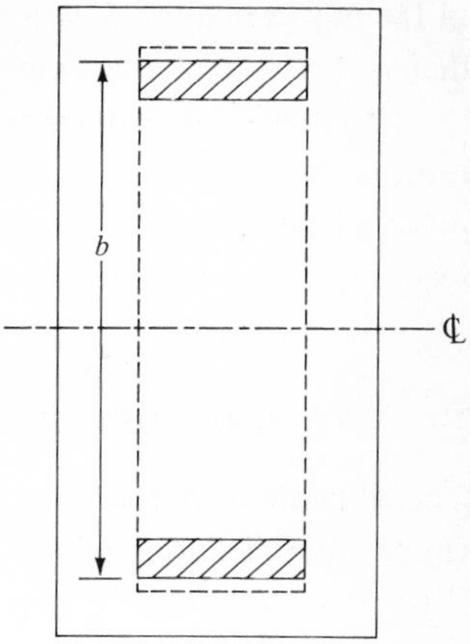

FIGURE 2-20
Auxiliary slits used to eliminate crystal tilt. Slit of height b (dashed outline) defines vertical divergence. For adjustment purposes this is covered by a plate which transmits only the narrow beams shown by shaded areas at top and bottom. Only one of these beams is used at a time, the other being blocked by an additional stop (not shown).

The crystal tilt is now adjusted until $\Delta\beta_{1p} = 0$. A mirror on one side of the crystal which can be observed with a telescope and scale will aid in interpolating to the correct position more quickly.

It is clearly advantageous to use a first-order position ($n = 1$) in this adjustment; this choice will minimize dispersion and hence observed line width. Consequently the angular error in $\Delta\beta_{1p}$ will also be minimized, as indicated by (2-171) and (2-172). In this adjustment it is, of course, necessary to measure accurately only very small changes in β; the absolute angle on the spectrometer circle is of no importance. With reasonable intensity it should be possible to reduce $\Delta\beta_{1p}$ to 0.1 second (of arc) or less.

In this adjustment it would be desirable to have a large value of b/L in order to minimize δ_1; however, b must be relatively small to keep the vertical-divergence correction within reasonable limits. Thus b/L may be of the order of 10^{-2}, and a value of $|\Delta\beta_{1p}|$ of 0.1 second (of arc) reduces $|\delta_1 - \Psi \sin \theta_n|$ to about 10 seconds (of arc) (5×10^{-5} rad), as shown by (2-198). It was indicated in Subsection V E that the telescope-level method should limit $|\Psi|$ to 2×10^{-4}. It follows that this may frequently be the limiting factor in eliminating tilt and that δ_1 will be of the order of 10^{-4} (20 seconds of arc) or less for typical Bragg angles. Because of the absence of a cross product in (2-191), the resulting error estimate will be even smaller than the 0.1 ppm given in Subsection V E.

The above discussion is slightly modified if the crystal is located between the slits, as discussed in Subsection III G. In this case the effective misalignment Ψ is related to the slit misalignment Ψ_s by (2-61). As a result, the term $-\Psi \sin \theta_n$ in (2-198) is replaced by $-\Psi_s \sin \theta_n - 2\sigma\delta_1 \sin^2 \theta_n$. In most applications we will have $2\sigma \sin^2 \theta_n \ll 1$, and the above conclusions will be qualitatively unchanged. If this condition does not hold, the equation must be modified accordingly; the sensitivity of the adjustment of δ_1 will be correspondingly reduced. The principle of this adjustment technique remains entirely valid.

It should be emphasized that, in contrast with the procedure of Section V, this is an x-ray technique for minimizing crystal tilts; it is not dependent on the employed plane's being precisely normal to the surface. In this respect it is similar to the x-ray technique recommended by Schnopper (1965).

E.   Alignment procedure: two-crystal case

The alignment of the two-crystal instrument in transmission can be performed by simple modifications of the techniques described previously. Again the three-point support of the rotating crystal is placed on the axis of rotation and the axis made vertical by the procedure described in Subsection V D. The first-crystal slide may be aligned and the x-ray source located by obvious modifications of the techniques of Subsection V E. (Since the crystal must now be parallel with the axis of the slide rather than normal to it, the right-angle prism is replaced by a mirror.) As the Bragg angles are usually small in transmission measurements, it may sometimes be feasible to leave the first crystal stationary and eliminate the slide.

It remains only to eliminate the tilts δ_1 and δ_2. For this purpose, the second crystal is *temporarily* used *in reflection*; hence $\delta_2 \approx 0$ as a result of the optical alignment. The detector slit is then covered with the plate shown in Fig. 2-20 and the tilt δ_1 minimized by the method described in the last section. In the present case it is best to perform this adjustment in the $(1,-1)$ position since this gives a narrow observed profile, with the width due only to the crystal windows. After the first crystal has been properly adjusted, the second crystal is rotated to give the $(1,-1)$ position with both crystals in transmission. The same process is repeated to minimize the tilt δ_2; this completes the alignment.

The theory of this adjustment requires the expression for β_{p-} in the $(n,-n)$ position. [Note that the sign convention for β_{p-} is opposite to that for the plus positions; see (2-107).] Equations (2-90) and (2-101) are substituted into (2-137); setting $m = n$ and $\eta = \theta_m = \theta_n$ then yields

$$
\beta_{p-} = - \int g\left(\psi\right) \left[-\tfrac{1}{2}(\psi^2 + \delta_2{}^2) \tan \theta_n + \frac{1}{\cos \theta_n} \right.
$$
$$
\times \left(-2\delta_1{}^2 \sin \theta_n + 2\psi\delta_1 - 2\delta_1\delta_2 \sin \theta_n + \psi\delta_2\right)
$$
$$
\left. + \tfrac{1}{2}(\psi^2 + \delta_1{}^2) \tan \theta_n - \frac{\psi\delta_1}{\cos \theta_n} \right] d\psi + \tfrac{1}{2} \psi^2 \tan \theta_n \qquad (2\text{-}199)
$$

or, since $g(\psi)$ is normalized,

$$
\beta_{p-} = \tfrac{1}{2}(3\delta_1{}^2 + 4\delta_1\delta_2 + \delta_2{}^2) \tan \theta_n - \frac{\delta_1 + \delta_2}{\cos \theta_n} \overline{\Psi} + \tfrac{1}{2}\overline{\psi^2} \tan \theta_n \qquad (2\text{-}200)
$$

With the help of (2-196) it follows immediately that

$$\Delta\beta_{p-} = -\frac{\delta_1 + \delta_2}{\cos\theta_n}(\Psi_{up} - \Psi_{low}) = \frac{b}{L\cos\theta_n}(\delta_1 + \delta_2) \qquad [(n,-n)\text{ case}] \qquad (2\text{-}201)$$

Thus the sensitivity of the tilt adjustment to $\Delta\beta_{p-}$ is similar to that given by (2-198) for the single-crystal case. Under the same assumptions, $\delta_1 + \delta_2$ would be limited to roughly 10 seconds (of arc); actually the precision should be higher in this case because of the narrowness of the $(1,-1)$ profile. However, the adjustment of δ_1 to zero obviously hinges on how nearly δ_2 vanishes in reflection, which in turn depends on the surface being parallel to the atomic plane involved. While errors will cumulate to some extent, it should be possible to keep both tilts within 10 to 20 seconds (of arc).

Even if the second slit is located at the detector, as discussed in Subsection IV G, equation (2-201) will remain valid for the $(n,-n)$ position. This results from the cancellation of the ψ^2 terms in (2-199). As a consequence, only the *change* in Ψ, i.e., b/L, appears in equation (2-201). Thus, although Ψ is now expressed in terms of Ψ_s by (2-141), this plays no role in the final result.

In describing the use of the $(n,-n)$ parallel position to adjust the tilt δ_1, we have implicitly assumed that there exists an atomic plane which is normal to the one used in transmission and which has the same grating constant. If this condition does not hold and it is necessary to use dissimilar planes, it is then desirable to choose an $(m,-n)$ position with low dispersion. Equation (2-201) must obviously be recalculated for this new situation; there will now be a small effect if the second slit is located at or near the detector.

In any event the principle of the adjustment remains the same. $\Delta\beta_{p-}$ is always directly proportional to b/L; the coefficient is at most a linear combination of Ψ_s, δ_1, and δ_2. If the slit misalignment Ψ_s and the tilt δ_2 (in reflection) can be made approximately zero, then it follows that $\Delta\beta_{p-}$ vanishes if and only if the tilt $\delta_1 = 0$. With Ψ_s and δ_1 now equal to zero, the second crystal can be employed in transmission and δ_2 eliminated by the same technique. While modifications in the form of (2-101) will alter the sensitivity somewhat, the basic theory of the method remains completely valid.

VII.   RELATIVE WAVELENGTHS

A.   Possible standards

Any field of physics where relative measurements can be made to higher accuracy than absolute ones almost inevitably produces its own unit, either explicit or implicit, through which such data can be compared. Perhaps the best-known example is the atomic mass unit, which is universally used for comparing atomic and nuclidic masses. In principle the kilogram would serve, but in practice the high accuracy of nuclidic-mass comparisons demands a suitable relative scale. Even in the field of electrical measurements, where the old International System of units was abandoned over twenty years ago, it is still necessary in the most precise work to distinguish between the NBS ampere and the absolute ampere. [See, for example, Taylor, Parker, and Langenberg (1969).]

Any standard intended for x-ray spectroscopy should obviously be of the same order as x-ray wavelengths, i.e., of atomic dimensions. There are at least three different types of candidates for this purpose, viz.:

1 The grating constant d_∞ of some selected species of crystal
2 The wavelength of some selected x-ray line
3 The wavelength of some selected γ-ray line

Historically the first unit to be employed specifically for x-ray spectroscopy was the x unit, which was introduced by Siegbahn. It gradually evolved from a submultiple of the meter to an empirical working standard based on the grating constant of calcite crystals. Eventually this accepted usage was stated as an explicit definition [see Siegbahn (1931)]; thus it was an example of the first type of unit discussed above. Its history has been reviewed in detail by Thomsen and Burr (1968).

The chief drawback of this standard arose from the variation in the grating constants of different calcite crystals, as noted by various experimenters, e.g., Bearden (1931a; 1965a). As a result some workers began to base their measurements on selected wavelengths as implicit working standards, primarily the Mo $K\alpha_1$ and Cu $K\alpha_1$ lines. Unfortunately it later developed that the numerical values commonly used for these two wavelengths were incompatible by about 20 ppm.

B.   The W $K\alpha_1$ wavelength standard

Both Sandström (1957) and Merrill and DuMond (1958) criticized the continued use of the old definition of the x unit and recommended the adoption of a wavelength as an x-ray length standard. By analogy, this proposal received further support in 1960 when the meter was redefined in terms of the orange-red krypton line rather than a material length standard.

Bearden (1964; 1967) published a new and comprehensive wavelength table, listing the "best values" of virtually all measured x-ray lines. As a preliminary step he felt it desirable to discard the x unit and adopt a new and more precisely defined standard, viz., a standard wavelength (Bearden, 1965b).

An ideal wavelength standard should satisfy the following criteria:

1 The width should be as narrow as possible to minimize statistical error of measurement, as indicated by equations (2-171) and (2-172).

2 The Bragg angles which may feasibly be employed should be of such magnitude that the angle $2\theta_n$ can be conveniently measured to high accuracy.

3 The wavelength should be suitable for transmission measurements in which the index-of-refraction correction vanishes.

4 The line should be highly symmetric and should be measurable by crystals with symmetric window functions. Thus the observed peak will coincide with the true peak, except for vertical-divergence and error terms.

5 The line should be obtainable from a convenient and readily available source which gives a highly reproducible wavelength, substantially free from chemical and isotopic shifts.

6 The line should lie in a conveniently accessible wavelength region.

On most counts a γ-ray line would make a highly desirable standard. In particular, line widths are of the order of 1 ppm (much smaller for Mössbauer radiation) and thus no more than 1% of those of x-ray lines. However, the actual observed width would be governed by the window of the crystal employed, which might limit the improvement to a factor of 3 or 4.

The crucial problem lies in obtaining a convenient γ-ray source with adequate intensity. The observed intensity will be roughly proportional to the geometrical or crystal window, whichever is narrower. (The results of Subsection V K are inapplicable here because of the narrow line width.) Equation (2-171) or (2-172) indicates that the statistical error will be reduced if w and I_p are decreased in the same proportion; hence it might appear desirable to turn to the narrow $(n, \pm n)$ two-crystal window $(n > 1)$. However, these equations do not include any background effect, which can create a severe signal-to-noise problem in the case of a γ ray. Hence Bearden finally concluded that the intensity problem was too formidable with any source of reasonable strength to warrant the adoption of a γ-ray wavelength standard at the present time.

He therefore turned back to x-ray wavelengths. While the Cu $K\alpha_1$ line (≈ 1.54 A*) has been extensively used as a working standard, it has a significant asymmetry. The Mo $K\alpha_1$ line (≈ 0.71 A*), which has also been widely employed, is free from this drawback and equally satisfactory on most other counts. The short wavelength W $K\alpha_1$ line (≈ 0.21 A*) is also highly symmetric and competitive with Mo $K\alpha_1$ in most respects; the problem quickly reduced itself to a choice between these two.

The larger Bragg angle was a factor in favor of the Mo $K\alpha_1$ line, although the measurements of Bearden et al. (1964) indicated, perhaps fortuitously, that the larger Bragg angle was not as important as it was once believed. In other respects the W $K\alpha_1$ choice seemed either equal or superior. It permitted measurements in transmission without difficulty, required only a readily available commercial source, and appeared more immune to chemical effects since the K and L levels lie deeper in heavier atoms. Furthermore it provided an excellent wavelength for preliminary testing of the crystals to be used in measurement. The wavelength is somewhat shorter than might be ideal for x-ray spectroscopy, but very well adapted to the γ-ray field. After weighing all these points, Bearden (1965b) chose the W $K\alpha_1$ line to define his wavelength unit.

It does not follow that an x-ray wavelength will always be the most logical choice, should the need for a distinct x-ray length standard continue far into the future. With a sufficient demand, the γ-ray intensity problem can probably be overcome even if measurements of the primary standard are restricted to nuclear reactor laboratories. Furthermore, with the improved silicon crystals now available, the possibility of returning to some crystal standard cannot be altogether disregarded. While crystal grating constants may continue to be less uniform than properly chosen x-ray wavelengths, the enormous difficulties in locating the peak to better than $w/10^3$ (see Subsection V I) tend to minimize this advantage if crystal variations can be held below 1 ppm. However, at the present writing it still seems best to retain the W $K\alpha_1$ wavelength standard.

C.   The A* unit

To complete the definition of a new wavelength unit, it is necessary to adopt some specific numerical value for the selected standard wavelength. The W $K\alpha_1$ wavelength might, for example, be taken as unity; this choice would yield a totally new unit of length, not likely to be confused with any other. On the other hand, the numerical value might be selected in order to make the new unit as nearly as possible equal to the old x unit or alternatively to the angstrom; either of these options would open a greater possibility of confusion between different units, but would avoid the use of new conversion factors in work of moderate accuracy.

After prolonged consideration Bearden (1965b) decided to approximate the angstrom as closely as possible. Despite the possible confusion of units which may occasionally result, he felt that this drawback was outweighed by the complications of introducing a totally new set of numerical values. Since there were already effectively three different working definitions of the old x unit, it seemed undesirable to perpetuate this quantity; hence he turned to the angstrom.

This choice had one major advantage. X-ray wavelengths are widely used in crystallography, and crystal dimensions are usually given in angstroms. Hence in calculations of moderate precision, crystallographers can employ wavelengths on this new scale without the need for conversion factors.

To indicate the similarity to the angstrom and yet distinguish the new unit as an independent entity, he termed it the "A*" unit. At the time the best available absolute value for the W $K\alpha_1$ wavelength was 0.209 010 0 A, as explained in Section VIII. Hence he defined the A* unit through the statement

This numerical value of the wavelength is now proposed for use with the W $K\alpha_1$ line to define *the x-ray wavelength standard* by the relation

$$\lambda_{\text{W } K\alpha_1} = 0.209 \, 010 \, 0 \text{ A*} \qquad (2\text{-}202)$$

In this definition a solid target of pure tungsten with natural isotopic abundances is assumed as the source; it is also understood that the wavelength is defined by the *peak* of the line profile. (However, due to the symmetry of the W $K\alpha_1$ line any other reasonable wavelength definition would yield the same result.)

While the above choice makes the conversion factor between the A* unit and the angstrom very close to unity, this conversion factor must remain an *experimentally determined* quantity. As better experimental information becomes available on the absolute wavelength of the W $K\alpha_1$ line, slight changes in this factor are to be expected. Such changes are no cause for alarm and require no modification of the A* unit. There is no more necessity to redefine the A* unit with each revision of the conversion factor than to redefine the atomic weight scale with each new value for Avogadro's number.

D. Secondary standards

As already indicated, the W $K\alpha_1$ line is near the short-wavelength end of the x-ray spectrum; it is thus not conveniently accessible to all x-ray spectroscopists or crystallographers. This factor makes it particularly important to have a set of reliable secondary standard lines spanning at least the most important portion of the x-ray spectrum. Given such a set, a spectroscopist can determine the wavelength of any other line by a crystal measurement through comparison either directly with the W $K\alpha_1$ line or with any of the secondary standards. The procedure is analogous to that of an optical spectroscopist who may measure a wavelength either directly in terms of the orange-red krypton line or in terms of a well-established secondary standard.

Such a secondary set has been supplied by the work of Bearden et al. (1964), who measured the wavelength ratios between the W $K\alpha_1$, Ag $K\alpha_1$, Mo $K\alpha_1$, Cu $K\alpha_1$, and Cr $K\alpha_2$ lines. These wavelengths span the range from roughly 0.21 to 2.3 A* and generally fulfill most of the requirements for a standard, as listed in Subsection VII B. The Cu $K\alpha_1$ line has a significant asymmetry, which may make it appear a questionable choice. However, it is easily obtained from a convenient target and has been very extensively employed in the past as a reference line; hence its inclusion seemed essential.

Five different crystals (one calcite, two quartz, and two silicon) were used in these measurements. This "fivefold evaluation" (five lines measured by five authors with five crystals) gave a set of data with a considerable degree of overdetermination. Each combination gave a λ/d_∞ ratio; twenty of the twenty-five possible combinations were actually measured. In each case a considerable number of curves were recorded, with angles being read to a precision of 0.1 second (of arc) or better. The results were then carefully corrected for circle calibration, temperature, and vertical divergence.

These data yielded twenty λ/d_∞ values. Probable errors were assigned to each one, depending on the amount and type of data represented; these errors varied from 2 to 4 ppm. Since only *ratios* were measured, a numerical value was then assigned to one wavelength (provisionally Mo $K\alpha_1$, although W $K\alpha_1$ would now be the obvious choice). This left nine unknowns (four wavelengths and five crystal grating constants) to be determined from twenty equations.

The obvious means of dealing with such an overdetermined set of data is a least-squares adjustment. [For a brief summary of this method see, for example, Bearden and Thomsen (1957).] Such an adjustment was carried out and yielded output values in generally good agreement with the input data and assigned errors; the actual value of χ^2 was 10.7, as compared with an expected value of 11.

It can be shown that the wavelength *ratios* resulting from this adjustment are independent of the particular numerical value assigned to the Mo $K\alpha_1$ line. Hence the output ratios may be used to obtain wavelengths in A* units on the basis of the defined numerical value for the W $K\alpha_1$ line, as explained in the preceding subsection. The results thus obtained are as follows:

$$
\begin{aligned}
\lambda_{\text{Ag } K\alpha_1} &= 0.559\ 407\ 5 \text{ A}^* \pm 1.3 \text{ ppm} \\
\lambda_{\text{Mo } K\alpha_1} &= 0.709\ 300 \text{ A}^* \pm 1.3 \text{ ppm} \\
\lambda_{\text{Cu } K\alpha_1} &= 1.540\ 562 \text{ A}^* \pm 1.2 \text{ ppm} \\
\lambda_{\text{Cr } K\alpha_2} &= 2.293\ 606 \text{ A}^* \pm 1.3 \text{ ppm}
\end{aligned}
\tag{2-203}
$$

The Mo $K\alpha_1$/Cu $K\alpha_1$ ratio is in good agreement with that subsequently found by Cooper (1965). At the present writing, it seems likely that the Ag $K\alpha_1$, Mo $K\alpha_1$, and Cu $K\alpha_1$ values may all have to be revised upward by one or two probable errors, even though their ratios remain substantially unchanged from those implied by (2-203).

E. Other wavelengths

Other prominent x-ray lines can be established in terms of the primary or secondary standards. This may be done either through a direct ratio measurement or, slightly less directly, by means of a crystal which has been calibrated in terms of one or more of the standards. Any of the three types of spectrometers discussed in Sections III and IV are suitable for such measurements.

For measurements of low-intensity lines the focusing curved-crystal spectrograph has many advantages. Such an instrument is best suited for measuring these wavelengths with respect to those of strong lines in the same spectral region. Photographic registration is usually employed, with two or more known lines recorded on each plate as calibration wavelengths.

This chapter will not treat this instrument in any detail. Descriptions and analyses have been given by Cauchois (1932), DuMond and Kirkpatrick (1930), DuMond (1947), and Inglestam (1937). Both Cauchois and Inglestam have made very extensive measurements of x-ray spectra with this type of spectrometer.

Measurements obtained by all the various methods outlined in this chapter form the input data for any wavelength table, such as those of Cauchois (1947), Sandström (1957), and Bearden (1964; 1967). Analysis of the data and adoption of recommended values call for a long and painstaking critical study. A discussion of the Bearden table, along with some selected wavelength values, is given in Appendix A.

VIII.   ABSOLUTE WAVELENGTHS

A.   Possible experimental methods

Despite the existence of a well-defined relative scale for x-ray wavelengths, it is obviously desirable to know them as accurately as possible in absolute terms, i.e., in terms of the angstrom or the meter. Indeed for many purposes, e.g., atomic constants experiments, such information is essential.

There are primarily three different approaches to the problem, viz.:

1   Direct measurements of x-ray wavelengths by means of a ruled grating.
2   Determination of some combination of atomic constants by both x-ray and non-x-ray methods. The ratio between the two results then gives the conversion factor, e.g., the conversion factor between A and A*.
3   Determination of crystal grating constants in terms of an optical wavelength with the aid of an x-ray interferometer and subsequent measurement of x-ray wavelengths with these crystals.

The first two approaches have been used a number of times, with estimated errors of the order of 10 ppm. The last method potentially promises a much higher accuracy, but is currently still in the development stage.

The results of such measurements are generally expressed in terms of a conversion factor. The ratio of a wavelength in angstroms to that in kilo-x-units has usually been denoted by Λ; i.e.,

$$\Lambda \equiv \frac{\lambda\ (A)}{\lambda\ (kxu)} \qquad (2\text{-}204)$$

(kxu is used rather than xu in order to make Λ close to unity.) In a similar way a conversion factor Λ^* is defined for the newer A* unit; i.e.,

$$\Lambda^* \equiv \frac{\lambda\,(A)}{\lambda\,(A^*)} \qquad (2\text{-}205)$$

This factor will be much closer to unity, the difference being of the order of 10 ppm.

All recent evaluations of the atomic constants have reviewed the available experimental data on these conversion factors and recommended best current values. The factor Λ was treated by Bearden and Thomsen (1957) and Cohen and DuMond (1965). The recent review of Taylor, Parker, and Langenberg (1969) includes both Λ and Λ^*.

B.   Ruled-grating experiments

Ruled-grating x-ray measurements are similar in principle to optical ones, but involve several highly important practical differences. Since the angles involved are usually small, it is customary to write the grating equation in terms of glancing angles (as in the Bragg law) rather than the usual optical ones; hence it becomes

$$n\lambda = d_G\,(\cos\zeta - \cos\xi_n) \qquad (2\text{-}206)$$

where n is the order of diffraction, λ the wavelength, d_G the grating space, and ζ and ξ_n the glancing angles for the incident and nth-order diffracted beams, respectively, as shown in Fig. 2-21.

An alternative form of the above equation is

$$n\lambda = 2d_G \sin\frac{\xi_n + \zeta}{2}\,\sin\frac{\xi_n - \zeta}{2} \qquad (2\text{-}207)$$

(This is more convenient for computational purposes since it avoids subtraction of two nearly equal quantities.) For small angles the expression (2-207) may be approximated as

$$n\lambda = \tfrac{1}{2}d_G(\xi_n{}^2 - \zeta^2) \qquad (2\text{-}208)$$

This simplified form is helpful for understanding the general features of x-ray gratings, but *not* sufficiently accurate for use in computing high-precision absolute-wavelength determinations.

The primary distinction between optical and x-ray use of diffraction gratings arises from the low reflectivity of all materials at x-ray wavelengths under normal conditions. Hence the only way of obtaining adequate intensity in the diffracted orders is to work in the region of total reflection. Since the index of refraction μ is slightly less than unity for most materials, this condition can be satisfied, but usually requires glancing angles of less than 1°. Consequently the dispersion is highly nonlinear, as follows from (2-208), and the qualitative features are quite different from those of the optical case.

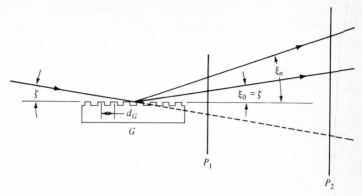

FIGURE 2-21
Schematic diagram of x-ray grating experiment, showing grating G with grating space d_G. X-ray beam is incident at glancing angle ζ; nth-order diffracted beam forms glancing angle ξ_n (ξ_0 corresponds to zero order, i.e., geometrical reflection). P_1 and P_2 represent photographic plates as used in experiments of Bearden and of Henins; dashed line indicates path of direct beam with grating removed.

The simple grating law (2-206) is strictly applicable only for Fraunhofer diffraction, i.e., for parallel incident and diffracted beams. Since no practical mirrors or lenses are presently available for x rays, the parallelism condition can be approximated only through large distances to source and image; strictly speaking, we always have a case of Fresnel diffraction. Fresnel corrections were investigated by Porter (1928), whose work was extended by Stauss (1929); their results indicated that Fresnel effects were of negligible importance in practical applications. However, their conclusion is dubious, because of an invalid assumption in Porter's analysis (a term in the argument of a cosine was dropped even though it may not be small compared to unity).

The Fresnel problem is eliminated by the use of a focusing concave-grating spectrograph, as used by Tyrén (1940) in a precision measurement of the Al $K\alpha_{1,2}$ doublet. The plates were calibrated by the lines of hydrogen-like spark spectra from highly ionized light atoms. These hydrogen-like wavelengths were calculated from theory, which did not include the then unknown Lamb shift. Subsequently this correction was considered and the results recalculated by Edlén and Svensson (1964). They obtained a value $\Lambda = 1.002\ 060 \pm 0.000\ 023$ A/kxu, based on the Al $K\alpha_{1,2}$ wavelength results of Nordfors.

While the Lamb-shift correction removes one problem regarding Tyrén's work, the Al $K\alpha_{1,2}$ wavelength in xu remains a significant source of possible error. The Nordfors wavelength which was employed differs from the recommended value in the Bearden (1964; 1967) table by about 20 ppm. Furthermore Sandström (1957) has pointed to the dangers of chemical effects on the surface of an aluminum target; he emphasized that in all future determinations of this type both the grating (absolute) and the crystal (x-ray scale) measurements should be made simultaneously with the *same source*. Hence there remains a substantial uncertainty as to the value of Λ^* implied by the Tyrén work.

In recent years Kirkpatrick, DuMond, and Cohen (1967) have attempted a new determination of Λ by Tyrén's method. Numerous difficulties were encountered, some still not explained satisfactorily; no definitive results were obtained. This method clearly presents formidable problems.

Bearden (1931*b*) performed a series of plane grating measurements of the Cu $K\alpha$, Cu $K\beta$, Cr $K\alpha$, and Cr $K\beta$ wavelengths. His method is shown schematically in Fig. 2-21. Two photographic plates P_1 and P_2 are used to record the various diffracted orders (including the zero order) and the direct beam (with grating removed). These plates are held by special holders which separate them by a precisely measured distance. From this length and the measured separations of the various lines on the plates, all the required angles may be calculated.

While his result remained for many years the most precise plane grating measurement, increasing accuracy of measurements on the x-ray scale suggested the need for a new determination of this type; it was also anticipated that such a result might permit x-ray experiments to carry greater weight in the evaluation of atomic constants. A. Henins (1971) carefully investigated some major modifications of the Bearden method, but finally returned to the same basic approach. There were primarily two significant changes. First, he developed a technique for producing a ruled grating with a known groove form (closely approximating the theoretical model). Second, he employed the longer wavelength (8.3 A) Al $K\alpha$ doublet, which permitted measurement of somewhat larger angles, while avoiding working too near the critical angle of the grating surface. By providing a hydrogen atmosphere for the entire x-ray path, he was able to eliminate any serious absorption effects without the mechanical strains and other complications of a vacuum system.

Measurements were made on 72 plates, more than half of which were measurable through the sixth order. A weighted average was used to compute the absolute wavelength of the Al $K\alpha_{1,2}$ doublet. Fresnel effects were carefully studied and calculated; such corrections were roughly 1 ppm, well within the experimental error.

A quartz-crystal measurement was made to determine the Al $K\alpha$ wavelength with respect to the Cu $K\alpha_1$ line for the specific x-ray source employed. It was assumed that the position of the photographic image, as read in the grating method, corresponded to the centroid of the crystal curve recorded for the doublet. The index of refraction of quartz at 8.3 A* introduced some uncertainty; an average of experimental and theoretically computed values for μ was used.

These measurements yielded a conversion factor

$$\Lambda = 1.000\ 009\ \frac{A}{A^*} \pm 7\ \text{ppm} \qquad (2\text{-}209)$$

The uncertainty quoted represents the probable error.

C. Atomic constants experiments

As previously noted, the atomic constants approach to the determination of Λ^* or Λ consists of measuring of some combination of constants by both x-ray and non-x-ray methods. Currently there are three possible candidates, which may be loosely termed as "annihilation wavelength," "h/e," and the "Avogadro-number experiments." These may be briefly summarized as follows:

1 Annihilation wavelength. When a positron-electron pair annihilates to form two photons, the energy relation is simply $2mc^2 = 2h\nu_a$, where m is the mass of the electron, c the velocity of light, h Planck's constant, and ν_a the annihilation frequency. It follows that the annihilation wavelength is given by

$$\lambda_a = \frac{c}{\nu_a} = \frac{h}{mc} \qquad (2\text{-}210)$$

which is simply the Compton wavelength of the electron. This quantity may be measured directly by crystal spectroscopy and compared with the results implied by non-x-ray data.

2 h/e experiments. The high-frequency limit of the continuous x-ray spectrum is also given by a simple energy equation $eV = h\nu_{hfl}$, where e and V denote the electronic charge and the accelerating voltage (both in mks units). Hence the short-wavelength limit becomes

$$\lambda_{swl} = \frac{c}{\nu_{hfl}} = \frac{1}{V}\frac{hc}{e} \qquad (2\text{-}211)$$

For a known voltage, λ_{swl} may be computed from a non-x-ray value of hc/e; if the wavelength is then also measured on the x-ray scale, the desired conversion factor may be obtained.

3 Avogadro-number experiments. The grating constant of a pure crystal may be calculated from non-x-ray data and compared with the value measured on the x-ray scale. For simplicity consider a cubic crystal with density ρ and unit cell of side d; the mass of the cell is thus ρd^3. On the other hand, if the unit cell contains effectively z molecules of molecular weight M, then its mass is also expressible as zM/N, where N denotes Avogadro's number. Equating these two expressions and solving for d yields

$$d = \left(\frac{zM}{\rho N}\right)^{1/3} \qquad (2\text{-}212)$$

If the cell structure and molecular weight are known and N is obtained from non-x-ray data, it is only necessary to measure the density to compute d.

All these methods are treated at greater length in reviews of atomic constants, such as those of Bearden and Thomsen (1957), Cohen and DuMond (1965), and Taylor, Parker, and Langenberg (1969). At the present time the third approach appears to offer the highest precision; the problems involved in this type of experiment have been discussed in detail by Deslattes et al. (1966).

D.   Avogadro-number measurements

Henins and Bearden (1964) measured the grating constants of high-purity silicon crystals in terms of the Cu $K\alpha_1$ and Cu $K\alpha_2$ lines. I. Henins (1964) also determined the densities of these crystals by precision hydrostatic weighing. Absolute wavelengths were computed from these data for each case, and agreed to the order of 1 ppm. The principal error sources were the molecular weight of silicon (due to variations of natural isotopic abundances) and the value of Avogadro's number.

Bearden (1965a) carried out a somewhat similar set of measurements with nine selected calcite crystals. Grating constants were measured with both the Cu $K\alpha_1$ and Ag $K\alpha_1$ lines, and the results were averaged. Since calcite is a rhombohedral rather than a cubic crystal, (2-212) had to be modified by a dimensionless geometrical factor. Since significant impurities were present in the crystals, these were analyzed and the results extrapolated by a least-squares analysis to a hypothetical crystal with zero impurity content. The principal uncertainties were due to the atomic weight of calcium and the value of Avogadro's number.

As a preliminary step toward his wavelength table, Bearden (1965a) made a new evaluation of the conversion factor Λ from all available precision data. While a number of different experiments were included, more than 80% of the weight rested on the silicon and calcite determinations described above; these two agreed within about 2 ppm. His recommended value was $\Lambda = 1.002\ 056 \pm 5$ ppm, based on $\lambda_{Cu\ K\alpha_1} = 1.537\ 400$ xu. This is obviously equivalent to a recommended absolute value for $\lambda_{Cu\ K\alpha_1}$.

When this result was combined with the wavelength ratios determined by Bearden et al. (1964), it gave the numerical value for the W $K\alpha_1$ line used in the definition of the A* unit, as given in (2-202). As a consequence, the initially recommended value of Λ^* became equal to unity. Of course, it remained an *experimentally determined quantity*, the initial experimental value being $\Lambda^* = 1.000\ 000 \pm 5$ ppm, as stated explicitly in the review of Thomsen and Burr (1968).

As indicated by equation (2-212), the Avogadro-number experiments depend on the use of a known value for N. Bearden obtained this datum from the then current evaluation of Cohen and DuMond. [For details of this adjustment, see Cohen and

DuMond (1965).] Subsequently, new developments in the field, particularly the ac Josephson-effect experiment, have led to a downward revision of the value of N by almost 60 ppm. As a result of this shift Taylor, Parker, and Langenberg (1969) concluded that the value of Λ^* implied by silicon- and calcite-crystal data described above was

$$\Lambda^* = 1.000\ 019_7 \pm 5.6\ \text{ppm} \qquad \text{(standard deviation)} \qquad \text{(2-213)}$$

E.   X-ray interferometer measurements

The availability of almost perfect crystals has recently opened the way for the development of an x-ray interferometer; the first such instrument was constructed by Bonse and Hart (1965). This principle is illustrated schematically in Fig. 2-22. Beam-splitter crystal C_1, "mirror" crystals C_2 and C_3, and analyzer crystal C_4 are all cut from the same block of perfect crystal. They are assumed to be perfectly aligned with atomic planes parallel to one another.

The x-ray radiation incident on C_1 is directed so as to satisfy the Bragg law for reflection by the internal plane normal to the surface. At C_1 the beam is effectively split into a reflected component I and a transmitted component II; these are in turn partially reflected by C_2 and C_3, respectively, and recombined at C_4. The detector D records a portion of the recombined radiation, as shown.

Assume that the crystals are initially aligned so that the atomic planes coincide, i.e., as though all were still part of the same perfect-crystal structure. The symmetry of the arrangement then assures equal path lengths for components I and II; hence

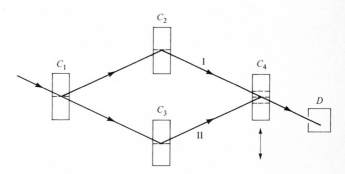

FIGURE 2-22
Schematic diagram of x-ray interferometer, with four crystals used in transmission. Ray I is internally reflected by C_1 and C_2 and transmitted by C_4; ray II is transmitted by C_1 and internally reflected by C_3 and C_4. Combined rays strike detector D, where intensity variations (i.e., fringes) are observed as C_4 is moved in the direction indicated by the arrows underneath it.

constructive interference results. Now assume that C_4 is moved normal to the internal reflecting planes, as indicated in the figure. If it is displaced by exactly one lattice spacing (i.e., grating constant), the new situation is identical to the former one, and constructive interference occurs once more. Between these two positions there is a point where the path of the ray II is altered by $\lambda/2$, while that of the interfering ray I is obviously unaffected; in this case destructive interference occurs. Thus, as C_4 is slowly displaced, a series of high- and low-intensity readings, i.e., interference fringes, will be recorded by detector D.

If the displacement of C_4 is measured by an optical interferometer (preferably using a laser source) and the x-ray fringes simultaneously recorded, the grating constant can then be directly computed in absolute units. The wavelength of the x rays is *not* involved in this calculation. In fact, it will be noted that the arrangement corresponds to the nondispersive $(n, -n)$ position of the two-crystal spectrometer. Thus a relatively wide band of wavelengths is effectively employed in the measurement. This minimizes the intensity problem, which is further alleviated by the Borrmann (1959) effect, i.e., anomalously high transmission of one component through almost perfect crystals.

The principle of the interferometer is more rigorously understood in terms of the fields in the various crystals. Such a description is outlined in the survey paper of Hart and Bonse (1970), who also review possible applications.

Three different groups are now working on precision measurement of silicon grating constants by the x-ray interferometer technique. Bonse, te Kaat, and Spieker (1971), Curtis et al. (1971), and Deslattes (1971) have all given preliminary accounts of their work. While the procedure may appear relatively simple from the outline given here, the small distances to be measured and the consequent demands on the stability of the apparatus make such experiments extremely difficult and painstaking. Thus far only very preliminary values have been reported, which are not sufficiently accurate to shed any new light on the value of Λ^*. However, these experiments hold great promise of yielding a substantially more accurate value in the near future and may indeed eventually eliminate the need of a separate unit for x-ray wavelengths.

F. Conversion factor

At the present writing two high-precision values of Λ^* are available, viz., the grating value of (2-209) and the Avogadro-number result of (2-213). The latter is stated with a standard deviation of 5.6 ppm, which implies a probable error of 3.8 ppm. Thus if the two values were statistically averaged with weights inversely proportional to the squares of the errors, the grating result would carry less than 25% of the weight.

However, as noted in Subsection VIII D, the Avogadro-number value has recently undergone a rather drastic change (of several probable errors) due to a revision in the value of N; it now seems likely that a further substantial shift in N will occur with the next evaluation of atomic constants. Because of this uncertainty and because the two values of Λ^* represent completely independent approaches, it now seems reasonable to give them equal weights and treat each as having a probable error of roughly 7 ppm.

With this assumption, we obtain an average value

$$\Lambda^* = 1.000\ 014 \pm 5\ \text{ppm} \qquad (2\text{-}214)$$

where the error represents a *probable error*.

IX. ATOMIC ENERGY LEVELS

A. Wavelength information

X-ray wavelengths are a measure of the separation between various energy levels and hence provide important information on atomic structure. The energy difference is, of course, given by the familiar relation $\Delta E = h\nu = hc/\lambda$, where h is Planck's constant and c the velocity of light. It is frequently desirable to express wavelengths on the x-ray scale and energies in terms of electron volts.

From (2-205) it is obvious that the wavelength in m is $\lambda = 10^{-10}\Lambda^*\lambda(\text{A}^*)$, while the separation in eV is simply $\Delta V = \Delta E/e$. Combining these expressions now yields

$$\lambda\,(\text{A}^*)\,\Delta V = \frac{10^{10}\ hc}{\Lambda^* e} \qquad (2\text{-}215)$$

where h, c, and e are all expressed in mks (SI) units.

Thus the energy difference in electron volts is inversely proportional to the wavelength, with the conversion factor given by (2-215). Taylor, Parker, and Langenberg (1969) recommended a value of $1.239\ 830\ 1 \times 10^4\ V \cdot \text{A}^* \pm 5.9$ ppm for this factor. The value of Λ^* recommended in Subsection VIII F of the present chapter is 5.7 ppm lower than the TPL value. When Λ^* is taken from (2-214), the conversion factor and its *probable error* become

$$\lambda(\text{A}^*)\,\Delta V = 1.239\ 837 \times 10^4\ \text{A}^* \cdot V \pm 6\ \text{ppm} \qquad (2\text{-}216)$$

Electronic transitions involved in emission-line production are shown schematically in Fig. 2-23 for a few selected cases. For example, the $K\alpha_2$ line corresponds to an *electronic* transition from the L_{II} shell to the K shell. (The energy level of the *atom* is usually described in terms of the level where the *vacancy* exists; thus the *atom* goes from a K to an L_{II} state in this example.) Hence the energy levels are related by

$$E_K - E_{L_{II}} = h\nu_{K\alpha_2} \qquad (2\text{-}217)$$

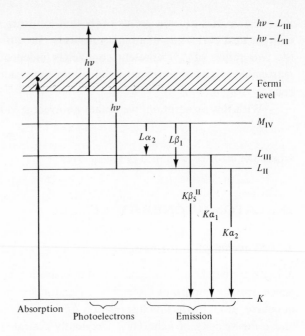

FIGURE 2-23
Schematic diagram illustrating the energy differences involved in emission lines, absorption edges, and photoelectron emission. Note that the arrows indicate direction of *electronic* transition, e.g., in the $K\alpha_2$ emission line process an electron goes from the L_{II} level to the K level. (Since x-ray states are usually described in terms of the *vacancy*, the *atom* is said to go from a K state to an L state in this example.) This diagram could apply to an atom such as molybdenum (Mo 42), where all the processes shown have actually been measured.

In most cases such energy differences can be obtained from wavelength data in more than one way. For example, these two states are also connected by a two-step path via the M_{IV} level; i.e.,

$$E_K - E_{L_{II}} = h(\nu_{K\beta_5}{}^{II} - \nu_{L\beta_1}) \qquad (2\text{-}218)$$

In the same way it is easy to trace two different paths between the K and L_{III} levels. Thus x-ray wavelength data usually furnish an overdetermined set of equations which fix the various energy-level differences.

B.   Absolute energy levels

It is also desirable to state the various energy levels on an absolute basis, i.e., with respect to some appropriate zero level. Since most x-ray lines are observed with solid targets, it is usually convenient to adopt the Fermi level for this purpose.

For many years x-ray absorption edges were used to relate the various x-ray states to the Fermi level; the L_{III} level was generally taken as the best choice to define the absolute scale [see, for example, Sandström (1957)]. The criterion for locating the edge spectroscopically was usually considered to be the inflection point of the absorption curve; this is nearly equivalent to the half-intensity point, i.e., the point with an intensity midway between those of the plateaus above and below the edge. This point was seldom obtained with a precision comparable to that of the best emission-line measurements. Furthermore it was not always clear how closely the resulting values were based on the true Fermi level.

The problem has now been very largely overcome by the photoelectron measurement of K. Siegbahn and collaborators. The methods and results have been summarized by Hagström, Nordling, and Siegbahn (1965); additional details on the β-ray spectrometer employed have been given by K. Siegbahn (1965). Bearden and Burr (1965; 1967) included a brief outline of these experiments in their evaluation of x-ray energy levels.

Briefly stated, the method is simply an x-ray application of Einstein's photoelectric equation. Radiation from a strong emission line of known wavelength is allowed to strike a target (converter); as a result, photoelectrons are emitted. The energy relation is approximately

$$E_{KE} = h\nu - E_B \qquad (2\text{-}219)$$

where ν is the frequency of the incident photon and E_{KE} the kinetic energy of the photoelectron. E_B represents the binding energy of the electron in question; e.g., it would be $E_{L_{II}}$ for an electron from the L_{II} level. The L_{II} and L_{III} photoelectron measurement processes are shown schematically in Fig. 2-23.

The kinetic energy of the photoelectron is determined by a β-ray spectrometer, i.e., by the magnetic field needed to produce a given radius of curvature. High-precision work demands great care in this measurement and requires a correction for the work function of the spectrometer slits; the details will not be discussed here.

Such measurements have been carried out for more than seventy elements. In many cases several levels were included; e.g., the L_I, L_{II}, and L_{III} levels were measured for molybdenum (Mo 42). Estimated errors were generally a few tenths of an electron volt; energy-level differences were determined to somewhat higher accuracy than absolute values with respect to the Fermi level. Such results are substantially more accurate than almost all absorption-edge data and have largely eliminated use of the latter in energy-level evaluations. However, Bearden and Burr (1965; 1967) found that, within their estimated errors, absorption-edge results were generally consistent with the newer photoelectron data.

C. Energy-level tables

A number of workers have prepared comprehensive tables of x-ray energy levels, e.g., Siegbahn (1931), Cauchois (1952; 1955), and Sandström (1957). In most instances one absorption edge, usually the L_{III} edge, was used to fix the zero of energy and other levels determined with respect to this one via the most accurate path supplied by emission-line data.

As indicated in Subsection IX A, the emission lines normally furnish an overdetermined set of equations. A further degree of overdetermination results when these are combined with photoelectron measurements. For example, Fig. 2-23 shows that the K energy level can be computed from either the L_{III} or the L_{II} photoelectron measurements; i.e.,

$$E_{L_{III}} + h\nu_{K\alpha_1} = E_K = E_{L_{II}} + h\nu_{K\alpha_2} \qquad (2\text{-}220)$$

In preparing an updated set of energy-level tables Bearden and Burr (1965; 1967) adopted a new approach. They normally retained the entire set of overdetermined data for each element and computed the various energy levels by a least-squares adjustment. [For a brief summary of the least-squares technique see, for example, Bearden and Thomsen (1957).] Their input data consisted principally of the emission wavelengths from the recent Bearden (1964; 1967) table and the photoelectron data of K. Siegbahn and collaborators described in Subsection IX B.

The least-squares method, of course, required the assignment of errors to each input item and yielded output errors for the resulting energy levels. The χ^2 tests indicated that the input errors were reasonable in the great majority of cases. A few input items were rejected and new evaluations carried out with the censored data for the elements in question.

A tabulation of selected energy levels, based on the Bearden and Burr tables, is given in Appendix B.

X. CONCLUSIONS

We have described the three types of x-ray crystal spectrometers best suited for high-precision measurements, viz., the Bragg spectrometer, the tube spectrometer, and the two-crystal spectrometer. Theoretical analyses have been given and alignment procedures outlined, with emphasis on the Bragg and two-crystal types. The problem of crystal selection and preparation has also been treated.

Under optimum conditions wavelengths can be measured on a relative x-ray scale (i.e., with respect to a standard x-ray wavelength) with a probable error of the order of 1 ppm. (The W $K\alpha_1$ line is taken as the standard wavelength to define the new A* unit.) Systematic errors are almost certain to be more critical than statistical ones; for this reason it is vital to vary as many experimental parameters as possible in high-precision measurements.

The conversion factor from the relative to the absolute wavelength scale is currently known with a probable error of about 5 ppm; the present value rests on an average of a recent ruled-grating experiment and two atomic constants determinations (Avogadro-number experiments). X-ray interferometer work currently under way may reduce the error by at least one order of magnitude.

Very recently the first instance of a nonlinear optical effect with x-rays was observed by Eisenberger and McCall (1971). It is too soon to predict what applications this may have for x-ray spectroscopy, but this could well be a development of major significance.

ACKNOWLEDGMENTS

The author wishes to thank numerous colleagues who have contributed in various ways to the preparation of this chapter. Dr. J. A. Bearden has reviewed the manuscript and offered numerous constructive suggestions; his active assistance in all phases was essential to the project. Most of the experimental techniques described, some heretofore unpublished, were developed in his laboratory at The Johns Hopkins University. Much of the background was provided by former members of his x-ray group; particular mention is due to Drs. A. F. Burr, Albert Henins, John G. Marzolf, William C. Sauder, and F. Y. Yap. The author is also grateful to Drs. R. D. Deslattes, John J. Merrill, and Barry Taylor for helpful discussions.

Finally he wishes to thank Mr. C. Millard LaPorte for preparing the figures, Mrs. George D. Rowe for typing the manuscript, and Mrs. John S. Thomsen for proofreading.

REFERENCES

The works listed below include only those specifically cited in this chapter; there is no attempt at complete coverage of the field. Siegbahn's (1931) book gives a comprehensive bibliography of x-ray spectroscopy up to that date; the Sandström (1957) article is more recent, but somewhat less inclusive. The Bearden (1964) wavelength table includes 385 references, most of which provided some numerical wavelength data; however, the majority of these were *not* listed in the Bearden (1967) review article. The book of Compton and Allison (1935) provides some additional references on various aspects of x-ray spectroscopy.

ALLISON, S. K. (1930): Experiments on the reported fine structure and the wave-length separation of the $K\beta$ doublet in the molybdenum x-ray spectrum, *Phys. Rev.*, **35**: 149–154.

―――― (1931): The resolving power attainable in x-ray spectroscopy by photographic methods, *Phys. Rev.*, **38**: 203–210.

―――― (1933): The natural widths of the $K\alpha$ x-ray doublet from 26 Fe to 47 Ag, *Phys. Rev.*, **44**: 63–72.

BATTERMAN, B. W., and COLE, H. (1964): Dynamical diffraction of x rays by perfect crystals, *Rev. Mod. Phys.*, **36**: 681–717.

BEARDEN, J. A. (1931*a*): Grating constant of calcite crystals, *Phys. Rev.*, **38**: 2089–2098.

―――― (1931*b*): Absolute wave-lengths of the copper and chromium K series, *Phys. Rev.*, **37**: 1210–1229.

―――― (1964): X-ray wavelengths, NYO-10586 (U.S. Department of Commerce, National Bureau of Standards, Clearinghouse for Federal Scientific and Technical Information, Springfield, Va.).

―――― (1965*a*): X-ray wavelength conversion factor Λ (λ_g/λ_s), *Phys. Rev.*, **B137**: 181–187.

―――― (1965*b*): Selection of W $K\alpha_1$ as the x-ray wavelength standard, *Phys. Rev.*, **B137**: 455–461.

―――― (1967): X-ray wavelengths, *Rev. Mod. Phys.*, **39**: 78–124.

―――― and BURR, A. F. (1965): Atomic energy levels, NYO-2543-1 (U.S. Department of Commerce, National Bureau of Standards, Clearinghouse for Federal Scientific and Technical Information, Springfield, Va.).

―――― and ―――― (1967): Reevaluation of atomic energy levels, *Rev. Mod. Phys.*, **39**: 125–142.

―――― and HENINS, A. (1965): Precision measurement of lattice imperfections with a photographic two-crystal method, *Rev. Sci. Instr.*, **36**: 334–338.

――――, ――――, MARZOLF, J. G., SAUDER, W. C., and THOMSEN, J. S. (1964): Precision redetermination of standard reference wavelengths for x-ray spectroscopy, *Phys. Rev.*, **A135**: 899–910.

――――, MARZOLF, J. G., and THOMSEN, J. S. (1968): Crystal diffraction profiles for monochromatic radiation, *Acta Cryst.*, **A24**: 295–301.

―――― and SHAW, C. H. (1935): Shapes and wavelengths of K series lines of elements Ti 22 to Ge 32, *Phys. Rev.*, **48**: 18–30.

―――― and THOMSEN, J. S. (1957): A survey of atomic constants, *Nuovo Cimento Suppl.*, **5**: 267–360; least-squares method, Appendix B, 349–353.

―――― and ―――― (1971): The double-crystal x-ray spectrometer: Corrections, errors, and alignment procedure, *J. Appl. Cryst.*, **4**: 130–138.

BONSE, U., and HART, M. (1965): An x-ray interferometer, *Appl. Phys. Lett.*, **6**: 155–156.

――――, TE KAAT, E., and SPIEKER, P. (1971): *Precision lattice parameter measurement by x-ray interferometry*, Proceedings of the International Conference on Precision Measurement and Fundamental Constants, August 1970, Government Printing Office, Washington, D.C.

BORRMANN, G. (1959): Röntgenwellenfelder, in O. R. FRISCH et al. (eds.), *Trends in atomic physics*, pp. 262–282, Interscience Publishers, Inc., New York.

BRAGG, W. H., and BRAGG, W. L. (1915): X rays and crystal structure, chap. 3, G. Bell and Sons, Ltd., London.

BROGREN, G. (1954): A note on the width of x-ray emission lines, *Arkiv Fysik*, **8**: 391–400.

CAUCHOIS, Y. (1932): Spectrographie des Rayons X par transmission d'un Faisceau non Canalisé à travers un Crystal Courbé, *J. Phys. Radium*, **3**: 320–336.

——— (1952): Les Niveaux d'Energie des Atomes Lourds, *J. Phys. Radium*, **13**: 113–121.

——— (1955): Les Niveaux d'Energie des Atomes de Numéro Atomique Inférieur à 70, *J. Phys. Radium*, **16**: 253–262.

——— and HULUBEI H. (1947): *Longeurs d'Onde des Émissions X et des Discontinuités d'Absorption X*, Hermann and Co., Paris.

COHEN, E. R., and DUMOND, J. W. M. (1965): Our knowledge of the fundamental constants of physics and chemistry in 1965, *Rev. Mod. Phys.*, **37**: 537–594.

COMPTON, A. H. (1917): The reflection coefficient of monochromatic x rays from rock salt and calcite, *Phys. Rev.*, **10**: 95–96.

——— (1931): A precision x-ray spectrometer and the wavelength of Mo $K\alpha_1$, *Rev. Sci. Instr.*, **2**: 365–376.

——— and ALLISON, S. K. (1935): *X rays in theory and experiment*, D. Van Nostrand Company, Inc., New York; spectroscopy in chap. 9; Lorentzian line profile in chap. 4, sec. 3; Darwin crystal window (general) in chap. 6; two-crystal-window width, p. 728; coefficient of reflection, pp. 394–399.

COOPER, A. S. (1965): The Mo $K\alpha_1$/Cu $K\alpha_1$ wavelength ratio, *Acta Cryst.*, **18**: 1078–1080.

CURTIS, I., HART, M., MILNE, A. D., and MORGAN, I. G. (1971): *A new determination of Avogadro's number*, Proceedings of the International Conference on Precision Measurement and Fundamental Constants, August 1970, Government Printing Office, Washington, D.C.

DARWIN, C. G. (1914): The theory of x-ray reflection, *Phil. Mag.*, **27**: 315–333 and 675–690.

DESLATTES, R. D. (1967): Single axis, two crystal x-ray instrument, *Rev. Sci. Instr.*, **38**: 815–820.

——— (1971): *Optical interferometry of the 220 repeat distance in a silicon crystal*, Proceedings of the International Conference on Precision Measurement and Fundamental Constants, August 1970, Government Printing Office, Washington, D.C.

———, PEISER, H. S., BEARDEN, J. A., and THOMSEN, J. S. (1966): Potential applications of the x-ray/density method for the comparison of atomic weight values, *Metrologia*, **2**: 103–111.

DUMOND, J. W. M. (1947): A high resolving power, curved-crystal focusing spectrometer for short wave-length x-rays and gamma-rays, *Rev. Sci. Instr.*, **18**: 626–638.

——— and HOYT, A. (1930): Design and technique of operation of a double crystal spectrometer, *Phys. Rev.*, **36**: 1702–1720.

——— and KIRKPATRICK, H. A. (1930): The multiple crystal x-ray spectrograph, *Rev. Sci. Instr.*, **1**: 88–105.

EDLÉN, B., and SVENSSON, L. A. (1964): The wavelengths of the Lyman lines and a redetermination of the x-unit from Tyrén's spectrograms, *Arkiv Fysik*, **28**: 427–447.

EISENBERGER, P., and MCCALL, S. L. (1971): X-ray parametric conversion, *Phys. Rev. Lett.*, **26**: 684–688.

FORSYTHE, G. E. (1957): Generation and use of orthogonal polynomials for data-fitting with a digital computer, *J. Soc. Ind. Appl. Math.*, **5**: 74–88.

HAGLUND, P. (1941): An experimental investigation into *L*-emission spectra, *Arkiv Mat. Astron. Fysik*, **A28** (8).

HAGSTRÖM, S., NORDLING, C., and SIEGBAHN, K. (1965): vol. 1, Appendix 2, pp. 845–862, in K. SIEGBAHN (ed.), *Alpha-, beta-, and gamma-ray spectroscopy*, North-Holland Publishing Company, Amsterdam.

HART, M. (1969): High precision lattice parameter measurements by multiple Bragg reflection diffractometry, *Proc. Roy. Soc. (London)*, **A309**: 281–296.

───── and BONSE, U. (1970): Interferometry with x rays, *Phys. Today*, **23** (8): 26–31.

HENINS, A. (1971): *Ruled grating measurement of the Al* $K\alpha_{1,2}$ *wavelength*, Proceedings of the International Conference on Precision Measurement and Fundamental Constants, August 1970, Government Printing Office, Washington, D.C.

HENINS, I. (1964): Precision density measurement of silicon, *J. Res. Natl. Bur. Std.*, **A68**: 529–533.

───── and BEARDEN, J. A. (1964): Silicon-crystal determination of the absolute scale of x-ray wavelengths, *Phys. Rev.*, **A135**: 890–898.

HRDÝ, J., HENINS, A., and BEARDEN, J. A. (1970): Polarization of the $L\alpha_1$ x rays of mercury, *Phys. Rev.*, **A2**: 1708–1710.

INGLESTAM, E. (1937): Die *K*-spektren der schweren elemente, *Nova Acta Reg. Soc. Sci. Upsalien.*, **10** (5): 13–15.

KIKUTA, S., and KOHRA, K. (1970): X-ray crystal collimators using successive asymmetric diffractions and their applications to measurements of diffraction curves. I. General considerations on collimators, *J. Phys. Soc. Japan*, **29**: 1322–1328.

KIRKPATRICK, H. A., DUMOND, J. W. M., and COHEN, E. R. (1967): *Remeasurement of the conversion constant* Λ, Proceedings of the Third International Conference on Atomic Masses, 1967, pp. 347–382, University of Manitoba Press.

KOHRA, K. (1962): An application of asymmetric reflection for obtaining x-ray beams of extremely narrow angular spread, *J. Phys. Soc. Japan*, **17**: 589–590.

LARSSON, A. (1927): Precision measurements of the *K*-series of molybdenum and iron, *Phil. Mag.*, **3**: 1136–1160.

MARZOLF, J. G. (1964): Angle measuring interferometer, *Rev. Sci. Instr.*, **35**: 1212–1215.

MERRILL, J. J., and DUMOND, J. W. M. (1958): Precision measurement of the *L* x-ray spectra of uranium and plutonium, *Phys. Rev.*, **110**: 79–84.

───── and ───── (1961): Precision measurement of *L* x-ray wavelengths and line widths for $74 \leq Z \leq 95$ and their interpretation in terms of nuclear perturbations, *Ann. Phys. (N.Y.)*, **14**: 166–228. Appendix B contains an algebraic error following equation (B-7c) which invalidates the resulting quadratic terms in the δ's. (Private communication, J.J.M.)

PARRATT, L. G. (1935): Precision of x-ray wavelength measurements, *Phys. Rev.*, **47**: 882–883.

PORTER, A. W. (1928): On the positions of x-ray spectra as formed by a·diffraction grating, *Phil. Mag.*, **5**: 1067–1071.

PORTEUS, J. O. (1962): Optimized method for correcting smearing aberrations: Complex x-ray spectra, *J. Appl. Phys.*, **33**: 700–707.

PRINS, J. A. (1930): Die Reflexion von Röntgenstrahlen an absorbierenden idealen Kristallen, *Z. Physik.*, **63**: 477–493.

RENNINGER, M. (1967): The asymmetric Bragg reflection and its application in double diffractometry, *Advan. X-Ray Anal.*, **10**: 32–41.

SANDSTRÖM, A. E. (1957): Experimental methods of x-ray spectroscopy: Ordinary wavelengths, in S. Flügge (ed.), *Handbuch der physik*, vol. 30, pp. 78–240, Springer-Verlag, Berlin; x unit and wavelength standard, p. 93; wavelength table, pp. 164–198; absolute grating measurements, pp. 239–240; absorption edges and energy levels, pp. 211–230.

SAUDER, W. C. (1966): General method of treating instrumental distortion of spectral data with applications to x-ray physics, *J. Appl. Phys.*, **37**: 1495–1507.

——— (1971): *A positron annihilation experiment for determining the fine structure constant*, Proceedings of the International Conference on Precision Measurement and Fundamental Constants, August 1970, Government Printing Office, Washington, D.C.

SCHNOPPER, H. W. (1965): Spectral measurements with aligned and misaligned two-crystal spectrometers. I. Theory of the geometrical window, *J. Appl. Phys.*, **36**: 1415–1423; II. Alignment, ibid., pp. 1423–1430; Erratum, ibid., p. 3692.

SCHWARZSCHILD, M. M. (1928): Theory of the double x-ray spectrometer, *Phys. Rev.*, **32**: 162–171.

SHACKLETT, R. L., and DUMOND, J. W. M. (1957): Precision measurement of x-ray fine structure; effects of nuclear size and quantum electrodynamics, *Phys. Rev.*, **106**: 501–512.

SIEGBAHN, K. (1965): vol. 1, chap. 3, pp. 79–202, in K. SIEGBAHN (ed.), *Alpha-, beta-, and gamma-ray spectroscopy*, North-Holland Publishing Company, Amsterdam.

SIEGBAHN, M. (1925): *The spectroscopy of x-rays*, translated by G. A. Lindsay, Oxford University Press, London.

——— (1929): On the methods of precision measurements of x-ray wave-lengths, *Arkiv Mat. Astron. Fysik*, **A21** (21): 1–20.

——— (1931): *Spektroskopie der Röntgenstrahlen*, 2d ed., Springer-Verlag, Berlin; x unit, pp. 42–44; vacuum spectrometer, pp. 112–117; energy-level table, pp. 346–352.

STAUSS, H. E. (1929): Errors in the use of gratings with x-rays due to the divergence of the radiation, *Phys. Rev.*, **34**: 1601–1604.

TAYLOR, B. N., PARKER, W. H., and LANGENBERG, D. N. (1969): *The fundamental constants and quantum electrodynamics*, Review of Modern Physics Monograph, Academic Press, New York; reprinted from *Rev. Mod. Phys.*, **41**: 375–496; electrical units, secs. II.B.3 and II.C.2; x-ray experiments and values of Λ and Λ^*, secs. II.C.6 and III.C (final values given in sec. VI.A).

THOMSEN, J. S. (1966): Calibration of a divided circle using two diametrically opposite pairs of microscopes, *Bull. Am. Phys. Soc.*, **11**: 388.

——— and BURR, A. E. (1968): Biography of the x unit—the x-ray wavelength scale, *Am. J. Phys.*, **36**: 803–810.

——— and YAP, F. Y. (1968): Effect of statistical counting errors on wavelength criteria for x-ray spectra, *J. Res. Natl. Bur. Std. (U.S.)*, **A72**: 187–205.

TYRÉN, F. (1940): Precision measurements of soft x-rays with concave gratings, *Nova Acta Reg. Soc. Sci. Upsalien.*, **12** (1).

WEISSKOFF, V., and WIGNER, E. (1930): Berechnung der natürlichen Liniebreite auf Grund der Diracschen Lichttheorie, *Z. Physik.*, **63**: 54–73.

WILLIAMS, J. W. (1932): A correction to wave-length measurements with the double-crystal spectrometer, *Phys. Rev.*, **40**: 636.

3

GRATING SPECTROMETERS AND THEIR APPLICATION IN EMISSION SPECTROSCOPY

John R. Cuthill

I. INTRODUCTION

The M spectra of the first transition-series metals and the L spectra of aluminum and magnesium, which have been studied rather extensively during the past two decades, involve the wavelength range from 50 to 500 A. In this range the grazing-incidence grating spectrometer is superior. The resolution of the grating spectrometer surpasses that of even the two-crystal spectrometer above 50 A. The wavelength cutoff at which total absorption occurs is a function of the refractive index of the material of the grating and the angle of incidence. However, grating spectrometers are commonly used in soft x-ray spectroscopy down to O K at 23.6 A (Holliday, 1967R). This is by no means a lower limit; Bearden (1931) carefully measured the diffraction angles of the Cr and Cu K lines from reflection gratings, to obtain a precise absolute determination of the wavelengths, which in turn could then be used to calculate the lattice spacings in crystals. Recently, Speer (1970) described a spectrometer that recorded on film spectral lines down to less than 1 A. Grating spectrometers would not ordinarily be used in this range, which is the realm of the crystal spectrometer. (However, one unique feature of a grating spectrometer that sometimes justifies coping with the severe problems of low grazing angle—i.e., a few minutes of arc, optical alignment, grating perfection, astigmatism, and intensity in this range—is that an entire broad spectrum can be recorded from a short-lived transient phenomenon if it is sufficiently intense and if film recording is used.)

The critical parameters and characteristics of grazing-incidence grating spectrometers will be discussed first; then the effects of such experimental conditions as specimen temperature, degree of vacuum, and type of specimen excitation will be discussed before reviewing soft x-ray spectra. Topics which have been covered in detail in the textbooks and reviews will be discussed only briefly here, leaving more space for detailed discussions of newer developments in soft x-ray spectroscopy. Over the years there have been many very good state-of-the-art review articles and books. A selection of these has been set apart under the heading of General Reviews, and they are identified by the R following the year in text citations. There have also been series of soft x-ray conferences over the years which have served to reveal the state-of-the-art as of their respective dates as well as the geographic distribution of interests. The longest series is the All Union X-Ray Spectroscopy Conference series sponsored by the Academy of Sciences of the U.S.S.R. The first conference was in 1955. The eighth conference was held in 1967. The proceedings of all the conferences have been published in the *Bulletin of the Academy of Sciences of the USSR, Physical Series*. A series of international character was initiated by Professor Parratt at Cornell University in 1962. The proceedings of the conference were unpublished, but the second in the series, entitled "X-Ray Spectra and Chemical Binding," was held at Karl Marx University, Leipzig, 1965, and the proceedings were published by the Physical Chemistry Institute of the University. The third in the series, entitled "X-Ray Spectra and the Electronic Structure of Matter," was held in Kiev in 1968. The proceedings were published by the IMF Academy of Sciences of the Ukrainian SSR, Kiev, 1969. The fourth in the series was held at the University of Paris in 1970, entitled "Single and Multiple Electronic Processes in the X-ray and UV Range." The proceedings were published as a supplement to the *Journal de Physique*, **32**: (1971) No. 10 (Colloque *C*-4). The fifth in this series was held at the University of Munich in September 1972. A new series was initiated at the University of Strathclyde, Glasgow, in 1967. The proceedings entitled *Soft x-ray band spectra and the electronic* *structure of metals and materials* was published by Academic Press, Inc., London, 1968. The proceedings of the second University of Strathclyde conference held in September 1971, entitled *Band-structure spectroscopy of metals and alloys*, will also be published by Academic Press, Inc. These conferences are not referenced again as such, but individual papers at the conferences and others are referenced under the appropriate topics and included in the tabulation of references at the end of the chapter.

II. THE GRAZING-INCIDENCE CONCAVE-GRATING SPECTROMETER

A. General

We will begin with a general description of the concave-grating spectrometer used in soft x-ray spectroscopy. The principal components and critical parameters of this type of spectrometer will be discussed in detail in subsequent sections of the chapter. The geometry of the grazing-incidence concave-grating spectrometer is shown in Fig. 3-1. The radius of curvature of the grating is equal to the diameter of the Rowland circle which, in the case of most soft x-ray spectrometers in use today, is in the range of 2 to 3 m. Electron excitation is almost universally used in the soft x-ray region. The resulting x rays emanating from the source slit S on the Rowland circle are directed to strike the grating at a fixed angle of incidence ϕ, generally in the range of 2 to 6° as measured from a tangent to the grating at its center. An x ray of wavelength λ is dispersed by the grating through an angle $\phi + \chi/2R$, as a function of the groove spacing σ, radius of the grating $2R$, and grazing angle of incidence ϕ, according to the grating equation in the form

$$\eta\lambda = \sigma\left[\cos\phi - \cos\left(\phi + \frac{\chi}{2R}\right)\right] \qquad (3\text{-}1a)$$

which is seen, again from Fig. 3-1, to be equivalent to the following two forms that are also commonly used:

$$\eta\lambda = \sigma(\cos\phi - \cos\psi_N) \qquad (3\text{-}1b)$$

$$= \sigma(\sin\alpha - \sin\beta) \qquad (3\text{-}1c)$$

η is the spectral order and is a positive integer. The minus sign before $\cos(\phi + \chi/2R)$ is used assuming that the dispersion angle is larger than the grazing-incidence angle, which is the case in soft x-ray spectroscopy. Note that this sign convention is opposite to that of Tomboulian (1957R) but follows the current practice of grating manufacturers.[1]

---

[1] *Diffraction grating handbook*, Bausch and Lomb, Inc., New York, 1970.

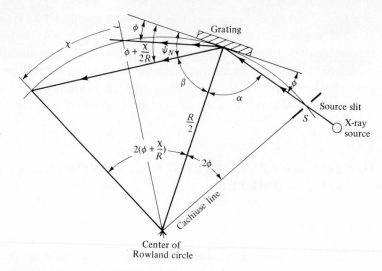

FIGURE 3-1
Geometry of the grazing-incidence concave-grating spectrometer.

The early investigators (Skinner, 1940R; Tomboulian, 1957R; Cauchois, 1948R; Curry, 1968; etc.) began with film recording which required that the film plate be bent to conform to the Rowland circle and cover the range of interest. However, because of a number of factors (poor sensitivity and narrow linearity range probably being among the most important), almost all investigators are using electronic recording today by scanning the spectral range of interest with a detector behind the analyzer slit S_A (see Fig. 3-2) moving around the Rowland circle. The intensity in the resulting spectrum is optimized for any particular level of resolution when the source slit S_S and analyzer slit S_A are of the same width.

It is now necessary to provide a drive mechanism to move that analyzer slit at a programmed rate. Grazing-incidence spectrometers for soft x-ray spectroscopy have essentially remained custom-built instruments, and probably the principal difference between designs has been in the mode of driving the analyzer slit. From the point of view of minimizing mechanical difficulties, pivoting a lead screw at one end about an axis through the grating is now considered to be the best design. The analyzer slit is mounted rigidly on a nut on the lead screw, which always lies along a chord of the Rowland circle which joins the grating and analyzer slit. This design can be used either in the case of the analyzer slit being confined by a radius arm pivoted at the center of the Rowland circle inside of the vacuum chamber (as in Fig. 3-2), or confined by a track which is an arc of the Rowland circle (as in Fig. 3-3). However, many successful spectrometers have been built with the axis of the

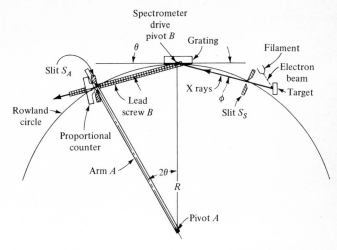

FIGURE 3-2
A spectrometer with detector on pivoted radius arm and pivoted lead screw.

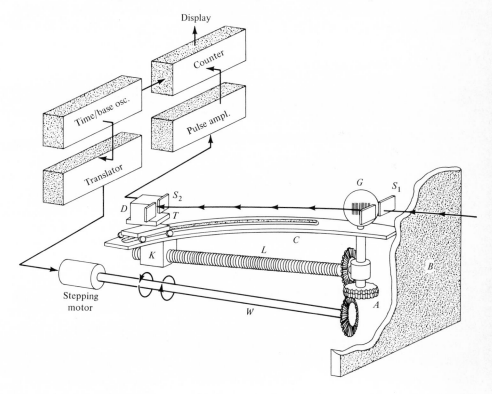

FIGURE 3-3
A spectrometer with detector on Rowland-circle track and pivoted lead screw.

FIGURE 3-4
2-m grazing-incidence grating spectrometer using film recording.

lead screw remaining in a fixed orientation while the analyzer slit is driven around the
Rowland-circle track by the lead screw nut through a system of linkages.

The optical path of a soft x-ray spectrometer must be in a high vacuum. This
requirement stems entirely from the high mass-absorption coefficient of air for x rays
in this wavelength region. As a result, a vacuum path without windows must be used.
Open electronic detectors must be used at least above 100 A, as well as open electron
guns for exciting the specimen, and a clean specimen surface free of oxide films.
These conditions require very good vacuums. The vacuum in most present-day soft
x-ray spectrometers is in the 10^{-6} to 10^{-8} torr range with a continuing trend to improve the vacuum. Grating spectrometers range in size from those on the order of a
few liters in volume to the mammoth instrument at Orsay with a volume of 400 liters
(Jaegle, 1965) (Fig. 3-4). The vacuum technique involved is that of good ultra-high-
vacuum-engineering practice so the vacuum-engineering consideration, though a
major part of any soft x-ray spectrometer design, will not be treated further in this
chapter.

B.   Focusing characteristics

A comprehensive review of the optics and production of diffraction gratings is given
by Stroke (1967R).

The theory of the focusing concave grating for optical spectroscopy was first
advanced by Rowland (1882; 1883). He showed that the various wavelength components in a point source S_S on the circumference of a circle of radius R (Rowland
circle) (see Fig. 3-5) will be diffracted by a concave grating of radius $2R$, tangent to
the Rowland circle at its center, and brought to an approximate focus at respective
points on the Rowland circle as a function of wavelength (*provided that the grooves in
the grating are equally spaced along the chord of the grating*), in accordance with the
grating equation (3-1).

The groove spacing's being uniform along the chord of the grating is the key to

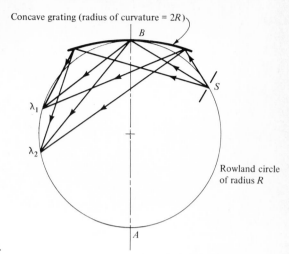

FIGURE 3-5
Focusing geometry of a concave grating.

success of the Rowland grating because this is the easiest geometry to design a ruling engine for; in fact, any of the ruling engines in use today can rule only grooves uniformly spaced along the chord of the grating. Also, the spherical surface is the easiest shape to polish to close tolerances. Equal spacing of the grooves along the chord of the grating causes the actual groove spacing on the face of the grating to become successively larger toward both edges of the grating in the plane of the Rowland circle. This increase in the groove spacing toward the edges of the grating is nearly compensated for, and the grating equation is nearly satisfied over the whole width of the grating by making the radius of the grating twice the radius of the Rowland circle. This compensating relationship can be seen with the help of Fig. 3-6. Referring to Fig. 3-6, after Strong and Madden (1958), consider a ray AB incident normal on the grating at its center and diffracted by the groove spacing σ at the center of the grating through an angle a along BC. In this case the angle of incidence is zero (normal incidence), and the grating equation becomes $\eta\lambda = \sigma \sin a$. Now consider the incident ray AD also reflected back to the point C on the Rowland circle. If the two rays AB and AD started out from point A in phase, they will be in phase again when they reach C if the grating equation is also satisfied for the ray path ADC. At D the grating spacing has increased to $\sigma/\cos \theta$, but the diffraction angle a has decreased almost in proportion. A decrease in the diffraction angle to a' almost compensates for this increase. We can see just how good this compensation is by calculating the difference in the phase retardation per groove between the center and the edge of the grating. This difference is readily seen, from applying the grating equation at points B and D, to be

$$\delta = \eta\lambda\sigma\left(1 - \frac{1}{\cos\theta}\frac{\sin a'}{\sin a}\right) \qquad (3\text{-}2)$$

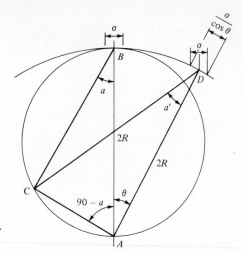

FIGURE 3-6
The Rowland grating principle, after
Madden and Strong.

For a grating with 1200 lines per mm, 3 cm wide, and a 2-m radius, this difference δ is calculated from (3-2) to be only $2 \times 10^{-5} \, \eta\lambda$, which is well within the Rayleigh criteria.

The term "approximate focus" was used above because in general the image will be strongly astigmatic; that is, the ray will be focused along the Rowland circle but will not be in focus in the direction perpendicular to the plane of the Rowland circle. The image of a point source will be an arc whose length will increase with an increase in the angle of incidence α (Fig. 3-5). In the grazing-incidence range of the usual soft x-ray spectrometer, the curved image will approximate the arc of a circle with radius approximating the radius of the grating, the curvature being concave toward the source slit. A common method of checking the alignment of a grating spectrometer is to place a mercury lamp at the position of the specimen and observe visually the green (5461-A) line brought to a focus at its proper position on the Rowland circle. The curvature in the image will be readily obvious.

The use of a mercury lamp at the position of the specimen and the observation of the green (5461-A) line brought to a focus at the proper location on the Rowland circle is a commonly used method of checking the alignment of grazing-incidence grating spectrometers. The detailed shape and intensity distribution of the astigmatic image as a function of the incidence and diffraction angles α and β, respectively, has been derived by Beutler (1945) and by Namioka (1959) for both a point source and a line source. Also, the analytical derivation of the concave-grating characteristics at grazing incidence is given by Mack, Stehn, and Edlén in their classic paper (1932). The mathematical derivation and resulting relationships will not be reproduced here. The reader is referred to these original papers or to a very good summary by Samson (1967R).

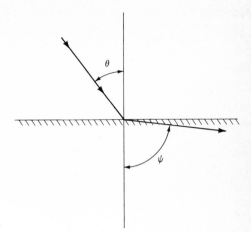

FIGURE 3-7
Snell's law in the x-ray region.

A stigmatic image is obtained with a conventional spherical grating only by using a geometry that diffracts the rays along the normal, or near the normal, to the grating (*AB* in Fig. 3-5). This can be done in the visible region of the spectrum. In the early days of x-ray spectroscopy attempts were made to diffract x rays at normal incidence. Compton and Doan (1925) were the first to obtain x-ray spectra from a ruled grating. Soon afterward O'Bryan (1931) obtained x-ray spectra at normal incidence down to 200 A.

C.   Spectral reflectance

For shorter wavelengths and usable intensities, we must resort to grazing-incidence geometry—with the attendant severe astigmatism—because for any angle of incidence there is a corresponding minimum x-ray wavelength, and x rays shorter than this minimum will not be reflected.

It is a little more direct to calculate the threshold angle of incidence θ_i corresponding to a particular x-ray wavelength λ_i'; i.e., at angles of incidence below θ_i an x ray of wavelength λ_i will not be reflected (although the cutoff as a function of angle is not sharp). θ_i is a function of the refractive index μ of the grating material which relates the angle of refraction ψ to the angle of incidence θ_i by Snell's law:

$$\frac{\sin \theta}{\sin \psi} = \mu \qquad (3\text{-}3)$$

where the angle of incidence θ and the angle of refraction ψ are defined in Fig. 3-7. Note that the refracted ray bends away from rather than toward the normal because the refractive index in the x-ray region is less than 1, contrary to that in the visible region. We will see (following Sproull, 1946) the reason for this presently.

Lorentz and Drude derived the following expression for μ, to explain for the first time on the basis of classical electron theory what had been previously a strictly empirical expression for diffraction in the visible region:

$$\mu^2 = 1 + \frac{e^2}{\pi m}\left(\frac{n_1}{v_1{}^2 - v_\lambda{}^2} + \frac{n_2}{v_2{}^2 - v_\lambda{}^2} + \cdots\right)$$

where e is the charge on an electron in esu, m is the mass of an electron, $n_1, n_2, \ldots$ are the numbers of electrons per cm^3 with respective natural oscillator frequencies $v_1, v_2, \ldots$, and v_λ is the frequency of the diffracted light. However, in the x-ray range the frequency v_λ is so much larger than $v_1, v_2, \ldots$ that they can be ignored; the Lorentz-Drude expression in the x-ray region thus becomes

$$\mu^2 = 1 - \frac{n}{\pi}\frac{e^2}{mv_\lambda{}^2}$$

Applying the binomial theorem, after Sproull, we obtain

$$\mu = 1 - \frac{n}{2\pi}\frac{e^2}{m}\frac{\lambda^2}{c^2} \qquad (3\text{-}4)$$

The criterion for the onset of total reflection is that the radiation gets diffracted just out of the surface; i.e., $\psi = 90°$ and $\sin \psi = 1$. Therefore substituting into Snell's law for this condition and for the value of μ gives

$$\sin \theta_i = 1 - \frac{n}{2\pi c^2}\lambda^2 = \cos \phi$$

where ϕ is the grazing angle of incidence. Making use of the trigometric identity, we find

$$1 - \cos \phi = 2 \sin^2 \frac{\phi}{2}$$

and noting that because ϕ is sufficiently small we can substitute $\sin \phi$ for $2 \sin (\phi/2)$, we get

$$\sin \phi = \frac{n}{\pi}\frac{e^2}{mc^2}\lambda \qquad (3\text{-}5)$$

Note that equation (3-5) is in the form given by Samson (1967R). Figure 3-8 is a plot of threshold angle versus wavelength for some commonly used grating surface materials. Figure 3-9 is a plot of experimentally observed reflectivities, for commonly used grating surface materials, as a function of wavelength and grazing-incidence angle.

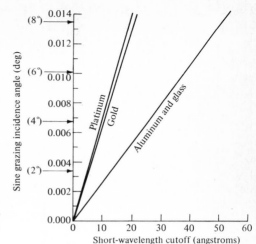

FIGURE 3-8
Threshold angle versus wavelength calculated from equation (3-5) for representative grating-surface materials.

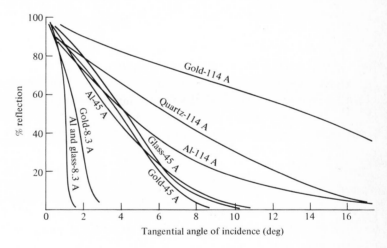

FIGURE 3-9
Relative reflectances of various grating coatings versus wavelength.

D.   Optimum grating width and groove length

Mack, Stehn, and Edlén (1932) extended the theory of the concave grating to the grazing-incidence case. They showed that, because the astigmatism is proportional to the fourth power of the width of the grating in the direction perpendicular to the grooves, there is an optimum grating width in terms of resolution to use in grazing-

Table 3-1 OPTIMUM GRATING WIDTH w_{opt} IN mm AT VARIOUS WAVELENGTHS FOR A 2-m-RADIUS GRATING, AS A FUNCTION OF GRATING SPACING σ AND ANGLE OF INCIDENCE θ (MEASURED FROM THE NORMAL)

w_{opt}

	$\sigma = 1/600$ mm			$\sigma = 1/1200$ mm		
λ, A	$\theta = 88°$	$\theta = 86°$	$\theta = 84°$	$\theta = 88°$	$\theta = 86°$	$\theta = 84°$
50	14.1	16.1	17.5	14.4	16.4	17.7
100	17.1	19.5	21.1	17.4	19.9	21.6
200	20.6	23.7	25.6	20.9	24.2	26.2
400	24.9	28.8	31.2	25.2	29.3	31.9

incidence spectrometers. This optimum width w_{opt} is appreciably less than that which can be used at normal incidence, and it is given by the formula

$$w_{opt} = 2.36 \frac{4\lambda\rho^3}{\pi(\tan\theta\sin\theta + \tan\theta'\sin\theta')} \tag{3-6}$$

where ρ is the radius of grating, θ is the angle of incidence from normal, and θ' is the corresponding angle of reflection for wavelength, from the grating equation. Table 3-1 shows some optimum widths per the Mack, Stehn, and Edlén criteria, calculated for a 2-m-radius grating of 1200 lines per mm and for grazing-incidence angles of 2, 4, and 6°, which approximates the geometry of most soft x-ray spectrometers in use today. For a grating of any other radius, multiply by

$$\frac{\rho}{2000} \tag{3-7}$$

where ρ is the radius of the grating in millimeters. In this range of geometry, the optimum grating width is virtually independent of the grating spacing.

Mack, Stehn, and Edlén show that extra length of ruling, unlike extra width of grating, is not positively harmful; but there is a practical limit beyond which the additional groove length does not increase the usable intensity. There is no gain in having the groove length longer than the analyzer-slit length.

E. Resolution

The Mack, Stehn, and Edlén criterion for resolution, upon which the foregoing optimum effective grating width is based, is that two emission lines of equal intensity are resolved when their wavelength difference $\Delta\lambda$ is such that the minimum total intensity midway between the lines is $8/\pi^2$ ($= 0.8106$) as great as the total intensity of both at the central maximum of either of the lines.

The definition of resolving power R is

$$R = \frac{\lambda}{\Delta\lambda} \qquad (3\text{-}8a)$$

The resolving power can be limited either by the grating or by the spectrometer slit width. The grating-limited resolving power is shown by Mack, Stehn, and Edlén to be

$$R_{\text{grating}} = 0.92(2w_{\text{opt}})\frac{m}{\sigma} \qquad (3\text{-}8b)$$

and the slit-limited resolving power to be

$$R_{\text{slit}} = \frac{0.91\,\rho\lambda m}{S\sigma} \qquad (3\text{-}8c)$$

where ρ is the radius of grating, m is the spectral order, λ is the wavelength, S is the slit width, and σ is the grating spacing. The resolution of the spectrometer is the lower of the two values.

F. Dispersion

"Dispersion" refers to the increment of Rowland-circle circumference per unit of wavelength or spectral energy, in contrast to "resolution" which indicates the allowable minimum spacing of two "just resolvable" spectral lines. Dispersion is a function of the Rowland-circle geometry only, and is independent of slit width, grating perfection, and general spectrometer alignment, whereas all these factors are important to the overall resolution capability of the spectrometer.

To find the dispersion, i.e., $\Delta\lambda/\Delta l$, in A/mm of Rowland-circle circumference, following Samson (1967R), let us write, referring to Fig. 3-10,

$$\frac{\Delta\lambda}{\Delta l} = \frac{\Delta\lambda}{\Delta\psi} \cdot \frac{\Delta\psi}{\Delta l}$$

Again from Fig. 3-10 we can see that

$$2R\,\Delta\psi = \Delta l$$

giving

$$\frac{\Delta\lambda}{\Delta l} = \frac{\Delta\lambda}{\Delta\psi} \cdot \frac{1}{2R}$$

In the limit we have

$$\frac{d\lambda}{dl} = \frac{d\lambda}{d\psi} \cdot \frac{1}{2R}$$

Using the grating equation in the form of (3-1b), we obtain

$$\frac{d\lambda}{dl} = \frac{\sigma \sin\psi}{2mR} \qquad (3\text{-}9)$$

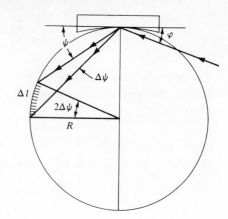

FIGURE 3-10
Dispersion relation.

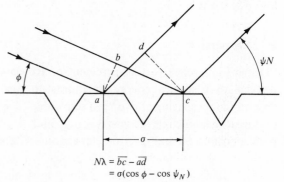

$$N\lambda = \overline{bc} - \overline{ad}$$
$$= \sigma(\cos\phi - \cos\psi_N)$$

FIGURE 3-11
Schematic of Siegbahn-type grating.

III. TYPES OF REFLECTION GRATINGS

A. Siegbahn-type grating

The so-called lightly ruled, or Siegbahn-type, grating, where the original undisturbed glass surface between the grooves is considered solely responsible for the reflection, has been the most widely used type of grating in the soft x-ray region until recently. The Siegbahn-grating geometry is shown in Fig. 3-11. Reinforcement occurs for a particular wavelength λ along wavefront cd when the difference in path length between bc and ad is an integral multiple of the wavelength λ, i.e., when $bc - ad = \eta\lambda$. Expressing bc and ad in terms of the cosines of the incident angle ϕ and the diffraction

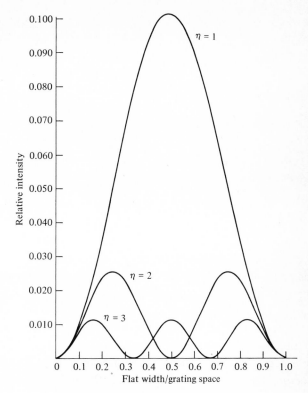

FIGURE 3-12
Relative intensity of spectral order versus flat-width-to-grating-spacing ratio.

angle ψ_N gives the grating equation (3-1b). In principle, the optimum situation is achieved when the net width of the undisturbed glass between two grooves is equal to the net width of a groove. For this case Sprague, Tomboulian, and Bedo (1955) showed that the first-order reflection would be a maximum, the second order would be canceled out, and the third-order intensity would be approximately one-tenth that of the first order. If the ratio of the net groove width to the grating spacing deviates from the ideal value of 0.5, the predicted effect on the relative intensities of the first-, second-, and third-order spectra is shown by the plot in Fig. 3-12. However, even if this idealized situation is achieved, the Siegbahn-type grating reflects specularly most of the incident light in the zeroth order. In actual practice the idealized case of zero second-order intensity is seldom achieved—probably because of scattering from the surface of the groove, and ploughed furrows of glass that protrude above the original surface (see Fig. 3-13) as well as the fact that the net groove width is not exactly one-half the grating spacing.

FIGURE 3-13
Electron photomicrograph of Siegbahn-
type rulings in glass showing the
ploughed-up furrows of glass on both
sides of each groove. Grooves are 30,000
per in.

The problem of ploughed furrows of glass being piled upon the original polished-glass surface on either side of the groove has recently been overcome by first coating the polished glass blank with a thin layer of gold. The rulings are then made through the gold layer into the glass (Bearden, 1969). Actually, a 30-A layer of molybdenum is first sputtered onto the glass to give good adherence for the subsequent 250- to 300-A layer of gold which is also applied by sputtering. Light rulings are made, cutting through the gold and molybdenum layer, and penetration is made into the glass to a depth of 200 to 300 A. The grooves are then etched to the desired width by flooding the surface with a dilute (1 to 2%) HF solution. The gold and molybdenum layers are then dissolved; the finished grating is coated with platinum.

B.   Laminar grating

A variation on the Siegbahn grating is the "laminar" or "phase" grating (Sayce and Franks, 1964; Franks and Lindsey, 1968; and Bennett, 1971). The principal difference between the Siegbahn and the laminar grating is that the grooves in the latter have flat bottoms and are of a controlled depth. The depth of the groove is made such that reflection from the bottom of the grooves, as well as from the lands, can be made to contribute to the intensity of the first-order spectrum. Grating efficiencies of 40% in first order are achieved, and the gratings can be used at grazing-incidence angles as low as 5 minutes of arc. This combination of properties has permitted spectra of short-lived transient phenomena extending down to 0.5 A to be recorded.

A groove depth is selected that will result in the x rays reflected from the bottom of the grooves being in phase with the x rays reflected from the lands. The intensity will peak at a particular wavelength as a function of the groove depth and grazing-incidence angle. Depending on the choice of grating parameters used, this peaking in grating efficiency as a function of wavelength could be rather sharp, which would be undesirable for the usual applications. However, by the proper choice of parameters and by varying the grazing-incidence angle, the grating efficiency in any particular case can be kept high over an appreciable wavelength range. In an example given by Bennett (1971), a grating with a groove depth of 0.01 times the grating spacing can be made to exhibit an efficiency in first order of greater than 30% over a wavelength range of 0.001 to 0.006 times the grating spacing by varying the grazing angle from 1° at the short-wavelength end to 4° at the long-wavelength end.

The laminar-grating concept is very old, but Franks at the National Physical Laboratory (NPL), in collaboration with Speer at Imperial College, London, only during the past 10 years has brought the concept into practical fruition. The key development in the production of a practical grating is the use of ion-beam etching rather than chemical etching of the glass to produce the flat-bottomed grooves of controlled depth. The steps in the production of the laminar grating are depicted in Fig. 3-14. The "chemical etching," noted under stage 3, is a solvent for the aluminum, to remove any traces of aluminum left by the ruling diamond at the bottom of the groove in order to expose the clear strip of glass. Grooves in the glass are then formed (Fig. 3-14d) by argon-ion etching.

C.   Blazed gratings

For more than three decades, the notion that only an original glass grating of the Siegbahn type could be used in the soft x-ray region had become so fixed that a blazed grating was never exploited. However, during the past decade Holliday has produced an abundance of spectra in the 20- to 100-A region using only blazed replica gratings, and today blazed replica gratings are widely used in grazing-incidence spectrometers. The advantage of the blazed grating is that by selecting the blaze angle, the bulk of the incident light, rather than being lost by specular reflection in the zeroth order, can be put into the first-order diffraction spectrum. However, the blazed grating inherently will exhibit a large variation in spectral response as a function of wavelength. The intensity will peak at the particular wavelength corresponding to the blaze angle, and will fall off to one-half the peak intensity at generally about two-thirds of the blaze wavelength on the short-wavelength side and again at about twice the blaze wavelength on the long-wavelength side. The geometry of the blazed grating is shown in Fig. 3-15. Similarly, as in the case of the Siegbahn grating, reinforcement occurs along the wavefront cd when the difference in path length between bc and ad is an integral multiple of the wavelength $N\lambda$, and likewise, as in the case of the Siegbahn grating, expressing bc and ad in terms of ϕ and ψ_N, respectively, again yields the grating equation in the same form. Also, specular reflection occurs when the angle of incidence S_i equals the angle of reflection S_r (assuming that S_i is larger than the critical angle so that total reflection occurs).

Now, however, this relation for specular reflection can be satisfied for any

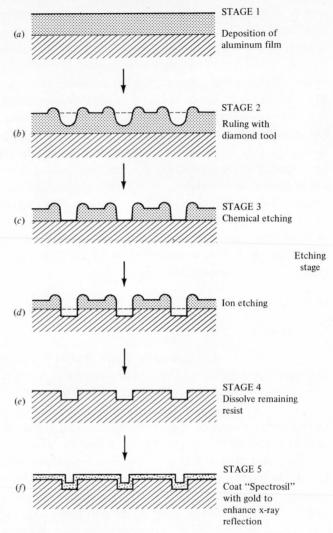

STAGE 1

Deposition of
aluminum film

STAGE 2

Ruling with
diamond tool

STAGE 3

Chemical etching

Etching
stage

Ion etching

STAGE 4

Dissolve remaining
resist

STAGE 5

Coat "Spectrosil"
with gold to
enhance x-ray
reflection

FIGURE 3-14
Steps in production of NPL laminar grating.

order η by constructing the appropriate blaze angle θ into the grating. Then the wavelength corresponding to maximum spectral intensity is computed from the grating equation, with ψ_N equal to $\phi + 2\theta$.

Blazed gratings are invariably ruled in aluminum deposited upon a glass blank. The grooves are wholly within the aluminum layer. Copies are made from the original by taking an impression in a thin layer of plastic backed up by a glass blank. The

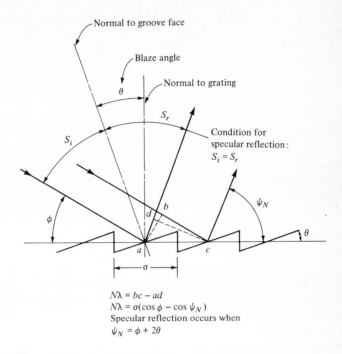

$N\lambda = bc - ad$
$N\lambda = \sigma(\cos\phi - \cos\psi_N)$
Specular reflection occurs when
$\psi_N = \phi + 2\theta$

FIGURE 3-15
Schematic of blazed grating.

grooved plastic surface thus formed is then coated by vapor deposition, generally with gold or platinum, for use in the soft x-ray region.

D.   Holographic gratings

The astigmatism in gratings at grazing incidence has been treated in detail by Beutler (1945), Namioka (1959), and by Kastner and Neupert (1963).   Beutler (1945) and Haber (1950) show how toroidal gratings, and Namioka (1961) ellipsoidal gratings, should result in reduced astigmatism.   Neither of these has been tried at grazing incidence although Ishii et al. (1971) reported the use of a toroidal mirror ahead of a conventional spherical grating at grazing incidence and showed a reduction of astigmatism in the soft x-ray region.   With the advent of the holographic gratings it appears to be possible to take better advantage of these nonspherical shapes to eliminate astigmatism.

Holographic gratings of good optical quality first became available in 1968.[1] The holographic grating is produced by projecting an interference fringe pattern

[1] Manufactured by the Jobin & Yvon Co., 26, Rue Berthollet, 94-Arcueil, France.

formed by two CW laser beams of slightly different wavelength onto a glass blank which has been coated with a thin film of photosensitive plastic (Labeyrie and Flamand, 1969; Maystre and Petit, 1971; Jenney, 1970). The glass blank is first polished either flat or to the desired surface contour to within one-quarter of a fringe. The coating of photosensitive plastic that is then applied is less than 10 microns thick.

Where the laser beams reinforce, producing bright fringes, the plastic is polymerized by the light. The unpolymerized plastic is then dissolved away with a suitable solvent, leaving a groove pattern in the plastic. The grating is then vacuum-aluminized in the conventional manner. Gold and platinum coatings also can be applied. The finished groove shape is between that of the conventionally ruled blazed grating and a sine wave. The gratings have been replicated using the same technique as for conventionally ruled gratings.

Because these gratings are not subject to the errors and limitations of ruling engines, they have some interesting properties and open up new possibilities (Cordelle et al., 1969). Rowland ghosts due to variation in groove spacing are virtually eliminated. The attainable resolution approaches 80 to 100% of the theoretical. The spherical aberration of gratings that focus on the Rowland circle can be reduced by a factor of 10 in comparison with conventionally ruled gratings, and gratings of special contour which focus off the Rowland circle can be made perfectly stigmatic at three wavelengths, which lie along a straight line, and which are only slightly astigmatic between these three points.

IV. DETECTORS

A. Photographic film

Before World War II photographic film was used almost exclusively. Film for use in the soft x-ray spectral range must have its layer of light-sensitive grains exposed directly to the radiation rather than embedded in a coating of gelatin as in the case of most photographic film. The big advantage of film recording is that the entire spectrum is recorded simultaneously. For certain applications, such as surveying the spectra of a transient phenomenon, photographic recording is the most practical method of detection. However, electronic detection has almost completely replaced photographic recording in soft x-ray emission spectroscopy because of the greater sensitivity of electronic detectors and the adaptability of electronic detection to the integration of a large number of counts to increase the signal-to-noise ratio. Photographic film has a very limited signal-integration capability before the response becomes nonlinear. Photographic film for the soft x-ray region is discussed by Tomboulian (1957R) and by Samson (1967R) and will not be discussed further here.

B.   Flow-proportional counters

Out to boron K, at 67 A, flow-proportional counters are commonly used, and in some cases are being used out to 150 A, while tolerating the fragile formvar (Baun and Fischer, 1970) or parlodian windows that must be used, because of the high counting rates and the energy-discriminating capability of the flow-proportional counter. The energy resolution of the flow-proportional counter, even at its best, does not begin to approach that of the grating spectrometer in the hard x-ray region (in the vicinity of 1 A), and is still poorer in the soft x-ray region. However, the energy discrimination is still adequate and indeed the most satisfactory method of blanking out interfering higher orders, when used as the detector in a dispersing spectrometer, either grating or crystal. The principal reason for the poorer discriminating ability in the soft x-ray region is that the combination of low-energy photons and low gas pressures in the tube (on the order of 100 mm Hg, absolute, because of the thin windows) results in most of the ionization event taking place close inside of the window and ions not reaching the central collector, or at least in numbers proportional to the energy of the original photon. Comprehensive discussions of the construction and characteristics of the flow-proportional counter are given by Samson (1967R), Blokhin (1957R), and by Spielberg (1967), and so no further discussion will be given here.

C.   Photoelectron multipliers

Above 150 A completely windowless detectors are necessary and are highly desirable even down to B_K (67 A). For about two decades the conventional photomultiplier structure of the Dumont no. 6292 type, with the glass envelope removed, was used (Tomboulian, 1957R). The dynodes were made of Be-Cu alloy which permitted periodic exposure of the dynode structure to air with some deterioration in signal output. In the last ten years the conventional photomultiplier structure has been largely replaced by the continuous-strip magnetic electron multiplier (MEM) and this in turn has been superseded by the curved-channel electron multiplier (CEM).[1] Meanwhile, the conventional discrete-dynode photoelectron-multiplier structure has been compacted into a wafer design that results in increased amplification while at the same time reducing background noise because of the shorter electron path length between dynodes. This wafer design of photoelectron multiplier is pictured in Fig. 3-16. It will find use in preference to the curved-channel electron multiplier when a large spot-sized source must be accepted by the detector. Probably the MEM is still the most commonly used type of detector in grating spectrometers. To optimize the quantum yield from this type of detector, it has become common practice to orient

[1] Manufactured by Bendix Corporation, Southfield, Mich., and by Mullard Ltd., London.

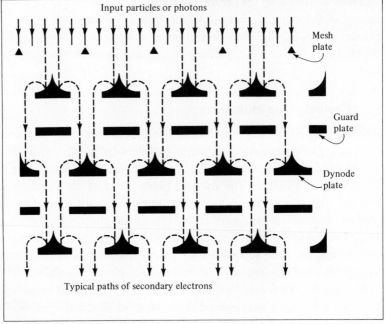

FIGURE 3-16
Wafter-type photoelectron multiplier showing longitudinal cross section through focused-mesh multiplier.

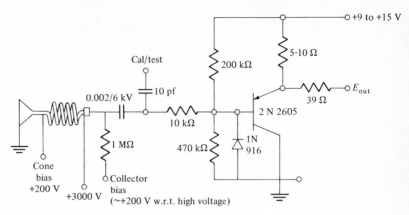

FIGURE 3-17
Schematic diagram of a typical CEM complete with a typical preamplifier circuit.

the MEM behind the analyzer slit such that the x-ray beam strikes the entrance grid of the MEM at an angle of 30° to the plane of the entrance grid of the MEM.

The CEM detector consists of either a lead or a vanadium oxide glass capillary tube, of 1-mm inside diameter and about 100 mm long (Weller and Young, 1970). The inside surface is made semiconducting by a hydrogen heat treatment. The resulting electrical resistance between the two ends of the tube is about 10^9 ohms. A voltage approaching 3000 volts is placed across the two ends of the tube. A photon entering the forward end of the tube will initiate a cascade of electrons down the length of the tube, multiplying upon each impact with the wall of the tube and resulting in an overall gain approaching 10^9. The final rectangular current pulse reaching the collector anode is about 0.6 ma and 25 nsec wide. The tube is curved to prevent scatter in the backward direction (Johnson, 1969). With the usual preamplifier circuitry of unity gain, the resulting final output pulse is generally in the range of 0.25 to 1.0 volt with 70% of the pulses being within approximately ±10% of the central value.

To accommodate line sources and larger spot-sized sources, a 45° glass cone of the same material as the tube is often attached to the front end, which gives an area of acceptance of up to 1 cm in diameter. The uniformity of response over this 1-cm-diameter cone is within 10%.

A schematic diagram of the CEM complete with a typical preamplifier circuit is shown in Fig. 3-17. It is generally advisable to use the CEM in the pulse-counting mode rather than the current mode to avoid effects of a change in gain due to aging and pumping off of adsorbed gas layers. A Faraday-cup-type collector should be used for the pulse-counting mode of operation, to catch backscattered electrons. Counting rates of several thousand per second can be handled by the CEM.

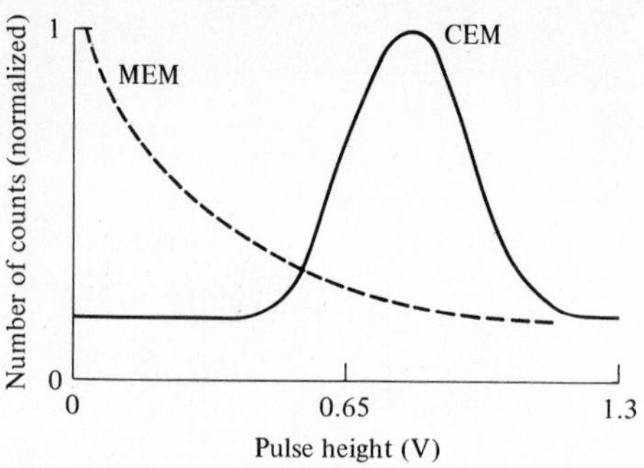

FIGURE 3-18
Comparison of the characteristic shapes
of the MEM and CEM pulse height
distributions.

Apart from its simplicity, the principal characteristics that make the CEM preferable to magnetic electron multipliers (MEMs) are (1) the narrow spread in output pulse size, and (2) the much better defined plateau in the curve of count rate vs. multiplier voltage. The characteristic difference between the pulse height distribution of the CEM and that of the MEM is shown in Fig. 3.18. The characteristics of a CEM are shown in Fig. 3-19. A typical count rate vs. multiplier voltage characteristic for the CEM in Fig. 3-19*a*, and a plot of the absolute quantum efficiency of the CEM for soft x-ray photons as a function of wavelength in Fig. 3-19*b*.

The small window area of the CEM is often a serious limitation that is only partially overcome by the mounting of a cone on the front of it. For applications requiring larger window areas, the so-called "Spiraltron electron multiplier" (SEM) has been constructed (Somer and Graves, 1969). This device is made up of Spiraltron matrix elements. Each matrix element consists of six CEM capillary tubes twisted around a central cone, as shown in Fig. 3-20*a*. These matrix elements are then stacked in close-packed array to give as large an effective window area as desired, as shown in Fig. 3-20*b*. Arrays with an effective window area of 1.5 × 1.5 cm have been constructed. However, in its present state of development the gain is somewhat less, the dark-count rate somewhat higher, and the spread in the pulse height distribution somewhat wider than that of the CEM.

V. EXCITATION METHODS

A. Fluorescent excitation

Although both fluorescent excitation (particularly Al *K* characteristic emission) and electron excitation are widely used in connection with crystal spectrometers in the soft x-ray region, grating spectrometers to date have employed electron excitation

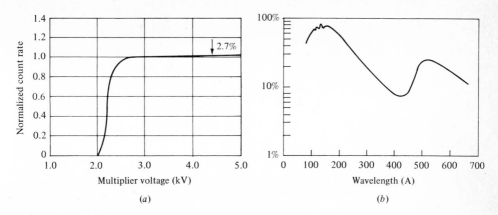

FIGURE 3-19
CEM operating characteristics. (*a*) Typical current rate versus multiplier voltage showing the well-defined plateau that is important in assuring stability; (*b*) absolute quantum efficiency for soft x-ray photons as a function of wavelength.

exclusively, the principal reason being intensity. For the same power going into the specimen, electron excitation produces an intensity of characteristic x-ray radiation, in the soft x-ray region, many times that produced by fluorescent excitation. Therefore electron excitation will be discussed in some detail. Fluorescent excitation is discussed in connection with crystal spectrometers. Other methods of excitation, although they have been little used to date, if at all, in connection with either grating or crystal spectrometers, will be discussed briefly because inherently they have some unique advantages in the soft x-ray region. Foremost among these peripheral methods are heavy-particle excitation (particularly protons) and radioactive sources.

B.   Electron excitation

Many spectrometers, particularly the earlier ones, employed a simple tungsten filament close to the face of the specimen. The specimen was mounted so that it could be periodically rotated for evaporation of a fresh layer of sample. There were two distinct objections to this design: (1) rapid contamination of the specimen surface by the spatter of tungsten from the filament, and (2) the restriction to pure metals that would yield good evaporated layers. Also, the electrons struck the specimen normal to the surface, and the x rays were taken off at an acute angle. This means that the high-energy electrons penetrated relatively deeply below the surface, making the oblique path length for the x-ray emission unduly long and resulting in considerable self-absorption.

In more recent spectrometers a focused electron beam, generally from a Pierce-type gun (Pierce, 1954), is directed obliquely at the specimen surface, and the x-ray

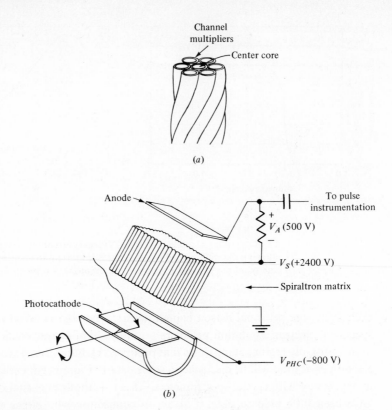

FIGURE 3-20
Spiraltron curved-channel electron multiplier. (a) Detail of matrix element;
(b) schematic of a large-window-area Spiraltron composed of a close-packed
array of the matrix elements shown in (a).

radiation is taken off perpendicular to the surface. This geometry minimizes self-
absorption. The electron beam is generally curved between the gun and the specimen
by electrostatic deflecting plates so that the specimen will not "see" the hot filament,
and thus avoid spatter. Common practice is to prepare the specimens outside of the
spectrometer and then do the final cleaning inside the specimen chamber, either by
argon-ion bombardment (Holliday, 1970R) or by mechanically scraping the specimen
(Crisp, 1961). In the argon-ion cleaning method, argon is let into the specimen chamber to a pressure of about 1 micron. A newer technique is to direct a beam of argon
ions on the specimen area to be cleaned. This is more effective with less argon in
the system to be pumped out. Argon-ion guns specifically for cleaning purposes are
commercially available. The beam of argon ions produced by electron excitation from
a separate filament strikes the specimen at a potential of about 1500 volts. Although
effective, objection has been raised that argon-ion bombardment severely distorts the

surface layer of the specimen. For this reason Crisp and others (Watson, Dimond, and Fabian, 1968) have used a technique wherein the surface is scraped with a knife blade manipulated from outside the specimen chamber.

The importance of the excitation geometry on self-absorption, mentioned previously, has been investigated experimentally by Bonnelle (1964). She maintained the x-ray take-off direction always perpendicular to the electron-beam direction and rotated the specimen through the angle shown in Fig. 3-21. The change in the amount of self-absorption in the Ni $L\alpha$ high-energy edge, as a function of angle of specimen orientation, and excitation energy, is dramatically shown in Fig. 3-22.

4000-volt electrons or even the 2500-volt electrons used by Cuthill et al. (1967) would seem extraordinarily excessive to excite photons of less than 100-eV energy, such as the M spectra of the first transition-series metals. Indeed it is, and as a result considerable multiple-ionization satellite structure is often produced on the high-energy side of the primary spectrum. However, if lower electron energies are used, the penetration depth will be too shallow, and the resulting x-ray intensity will be severely reduced. The intensity of the resulting boron K ($\lambda = 67$ A), carbon K ($\lambda = 44$ A), and silver $M_{\mathrm{IV,V}}$ ($\lambda = 40$ A) characteristic emission and the depths of electron penetration into the respective targets vs. electron energy were investigated experimentally by Hoffman, Wiech, and Zöpf (1969). Their measurements were made over the range of electron energies of 1 to 4 keV, and for the geometry shown in Fig. 3-23. Their results are plotted in Fig. 3-24 in a different form from that given by the authors. The maximum effective penetration depth vs. electron energy is plotted for each of the four materials. The targets were formed by vapor which deposited successively increasing thicknesses of the respective materials onto tungsten. The respective thicknesses of the evaporated layers were measured by optical interferometry. The maximum effective depth was then taken to be the value of the deposited film thickness beyond which no further increase in the x-ray intensity was observed with further increase in deposited film thickness.

The situation is considerably better in the crystal-spectroscopy range, say the Al K spectrum with its high-energy emission edge about 1830 eV, and the L spectra of the first transition-series metals in the 900-eV range. In this range Liefeld and his students (Liefeld, 1968) have been able to show that by reducing the excitation voltage from a considerable excess above to a value just above threshold this high-energy satellite structure could be avoided.

C. Proton excitation

Proton excitation is an interesting alternative to electron excitation. Whereas fluorescent excitation efficiency decreases with an increase in wavelength, both electron and proton excitation efficiencies increase with increase in wavelength. Proton excitation

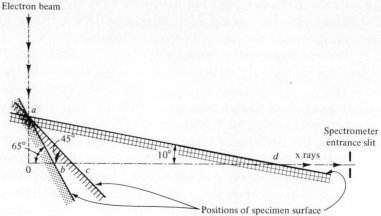

FIGURE 3-21
Excitation geometry employed by Bonnelle (Thesis, University of Paris, 1964) in testing for evidence of self-absorption in the resulting emission spectrum.

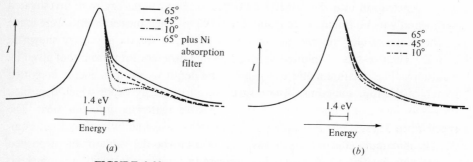

FIGURE 3-22
Self-absorption effects in the Ni Lα emission spectrum as a function of the variation in excitation geometry shown in Fig. 3-21. (a) 2.5 keV; (b) 1.1 keV.

offers the further advantage that the Bremsstrahlung background is greatly reduced. The background is reduced by the ratio of the square of the ratio of electron mass to proton mass. Some pioneering work has been done using proton excitation in connection with both crystal and grating spectrometers, notably by Sterk and his coworkers (Sterk, Marks, and Saylor, 1967) and others (Birks et al., 1964). Sterk, Marks, and Saylor (1967) find for the efficiency of x-ray production:

$$N = K_{K,L,M} z^{-6.52} e^{0.062 E_P} \qquad (3\text{-}10)$$

where N is the x-ray yield in photons per proton, $K_K = 0.192$, $K_L = 10^2$, $K_M = 3.4 \times 10^2$, and E_P is the proton energy, in keV.

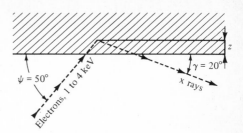

FIGURE 3-23
Excitation geometry employed by Hoffmann, Wiech, and Zöpf (1969) to investigate the effective specimen depth in soft x-ray emission as a function of excitation energy, specimen atomic number, and wavelength.

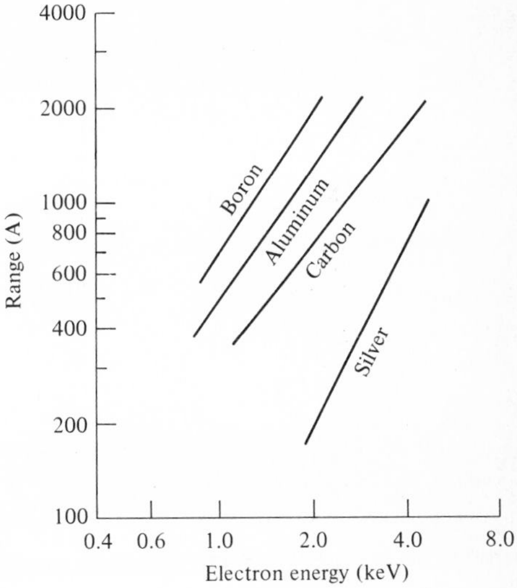

FIGURE 3-24
Maximum effective specimen depth and the corresponding excitation energy for boron K ($\lambda = 67$ A), carbon K ($\lambda = 44$ A), and silver M ($\lambda = 40$ A).

Proton excitation has been very little used in comparison with electron excitation because of the rather complex instrumentation required compared with the simple hot-wire thermionic emitter used as a source for electron excitation. A schematic diagram of a comparatively simple proton source is shown in Fig. 3-25. Birks et al. (1964) concluded that the disadvantages outweigh the advantages of proton excitation in the harder x-ray region (1 A). However, ion excitation has been found useful for identifying satellite structure (Nagel, 1971).

D. Radioactive sources

A number of radioactive sources, including α, β, and γ emitters, are being used in commercial nondispersive analytical instruments. This type of source makes possible portable instruments that have found much use in field applications, including lowering them down boreholes in oil and mineral exploration. However, the flux from

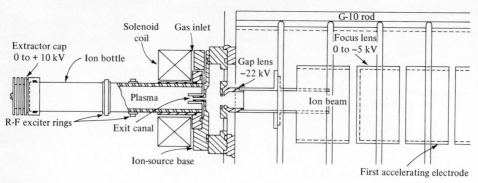

FIGURE 3-25
A typical proton generator suitable for soft x-ray excitation.

any of these sources has not been sufficient to use them in connection with a dispersive-type instrument, either grating or crystal, except in one special case. This is the case of K-capture sources where the radioactive source is the specimen. The only study thus far has been on Fe[55] and happens to have been done on a crystal spectrometer (Schnopper, 1968). The Fe[55] isotope actually emits the manganese characteristic x-ray spectrum.

Schnopper compared the Mn $K\alpha_{1,2}$ and Mn $K\beta_{1,3}$ spectra from the Fe[55] K-capture source with the same spectrum from bulk manganese metal obtained by conventional electron excitation, and he found no significant difference. Because of the low intensity no work has yet been done in the soft x-ray range. Although the work of Schnopper is outside of the x-ray range with which this chapter is concerned, it is worthwhile to note the importance of his findings. Of course, the K-capture process gives a clean single vacancy in the K shell, and so the fact that there was no significant difference in the spectra resulting from the two different excitation processes indicates that speculation about electron excitation causing a significant production of double vacancies was not borne out.

VI. SOFT X-RAY EMISSION BAND SPECTRA

In this section we will consider the instrumental correction factors involved in relating the observed spectrum to a measure of specimen parameters such as bandwidth, electron density of states, etc., which is the usual goal of any soft x-ray spectroscopic study. For an index, by material studied, to what soft x-ray spectral data are available in the literature, the reader is referred to a number of extensive compilations: Faessler, 1955; Yakowitz and Cuthill, 1962; Baun, 1972; Carter et al., 1970, 1971; McAlister et al., 1973.

The soft x-ray valence-band emission spectrum, as obtained from the spectrometer, can be considered to be a distorted representation of the distribution of occupied electronic energy states. In the sense used here, the "valence band" is considered to be all the electrons beyond the outermost normally filled subshell. The factors responsible for the distortion fall into two categories: (1) instrumental factors, i.e., distortion of the spectrum traceable directly to factors inherent in the experimental method used to obtain the spectrum, and (2) distortion factors inherent in the electronic structure of the sample, such as inner-level width, variation in transition probability, transition lifetimes, multiple-vacancy effects, etc. These two categories of distortion factors are virtually independent of each other. The latter group will be considered in detail in other chapters together with the interpretation of the spectra in the light of modern theories of the solid state. The discussion in this section will be confined to illustrations of representative spectra and effects of the more important instrumental factors on the spectra. These are: (1) instrumental resolution, (2) counting error, (3) energy-dispersion variation with wavelength, (4) specimen surface contamination, (5) self-absorption, (6) specimen temperature, and (7) excessive excitation energies. It should be emphasized that these various experimental factors are not necessarily listed in the order of their importance; in fact, they will vary depending upon the particular instrument and the method of operating it. Usually, distortions in a spectrum will be due to a number of these factors, whose individual effects cannot be resolved very precisely. Some of these experimental factors, such as resolution, dispersion, and self-absorption, have been discussed earlier in this chapter under the various appropriate topics. One of the experimental factors that is worthy of some extended discussion is temperature because (1) temperature can affect the spectrum in a variety of ways and (2) temperature is one of the factors that is frequently ignored.

Until recently, spectra have been reported, generally, with the implication that they were room-temperature results, but many of these spectra were from vapor-deposited material on a water-cooled, stainless-steel specimen post, and there is good reason to believe that the actual effective specimen temperature was considerably above room temperature. The lithium K spectrum shown in Fig. 3-26 is an interesting example. There have been numerous theoretical papers explaining the peaking of the spectrum which takes place abnormally far below the high-energy edge (compare with sodium for instance). However, McAlister (1969) has shown that the broadening of the K level by the phonon spectrum, in accordance with the following Overhauser relation, can account for the entire effect seen in the lithium spectrum if it is assumed that the effective specimen temperature was in the neighborhood of 162°C.

$$\langle (\mathrm{eV})^2 \rangle = \frac{N E_F{}^2 k\theta}{6\rho\mu^2}\left(1 + \frac{8T}{3\theta}\right) \qquad (3\text{-}11)$$

where $\langle (\mathrm{eV})^2 \rangle$ represents the root mean square of the thermal vibration energy of the

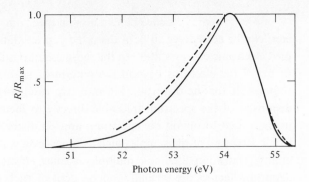

FIGURE 3-26
Lithium K emission spectrum; calculated spectrum (solid curve) and experimental spectrum (dashed curve).

$1s$ electrons, N is the number of atoms per cm^3, ρ is the density in g/cm^3, μ is the longitudinal sound velocity in cm/sec, θ is the Debye temperature, k is the Boltzmann constant, T is the absolute temperature, and E_F is the Fermi energy.

Thermal broadening of the high-energy edge was recognized early in the development of valence-band soft x-ray spectroscopy. This broadening follows the Fermi-Dirac distribution

$$f(E) = \frac{1}{1 + e^{(E-E_F)/kT}} \qquad (3\text{-}12)$$

where k is Boltzmann's constant ($= 7.244 \times 10^{-5}$ eV/°K). Skinner (1940R) demonstrated phenomenal agreement between the high-energy widths observed in the sodium L_{III}, magnesium L_{III}, and aluminum L_{III} spectra, and calculated edge widths. Skinner's results are shown in Fig. 3-27.

The third direct way in which specimen temperature can affect the spectrum is the usual expansion of the crystal lattice with increase in temperature. An increase in the volume of the crystal unit cell in real space corresponds to a constriction of the Brillouin zone in reciprocal space. It would be expected that the energy bands would cut the Fermi surface at different points and there would be shifts in the fine structure of the density-of-states distribution throughout the band. This type of effect has been shown in the comparison of the absorption spectra of aluminum at two different temperatures by Jope (1970).

The major cause of distortion from the true density of states, after transition-probability variation and satellites are accounted for (see Chap. 4), is the smearing of the spectrum due to the width of the inner level and the spectrometer slit width. It is appropriate to use the Lorentzian distribution function

$$L(E - E_0) = \frac{1}{\delta^2 + (E - E_0)^2} \qquad (3\text{-}13)$$

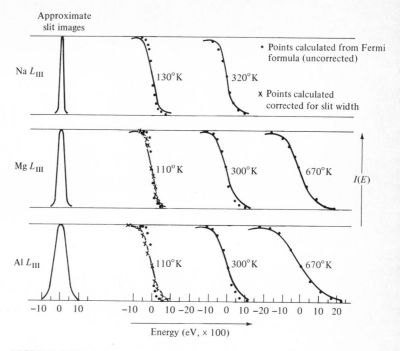

FIGURE 3-27
Thermal broadening of high-energy edges in the Na L_{III}, Mg L_{III}, and Al L_{III} spectra.

where δ is the half-width at half-maximum for the smearing effect of the inner-level width and the Gaussian distribution function

$$G(E - E_0) = e^{-\alpha(E - E_0)^2} \qquad (3\text{-}14)$$

for the smearing effect of the spectrometer slit width upon the spectrum.

When the Lorentzian and Gaussian functions are normalized to unit height and unit area and compared, as is done in Fig. 3-28, it is seen that the Lorentzian is the narrower at half-maximum but has broader tails than the Gaussian.

It has become customary (Tomboulian, 1957R) to relate the observed intensity to the density of electron states $N(E)$ through the various correction factors, in the form

$$I_{\text{photon}}(E) \propto v[T(E)G(E)L(E)P(E)N(E)] \qquad (3\text{-}15)$$

which make them appear, to be multiplication factors whereas in fact they are each integral smearing functions which operate independently upon the true density of states. The transition probability $P(E)$ is given by a matrix element of the electron momentum. Often, $P(E)$ is expressed in terms of only the radial part of the electron

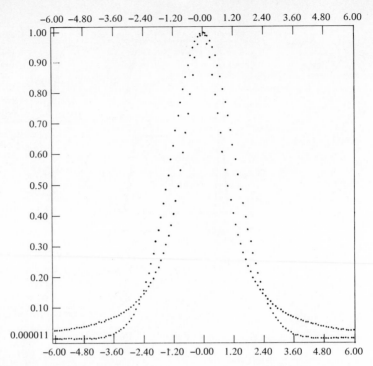

FIGURE 3-28
Comparison of the Lorentzian and Gaussian distribution functions normalized
to unit height and unit area.

wavefunctions, which allows an additional v^2 to be factored out (see Chap. 4 for a
detailed discussion of the transition probability). Now our expression becomes

$$I_{\text{photon}}(E) \propto v^3[T(E)G(E)L(E)P(E)N(E)] \qquad (3\text{-}16)$$

In this form, the intensity is assumed to be expressed in photon counts per unit time
per *unit energy*. Digital counting of photons is the most common intensity recording
method today in the soft x-ray grating spectroscopy range. However, the unit counting interval of spectral width is generally a unit of wavelength because this is as a unit
of distance around the Rowland circle and the spectrum is being mechanically
scanned. Therefore, the intensity generally being recorded is

$$I\lambda_{\text{photons}}(E)\, d\lambda$$

i.e., the number of photons per unit time over an interval $d\lambda$ centered at wavelength
λ. To convert to $I_{\text{photon}}(E)\, dE$ to be compatible with the units of $N(E)$, the distribution
of electron energy states necessitates multiplying the right-hand side by an additional

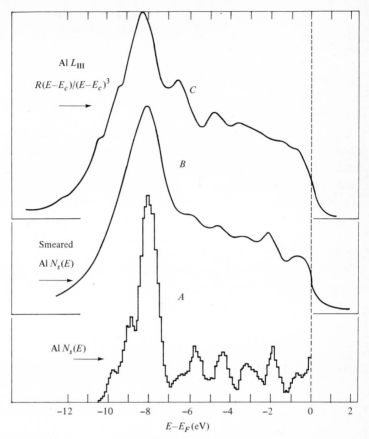

FIGURE 3-29
Aluminum L_{III} emission spectrum from $AuAl_2$.

v^2, which arises from the reciprocal relationship between λ and E; i.e., $\lambda \approx 1/v$, giving $d\lambda \approx dv/v^2$. Therefore, when the photon-counting interval has been an interval of wavelength, the final expression relating spectral intensity to $N(E)$ becomes

$$I\lambda_{\text{photon}}(E) = v^5 [T(E)G(E)L(E)P(E)N(E)] \qquad (3\text{-}17)$$

Of course, it is quite conceivable that the intensities could be measured directly in number of photons per unit *energy* interval in which case this last v^2 factor would *not* apply. For instance, a flow-proportional counter is energy discriminating (see Section IV on detectors) and therefore can be set to count photons over a narrow energy interval which can be scanned electrically over the width of the spectrum of interest.

Intensities may be recorded in energy units rather than in number of photons. In such cases another factor of v must be added to the coefficient in front of the integral. An example is recording current which is often done when the counting rate is extremely high—very seldom the case in soft x-ray spectroscopy in the grating-spectroscopy range.

The problem now is to unfold the effect of the smearing functions from the measured intensity distribution to yield a representation of the density of electron energy states. There have been many unfolding procedures developed (Porteus, 1961; Jones and Misell, 1967; Farach and Teitelbaum, 1967; Dotti, 1967) for removing this smearing from the spectrum, but usually one wants to compare the soft x-ray results obtained experimentally with the theoretical density of states obtained from a band calculation, and it is more convenient to fold into the calculated density of states the smearing that is present in the experimentally determined spectrum. See, for example, Fig. 3-29 showing the aluminum L spectrum from $AuAl_2$ (Williams et al., 1972) where smearing functions of the foregoing type were folded into a theoretical density of states labeled A to produce the "spectrum" labeled B. The latter is now comparable with spectrum C which is the experimentally determined spectrum after it has been corrected for instrumental dispersion.

GENERAL REVIEWS

BAUN, W. L. (1968R): Instrumentation, spectral characteristics, and applications of soft x-ray spectroscopy, in E. G. BRAME (ed.), *Applied spectroscopy reviews*, vol. 1, pp. 379–432, Marcel Dekker, Inc., New York.

BLOKHIN, M. A. (1957R): *The physics of x-rays*, 2d ed. (State Publishing House of Technical-Theoretical Literature, Moscow), English translation by U.S. Atomic Energy Commission, document AEC-tr-4502.

CAUCHOIS, Y. (1948R): *Les spectres de rayon x et la structure électronique de la matière*, Gauthier-Villars, Paris.

HOLLIDAY, J. E. (1967R): Soft x-ray emission spectroscopy in the 10 to 150 A region, in E. F. KAELBLE (ed.), *Handbook of x-rays*, pp. 38-1–38-41, McGraw-Hill Book Co., New York.

———— (1970R): Soft x-ray spectroscopy in metals research, in R. F. BUNSHAH (ed.), *Techniques of metals research*, vol. 3, part 1, pp. 325–418, John Wiley & Sons, Inc., New York.

SAMSON, J. A. R. (1967R): *Techniques of vacuum ultraviolet spectroscopy*, John Wiley & Sons, Inc., New York.

SHAW, C. H. (1956R): The x-ray spectroscopy of solids, in *Theory of alloy phases*, pp. 13–62, American Society for Metals, Metals Park, Cleveland.

SKINNER, H. W. B. (1940R): The soft x-ray spectroscopy of solids, *Phil. Trans. Roy. Soc. London, Ser. A*, **239**: 95–112.

STROKE, G. W. (1967R): Diffraction gratings, in s. FLÜGGE (ed.), *Encyclopedia der physik*, vol. 29, pp. 426–735, Springer-Verlag, Berlin.

TOMBOULIAN, D. H. (1957R): The experimental methods of soft x-ray spectroscopy and the valence band spectra of the light elements, in s. FLÜGGE (ed.), *Handbuch der physik*, vol. 30, pp. 246–304, Springer-Verlag, Berlin.

ZIMKINA, T. M., and FOMICHEV, V. A. (1971R): *Ultra-soft x-ray spectroscopy* (in Russian), Leningrad University Publishers, Leningrad.

REFERENCES

BAUN, W. L. (1972): *Soft x-ray bibliography*, Air Force Materials Laboratory, Wright Patterson Air Force Base, Dayton, Ohio.

—— and FISCHER, D. W. (1970): Soft x-ray spectroscopy as related to inorganic chemistry, in C. N. RAO and J. R. FERRARO (eds.), *Spectroscopy in inorganic chemistry*, vol. 1, pp. 209–246, Academic Press, Inc., New York.

BEARDEN, J. A. (1931): Absolute wavelengths of the copper and chromium *K* series, *Phys. Rev.*, **37**: 1210–1229.

—— (1969): Ruled gratings for x-ray and ultraviolet spectra, AD700352 (U.S. Department of Commerce, National Bureau of Standards, National Technical Information Service, Springfield, Va.).

BENNETT, J. M. (1971): "Physical structure and diffraction performance of laminar x-ray gratings," Thesis, Imperial College of Science and Technology, London.

BEUTLER, H. G. (1945): The theory of the concave grating, *J. Opt. Soc. Am.*, **35**: 311–350.

BIRKS, L. S., SEEBOLD, R. E., BATT, A. P., and GROSSO, J. S. (1964): Excitation of characteristic x-rays by protons, electrons, and primary x-rays, *J. Appl. Phys.*, **35**: 2578–2581.

BONNELLE, C. (1964): "Contribution à l'étude des metaux de transition du premier groupe, du cuivre et de leurs oxydes par spectroscopie x dans le domaine de 13 à 22 A," Thesis, University of Paris.

CARTER, G. C., KAHAN, D. J., BENNETT, L. H., CUTHILL, J. R., and DOBBYN, R. C. (1970): *The NBS alloy data center: Author index*, NBS-OSRDB-70-2 (U.S. Department of Commerce, National Bureau of Standards, National Technical Information Service, Springfield, Va.).

——, ——, ——, ——, and —— (1971): *The NBS alloy data center: Permuted materials index*, Natl. Bur. Stand. Spec. Publ. 324, Washington.

COMPTON, A. H., and DOAN, R. L. (1925): X-ray spectra from a ruled reflection grating, *Proc. Natl. Acad. Sci. U.S.*, **11**: 598–601.

CORDELLE, J., FLAMAND, J., PIEUCHARD, G., and LABEYRIE, A. (1969): Aberration-corrected concave gratings made holographically, *Opt. Instr. Tech.*, Oriel Press, Lourdes.

CRISP, R. S. (1961): "The soft x-ray emission spectra of the light elements and some alloys," Thesis, University of Western Australia.

CURRY, C. (1968): Soft x-ray emission spectra of alloys and problems in their interpretation, in D. J. FABIAN (ed.), *Soft x-ray band spectra and the electronic structures of metals and materials*, pp. 173–189, Academic Press, Inc., New York.

CUTHILL, J. R. (1970): A soft x-ray spectrometer with improved drive, *Rev. Sci. Instr.*, **41**: 422–423.

———, MCALISTER, A. J., DOBBYN, R. C., CARTER, G. C., and KAHAN, D. J. (1972): *Critical review of soft x-ray emission spectroscopy of metals and alloys* (in preparation).

———, ———, WILLIAMS, M. L., and WATSON, R. E. (1967): Density of states of nickel: Soft x-ray spectrum and comparison with photoemission and ion neutralization studies, *Phys. Rev.*, **164**: 1006–1017.

DOTTI, D. (1967): CSUP, a computer program to reconstruct a physical distribution from counting spectra distorted by the instrumental resolution, *Nucl. Instr. Methods*, **54**: 125–131.

FAESSLER, A. (1955): Spektroskopie der valenzelectronenbänder, *Landolt-Börnstein tables*, 6th ed., vol. 1, part 4, pp. 769–868, Springer-Verlag, Berlin.

——— and WIECH, G. (1973): *Landolt-Börnstein tables, new series*, Springer-Verlag, Berlin.

FARACH, H. A., and TEITELBAUM, H. (1967): Spectroscopic line analysis using a Gaussian and Lorentzian convolution technique, *Can. J. Phys.*, **45**: 2913–2921.

FRANKS, A., and LINDSEY, R. (1968): Dispersion of 1A x-rays with NPL x-ray gratings, *J. Sci. Instr.*, (2) **1**: 144.

HABER, H. (1950): The torus grating, *J. Opt. Soc. Am.*, **40**: 153–165.

HOFFMANN, L., WIECH, G., and ZÖPF, E. (1969): Zur Tiefenverteilung der durch Electronenstoss angeregten charakteristischen Röntgenstrahlung von Bor, Kohlenstoff, Aluminum und Silber, *Z. Physik.*, **229**: 131–142.

ISHII, T., AITA, O., ICHIKAWA, K., and SAGAWA, T. (1971): Use of a toroidal mirror in the soft x-ray optical system, *J. Appl. Phys. (Japan)*, **10**: 637–642.

IAEGLE, P. (1965): "Contribution à la spectrographie dans l'ultraviolet de la région de Holweek," Thesis, University of Paris.

JENNEY, J. A. (1970): Holographic recording with photopolymers, *J. Opt. Soc. Am.*, **60**: 1155–1161.

JOHNSON, M. C. (1969): A secondary standard vacuum ultraviolet detector, *Rev. Sci. Instr.*, **40**: 311–315.

JONES, A. F., and MISELL, D. L. (1967): A practical method for the deconvolution of experimental curves, *Brit. J. Appl. Phys.*, **18**: 1479–1483.

JOPE, J. A. (1970): Temperature displacement of x-ray absorption fine structure in aluminum, *J. Phys. C.: Metal Phys. Suppl.*, No. 1, pp. 21–23.

KASTNER, S. O., and NEUPERT, W. (1963): Image construction for concave gratings at grazing incidence, by ray tracing, *J. Opt. Soc. Am.*, **53**: 1180–1184.

LABEYRIE, A., and FLAMAND, J. (1969): Spectrographic performance of holographically made diffraction gratings, *Optics Commun.*, **1**: 5–8.

LIEFELD, R. J. (1968): Soft x-ray emission spectra at threshold excitation, in D. J. FABIAN (ed.), *Soft x-ray band spectra and the electronic structures of metals and materials*, pp. 133–149, Academic Press, Inc., New York.

MCALISTER, A. J. (1969): Calculation of the soft x-ray K-emission and absorption spectra of metallic Li, *Phys. Rev.*, **186**: 595–599.

MACK, J. E., STEHN, J. R., and EDLÉN, B. (1932): On the concave grating spectrograph, especially at large angles of incidence, *J. Opt. Soc. Am.*, **22**: 245–264.

MAYSTRE, D., and PETIT, R. (1971): Essai de determination theorique du profil optimal d'un reseau holographique, *Optics Commun.*, **4**: 25–28.

———, DOBBYN, R. C., CUTHILL, J. R., and WILLIAMS, M. L. (1973): Soft x-ray emission spectra of metallic solids: Critical review of selected systems and annotated spectral index, Natl. Bur. Std. Spec. Publ. 369.

NAGEL, D. J. (1971): Ion-impact excitation of x-ray spectra for electronic structure studies, in D. J. FABIAN (ed.), *Band structure spectroscopy of metals and alloys*, Academic Press, Inc., New York.

NAMIOKA, T. (1959): Theory of the concave grating I, *J. Opt. Soc. Am.*, **49**: 446–465.

—— (1961): Theory of the ellipsoidal concave grating I, *J. Opt. Soc. Am.*, **51**: 4–12.

O'BRYAN, H. M. (1931): Reflecting power and grating efficiency in the extreme ultraviolet, *Phys. Rev.*, **31**: 32–40.

PIERCE, J. R. (1954): *Theory and design of electron beams*, 2d ed., pp. 173–193, McGraw-Hill Book Co., New York.

PORTEUS, J. O. (1961): Optimized method for correcting smearing aberrations: Complex x-ray spectra, *J. Appl. Phys.*, **33**: 700–707.

ROWLAND, H. A. (1882): Preliminary nature of the results accomplished in the manufacture and theory of gratings for optical purposes, *Phil. Mag.*, **13**: 469–474.

——— (1883): On concave gratings for optical purposes, *Phil. Mag.*, **16**: 197–210.

SAYCE, L. A., and FRANKS, A. (1964): N.P.L. gratings for x-ray spectroscopy, *Proc. Roy. Soc. London*, **A282**: 353–357.

SCHNOPPER, H. W. (1968): Atomic readjustment to an inner-shell vacancy: Manganese K x-ray emission spectra from an Fe^{55} K-capture source and from the bulk metal, *Phys. Rev.*, **154**: 118–123.

SOMER, T. A., and GRAVES, P. W. (1969): Spiraltron matrices as windowless photon detectors for soft x-ray and extreme UV, *IEEE Trans. Nucl. Sci.*, **16**: 376–380.

SPEER, R. J. (1970): Grating studies at x-ray wavelengths, *Advan. X-Ray Anal.*, **13**: 382–389.

SPIELBERG, N. (1967): Characteristics of flow proportional counters for x-rays, *Advan. X-Ray Anal.*, **10**: 534–545.

SPRAGUE, G., TOMBOULIAN, D. H., and BEDO, D. E. (1955): Calculations of grating efficiency in the soft x-ray region, *J. Opt. Soc. Am.*, **45**: 756–761.

SPROULL, W. T. (1946): *X-rays in Practice*, pp. 97–103, McGraw-Hill Book Co., New York.

STERK, A. A., MARKS, C. L., and SAYLOR, W. P. (1967): Production efficiencies of x-ray emission spectra by proton bombardment, *Advan. X-Ray Anal.*, **10**: 399–408.

STRONG, J. (1958): *Concepts of classical optics*, pp. 597–615, W. H. Freeman and Company, Publishers, San Francisco.

WATSON, L. M., DIMOND, R. K., and FABIAN, D. J. (1968): Soft x-ray emission spectra of magnesium and beryllium, in D. J. FABIAN (ed.), *Soft x-ray band spectra and the electronic structures of metals and materials*, pp. 45–58, Academic Press, Inc., New York.

WELLER, C. S., and YOUNG, J. M. (1970): Photomultiplier yield of a channel electron multiplier in the 304–493 A wavelength range, *Appl. Opt.*, **9**: 505–506.

WILLIAMS, M. L., DOBBYN, R. C., CUTHILL, J. R., and MCALISTER, A. J. (1972): Soft x-ray emission spectrum of Al in AuAl$_2$, *Proceedings of third IMR symposium, electronic density of states* (Natl. Bur. Std. Spec. Publ. 323).

YAKOWITZ, H., and CUTHILL, J. R. (1962): Annotated bibliography on soft x-ray spectroscopy, Natl. Bur. Std. Monograph 52.

4

THEORY OF EMISSION SPECTRA

G. A. Rooke

I. INTRODUCTION

This chapter describes the interpretation of the shape and character of x-ray emission spectra in terms of the electronic properties of the emitting substances.

A phenomenological approach has been adopted in order to fulfill two objectives: (1) to give the cursory reader a brief review of the subject and (2) to give the serious student a unified introduction to the considerable amount of further reading which he will undoubtedly be required to do. It is not a comprehensive bibliography, and no attempt has been made to interpret or describe a comprehensive selection of spectra; instead, references and spectra have been selected purely to illustrate points made during the discussion.

As a crude approximation, the shape of the spectra can be considered to be proportional to the product of two functions of the electronic properties: the density of states and the transition probability. Many other articles have given detailed descriptions of the concepts of the density of states, but mostly they have given only a limited treatment of the concepts incorporated in the transition probability. A whole chapter of Wilson's book (1936) *Theory of metals* was devoted to this type of description, but unfortunately, this was omitted from later editions. In this chapter, both functions are described in detail.

Several useful reviews discuss the interpretation of the spectra. The following is a selection of the more important ones: Cauchois (1948), Parratt (1959), Tomboulian (1957), and Blokhin (1957). A concise introduction to the subject is provided by a review by Skinner (1940). An excellent bibliography of the literature up to 1960 has been published by Yakowitz and Cuthill (1962), and a revised edition of this is being prepared. The proceedings of two recent conferences, Fabian (1968) and Nemoshkalenko (1969), summarize the developments that have occurred since these reviews were written.

II.   WAVEFUNCTIONS, QUANTUM NUMBERS, AND SPHERICAL HARMONICS

A.   One-electron wavefunctions

It can be assumed that all matter consists of positive atomic nuclei and negative electrons held together by the attraction between the unlike charges. Because the atomic nuclei are much heavier than the electrons, they accelerate relatively slowly, and an attendant cloud of electrons will always travel with them to keep the total moving charge roughly neutral. On the other hand, when an electron moves, the nuclei will be too heavy to follow, and it will appear as though the electron can move independently of the nucleus and that the nuclei remain stationary. The ability to treat the electrons as though the nuclei are stationary provides a tremendous simplification in the mathematics required to determine the behavior of the electrons. Corrections for this assumption can be made later by considering electron-phonon interactions; *phonons* are collective vibrations of the nuclei with their attendant clouds of electrons (atoms), and they form the dominant mode of atomic movement.

When studying the electronic structure, it is further assumed that the motion of any individual electron is independent of the momentary position of any other electron; any electron is assumed to experience only the time-average effect of the other electrons. Because most of the electrons are attracted to the vicinity of the positive charges, their average position is often near the center of the atoms. The attractive potential of the positive charge is effectively reduced, the results being called the *screened* or *effective potential*. Correction for the assumption that electrons move independently is made in two parts: corrections for the electron-electron interactions and corrections for plasmon effects; *plasmons* are collective oscillations of the electrons.

For each point in space there is a probability that an electron will be in the vicinity of that point during some small time interval. The probability is proportional to the electron density at that point. In some regions, often those close to the positive charges, the probability of finding an electron is higher than in others. Unless there is a physical discontinuity, the variations of the probability must be smooth and continuous. If one followed a path that formed a continuous loop, one would find that this probability varied continuously, slowly rising and falling in succession until, on completion of the circuit, it had returned to its original value. Of course, the probability must always be positive or zero, since negative probabilities are meaningless.

Since the probability is positive, it can always be expressed as the square of a function that can be either positive or negative. It is not necessary for this function to be real since any complex number when multiplied by its complex conjugate gives a positive real number. Therefore, the field of probabilities can be represented by a field of complex numbers whose amplitude squared equals the probability and whose phase is indeterminate. The field of complex numbers is called the *wavefunction* and is the foundation of quantum mechanics. The electronic structure of matter is predicted by determining the form of these wavefunctions.

B. Group theory of an isolated atom

The character of the wavefunction is determined by the symmetry of the system. As an example, it is proposed to examine the case of an isolated atom with full spherical symmetry. It is required to find the character of the continuous functions that can be placed around this atom. In the following discussion, it is convenient to assume that the wavefunctions are real, although this assumption does not seriously affect the conclusions reached.

Without specifying any numerical magnitudes, it is possible to say that any function may consist of regions where it is positive and regions where it is negative and that these regions (antinodes)[1] will be distributed in some regular manner around the center of symmetry.

Although the antinodes may be associated with either positive or negative wavefunctions, in physical reality where one is considering only the squares of the wavefunctions, the signs are chosen quite arbitrarily. (They certainly have no connection with electrical charge.) Because they are chosen arbitrarily, it is possible to consider specifically only those of one sign, the positive sign say, and to assume that the conclusions would not be different if one had considered those of opposite sign.

C. Angular symmetry

The number of distinct positive antinodes must be integral and can vary from zero to infinity; this number is called the *azimuthal quantum number l*. Wavefunctions with $l \geq 1$ have l positive antinodes and usually an equivalent number of negative antinodes. The function with $l = 0$ has no clearly defined antinodes; that is, it is entirely positive or entirely negative. For historic reasons, states with different values of l have been represented by different letters of the alphabet as follows:

$$l = 0 \quad 1 \quad 2 \quad 3 \quad 4 \quad 5 \quad 6 \quad 7 \quad \cdots$$
$$s \quad p \quad d \quad f \quad g \quad h \quad i \quad j \quad \cdots$$

States with $l = 0$ are said to be *s*-like, and so forth.

---

[1] The word "antinode" is an unfortunate choice here because it is being used in its wave sense. In the usual meaning it represents node-like shapes.

The positive antinodes will orient themselves symmetrically around the center of the atom, and the negative regions will fill in between them. Functions are said to be *symmetrical* with respect to an operation if after the operation they look the same as before it. For an isolated atom we are interested in symmetry with respect to all proper rotations and improper rotations, i.e., rotations combined with reflections in a plane perpendicular to the axis of rotation.

The representation of symmetrical functions is always simplified by defining them with respect to their symmetry axes. After one function is defined, it can easily be imagined that other related functions will align themselves symmetrically with respect to it. Therefore, the simplest way of describing sets of symmetrical functions is to choose an axis of symmetry of one of the functions, call it the z axis, and describe the symmetrical alignment of the other functions with respect to this axis.

The rotational symmetry around this z axis can be used to uniquely define any specific symmetrical orientation of a function to this axis. If on rotating the function through $360°$ around the z axis it has had its original appearance m times, it is said to have *m-fold symmetry*, and it is given a magnetic quantum number of $\pm m$. The functions represented by $+m$ and $-m$ are distinguished by each having nodes where the other has antinodes. The function with $m = 0$ is symmetrical for all rotations about this axis.

Functions with m-fold symmetry around the z axis, when viewed along the z axis, have m positive real antinodes jutting out from the axis. Therefore, it is impossible for $|m|$ to exceed l. It can be shown that m can have all integral values from $-l$ to $+l$ and hence that there are $(2l + 1)$ distinct symmetrical orientations of the function around the chosen axis of symmetry, that is, around the z axis.

As specific examples, I will describe the functions with l equal to 0, 1, and 2 which are shown in Fig. 4-1. The function with $l = 0$ has full spherical symmetry like a sphere. Any axis is therefore an axis of complete rotational symmetry, and m is given the value of zero.

The function with $l = 1$ has one positive antinode and one negative antinode. Since these antinodes represent regions of high charge density, they will repel one another to positions on opposite sides of the atom. The function will appear like a dumbbell with one ball positive and the other negative. Three symmetrical orientations are possible; that with $m = 0$ has the dumbbell lying along the chosen z axis while those with $m = \pm 1$ will lie at right angles to it and at right angles to each other.

The function with $l = 2$ has the form of a cross with alternate arms positive and negative. The orientation with the positive arms lying along the z axis is labeled with $m = 0$; it has full rotational symmetry only if we assume that the two negative nodes combine into a doughnut shape, and since we are really defining the functions by only the number and orientation of the positive nodes, this assumption is allowable. The orientations labeled $m = \pm 1$ have the z axis lying between the arms of the cross, while

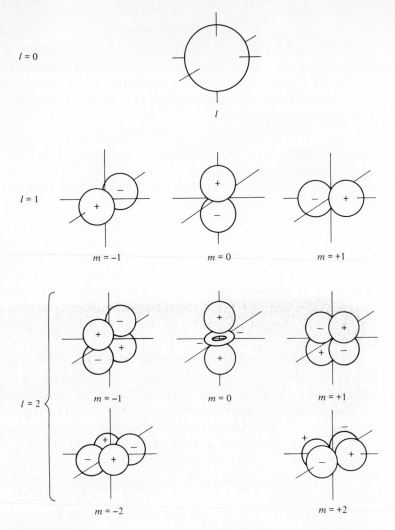

$l = 0$

l

$l = 1$

$m = -1$ $m = 0$ $m = +1$

$l = 2$ $m = -1$ $m = 0$ $m = +1$

$m = -2$ $m = +2$

FIGURE 4-1
Symmetrized set of spherical harmonics (schematic) illustrating the symmetries of the functions.

the orientations with the z axis perpendicular to the plane of the cross are labeled with $m = \pm 2$.

If one had carried out the above calculations using complex numbers for which one has positive and negative antinodes for both the real and the imaginary parts, the solutions would have been slightly different, but not significantly so. If one had used relativistic theory and complex numbers, twice as many symmetrical orientations would have been found: two for each set of quantum numbers described above. The two orientations with the same quantum numbers are labeled with $s = \pm 1/2$ to distinguish them; s is called the *spin quantum number*.

D.   Radial symmetry

Finally, functions may be characterized by a fourth quantum number that is related to the radial variation of the function. The number of positive antinodes found when moving from infinity to infinity diametrically across the atom can vary from zero to infinity; again, the wavefunction that has only one sign is counted as having no positive antinodes. The number of positive antinodes is equated to $n - l - 1$, where n is the principal quantum number. This quantum number can have any integer value greater than 1. States with different principal quantum numbers have been allocated different letters of the alphabet as follows:

$$n = 1 \quad 2 \quad 3 \quad 4 \quad 5 \quad 6 \quad \cdots$$
$$ K \quad L \quad M \quad N \quad O \quad P \quad \cdots$$

States with $n = 1$ are said to be in the K shell. When quoting the lowercase letters that represent the values of l, the numerical version of the principal quantum number is used; for instance, the state with $n = 2$ and $l = 1$ is said to be a $2p$ state.

These four quantum numbers can specify completely the character of the functions as far as x-ray spectroscopy is concerned. They all result from the symmetry and the continuous nature of the function, and they are not directly dependent on any quantum conditions. However, they are closely tied to the quantum nature of the system.

E.   The wave equation

We have seen that it is possible to consider the electrons as though they are independent particles; this implies that each electron will have a well-defined energy and well-defined momentum. The relationship between these physical quantities and the wavefunction is the fundamental assumption of quantum mechanics.

The total energy of a particle E is expressed as a function of the momentum p, the electronic mass μ, and the potential energy V:

$$E = \frac{p^2}{2\mu} + V(r) \qquad (4\text{-}1)$$

It is postulated that the momentum p should be replaced by a differential operator $i\hbar\nabla$. Equation (4-1) then contains an operator, and this term needs a function to operate on. To overcome this, the equation is multiplied by the wavefunction $\psi(r)$ and becomes

$$E\psi = \frac{-\hbar^2}{2\mu} \nabla^2 \psi + V\psi \qquad (4\text{-}2)$$

$\hbar$ is Plank's constant divided by 2π, and ∇ is the differential operator $\partial/(\partial x)\hat{\mathbf{x}}$, where $\hat{\mathbf{x}}$ is the unit vector in the direction of the greatest increase of p.

When the potential is spherically symmetrical, it is possible to write ψ as a product of three terms, each of which is dependent on only one of the variables r, θ, and φ:

$$\psi(r,\theta,\varphi) = \frac{1}{2\pi} R(r)\Theta(\theta)\Phi(\varphi) \qquad (4\text{-}3)$$

The Schrödinger equation (4-2) can then be split into three separate equations, each containing only one of the three variables:

$$\frac{1}{\Phi} \frac{\partial^2 \Phi}{\partial \varphi^2} = -a^2 \qquad (4\text{-}4)$$

$$\frac{1}{\Theta \sin \theta} \frac{\partial}{\partial \theta}\left(\sin \theta \frac{\partial \Theta}{\partial \theta}\right) - \frac{a^2}{\sin^2 \theta} = c \qquad (4\text{-}5)$$

$$\frac{\hbar^2}{2\mu} \frac{1}{r^2} \frac{\partial}{\partial r}\left(r^2 \frac{\partial R}{\partial r}\right) + \left(E - V - \frac{c}{r^2}\right) R = 0 \qquad (4\text{-}6)$$

F. Spherical harmonics

Equations (4-4) and (4-5) have continuous solutions only when $a = m$ and $c = -l(l + 1)$, where m and l are the quantum numbers discussed above. The solutions are of the form

$$\Phi_m = \frac{1}{\sqrt{2\pi}} e^{im\varphi} \qquad (4\text{-}7)$$

and

$$\Theta_{l,m} = \sqrt{\frac{(2l + 1)(l - m)!}{2(l + m)!}}\, P_l^m (\cos \theta) \qquad (4\text{-}8)$$

where $P_l^m (\cos \theta)$ are functions known as the *associated Legendre polynomials*.

The product of the two angular terms

$$Y_l^m(\theta,\varphi) = \Theta_{l,m}\Phi_m \qquad (4\text{-}9)$$

is called a *spherical harmonic*. Spherical harmonics are functions that have angular dependences that are the same as those described in the beginning of this section as being necessary for the wavefunctions of isolated atoms.

Wavefunctions that do not result from spherical potentials will not be simply a product of a radial function and a spherical harmonic. However, it is interesting to note that any function can always be uniquely represented by a sum of such terms.

$$\psi(r,\theta,\varphi) = \sum_{l,m} a_{l,m}(r) Y_l^m(\theta,\varphi) \qquad (4\text{-}10)$$

This is possible because the spherical harmonics form a complete set of orthonormal functions.

Orthonormality is defined by the conditions

$$\int Y_{l'}^{m'} Y_l^m \cos \theta \, d\theta \, d\varphi = \begin{cases} 0 & l' \neq l \text{ or } m' \neq m \\ 1 & l' = l \text{ and } m' = m \end{cases} \qquad (4\text{-}11)$$

This condition ensures the uniqueness of the coefficients $a_{l,m}(r)$ in the expansion. For the set to be complete, all possible values of l and m must be included.

The recurrence relationships are important for the derivation of selection rules. The most important formula is

$$\Theta_{m,l} \cos \theta = \sqrt{\frac{(l+1)^2 - m^2}{(2l+1)(2l+3)}} \, \Theta_{m,l+1} + \sqrt{\frac{l^2 - m^2}{(2l+1)(2l-1)}} \, \Theta_{m,l-1} \qquad (4\text{-}12)$$

Brief descriptions of spherical harmonics and associated Legendre polynomials are given by most comprehensive books on quantum mechanics like Schiff (1955) and Messiah (1964), as well as in the mathematical literature.

The discussion of the possible types of wavefunction that an electron can have, once the symmetry of its environment has been defined, is the province of group theory. The detailed arguments are necessarily more mathematical than could ever be included here. A good treatise on the application of group theory to atoms, molecules, and solids has been given by Tinkham (1964), but this is only one of many equally good books on the topic. Seitz (1940) includes a good discussion using the more intuitive approach while two papers by Altmann and Cracknell (1965) and Altmann and Bradley (1965) provide the necessary information to enable one to apply the concepts discussed by Tinkham to simple solids.

G. The energy of states

Equation (4-6) shows that the energy of the particle depends on the radial distribution of the wavefunction and hence on the charge density. We might expect that the $1s$ states have the lowest energy because they have no regions of zero charge density, and the electrons can distribute themselves so that they are on the average as far from one another as possible.

For Coulomb potentials, our expectations are proved true. In this case, if the nucleus has a charge Ze, it is possible to find solutions to equation (4-6) only if the energy is given by

$$E = \frac{-\mu e^4 Z^2}{2n^2 \hbar^2} \qquad (4\text{-}13)$$

where n is the principal quantum number. It can be seen that the various states have discrete energy levels that change with n.

Electrons tend to fill the states of lowest energy, but the repulsive forces and the nature of the charge distributions defined by the wavefunctions are such that one and only one electron can fill each state that is distinguished by a unique set of four quantum numbers. States with lowest n will be filled in preference to other states. As there are only a finite number of electrons in an atom, there will be a value of n that corresponds to the electrons with highest energy. Electrons with this value of n are called the *valence electrons*, and others, which will require a significantly greater energy to become excited to normally unoccupied states, are called the *core electrons*.

When the calculations are carried out fully, allowing for relativistic effects, electron-electron interactions, spin-orbit coupling, etc., it is found that although the energy depends mostly on n, it also depends on l, m, and s. All states with the same n have energies that are similar to one another, but each state has a slightly different energy from the others. Within this set of states with the same n, all states with the same l have similar energies, but they are slightly different because the m and s quantum numbers are different. For atomic states not in magnetic fields, the energy differences between states with the same values of n and l are often negligibly small, especially for the core states. The elements most sensitive to the values of these quantum numbers are the ferromagnetic metals Fe, Co, and Ni and to a lesser extent the other transition metals.

H.   The momentum of states

Because of the spherical symmetry, we would expect the angular momentum to be conserved and therefore to have a well-defined value for each state. Also, since angular momentum is an angular quantity, it would be expected to involve the angular quantum numbers l and m. Indeed, it can be shown that the total angular momentum is given by $\sqrt{l(l + 1)}\hbar$, and the component of angular momentum in the direction of the z axis, along which the functions have aligned themselves, is given by $m\hbar$. The spin s modifies the angular momentum, but this effect can be ignored in many cases. When l is modified by s, it is relabeled j, which is only allowed to have half-integer values.

Linear momentum is not conserved within an atom because the electrons are not in a field-free region of space. Therefore discussions about linear momentum are not relevant in this section.

III.   THEORY OF ELECTROMAGNETIC RADIATION

The theory of electromagnetic radiation in the x-ray wavelength region can be explained adequately only in terms of the theory of electrodynamics. [See Schiff (1955) and Messiah (1964).] However, the complexity is beyond the scope of this chapter, and only an outline of the essential character of the theory can be given. Only emission spectra are discussed.

A.   Excited states

Emission of radiation can occur whenever the electronic states of an atom have been excited. Within the independent-electron approximation, excited states are considered to be created when one electron has been excited from its normal state to a normally empty state at a higher energy. In reality, this situation has created two excited states that can decay independently of one another. The first excited state is the vacancy into which other electrons in states of higher energy can fall, and the other excited state is the electron in the normally unoccupied state of higher energy which may fall to other unoccupied states of lower energy. Both of these excited states may radiate the excess energy in the form of photons as they decay. The energy of the photon radiated is given by the energy difference between the initial and final states of the electron.

There are normally several electrons that are capable of filling a vacancy, and there are many states into which the excited electron may fall. The transitions occur according to the statistical probabilities associated with the respective transitions.

B.   Classical transition probabilities

In order to derive the "Golden Rule" for the transition probability, a perturbation approach can be taken with the Maxwell equations as the starting point. The Maxwell equations of motion for the electromagnetic field are:

$$\mathbf{V} \times \mathbf{E} + \frac{1}{c}\frac{\partial \mathbf{H}}{\partial t} = 0 \qquad (4\text{-}14)$$

$$\mathbf{V} \cdot \mathbf{H} = 0 \qquad (4\text{-}15)$$

$$\mathbf{V} \times \mathbf{H} - \frac{1}{c}\frac{\partial \mathbf{E}}{\partial t} = \frac{4\pi \mathbf{J}}{c} \qquad (4\text{-}16)$$

$$\mathbf{V} \cdot \mathbf{E} = 4\pi\rho \qquad (4\text{-}17)$$

where $\mathbf{H}$ is the magnetic field vector, $\mathbf{E}$ the electric field vector, $\mathbf{J}$ the current density, ρ the charge density, and c the velocity of light.

Taking the curl of equation (4-16) and then substituting equations (4-14) and (4-15) into it will yield

$$\nabla^2 \mathbf{H} - \frac{1}{c} \frac{\partial^2 \mathbf{H}}{\partial t^2} = - \frac{4\pi}{c} \nabla \times \mathbf{J} \qquad (4\text{-}18)$$

If we assume that $\mathbf{E}$, $\mathbf{H}$, and $\mathbf{J}$ are all periodic functions of time and that they all have the same frequency ω, the time derivative can be taken, and equation (4-18) can be written

$$(\nabla^2 + \mathbf{k}^2)\mathbf{H}(\mathbf{r}) = - \frac{4\pi}{c} \nabla \mathbf{J}(\mathbf{r}) \qquad (4\text{-}19)$$

where $\mathbf{k} = \omega/c$. This equation can be solved to find $\mathbf{H}$ in terms of $\mathbf{J}$, a nontrivial procedure which yields

$$\mathbf{H}(\mathbf{r}) = \frac{1}{c} \int \frac{\nabla \times \mathbf{J}(\mathbf{r}')}{(\mathbf{r} - \mathbf{r}')} \exp\left(i\mathbf{k} \cdot |\mathbf{r} - \mathbf{r}'|\right) d\tau' \qquad (4\text{-}20)$$

A similar expression describing $\mathbf{E}$ in terms of $\mathbf{J}$ can be found.

The time average of the energy flow associated with an electromagnetic field is given by the time average of the Poynting vector:

$$\mathbf{P}(\mathbf{r}) = \frac{c}{2\pi} \text{Re} \left[\mathbf{E}(\mathbf{r}) \times \mathbf{H}(\mathbf{r})\right] \qquad (4\text{-}21)$$

where Re denotes "the real part of." Substituting the expressions for $\mathbf{H}$ and $\mathbf{E}$ into this equation and simplifying it—a process requiring considerable mathematical juggling—we find the energy flux given by

$$\mathbf{P}(\mathbf{r}) = \frac{\mathbf{k}^2}{2\pi c} \left| \int \mathbf{J}_{l,\mathbf{k}}(\mathbf{r}') \exp\left(i\mathbf{k} \cdot |\mathbf{r} - \mathbf{r}'|\right) d\tau' \right|^2 \qquad (4\text{-}22)$$

where $\mathbf{J}_{l,\mathbf{k}}$ is the component of $\mathbf{J}$ perpendicular to $\mathbf{k}$.

C.   Quantized transition probabilities

So far the theory has been purely classical. Now we must add in the quantum assumption that the momentum $\mathbf{P}$ can be replaced by $i\hbar\nabla$. This introduces wavefunctions into the energy equation and into equation (4-22) above.

The current density $\mathbf{J}$ can be represented by the product of the probability density $|\psi|^2$, the charge e of an electron, and the velocity $\mathbf{v}$ of the electron. However, we are interested in transitions between the states q and q'[†] so that the probability density or probability of finding an electron at the point $\mathbf{r}$ should be replaced by $\psi_{q'}^* \psi_q$. The velocity is given by the cross product between the radial position vector $\mathbf{r}$ and the angular frequency vector ω.

[†] q stands for the set of quantum numbers n, l, m, and s.

The total energy radiated per second is found by substituting the current density into equation (4-22). If we divide this by the energy per photon $\hbar\omega$, the number of photons radiated per second is given by

$$W_{q'q}(\omega,j) = \frac{e^2\omega}{hc^3 2\pi} \left| D_{q'q} \right|^2 \qquad (4\text{-}23)$$

where

$$|D_{q'q}| = \int \psi_{q'}^* \omega \cdot \mathbf{r}_j \exp(i\mathbf{k} \cdot \mathbf{r}) \psi_q \, d\tau \qquad (4\text{-}24)$$

and $\mathbf{r}_j$ is the component of $\mathbf{r}$ in the direction $\mathbf{j}$ of the polarization.

The integral in equation (4-24) is a measure of the overlap of the two wave-functions $\psi_{q'}$ and ψ_q. However, it is a weighted average of this overlap. Only those parts of the waves that are rotating around the center with angular frequency ω, that have the same components of the radius along the direction of the polarization, and that have the same wave number $\mathbf{k}$ are considered in the average. It measures the probability that an electron in one state is behaving so much as though it were in the other state that, immediately after the transition, apart from losing the appropriate amount of energy and momentum, it is exactly as it was immediately before the transition.

To really represent the transition probability, this function has to be squared as in equation (4-24). The reason for this is that we want a function that looks something like $\psi_q^*(\mathbf{r})\psi_q(\mathbf{r})\psi_{q'}^*(\mathbf{r})\psi_{q'}(\mathbf{r})$ which is the probability that both the electron and the vacancy are at the same point in space at the same time.

The angular frequency ω of the radiation is related to the energy of the photon and hence to the energy difference between the two states:

$$\hbar\omega = E = |E_q - E_{q'}| \qquad (4\text{-}25)$$

This is a further consequence of having made the quantum assumption that we could replace $\mathbf{P}$ by $i\hbar\nabla$.

D. The dipole approximation

The exponential term in equation (4-24) can be expanded as a power series

$$\exp(i\mathbf{k} \cdot \mathbf{r}_j) = 1 + i\mathbf{k} \cdot \mathbf{r}_j + \frac{(i\mathbf{k} \cdot \mathbf{r})^2}{2!} + \frac{(i\mathbf{k} \cdot \mathbf{r})^3}{3!} + \cdots \qquad (4\text{-}26)$$

$\mathbf{r}_j$ is typically of the order of 10^{-8} cm for the valence electrons and less for the core electrons. $\mathbf{k} = 6 \times 10^6$ for x rays with a wavelength of 100 A, and it is 6×10^8 when the wavelength is 1 A. For 100-A x rays, the terms in equation (4-26), other than the first term, are small, and the bulk of the radiation can be described by omitting the exponential term in equation (4-24); this is called the *dipole approximation*.

The shorter wavelength x rays involve the inner-core electrons for which r_j is of the order of 10^{-9} cm. In this case the approximation is not very good, but the conclusions reached by making it should at least explain the dominant effects seen in the spectra.

If the second term in equation (4-26) adds a small contribution to the transition probability, this contribution is called the *quadrupole radiation*.

ω is independent of position and can be taken outside of the integral in equation (4-24). It is replaced by the radiation frequency $v = \omega/(2\pi)$. Making these assumptions and changes to equations (4-23) and (4-24) produces

$$W_{q',q}(\omega,j) = \frac{4\pi^2 e^2}{\hbar c^3} v^3 |r_{q'}{}^q|^2 \qquad (4\text{-}27)$$

and the term $r_{q'}{}^q$, which is often called the *matrix element* of the transition probability, is given by

$$r_{q'}{}^q = \int \psi_{q'}^* \mathbf{r}_j \psi_q \, d\tau \qquad (4\text{-}28)$$

Of course, the total transition probability at frequency v is given by the summation over all states q and q' that have the same energy difference E

$$I(v) \propto \sum_q \sum_{q'} \delta(E_q - E_{q'} - hv)v^3 |r_{q'}{}^q|^2 \qquad (4\text{-}29)$$

where δ is the Dirac delta.

E.   The uncertainty principle

One of the consequences of the quantum assumption is the creation of uncertainty in some of the values of the physical parameters of the system. After the assumption is made that the states of matter can be represented by a wavefunction and the Schrödinger equation is solved to find these wavefunctions, a problem occurs when determining their physical properties.

To find the wave number $\mathbf{k}$, we must sample the wave at a series of points over a distance x. The greater the distance over which the samples are taken, the more accurate the determination of the wave number. Of course the wave number so determined is only the average of the wave number, where the average is taken over the distance x. It is therefore seen that there is an uncertainty in $\mathbf{k}$ that can be decreased only by increasing the uncertainty in the position at which it really has this average value. This can be expressed by the equation

$$\delta x \, \delta k \sim 1 \qquad (4\text{-}30)$$

In a similar way, there is a balance between the knowledge of the frequency of the wave and the time at which it has that frequency. The wave must be sampled for a finite time before its frequency v and hence its energy E can be determined. The appropriate uncertainty relationship can be written as

$$\delta t \, \delta v \sim 1$$

or as

$$\delta t \, \delta E \sim \hbar \qquad (4\text{-}31)$$

F. Lifetime effects on the transition probabilities

An electron lies in an excited state before and after the emission of a photon, and there is only a short time before these states decay. The short lifetime δt of an excited state limits the time available to determine the energy of that state; therefore, there is an uncertainty in the energy: $\delta E \sim \hbar/\delta t$. When the same states on different atoms are excited, they have a mean energy E, but they may have individual energies scattered about this by an amount of the order of δE. Energy conservation is maintained by the deexcitation process which will have corresponding fluctuations in the outgoing energies, thus giving the spectral line a natural line width.

   If the decay of the excited states takes place by only one type of radiation process, the average lifetime of the states will be inversely proportional to the transition probability. Hence, the line width of the spectral line will be proportional to the transition probability.

   If several processes contribute to the decay of an excited state, the lifetime will be appropriately shortened, and this will produce a corresponding increase in the uncertainty of the energy; the line width is then directly proportional to the sum of the transition probabilities. For atoms with high atomic number, and hence a large number of electrons that may fill a vacancy in a core state, the lifetime of the excited state is small, and the radiation from any one line is spread over a wide energy range. This reduces the intensity per unit energy range at the peak intensity of the line, thus making the line spectra from these elements apparently weak compared to the background radiation.

   In reality, the energy of the radiated photons will have uncertainties δE due to the lifetime of both the initial and the final states, both of which are excited states. The probability that a photon will have an energy hv different from E, the average energy difference between the two states, is proportional to

$$\frac{\delta E/2}{(E - hv)^2 + (\delta E/2)^2} \qquad (4\text{-}32)$$

This formula has been derived from quantum electrodynamics, and the derivation is given in Messiah (1964); an intuitive derivation is given in Schiff (1955). The discussion of the broadening of spectra is taken up in more detail in Chap. 5.

The product of equations (4-29) and (4-32) gives the total probability that all pairs of states with an energy difference E will radiate with frequency v as

$$I(v) \propto \sum_q \sum_{q'} \frac{(E_q - E_{q'} - hv)v^3|r_{q'}{}^q|^2(\delta E/2)}{(\delta E/2)^2 + (hv - E)^2} \qquad (4\text{-}33)$$

If several processes compete for the decay of an excited state, besides decreasing the lifetime of the state they decrease the probability that the excited state will decay by any one process. This reduces the probability given in equation (4-33) by a fraction equal to the ratio of the transition probability given in that equation to the sum of the transition probabilities of all the competing processes. It is normally assumed that this fraction will be a constant over any small spectral region, and it is only to be considered when comparing absolute intensities of spectral features from different regions of the spectrum.

Often the corrections for line broadening can be ignored, particularly when δE is small compared to hv. Equation (4-29) is normally used unless these broadening effects are known to be large.

IV.   LINE EMISSION SPECTRA

In this section we are interested in the transitions between one core state and another. Electrons may be knocked out of a core state by any method such as electron bombardment or by irradiation with x rays. An electron can then fall from other core states into the vacant state, giving up its excess energy to a photon. This energy is normally in the x-ray region (> 20 eV). If the initial vacancy lies in the K shell, the spectrum is said to be a *K spectrum*, and so on.

A.   Matrix elements of hydrogen-like atoms

Because both states lie in the core, they are in an almost spherically symmetrical environment where the effects of the surrounding, apparently neutral atoms are minimal. As we have seen for the isolated atom, the wavefunctions may be expressed as a product of a radial function and a spherical harmonic:

$$\psi_{n,l,m} = R_{n,l}(r)\Theta_{l,m}(\theta)e^{im\varphi} \qquad (4\text{-}34)$$

There will be a slight change in the energies of the core states on going from atoms to solids, but these changes will be small and will consist of a general shift of all the energy levels up or down by nearly the same amount. This movement of the energy of the core states enables us to obtain meaningful data for chemical analysis from the results of electron spectroscopy. This aspect is discussed further in Chap. 9.

Substituting these wavefunctions into equation (4-28) produces

$$Z_{n',l',m'}^{n,l,m} = \int_0^\infty R_{n,l}(r)R_{n',l'}(r)r^3 \, dr \times \int_0^\pi \Theta_{l,m}(\theta)\Theta_{l',m'}(\theta) \sin\theta \cos\theta \, d\theta$$

$$\times \frac{1}{2\pi}\int_0^{2\pi} e^{i(m-m')\varphi} \, d\varphi \tag{4-35}$$

and similar terms for the components of the radiation that are polarized in the x and the y directions.

B.   Selection rules

In equation (4-35) the integral over φ vanishes unless

$$\Delta m = m - m' = 0 \tag{4-36}$$

in which case it equals 1.

The integral over θ can be evaluated by replacing $\Theta_{l,m}(\theta) \cos\theta$ by the recurrence relationship (4-12) and then using the orthonormality relationship (4-11). The integral is zero unless

$$\Delta l = l' - l = \pm 1 \tag{4-37}$$

in which case it equals 1.

The values of $Z_{n',l',m'}^{n,l,m}$ are given by

$$Z_{n',l-1,m}^{n,l,m} = \frac{l^2 - m^2}{(2l+1)(2l-1)} R_{n',l-1}^{n,l}$$

$$Z_{n',l+1,m}^{n,l,m} = \frac{(l+1)^2 - m^2}{(2l+3)(2l+1)} R_{n',l+1}^{n,l} \tag{4-38}$$

$$Z_{n',l',m'}^{n,l,m} = 0 \text{ for all other } l' \text{ and } m'$$

Here

$$R_{n',l'}^{n,l} = \int_0^\infty R_{n,l}(r)R_{n',l'}(r)r^3 \, dr \tag{4-39}$$

Repeating the above analysis for the x and the y components of polarization of the radiation produces a similar set of equations, provided that

$$\Delta m = \pm 1 \tag{4-40}$$

and

$$\Delta l = \pm 1 \tag{4-41}$$

The analysis is not quite so simple in this case, and the details are given by Bethe and Salpeter (1957).

Summing up, for any polarization only the following transitions are allowed:

$$\Delta l = \pm 1 \qquad (4\text{-}42)$$

and

$$\Delta m = 0 \text{ or } \pm 1 \qquad (4\text{-}43)$$

These conditions are called the *selection rules*.

C. Transition probabilities of hydrogen-like atoms

Because the energy dependence of states on m and s is normally negligible, transitions between all states n',l' and all states n,l will create photons having the same energy. The contribution to the spectrum by each state is found by adding the contributions from all the polarizations x, y, and z. Then by summing over all the states that can contribute to the same line, as required by equation (4-29), the total intensity of that line can be found. The summation over s just multiplies every state by 2 in the approximation that we have been using. The summation over m gives

$$\sum |r^{n,l}_{n',l-1}|^2 = \frac{l}{2l-1}(R^{n,l}_{n',l-1})^2 \qquad (4\text{-}44)$$

and

$$\sum |r^{n,l}_{n',l+1}|^2 = \frac{l+1}{2l+3}(R^{n,l}_{n',l+1})^2 \qquad (4\text{-}45)$$

where l is the azimuthal quantum number of the final state.

These two relations have resulted from the assumption that the potential experienced by the core electrons was spherically symmetrical. This proves to be a valid assumption in practice. However, if it is not considered to be a valid assumption, the wavefunction should be expressed as the sum of spherical harmonics. The coefficients of the spherical harmonic to which the real function is a good approximation will have a value which is large compared to all the other coefficients. If the sum of spherical harmonics is substituted into equation (4-28), there will be one pair of equations like (4-44) and (4-45) for every value of l in the summation. However, since the transition probability depends on the square of these terms, the size of these other contributions will be exceedingly small.

D. Energy levels

The energy of the core states given in equation (4-13) was derived assuming that the potential was Coulombic, that is, that it varies as $1/r$. This assumption is accurate only for an isolated hydrogen atom, but is nevertheless found to give a rough guide to the relative spacing of the energy levels of core states within any atom. For instance, it shows the variation with the square of the atomic number that was first determined experimentally; however, it is not sufficient to give an accurate estimation of the energy of a state. To make these calculations, more sophisticated approximations must be made; for example, see Hermann and Skillman (1963).

A clear, detailed description of the use of atomic wavefunctions for determining the characteristics of the appropriate spectra is given by Bethe and Salpeter (1957). A more limited description of the origin of the selection rules is given by Schiff (1955). Chapter 2 of this volume gives a further discussion of the measurement of the atomic energy levels.

V. VALENCE BANDS IN SOLIDS

In order to discuss the theory of solids, we should look a little more closely at the behavior of the electrons in atoms.

A. Standing waves

Classically, electrons occupy only regions for which $E > V$. They slow down as V approaches E and when the two are equal, they are reflected so that they can accelerate away from the potential wall. Quantum mechanically, the behavior of the electrons is given by equations like (4-6), whose solution is a wave which, in general, represents a moving electron.

In the normal state, when no current is flowing, there are as many electrons (and hence waves) flowing in one direction as in the other. The distribution of electron density (square of the wave amplitude) is not changing with time, so that the pairs of waves moving in opposite directions to one another must form standing waves. These standing waves must have an integral number n of antinodes between the points where reflection occurs; n is the principal quantum number.

We should impose a further artificial restriction here. As in Section II, we should concern ourselves with only the positive antinodes and count only those solutions that have as many negative antinodes; that is, the number of antinodes must be even.[1]

Because the boundaries are not perfect reflectors, the waves are not reflected exactly at the boundaries but will be subject to boundary or end corrections similar

---

[1] The reason for this assumption is associated with the spin states and the way we intend to include them in the theory. If we do not do the theory correctly, that is, with the full relativistic treatment, we must expect that the results will occasionally be a little nonphysical. This assumption just prevents some of the nonphysical character affecting the conclusions.

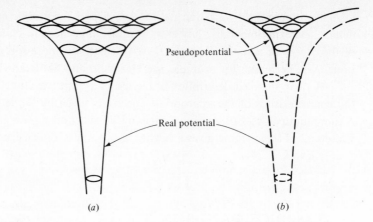

(a)   (b)

FIGURE 4-2
(a) Standing waves in a real potential (schematic) illustrating the formation of quantized states; (b) standing waves in a pseudopotential obtained by assuming that the electrons in the lowest two states of (a) screen the real potential well.

to those required by sound waves in an organ pipe. Therefore, the effective boundaries are a little distance away from the classical boundaries. Usually, the effective boundaries are some small fraction of a wavelength outside the classical boundary, and the corrections are trivial.

B. Energy levels in atoms

In the model of an isolated atom, the potential extends to infinity at $E = 0$, and it is possible to fit an infinite number of positive antinodes at this energy. Other states with lower energy have finite numbers of nodes spread across a finite region of space. As the number of antinodes increases, the rate of change of phase must increase to fit the discrete number of extra waves between the boundaries. This requires discrete steps in the energies of states with successive numbers of positive antinodes or quantum numbers.

For the one-dimensional case, the creation of these successive states is shown pictorially in Fig. 4-2a. This picture does not possess a singularity in the potential at the center of the atom like that at the center of the Coulomb potential; this singularity is not a physical reality,[1] because the electrons are repelled from the center of the nucleus by making physical contact with the nucleons.

From this model, it can be seen that electrons with lower energy and hence a

_____

[1] No singularities can exist in physical reality; they appear as mathematical conveniences or inconveniences, depending on one's point of view.

smaller quantum number n lie, on the average, closer to the nucleus than those with larger quantum numbers.   The electrons with the same value of the principal quantum number n lie at comparable distances from the nucleus; this cannot be seen from the one-dimensional model but can be shown by plotting the wavefunctions given in Section II.

C.   Pseudopotentials

In Section II, it was stated that we were assuming that the electron was independent of the momentary position of any other electron and that it experienced only the time-average effect of the other electrons. We then used the Coulomb potential as a model potential, but this potential does not include the effect of the repulsion of the other electrons. This time-average repulsion of the inner levels can be included by assuming the Pauli exclusion principle; that is, no more than one electron may fill any one state. Remembering that electrons try to fill the states of lowest energy first, we see that the effect of this repulsion by the inner levels is to force some of the electrons to fill states with higher energy than that of the ground state and, consequently, to be not so tightly bound to the atom.

There is an alternative way of allowing for the repulsion of the core electrons. Because they lie largely inside the territory of the electron under consideration and because the total charge distribution of a full set of states with the same value of n is spherically symmetrical, they can be considered to have reduced the atomic number Z by the number of electrons in the inner shells. Substituting this effective potential or pseudopotential into the radial equation and pretending that the first valence electron has an energy appropriate to a state with n equal to 1 should predict the correct energy of this state. This model is depicted in Fig. 4-2b.

The latter method generates wavefunctions that are not realistic. Once the outer electrons are in the vicinity of the inner electrons, the pseudopotential is no longer exact, and the wavefunctions are not changing phase rapidly enough. Thus the pseudo-wavefunctions are too smooth in this region. The pseudo-wavefunctions are adequate for describing the character of the spectra, but if careful and detailed calculations are required, more accurate wavefunctions will be needed.

D.   Energy levels in solids

We must now apply the above concepts to a solid. A solid consists of an array of ions in which each contributes its effective potential to the total potential. When all the contributions are added, the average potential is reduced and smoothed out as shown in Fig. 4-3a. The potential not near the surface of the solid can be approximated by a smoother potential as shown in Fig. 4-3b.

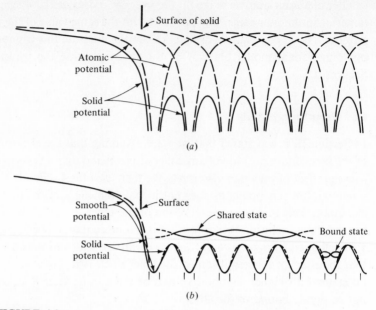

FIGURE 4-3
(a) How the potentials for atoms add together to produce the potential for a solid; (b) how, except near the centers of the atoms, the potential for a solid can be well represented by a smooth, almost sinusoidal potential.

At the surface of the solid, the potential may increase very rapidly, but within an atomic spacing or two outside the surface it will increase approximately as the inverse of the distance from the last atom, much as in the atomic case. This forms a potential wall around the solid, and the electron energy must exceed $E = 0$ in order to escape from the solid.

Provided that the required number of antinodes can be created both in the small potential wells of each atom and with an energy less than V_a, the valence electrons are attached more or less separately to each individual atom. If all the valence electrons can be accommodated in this way, the potential at any one site differs from that of the isolated atom only because of the effect of the polarization of the neighboring neutral atoms. This is the origin of the van der Waals forces which explain the behavior of those rare gases for which the pseudopotentials are moderately large. In certain solids, it is possible for some of the valence electrons to bond in this way while others are shared between all the atoms; the electrons held at individual atomic sites are called *bound electrons*.

If the potential wells are not large enough to form bound states, the electrons will try to form states with each ion contributing to the phase change. The bound state with one positive antinode in the vicinity of one atom would require sufficient energy to give a phase change of 2π in the neighborhood of that one atom. But the shared electron requires a phase change of only $2\pi/N$ in each atomic cell, since every cell contributes to the total phase change of 2π; N is the number of atoms across the solid. Therefore, the shared electron has a lower energy than the bound electron, and this contributes to the binding of the solid. However, there are now N electrons to be shared, and the wavefunction for each must have one more positive antinode than the one before it. Consequently, each will require a slightly greater energy than the one before it, and therefore it will contribute less to the binding energy of the solid. Some electrons may have a higher energy than in the atomic state and thereby oppose binding; they are then called *antibonding electrons*. The highest energy normally possessed by an electron (at the absolute zero of temperature) is called the *Fermi energy*, and states above this energy are not normally occupied. Provided that the total of the contributions to the binding energy is favorable, the solid state will be formed. The state of lowest energy will have an energy close to V_0.

Between the potential wells the waves will cease to change phase, but will still have some amplitude. This tunneling of the electrons implies that there is some probability that the electron lies in the region that would have been forbidden classically. Crudely speaking, the end corrections mentioned earlier have allowed the electrons to penetrate the barrier sufficiently to link up in the middle of it.

E.   Density of states

The spacing of successive energy levels for the shared electrons is much smaller than that for the bound states. As the number N approaches infinity, the energy spacing becomes infinitely small, and we have an almost continuous band of energies. In practice, N is sufficiently large to enable us to assume that the band is continuous; theories based on this assumption are called *band theories*.

The density of states per unit energy range per unit volume is finite and inversely proportional to the spacing of the states. It changes regularly for most energies above V_a. Between V_a and V_0 the states become relatively more spread out, and their energies are difficult to calculate.

F.   Free-electron model

If we make the assumption that the potential $\bar{V}$ is completely flat even at the surface of the solid, we can solve the Schrödinger equation.[1] This model is the basis of the free-electron model of the electronic structure.

---

[1] There are only a few potentials for which we can solve the Schrödinger equation exactly. They are all highly symmetrical potentials such as the spherically symmetrical potential, which has full rotational symmetry, and the flat potential, which has full translational symmetry.

The Schrödinger equation (4-2) can again be separated into three independent equations. In this case the three equations are all similar:

$$\frac{\partial^2 \psi_x}{\partial x^2} - \frac{2\mu}{\hbar^2} E_x \psi_x = 0 \qquad (4\text{-}46)$$

and two equations where x is replaced by y and z.

$$E = E_x + E_y + E_z \qquad (4\text{-}47)$$

This equation has a solution

$$\psi_x \propto e^{i\mathbf{k}_x \cdot \mathbf{x}} \qquad (4\text{-}48)$$

where

$$\mathbf{k}_x = \pm \sqrt{\frac{2\mu E_x}{\hbar^2}} \qquad (4\text{-}49)$$

The vector $\mathbf{k} = (\mathbf{k}_x, \mathbf{k}_y, \mathbf{k}_z)$ is the wave vector, and its magnitude is the wave number

$$k = \pm \frac{2\pi}{\lambda} \qquad (4\text{-}50)$$

G.   Free-electron density of states

In the one-dimensional case, we have seen that the boundary conditions require that there be an integral number of wavelengths across the solid:

$$n\lambda = L = Na \qquad (4\text{-}51)$$

where a is the atomic spacing and L is the linear dimension of the solid. Thus,

$$k_x = \frac{2\pi n}{aN} \qquad (4\text{-}52)$$

The value of $n = N$ corresponds to the state that has the highest energy when one electron from each atom has been allocated to the band, and this condition is always represented by the constant value of $k = 2\pi/a$.[1] This state always has the same energy no matter what the value of N. If N is increased, the states are more closely spaced to allow for the increased number, but the energy spread required to house one electron per atom is unchanged; that is, the bandwidth is independent of the size of the solid. The free-electron model requires that both N and n be infinite in size, but the ratio of n to N remains finite and proportional to k, where k is still well defined by equation (4-49), but it is then a continuous function.

If the wave vector $\mathbf{k} = (\mathbf{k}_x, \mathbf{k}_y, \mathbf{k}_z)$ is plotted in a three-dimensional space, the space is called $\mathbf{k}$ space. The points of the vectors in $\mathbf{k}$ space are positioned according

---

[1] When spin is also considered, this value of k corresponds to two electrons per atom.

to (4-52). This means that given any volume of $\mathbf{k}$ space, it will always contain the same density of points; the density of states in $\mathbf{k}$ space is constant.

The energy of any state is given by

$$E = \overline{V} + \frac{\hbar^2}{2\mu}(\mathbf{k}_x{}^2 + \mathbf{k}_y{}^2 + \mathbf{k}_z{}^2)$$

$$= \overline{V} + \frac{\hbar^2}{2\mu}|\mathbf{k}|^2 \qquad (4\text{-}53)$$

All states with the same energy lie on the sphere defined in $\mathbf{k}$ space by equation (4-53), with E a constant equal to the energy being considered.

The number of states with energy between E and $E + dE$ is the number of states lying between the two corresponding spheres. The thickness of the shell is the modulus of the vector $d\mathbf{k}$ found by taking the differential of equation (4-53). Because $\mathbf{k}$ and $d\mathbf{k}$ are both radial vectors,

$$d\mathbf{k} = \frac{\mu}{\hbar^2}\frac{dE}{\mathbf{k}} \qquad (4\text{-}54)$$

The volume of the spherical shell is then

$$\frac{4\pi\mu}{\hbar^2}k\,dE$$

so that the number of states per unit volume per unit energy range is given by

$$N(E) = 4\pi\left(\frac{2\mu}{\hbar^2}\right)^{3/2}E^{1/2} \qquad (4\text{-}55)$$

The extra factor of 2 allows for the spin. The density of occupied states also rises as $E^{1/2}$, but it falls to zero for all states above the Fermi energy.

VI. BAND SPECTRA FROM FREE ELECTRONS

In Section III we were interested in core-core transitions, whereas in this section we are interested in transitions of electrons from the valence states to core states. The valence states are represented by the free-electron wavefunctions. This is a reasonable first-order approximation for metals because their pseudopotentials are small.

A. Free-electron band spectra

If a vacancy is created in a core state, the excited state produced has a well-defined energy, provided, of course, that we assume that its lifetime is very long. If an electron from the valence band now falls into this vacancy, it will radiate an x-ray photon with a frequency that depends on the energy difference between the two states. Valence electrons from other parts of the band will create photons with different frequencies. If enough electrons make similar transitions to the one type of core state, the radiation will appear across a whole band of frequencies, and the spectrum is called a *band spectrum*. The width of the band, in terms of the range of energies of the photons, equals the width of the valence band.

We have seen that any wavefunction can be expressed as the sum of spherical harmonics. If, in order to evaluate the transition probability, the valence-state wavefunction expressed as such a sum is substituted into equation (4-28), the integral can be split into a sum of integrals. Each integral in the sum will involve the wavefunction of the core state and one of the series of spherical harmonics from the valence-state wavefunction. Only the integrals involving the states whose azimuthal quantum number *l* differs by 1 from that of the core state will be nonzero. If the core state is an *s* state, only the *p*-like electrons can contribute to the transition probability; but if the core state is a *p* state, both *s*-like and *d*-like electrons can make the transition.

For the free-electron model the valence-state wavefunction is proportional to $\exp{(i\mathbf{k} \cdot \mathbf{x})}$. When this is expanded as a series of spherical harmonics, only those harmonics with $m = 0$ have nonzero coefficients, and they can be added together to give

$$e^{i\mathbf{k} \cdot \mathbf{x}} = \sum_l (2l + 1)^{1/2} i^l j_l(|\mathbf{k}|r) Y_l^0(\theta, \varphi) \qquad (4\text{-}56)$$

The terms $j_l(|\mathbf{k}|r)$ are spherical Bessel functions.

B.   The symmetry of the wavefunctions

We can anticipate the form of these functions by considering the symmetry of the wavefunctions. When a node is formed in three dimensions, it is really a two-dimensional surface that separates two three-dimensional volumes in which the wavefunctions have opposite signs. In the solid, these surfaces can be envisaged as planes, although their exact shape will depend on the exact shape of the solid. For the following discussion it is adequate to imagine that Fig. 4-3 is a cross section along some line of symmetry in the crystal.

The state with $\mathbf{k} = (0,0,0)$ has a wavefunction that has constant amplitude at all points in real space. Therefore, because the sign is the same in all directions away from any one atom, the wavefunction expressed as a sum of spherical harmonics centered on that atom will contain only those terms corresponding to states with $l = 0$. The wavefunction is said to be *s-like*.

As planes of nodes are introduced with larger $\mathbf{k}$, there is an increased probability that atoms will find themselves with a region of positive wavefunction on one side and a negative region on the other side. If the wavefunction were expanded as a sum of a series of spherical harmonics about these atoms, the harmonics with $l = 1$ would predominate. Other atoms would exist for which this wavefunction would appear to be almost constant on opposite sides. The expansion around these atoms would have the largest contributions from the harmonic with $l = 0$, but the lack of complete symmetry of the wavefunction would be due to a small contribution from states with $l = 1$. On the average, states with this energy would have largely the character of s-like states with some p-like character. The percent of p-like character would increase with energy, up to the point where the function is as much p-like as s-like.

When the number of nodal planes has increased even more, the junctions of two or more planes will make the symmetry of the wavefunctions at these junctions much more complex. Atoms near these points in space have d-like symmetries and sometimes contributions from even higher harmonics. The introduction of these higher harmonics is not simple.

When the number of nodes along any line of a crystal equals the number of planes of atoms along that line, the wavelength along that line equals twice the atomic spacing. This is the case for which $k = \pi/a$ or $n/N = 1$. If one of the nodal planes falls on a plane of atoms, every plane of atoms will correspond to a nodal plane; in this case the wavefunction is entirely p-like. On the other hand, if one atom lies between two nodal planes, all other atoms are similarly positioned, and the wavefunction is entirely s-like at all atomic sites. Therefore, states with this **k** vector may sometimes appear to be entirely p-like or entirely s-like. If the state does not have such a **k** vector, it will appear more like a mixture of both s and p states.

C. Transition probabilities

In the calculation of transition probabilities, the extra term **r** in equation (4-28) shows that the electrons that can contribute most are those that have large radii and yet still overlap the core. In solids, for which the atoms are packed so that the ionic cores just touch one another, the largest contributions come from electrons that lie in the region where $r \approx a/2$ and a is the average atomic spacing.

If it is assumed that the radial variation of the wavefunction is small for this region, $R_{n,l}(r)$ in equation (4-39) can be replaced by $R_{k,l}(a/2)$. The quantum number n is a measure of the number of nodes across the potential and is therefore replaced by **k**. This radial term can then be removed from the integral in equation (4-39), leaving the remaining integral independent of **k**. Examination of equation (4-56) shows that the radial term for free electrons is just the appropriate spherical Bessel function, $j_l(ka/2)$.

To calculate the transition probabilities, equation (4-29) should be used. The core state is assumed to consist of one single state so that the summation over q becomes trivial. The summation over q' is really the summation over the three quantum numbers k, l, and m, but the summation over the three polarizations still turns out to be given by equations (4-44) and (4-45). The summation over l' is reduced to the summation of the two terms $l + 1$ and $l - 1$ given by these equations. The summations are limited by the Dirac delta function to the summation over all states $\mathbf{k}$ that have the same energy which differs from that of the core state by $h\nu$. This limits the summation to all states for which $|\mathbf{k}|$ is a constant.

Because $j_{l'}(ka/2)$ depends only on $|\mathbf{k}|$, the matrix elements $|r_q^{q'}|^2$ can be removed from the summation in equation (4-29). The summation then becomes simply the number of states at the energy $E = \hbar^2 k^2/(2\mu)$; this is the density of states. The total transition probability then becomes

$$I(\nu) \propto \nu^3 N(E) \sum_{l'} f_{l'} j_{l'}^2 \, (ka/2) \qquad (4\text{-}57)$$

where $f_{l'}$ is the fraction from either equation (4-44) or (4-45) and the summation is over only the two states $l + 1$ and $l - 1$. The energy dependence of the spectra can be determined by noting that $N(E) \propto E^{1/2}$ and by noting the shape of the spherical Bessel functions; the first three of these are plotted in Fig. 4-4a.

D.   Band shapes

The square of the Bessel function for s-like states is almost constant near $k = 0$. This reflects the fact that, with few nodal planes, only a very small proportion of the wavefunction will ever appear to be p-like to any atom. This means that the density of s-like states near the bottom of the band will rise like $E^{1/2}$, this being the product of the density of states and the percent of states at each energy that are s-like. Toward higher energies, the density of s-like states decreases, and as the energy is increased even further, it oscillates through successive maxima and minima.

The corresponding curve for the p-like states rises proportionally to k^2 near $k = 0$. This occurs because the number of atoms having a nodal plane superimposed on them is directly proportional to the number of nodal planes and hence to k. Therefore, the density of p-like states rises like $E^{3/2}$. Toward higher energies it also oscillates through successive maxima and minima, but it reaches its first peak at the same energy as the density of s-like states reaches its first minima, that is, where there

FIGURE 4-4
(a) Plot in $\mathbf{k}$ space of the first three spherical Bessel functions representing the contributions to the transition probability of s-, p-, and d-like electrons; this is only approximately valid near $k = 0$. (b) The sum (crosses) and difference (dashes) of the first three spherical Bessel functions centered on $k = 0$ and these three functions centered on the lattice vector $\mathbf{k} = \mathbf{g}$; these combinations approximately represent the contributions to the transition probability of the s-, p-, and d-like electrons near $\mathbf{k} = \mathbf{g}/2$.

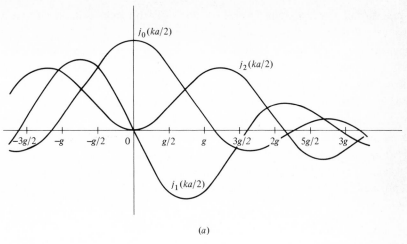

(a)

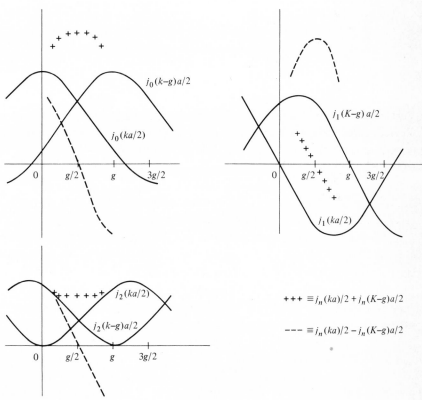

$+++ \equiv j_n(ka)/2 + j_n(K-g)a/2$

$--- \equiv j_n(ka)/2 - j_n(K-g)a/2$

(b)

are sufficient nodal planes for each atom to lie on one. This occurs at the energy given by $E = \hbar^2(\pi/a)^2/(2\mu) = 50.05/a^2$ (eV).

The spherical Bessel functions for the d-like states increase with the probability of two nodal planes intersecting on or near an atom. This increases as the square of the number of nodal planes and hence as the square of k. This causes the density of d-like states to rise as $E^{5/2}$ and severely limits the number of d-like electrons at low energies. The discussion of the behavior of the d-like states at higher energy is normally more complicated than can be described within the framework of the free-electron theory, and so it is left until the next section.

The percent of states having higher symmetries increases even more slowly and is normally negligibly small for energies below the Fermi energy. They will not be discussed in this chapter.

Because there is a finite number of electrons there is a maximum energy—the Fermi energy—above which the states are not normally filled. It causes the sharp cutoff, known as the *Fermi edge*, at the high-energy end of the band. The energy can be calculated by integrating the density of states from the bottom of the band until the integral equals the number of available states. The Fermi energy is found to be

$$E_F = \overline{V} + \frac{50.05}{a^2} \frac{Z}{2} \qquad \text{eV} \qquad (4\text{-}58a)$$

where Z is the valence of the atoms, $\overline{V}$ is the average energy of the bottom of the band, and the energies are given in electronvolts. The density of states, density of s-like states, and density of p-like states are shown in Fig. 4-5a.

E. Plasmon satellites

In Section I of this chapter it was assumed that the electrons could move independently of atoms and other electrons. The interactions between the electrons will be discussed in the next chapter, but some discussion of these effects is essential in this section. A useful introductory discussion of these interactions is given by Pines (1955).

The electrons can be set into a collective oscillation which is called a *plasmon*. These plasmons have an energy of formation which, in first-order theory, depends only on the density of the electrons, which is in turn related to a and Z.

$$E_p = \hbar\omega_p = \frac{47.11}{a} Z^{1/2} \qquad \text{eV} \qquad (4\text{-}58b)$$

Normally, electrons do not have enough energy to excite these modes of oscillation, but when an x ray is created, the photon can come away from the solid leaving the valence electrons excited in this way. The photons produced have their energies reduced by $\hbar\omega_p$ and form a satellite spectrum at an energy $\hbar\omega_p$ below the parent

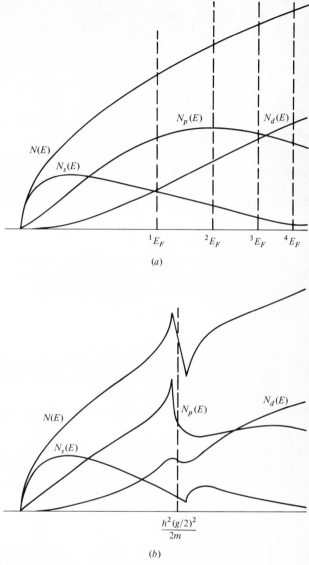

^1E_F ^2E_F ^3E_F ^4E_F

(a)

$$\frac{h^2(g/2)^2}{2m}$$

(b)

FIGURE 4-5
(a) Density of states for free electrons, showing the contributions from the s-, p-, and d-like electrons; (b) pair of discontinuities that appear in the density of states from a critical point that is the point of closest approach of a Brillouin-zone boundary to the origin.

spectrum. The shape of these plasmon satellites is not identical to the parent spectrum because the energy of the plasmons is not exactly $\hbar\omega_p$. These satellites were first observed in the $L_{II,III}$ spectra from Na, Mg, and Al (Rooke, 1968) and have subsequently been observed in the $L_{II,III}$ spectrum of Si (Fomichev[1]) and the K spectrum of Be (Watson, Dimond, and Fabian, 1968).

F.   Auger broadening

Another effect of the interactions between electrons is called the *Auger effect*. When an electron has made a transition from the valence band, it leaves a vacancy or hole in its original state; that is, it leaves the valence band excited. This excited state has a finite lifetime in the same way as a vacancy in the core state. Therefore, it contributes to the lifetime broadening.

Holes can be filled by any electron whose energy is greater than that of the hole. This transition could radiate, but it is more probable that it is a nonradiative Auger transition. The electron that makes the transition to the hole collides or interacts with a second electron. Provided that the second electron can accept the energy that the first requires to emit, the process is possible. The second electron can accept the energy if this energy raises it to a level above the Fermi energy where there are empty states it can fill. Otherwise the Pauli exclusion principle prevents the transition.

Holes near the bottom of the band have more electrons to fill them, and many of these electrons create large excesses of energy to be dissipated; therefore there is a large percentage of the other electrons that could be excited to states above the Fermi energy. Holes near the top of the band have only a few electrons to fill them, and these electrons have only a small amount of energy with which to excite the other electrons to states above the Fermi energy. Therefore, holes near the bottom of the band have lifetimes that are considerably shorter than those for holes near the top.

The band spectrum is considerably broadened at the low-energy, but not so severely broadened at the high-energy end. This explains the large low-energy tails seen on the spectra of many metals. For insulators, the electrons have to be excited to energies above the band gap (see below), and therefore fewer Auger transitions are possible. The spectra from insulators show much smaller low-energy tails.

If the lifetime of the holes decreases rapidly as we move from the Fermi energy toward lower energy, states near the Fermi energy, which are not greatly broadened, will create a peak in the spectrum at this edge. This possibility has been discussed by Pirenne and Longe (1964).

To predict spectra that include the effects of the lifetime broadening, equation

---

[1] V. A. Fomichev, University of Leningrad, unpublished. I had noticed the same small features on spectra taken by R. S. Crisp, University of Western Australia, but these, like Fomichev's, were sufficiently small to make me hesitate to publish them.

(4-33) should be used. When summations that are equivalent to those used to derive equation (4-57) have been made, the equation becomes

$$I(v) \propto v^3 \int_{E_0}^{E_F} \frac{N(E) \; \delta E/2 \sum_l f_l j_l^2(E, a/2) \; dE}{(\delta E/2)^2 + (hv - E)^2} \tag{4-59}$$

An elementary discussion of the derivation and application of the lifetime has been given by Blokhin and Sachenko (1960). More recent descriptions of these effects have used many orders of perturbation theory and are described in the next chapter.

G. Hole in the core

Recent work by Mahan (1967) and by Mizuno and Ichikawa (1968) has shown the importance of the effect of the valence states of the hole that exists in the core state. This hole causes a redistribution of the charge which is reflected in the renormalization of the wavefunction. Effectively, this renormalization increases the probability that s-like electrons near the Fermi surface will make the transition and decreases the probability that the corresponding p-like electrons will make it. This produces a peak at the Fermi edge in L spectra and a decrease toward the edge of K spectra. This effect is probably larger than any effect resulting from the interaction between valence electrons alone.

H. Thermal effects

The final many-body interaction is that between the valence electrons and phonons. Electrons can be excited to states above the Fermi energy by absorbing phonons. Because the energy of the phonons is only a small fraction of an electronvolt, however, only those electrons that are very close to the Fermi surface are affected. The intensity should be multiplied by the term

$$\frac{1}{\exp\left[(E - E_F)/kT\right] + 1} \tag{4-60}$$

which is derived from Fermi-Dirac statistics [for a simple derivation see Dekker (1962)]; here, k is Boltzmann's constant and T is the absolute temperature. This function causes a slight decrease in the density of states on the low-energy side of the Fermi edge and creates a small tail of states on the high-energy side. This thermal broadening of the Fermi edge has been verified experimentally by Skinner (1940).

I. Experimental spectra

In the preceding sections, the general properties of the band spectra have been predicted from the free-electron theory. These results are now compared against the experimental spectra from the simple metals. The spectra from the heavier metals

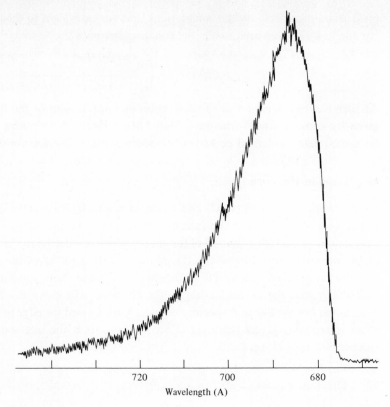

FIGURE 4-6
Lithium *K* emission spectrum. (*After Crisp and Williams*, 1960.)

and from nonmetals are not free-electron-like, and a comparison to the spectra from these elements is deferred to the next section. The comparisons made here are restricted to the parent bands because the plasmon satellites will be described in detail in the next chapter.

The lithium *K* emission spectrum has been recorded by Crisp and Williams (1960) and is shown in Fig. 4-6. It has the low-energy tail due to the Auger broadening, and it rises from the bottom of the band as $E^{3/2}$ like the density of *p*-like states. At the Fermi edge the intensity falls to about 0.75 of its peak height. This fall in intensity is most probably due to the effects of the hole in the core state. In general, the shape is well explained by the free-electron theory.

The beryllium *K* emission spectrum has been measured by several investigators, and a comparison with theory has been given by Sagawa (1968). The density of states is far from free-electron-like, and, in general, the shape of the spectrum reflects the shape of the density of states. The bottom of the spectrum rises from the Auger tail more like the $E^{3/2}$ of the density of p-like states than like the $E^{1/2}$ of the total density of states; this is as expected. Near the top of the band the intensity falls sharply. This appears to follow the density of states, but it is probably partly due to the effect of the hole in the core.

The sodium $L_{II,III}$ emission spectrum (Skinner, 1940) shows a parabolic form rising from the predicted low-energy Auger tail. A small peak near the Fermi edge could be due either to electron-electron interactions or to the perturbation caused by the hole in the core; it is felt that it is mostly due to the latter. The K spectrum has been reported by Skinner (1940) to look rather like the lithium K spectrum, as we would expect.

The free-electron $L_{II,III}$ emission spectrum of magnesium, as predicted by the theory, rises from the low-energy tail like $E^{1/2}$ but reaches a maximum before the Fermi edge. Since there are two electrons per atom, it is expected that the spectrum would fall to a minimum somewhere near the Fermi edge following the free-electron density of s states; the d states are not important for magnesium. In general, this behavior is observed in the measured spectrum (Watson, Dimond, and Fabian, 1968). There is a high-energy peak which may be partly due to the hole in the core state but which is mostly due to band-structure effects which will be explained in Section VII; other fine structure due to this same effect is also observed.

The K emission spectrum of magnesium has been reported by Baun and Fischer (1965), and it shows the expected $E^{1/2}$ dependence and the low-energy tail. The intensity increases slowly toward the Fermi edge, but there is insufficient resolution to distinguish any fine structure.

The aluminum $L_{II,III}$ emission spectrum has been thoroughly discussed by Rooke (1968); see Fig. 4-7. It shows the $E^{1/2}$ rise from the Auger tail, and then it falls to a minimum at an energy approximately given by

$$E = \frac{\hbar^2}{2\mu}\left(\frac{\pi}{a}\right)^2 \qquad (4\text{-}61)$$

in the same way as the magnesium spectrum. Because the bandwidth for aluminum is greater than for magnesium, however, the region of increasing intensity near the Fermi edge is much broader and larger than for the magnesium spectrum. Some fine structure is again seen, and this will be described in the next section.

The aluminum K spectrum, which was measured by Fischer and Baun (1965), rises to a maximum at an energy roughly corresponding to the minimum in the $L_{II,III}$ spectrum, and this is rounded off to the Fermi edge by the resolution. Because of this lack of resolution in the hard x-ray region, no fine structure is seen in this spectrum. However, if the theoretical spectrum is given the same broadening as the experimental spectrum, their shapes are very similar.

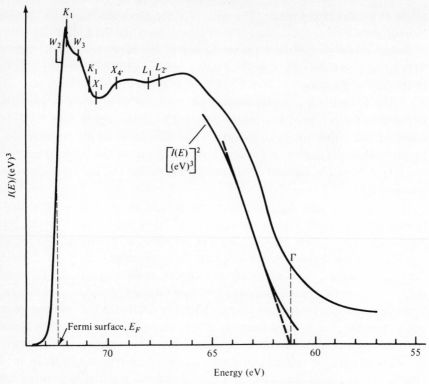

FIGURE 4-7
Aluminum $L_{II,III}$ emission spectrum. (*After Rooke*, 1968.)

Only two spectra have been obtained from liquid metals; Catterall and Trotter (1958) have reported the $L_{II,III}$ emission spectrum from liquid aluminum, and Fischer and Baun (1965) have reported the K emission spectrum from liquid aluminum. Both spectra were remarkable because they were so similar to the respective spectra from the solid. It had been suspected that aluminum would be very much free-electron-like, and these spectra were thought to give evidence to the contrary. However, it is now realized that the pure-metal spectrum is very nearly free-electron-like itself and that no contradictions may occur. Rooke (unpublished) has calculated the shape of the free-electron spectrum for a trivalent metal with the appropriate lattice constant and has shown that the predicted shape is slightly closer to that obtained from the liquid metal than that obtained from the solid. However, it was necessary to assume that the small peak near the Fermi edge had the same origin as the small peak near the Fermi edge in the sodium $L_{II,III}$ spectrum. If this is so, a significant portion of the high-energy peak in the solid spectrum is due to the hole in the core;

this may well be true as any calculations have not been sufficiently accurate to make significant tests.

The $L_{II,III}$ and the $M_{IV,V}$ emission spectra of potassium have been recorded by Kingston (1951) and by Crisp (1960), respectively. They both show the $E^{1/2}$ shape, but the high-energy peak seen in sodium is not observed. Other metals are not free-electron-like, largely due to the presence of d-like states. They will be discussed briefly in the next section.

VII. BAND SPECTRA FROM NEARLY FREE ELECTRONS

The assumption that the potential V is equal to a constant V, which is the basis of the free-electron model, is rather crude. We have seen earlier that the effective potentials are small but not normally negligible and that, in a more realistic model of a solid, the potential has the periodicity of the crystal lattice.

In this section, we consider the pseudopotential method, which, together with other related methods, has been most successful in predicting a wide range of properties for a large number of elements. This method is becoming increasingly important in making realistic calculations of the properties of materials.

A. Fourier transforms of the potential

The infinite set of all plane waves forms a complete orthonormal set just like the set of all spherical harmonics. Therefore, any function can be expanded as a sum of plane waves:

$$f(\mathbf{r}) = \sum_{\mathbf{q}} a(\mathbf{q})e^{i\mathbf{q}\cdot\mathbf{r}} \qquad (4\text{-}62)$$

where $\mathbf{q}$ is a vector in $\mathbf{k}$ space.

The set of coefficients $a(\mathbf{q})$ can be plotted as a function of $\mathbf{q}$ in $\mathbf{k}$ space, and this function is called the *Fourier transform of the function $f(\mathbf{r})$*. These transforms can be calculated from the function by the formula

$$a(\mathbf{q}) = \Omega^{-1} \int f(\mathbf{r})e^{-i\mathbf{q}\cdot\mathbf{r}} \, d\tau \qquad (4\text{-}63)$$

where Ω is the volume of the solid.

Except for certain minor modifications, the potential of a crystal is the sum of the potentials of the individual atoms:

$$W(\mathbf{r}) = \sum_{j} w(\mathbf{r} - \mathbf{r}_j) \qquad (4\text{-}64)$$

where r_j is the position of the jth atom. We will write the Fourier transform of this crystal potential as $\langle k + q|W|k\rangle$.[1] Then

$$\langle k + q|W|k\rangle = \Omega^{-1} \int \sum_j w(\mathbf{r} - \mathbf{r}_j)e^{-i\mathbf{q}\cdot\mathbf{r}}\, d\tau$$

$$= \frac{1}{N} \sum_j e^{-i\mathbf{q}\cdot\mathbf{r}_j} \frac{N}{\Omega} \int w(\mathbf{r} - \mathbf{r}_j)e^{-i\mathbf{q}\cdot(\mathbf{r}-\mathbf{r}_j)}\, d\tau$$

$$= S(\mathbf{q})\langle k + q|w|k\rangle \qquad (4\text{-}65)$$

B.   The structure factor

The structure factor, which is the same as the one used in diffraction theory (see the companion volume "X-ray diffraction"), is defined as

$$S(\mathbf{q}) = \frac{1}{N} \sum_j e^{-i\mathbf{q}\cdot\mathbf{r}_j} \qquad (4\text{-}66)$$

It is really just the Fourier transform of the crystal lattice. The crystal lattice $\chi(\mathbf{r})$ is the function that has a delta function at every lattice point (atomic site is a simple crystal structure) $\mathbf{r}_j$ and is zero elsewhere. Substituting this function into the expression (4-63) used for finding the Fourier transforms gives

$$S(\mathbf{q}) = \Omega^{-1} \int \chi(\mathbf{r} - \mathbf{r}_j)e^{-i\mathbf{q}\cdot\mathbf{r}}\, d\tau$$

$$= \frac{1}{N} \sum_j e^{-i\mathbf{q}\cdot\mathbf{r}_j} \qquad (4\text{-}67)$$

in accordance with the definition. The structure factors are well known for most crystal structures.

The structure factor $S(\mathbf{q})$ is zero for most $\mathbf{q}$, but at some specific values of $\mathbf{q} = \mathbf{g}$ it has a delta function with a certain magnitude. These points form a lattice in $\mathbf{k}$ space often called the *reciprocal lattice*.[2] Each point represents a stack of parallel planes in the real crystal lattice, and the magnitude of the vector to the points in the reciprocal lattice is proportional to the reciprocal of the interplanar spacing of the corresponding stack. In the one-dimensional case, with the atoms spaced a distance a_0 apart, $|\mathbf{g}|$ has values of $2\pi n/a_0$, where n is any integer.

---

[1] This notation for the transform has been used to give the function the appearance of the matrix elements that are used by Harrison (1966). The "k" appears to be redundant, but it does serve to remind us that every state **k** may require a different pseudopotential and that it is only an approximation to assume that all valence electrons experience the same effective potential. This is an approximation that we introduce to allow us to use perturbation theory.

[2] Solid-state physicists normally define reciprocal-lattice vectors in units of $2\pi/\lambda$, whereas x-ray crystallographers use units of $1/\lambda$.

C.   The form factor

The term $\langle \mathbf{k} + \mathbf{q} | w | \mathbf{k} \rangle$ is called the *form factor* and is defined by

$$\langle \mathbf{k} + \mathbf{q} | w | \mathbf{k} \rangle = \Omega_0^{-1} \int w(\mathbf{r}) e^{-i\mathbf{q} \cdot \mathbf{r}} \, d\tau \qquad (4\text{-}68)$$

where the integral is taken over the volume Ω_0 of one cell of the crystal; it is the Fourier transform of the effective potential of a single atom in simple structures containing one atom per unit cell.

It is the estimation of these form factors that is difficult. Normally, one postulates a model and then modifies it so that it correctly predicts the energies of atomic spectral lines in the optical region. The potential is normally found to be small and roughly equal to the effective potential as a result of the assumption that the core electrons screen the nucleus and leave an effective potential equal to the product of the valency of the atom and the electronic charge. The other valence electrons will screen the core, particularly in metals where the electrons can move easily. This smooths out high-frequency ripples in the potential, thus decreasing the coefficients of the form factor that have high values of $\mathbf{q}$.

Examination of the tables of form factors given by Harrison (1966) shows that they are large and negative at low $\mathbf{q}$. They increase with $\mathbf{q}$ to a maximum positive value at about $\mathbf{q} = 2\mathbf{k}_F$, and then they oscillate between small positive and negative values above this; they become negligibly small at high $\mathbf{q}$. Because of the structure factor, only those values of the form factor for which $\mathbf{q}$ is a lattice vector $\mathbf{g}$ will affect the properties of the solid. These values are normally either negative or small.

A positive component of the pseudopotential implies that the periodic potential, as represented in Fig. 4-3*b*, has its sign reversed. It attracts electrons less strongly to regions close to the ions than to the regions between the ions.

Because the coefficients at high $\mathbf{q}$ are negligibly small, the crystal potential can be written

$$V = W(\mathbf{r}) = \sum_{\mathbf{g}} S(\mathbf{g}) \langle \mathbf{k} + \mathbf{g} | w | \mathbf{k} \rangle e^{i\mathbf{g} \cdot \mathbf{r}} \qquad (4\text{-}69)$$

where the summation is taken over only a few lattice vectors close to the origin in $\mathbf{k}$ space.

D.   Solution of the Schrödinger equation

The wavefunctions and energies of the states in the band can be calculated by substituting this potential into the Schrödinger equation (4-2). The wavefunction is expanded as a series of plane waves:

$$\psi_{\mathbf{k}}(\mathbf{r}) = \sum_{\mathbf{g}} a_{\mathbf{k}}(\mathbf{g}) e^{i(\mathbf{g}+\mathbf{k}) \cdot \mathbf{r}} \qquad (4\text{-}70)$$

and the coefficients are found to first order by perturbation methods.

$$a_{\mathbf{k}}(\mathbf{g}) = \frac{\langle \mathbf{k} + \mathbf{g} | W | \mathbf{k} \rangle}{[\hbar^2/(2\mu)](k^2 - |\mathbf{k} + \mathbf{g}|^2)} \tag{4-71}$$

Second-order perturbation theory gives

$$E = \overline{V} + \frac{\hbar^2 k^2}{2\mu} + \sum_{\mathbf{g}}{}' \frac{\langle \mathbf{k} + \mathbf{g} | W | \mathbf{k} \rangle \langle \mathbf{k} | W | \mathbf{k} + \mathbf{g} \rangle}{[h^2/(2\mu)](k^2 - |\mathbf{k} + \mathbf{g}|^2)} \tag{4-72}$$

The prime on the summation indicates that the term with $\mathbf{g} = 0$ is not included.

The form factors that are most important are those for which $k^2 = |\mathbf{k} + \mathbf{g}|^2$. This condition occurs when the projection of $2\mathbf{k}$ onto $\mathbf{g}$ equals $-\mathbf{g}$. Because the reciprocal lattice is always centrosymmetric, this condition requires that $\mathbf{k}$ lie on a plane that is the perpendicular bisector of $\mathbf{g}$. These planes are called *Brillouin-zone boundaries*, and the volumes in $\mathbf{k}$ space that they enclose are called Brillouin zones.

When $\mathbf{k}$ does not lie near a Brillouin-zone boundary, all the coefficients given by equation (4-71), except that for which $\mathbf{g} = 0$, are small. The last term in equation (4-72) is also very small. The wavefunction obtained from equation (4-70) is then a plane wave, like the wavefunction for free electrons, and the energy is given by equation (4-53). Therefore, when $\mathbf{k}$ is not at a Brillouin-zone boundary, the contribution of each state to a spectrum is similar to that from free electrons. This explains the free-electron-like shapes of most light-metal spectra.

When $\mathbf{k}$ lies on a Brillouin-zone boundary, equations (4-71) and (4-72) have nonphysical singularities, and a different set of approximations must be made. In the simplest cases, the wavefunction becomes

$$\psi_{\mathbf{k}}(\mathbf{r}) \propto e^{i\mathbf{k} \cdot \mathbf{r}} \pm e^{i(\mathbf{k}-\mathbf{g}) \cdot \mathbf{r}} \tag{4-73}$$

and the energy is given by

$$E = \overline{V} + \frac{h^2}{2\mu} \left(\frac{\mathbf{g}}{2}\right)^2 \pm \langle \mathbf{k} - \mathbf{g} | W | \mathbf{k} \rangle \tag{4-74}$$

These expressions give prominence to the terms associated with $\mathbf{g}$, without creating singularities.

The symmetry of the wavefunction given by equation (4-73) can be seen by adding the two sets of spherical Bessel functions. These functions are shown as a function of $\mathbf{k}$ in Fig. 4-4*a*. Figure 4-4*b* shows the same functions drawn again, and another set centered on $\mathbf{g}$. If they are subtracted, the even functions *s* and *d* become zero at $\mathbf{k} = \mathbf{g}/2$, and so the function is purely *p*-like. If they are added, the odd function *p* becomes zero at $\mathbf{k} = \mathbf{g}/2$, and the function is purely *s*- and *d*-like. This is just the three-dimensional analog of the example given earlier for the one-dimensional case with $\mathbf{k} = \pi/\mathbf{a}$. Of course, if $\mathbf{k}$ is close to two or more Brillouin-zone boundaries, the situation is more complex.

All electrons have a kinetic energy of the order of $\hbar^2 k^2/(2\mu)$ above the average potential that they experience. We have seen that the wavefunctions of s-like states have nodes at the ionic sites and that the wavefunctions of p-like electrons have nodes lying between the ions. As the nodes in the wavefunctions are associated with regions of high electronic charge density, s-like electrons mostly experience the potential near the ions while the p-like electrons mostly experience the potential between the ions. If the form factor is positive, the potential near the ions, which is given by $\overline{V} - |\langle \mathbf{k} + \mathbf{g}|W|\mathbf{k}\rangle|$, is less than the potential between the ions, which is given by $\overline{V} + |\langle \mathbf{k} + \mathbf{g}|W|\mathbf{k}\rangle|$. In this case, the s-like electrons have the lower of the two energies given by equation (4-74). If the form factor is negative, the s-like electrons have the upper of the two energies in that equation.

Thus the effect of the small periodic potentials is to move some of the states that lie close to the Brillouin-zone boundaries to lower energy and to move the others to high energy and, at the same time, to change the symmetries of the wavefunctions so that those that move down in energy have a different symmetry from those that move up in energy. Second-order perturbation theory is more complex, causing terms involving other values of $\mathbf{g}$ to be included in equation (4-74).

E. Band spectra and the density of states

The shifting of some states to different energies changes the density of states. For free electrons, the thickness in $\mathbf{k}$ space of a shell of states in the energy range E to $E + dE$ was found by differentiating equation (4-53). For states in a small periodic field, we should differentiate equation (4-72) instead. As this equation does not contain $\mathbf{k}$ explicitly, it is only possible to write

$$dE = \nabla_{\mathbf{k}}(E)\, d\mathbf{k} \qquad (4\text{-}75)$$

The number of states in the shell is given by

$$N(E)\, dE = 8\pi \int \frac{d^2\mathbf{k}}{\nabla_{\mathbf{k}}(E)}\, dE \qquad (4\text{-}76)$$

and the density of states becomes

$$N(E) = 8\pi \int \frac{d^2\mathbf{k}}{\nabla_{\mathbf{k}}(E)} \qquad (4\text{-}77)$$

The integral is taken over the surface in $\mathbf{k}$ space on which all states have the same energy.

Similarly, taking equation (4-29) and summing or integrating over all states that have an energy that differs from the core state by an amount $h\nu$ give the intensity at this energy in the band; it is assumed here that the lifetime broadening is negligible.

$$I(\nu) \propto \nu^3 \int \frac{|\mathbf{r_k}^q|^2 \, d^2\mathbf{k}}{\nabla_{\mathbf{k}}(E)} \qquad (4\text{-}78)$$

If the matrix element $|r_{\mathbf{k}}{}^{q}|$ were independent of $\mathbf{k}$, it could be removed from the integral, and this equation would become

$$I(v) \propto v^{3}|r_{\mathbf{q}'}{}^{q}|^{2}N(E) \qquad (4\text{-}79)$$

However, this is not always valid; some states with a certain value of $\mathbf{k}$ could be almost entirely p-like, while other states with the same energy but at different $\mathbf{k}$ might be almost entirely s-like. If this occurs, the matrix element is highly dependent on the wave vector $\mathbf{k}$, and equation (4-79) becomes a very poor approximation.

The similarity of equations (4-77) and (4-78) occurs because of the gradients in their denominators and because the integrations are stopped at the Fermi energy. The latter causes both functions to have a Fermi edge. The effect of the former is more difficult to see; we shall first study the effect of the gradient on the density of states.

F.   Brillouin-zone effects

The first Brillouin zone is that volume of $\mathbf{k}$ space that can be reached from the origin without crossing a zone boundary. If we assume that the Fourier coefficients of the effective potential are negligibly small, the situation is the same as that for free electrons. The density of states rises as $E^{1/2}$ until the first vector touches a Brillouin-zone boundary; this is said to be a *critical point* in $\mathbf{k}$ space. The density of only those states that lie in the first Brillouin zone begins to fall, creating a discontinuity at the energy corresponding to this critical point. As other boundaries are met, other critical points occur. Two Brillouin-zone boundaries meet along a straight line, and the point of closest approach of this to the origin is another critical point. When this line reaches a third boundary, there is a corner in the first zone, and this critical point represents the largest possible $\mathbf{k}$ vector that can be fitted into the first zone. For each of these critical points there is a discontinuity at the corresponding energy in the density of states of the first zone.

When the Fourier coefficients of the potential are not negligible, all states in the first Brillouin zone that lie close to a zone boundary are moved down in energy by the last term in equation (4-72). Because the discontinuities are caused by states at the zone boundaries, the energies at which the discontinuities occur are also lowered.

In the second and higher zones, there are similar discontinuities, which in the free-electron case exactly counteract those in the first zone so that the total density of states is a smooth continuous curve. However, when the coefficients of the potential are not negligible, those discontinuities that correspond to discontinuities in the first zone move up in energy. Thus, in a realistic density of states, the discontinuities no longer counteract one another. Figure 4-5b shows a sketch containing a pair of discontinuities resulting from the critical point that is the point of closest approach of a

Brillouin-zone boundary to the origin. The discontinuities are spaced $2\langle \mathbf{k} + \mathbf{g}|W|\mathbf{k}\rangle$ apart, as required by equation (4-74). States inside the first zone that lie close to the critical point have energies close to that of the lowest-energy critical point. States in the second zone that lie close to the critical point will not be occupied unless there are electrons with energy above that of the other critical point.

The shape of the features resulting from the critical point is determined by the gradient in equation (4-77) and by the way the integral is taken in that equation. Because of the similarity between this equation and equation (4-78), similar features will be expected to appear in the spectra. However, when states are shifted in energy, their symmetries are also changed. If the states associated with one discontinuity are entirely s-like, it will not be observed in K spectra but will be exaggerated in the $L_{II,III}$ spectra. Thus the transition probabilities limit the number of discontinuities that can be observed in any one spectrum. The symmetry of the wavefunctions can be determined by using group theory and tables such as those given by Altmann and Cracknell (1965) and Altmann and Bradley (1965).

The bandwidth is also affected by the periodicity of the potential. If a band gap occurs at the Fermi energy, more filled states have their energy reduced than have it increased. This reduces the bandwidth by a few percent and increases the binding energy.

The crystal structure for which the binding energy is largest is the most stable. In the last couple of years, pseudopotential theory has been successfully used to predict the stable-crystal structures for many substances; the method of performing these calculations has been described by Harrison (1966).

G. Light-metal spectra

The discontinuities associated with the critical points in the Brillouin zone are not seen in the emission spectra from monovalent metals because they all occur at energies above the Fermi edge. Sagawa (1968) has described the discontinuities in the beryllium K spectrum; Watson, Dimond, and Fabian (1968) have described them in the magnesium $L_{II,III}$ spectrum; and Rooke (1968) has described them in the aluminum $L_{II,III}$ spectrum. Except for the band limits, they are always extremely small, and the resolution of the spectra is such that it is difficult to measure the coefficients of the pseudopotential that are involved. The Fermi surface techniques, such as the de Hass-van Alphen effect, are vastly superior for this purpose although they cannot show that these coefficients are almost independent of $\mathbf{k}$; verification of this important assumption has been obtained from the x-ray spectra listed above. The resolution of the hard x-ray spectra is so poor that, other than those at the band limits, the discontinuities have not been seen.

It has been assumed that all core electrons are equally effective in screening the nucleus. This is not entirely accurate when the core electrons have different symmetries. The plane waves used as a basis for expanding wavefunctions, as in equation (4-70), should be modified so that they are orthogonal to the core states as well as to one another. When these orthogonal plane waves are used, it is found that the core electrons are most effective in screening those valence electrons that have the same symmetry as the core state. If there are no core states with a specific type of symmetry, the valence electrons with that type of symmetry are less well screened than the others.

One is not surprised to learn that throughout the valence band of lithium there are more p-like states than in the free-electron band, or that beryllium does not have a free-electron-like density of states; none of the elements in this period have any p states in their cores. For the same reason, boron is a semimetal, and carbon forms extremely strong bonds when it has the diamond-crystal structure.

H.   Insulators and semiconductors

The band gaps of some elements can be so large that there exists a band of energies in which there are no allowed states. If the Fermi edge lies in this forbidden band, the element is either an insulator or a semiconductor. It is an insulator if the small amounts of kinetic energy normally given to valence electrons by thermal or other means are insufficient to excite them to states above the band gap. If the gap exists but is so small that a limited number of electrons can be excited to the empty states above the Fermi energy, the substance is a semiconductor.

The binding energy of an element may be reduced by having a band gap at the Fermi edge, because some electrons are moved to lower energy while there are no electrons in those states whose energy is raised. This contribution to the binding energy will depend on the size of the band gap. There may be opposition to the band structure because large band gaps are usually associated with local high electron densities which oppose binding. Only elements toward the right-hand side of the periodic table have large enough pseudopotentials to become insulators, and only those near the center of the table tend to become semiconductors. Band gaps can occur at the Fermi edge only if there are two electrons per cell in the crystal. Those elements with an odd number of electrons per atom can become insulators by taking crystal structures that have two atoms per cell.

Although, in general, pseudopotential methods are suitable for calculating the properties of these elements, perturbation methods are not. Tight-binding methods have been used for most of the calculations for these materials to date. The wavefunction, instead of being expanded as a series of plane waves or orthogonal plane waves, is expanded as a series of atomic orbitals, i.e., wavefunctions possessed by isolated atoms.

The spectra of these elements tend to have the appearance of a set of broad overlapping lines. No discontinuities caused by critical points in the Brillouin zones have been detected, although where band-structure calculations exist, the peaks in the spectra correspond to peaks in the density of states; the peaks in the density of states are associated with several discontinuities. Where there is lack of agreement, it is probably due either to the crudeness of the estimation of the band structure or to the lack of resolution of the spectra. As mentioned above, the Auger tails are noticeably smaller for insulators due to the band gap which restricts the number of electrons that can be excited to states above the Fermi surface.

I. Spectra from *d*-band metals

The final class of elements to be discussed is the metals that have a considerable number of *d*-like electrons in the valence band. Two factors combine to give the *d*-like electrons a peculiar character. First, the Fourier coefficients of the potential increase with the number of the period in the periodic table; second, the term $c = -l(l + 1)$ in equation (4-6) acts like a negative potential and attracts the electrons to the ions. The total effect of these two factors is that the *d*-like electrons are localized around their individual atoms.

The wavefunctions approximate atomic orbitals as though the atoms were isolated. However, the effects of the little tunneling that does occur between sites and the effects of the interactions with the *s*-like and *p*-like electrons, which often lie in the same energy range, perturb the wavefunction so that it is more accurately represented by a sum of a series of atomic orbitals. In general the *d*-like states that have the lowest energy are mixed with *s*-like states, while the *d*-like states that have the highest energy are mixed with *p*-like states, although there are exceptions to this.

The density of the *d*-like states consists of a set of narrow bands closely spaced and somewhat reminiscent of a set of broad atomic lines. From the band-structure point of view, it consists of a narrow band that is capable of holding 10 electrons per atom, and these states will occupy five complete Brillouin zones. The free-electron-like theory is not applicable. The band gaps are large when compared with the band-width, and they break the band into a series of between two and five almost-distinct broadened lines.

The recording of the spectra from the heavy metals is more difficult than for the light metals. Self-absorption, unidentified satellites, increased background radiation, and increased broadening all change the shape of the spectra and lead to uncertainties in finding its true shape. [See Liefeld (1968).] In the last ten or more years, much of this work has been carried out in the U.S.S.R. I know of no review articles describing this work in Western journals, although the Ukrainian Academy of Science [see Nemoshkalenko (1968) and Nemnonov (1968)] has published a booklet on this topic. These authors also have published numerous articles in the literature. The school headed by Nemnonov compares its results with densities of states calculated by band-structure techniques, such as those published by Mattheiss (1964). The school led by Nemoshkalenko interprets its results in conjunction with results from other experimental techniques such as the Mössbauer effect. Their discussion is largely centered on the atomic orbital approach and the percentages of the states that have predominantly s, p, or d character. Many x-ray spectroscopy groups in the Western world have measured spectra from these metals, and they have interpreted their results mostly by comparing them with the density of states, although Cuthill et al. (1968) made some attempts to introduce the effects of the transition probabilities.

It is hoped that the recent extension of the pseudopotential theory to these metals by Fong and Cohen (1970) will simplify the calculation of their various properties sufficiently to enable their x-ray spectra to be more accurately predicted. The resolution of the spectra will always limit the information that they can provide, but even this small amount of information can be useful, particularly when studying alloys.

VIII. BAND SPECTRA FROM COMPOUNDS AND ALLOYS

When spectra can be obtained from more than one element in the solid, one fact becomes quite obvious: the properties of the electrons in the neighborhood of the atoms of one element are considerably different from their properties in the neighborhood of the atoms of the other elements. Often the spectra from the individual elements are more akin to spectra from the respective pure element than those from other elements in the alloy or compound.

This ability to sample the electronic states locally provides x-ray spectroscopy with a distinct advantage over other forms of spectroscopy. It results from the inclusion of the wavefunction of the core state into the integral used to calculate the transition probability; see equation (4-28). Of course, even if the valence electrons were behaving similarly in the local neighborhoods of both elements, the transition probabilities associated with the different core states excited would produce different spectra. However, the evidence suggests that some of the discrepancies are due to the behavior of the valence states.

A. Basic processes in the creation of alloys

It is instructive to examine the changes that occur in the electronic properties when one element is alloyed or compounded with another. Although all the effects occur simultaneously, they are discussed below successively. The alloy being created will be assumed to be one in which all the elements occupy cells that have the same size.

As a first step, each element is forced to conform to the crystal structure of the alloy. The average electron density is kept equal to that of the natural crystal of the element. The discontinuities in the spectra depend on the wavefunction and the energy of each state. These depend directly on the form factors $\langle \mathbf{k} + \mathbf{g} | W | \mathbf{k} \rangle$ associated with the set of reciprocal-lattice vectors $\mathbf{g}$ for which the structure factor is nonzero. Changing the crystal structure changes the reciprocal lattice and hence the form factors. Thus, the bands for the new crystal structure have a new set of band gaps that occur at different energies $\hbar^2(\mathbf{g}/2)^2/(2\mu)$, and the states that lie close to these will have changed symmetries. Thus the shape of the spectra will be changed from that of the pure metal in any energy region that lies close to the energies of the discontinuities in either the pure-element spectra or the alloy spectra.

Away from the energies affected by the discontinuities, the pure-element and the alloy spectra will have the same basic shape. These basic shapes are determined by the form factors. For instance, the form factors for aluminum are small, and the bands have a basic free-electron-like shape, whereas beryllium has large form factors and its bands do not have a free-electron-like shape. The form factors are determined by the way the nucleus is screened by the core electrons and by the other valence electrons. Since this screening is unlikely to change and the new values of $\mathbf{g}$ are unlikely to be grossly different from the old ones, the valence electrons will always retain the tendency to heap themselves in a characteristic way. Brewer (1967) has based a theory of alloying on this property of the electrons.

The second step to be considered in the alloying process is the change of volume, so that the new cells have the size that they will have when they are placed in the compound or alloy. A change in volume creates a change in the atomic spacing a. Examination of equations (4-52), (4-53), and (4-57) shows that, in general, the energy of states varies as $1/a^2$ and that the bandwidth is decreased if the volume increases. This change of bandwidth is normally only about 1% of the total bandwidth and is barely detectable under even the best experimental conditions.

Because each cell is neutral, the atom in it could be replaced by an atom of the other element without appreciably changing the potential in the neighboring cells. However, at the cell boundaries, the electron density would be discontinuous, particularly if the neighboring atoms have different valences. The electrons will move to smear out the discontinuities, thereby reducing the energy of the electron-electron interactions. The final distribution of electrons will be a compromise between having neutral cells and having a smooth density distribution of the electrons.

In the next chapter it will be shown that for metals, the electrons can screen a point charge so that its effect on the potential at distances of the order of an atomic radius from the point charge is negligible. Thus, in order to screen a trivalent ion within an atomic radius, the electron density in its own cell must be sufficient for the cell to contain about three electrons. The nature of the wavefunctions required to do this screening will be discussed below.

B.   Spectra from d-band metals in alloys

At this stage it is possible to explain the general features of spectra from d-band metals in alloys. Because the d-like electrons are localized, their potentials are less affected by the neighboring cells. The general form of the potential will be unchanged, and the spectra will retain the basic structure of a narrow band that looks like a series of broadened spectral lines. The energy at which this narrow band occurs will not be changed. The energies of the discontinuities will be changed, but since these are smeared out by the large broadening that is present in the heavy-metal spectra, this effect will not be seen. Because of these considerations, it would be surprising if the spectra from d-band metals changed a lot when their spectra are recorded from alloys.

This lack of change has been reported by a number of authors; for example, see the spectra from nickel-zinc alloys reported by Appleton and Curry (1967). In general it has been found that if two d-band metals with different valences are alloyed, instead of the width of the d band changing according to the average number of d electrons per atom, the spectra are more consistent with having two d bands, each roughly the same shape as for the appropriate pure metal, and each will have an intensity proportional to the number of atoms of that type in the alloy. When the d states on the different atoms have the same energy as one another, it is no longer possible to draw conclusions from the spectra, and the theory suggests that the bands are no longer separate. For a discussion of the theory associated with this type of band, see Beeby (1964).

Because of the problems of observing the spectra of heavy metals, it is difficult to observe the behavior of the d bands. When aluminum is alloyed with d-band metals, there is enough overlap between the aluminum core and the d-like electrons to create large changes in the aluminum spectra; see Watson et al. (1972). This may be a useful way of sampling the behavior of the d-like electrons without the effects of broadening. The intensity of these peaks is surprisingly high, and these spectra may provoke considerable theoretical interest.

C.   Spectra from light metals in alloys

To discuss the behavior of the light-metal spectra from alloys with other light metals, further consideration must be given to the behavior of the band gaps. Because the s-like and p-like electrons are not localized, they are spread out through the crystal, and any one electron experiences the average potential in the regions where its wave-function is large. Therefore the form factors appropriate to a specific reciprocal-lattice vector will be the weighted average of the form factors for the individual elements. The weighting factors are proportional to the number of atoms of that type in the crystal. The effects of alloying on the band gaps are discussed by Harrison (1966).

If the alloy is ordered, superlattice band gaps will occur; these should be particularly noticeable in the *s-p* bands. Harrison shows that the appropriate form factor is the difference between the equivalent form factors for the pure elements. No fully ordered alloys have yet been measured under conditions that would have guaranteed that the sample had remained ordered, and these band gaps have not been observed.

It is expected that the spectra from the light metals will retain their basic shape in that part of the band that does not contain discontinuities; that is, the bottoms of the bands are expected to retain approximately the shape of the pure-metal spectra. In those regions that are affected by the discontinuities, it is expected that the band gaps and the transition probabilities will change very slowly across any one phase, but that on going from one phase to another there will be a bigger change. For most concentrations, it is expected that the changes are insufficient to alter the general shape of the band; for example, it is expected that the aluminum spectra from light-metal alloys will always retain the characteristic free-electron shape described in Section VI.

This behavior has been observed by Dimond (1970) in a series of aluminum-magnesium alloys. It was observed that the spectra from either element, when recorded from a considerable range of concentrations, changed slowly but never differed markedly from the pure-metal spectra. However, on very close examination, it is possible to see that there is a distinct difference between spectra from different phases.

Considerable controversy has arisen about the fact that the bottoms of the bands remain similar to those in the pure metal. This causes the individual spectra to retain shapes in which the majority of their intensity is spread over the same energy range as for the pure metals, giving the impression that the bandwidths are the same as for the pure metals. Theories based on perturbation methods do not predict this, but, as Friedel (1954) has pointed out, this is most probably due to the fact that this method is not suitable for treating the states that lie near the extremities of the bands. Other theories such as those proposed by Beeby (1964) and by Soven (1969) suggest that these results are consistent with the theory.

D. Spectra from compounds

Compounds are formed from elements that have large form factors because the valence electrons screen the core states less effectively. These large form factors cause the electrons to be localized, either on one of the atoms, as in the case of ionic compounds, or between the atoms, as in the case of covalent compounds. It is still possible to treat compounds using band-structure techniques; Bonnelle (1968) has compared some spectra from compounds with their band structures. Holliday (1968) has interpreted x-ray spectra in terms of bonding theory, where the features in the spectra are correlated with the formation of various types of chemical bonds.

These two extremes are probably both valid and useful, but one cannot help hoping that the interpretation of the spectra of alloys and compounds could be consistently achieved using a theory as simple as the pseudopotential theory for the pure metals. Much work is being done to develop this technique by Blandin (1967), Pick (1967), and others. If this approach is successful, the ease of doing the theory should be such that experimentalists may well be able to handle the theory required to predict the results of their own measurements.

IX.   CONCLUSIONS

It is possible that most of the spectra that have been recorded to date could be interpreted with the theory described in this chapter. However, a comparison between the experimental spectra and detailed theoretical calculations has been made for only a few elements. The use of pseudopotential theory has simplified the mathematics so that many more calculations can be expected in the future. Another factor that will stimulate interest in the field is the growing awareness among theoreticians that when calculating band structures, there are sufficient data generated to enable them to predict x-ray spectra with very little extra effort.

The spectra from alloys and compounds are still not well understood, but recent developments in the theory of these should improve this situation. Again, pseudopotential theory should assist with this, particularly now that this theory is being successfully applied to the d-band metals.

The many-body interactions, particularly electron-electron interactions, are also seen to be very important. Besides creating satellites, they play a large part in determining the shape of the main emission bands. They are directly involved in the prediction of the screening of the potentials that should be used when doing calculations of the properties of alloys.

Absorption spectra and optical properties are also slowly being interpreted in the same way as the emission spectra. For a while, it was thought that optical spectra would require a different form of the theory that involved many-body interactions in the fundamental transition processes as well as in the determination of the lifetime broadening. However, recent indications suggest that this may not be necessary; see Cardona (1970). Absorption spectra are discussed in Chap. 6, but it should be noted here that Fano and Cooper (1968) have found that many of the features of atomic absorption spectra can be explained in exactly the same way as the emission spectra have been interpreted in this chapter. An explanation put forward by Lytle (1966) for the extended fine structure in the absorption spectra from solid metals is based on the oscillation with energy of the probability of the states having a particular symmetry; we have seen that this type of oscillation is predicted by the free-electron theory.

There are many other methods of obtaining related spectra, and of these ultra-violet photoelectron emission, x-ray photoelectric emission, and ion-neutralization spectroscopy are the most important. It is anticipated that with time these spectra will also be interpreted in the same way, although the results from the ultraviolet photoelectron spectroscopy suggest that many-body interactions may play a fundamental role in some solids. For all methods of spectroscopy, the interpretation of the spectra from alloys and compounds provides the greatest difficulties.

REFERENCES

ALTMANN, S. L., and BRADLEY, C. J. (1965): Lattice harmonics. II. Hexagonal close-packed lattice, *Rev. Mod. Phys.*, **37**: 33–45.

―――― and CRACKNELL, A. P. (1965): Lattice harmonics. I. Cubic groups, *Rev. Mod. Phys.*, **37**: 19–32.

APPLETON, A., and CURRY, C. (1967): Soft x-ray emission spectra of some binary alloys, *Phil. Mag.*, **16**: 1031–1037.

ASHCROFT, N. W. (1968): Electron-ion pseudo-potentials in the alkali metals, *J. Phys. Chem.*, **1**: 232–243.

BAUN, W. L., and FISCHER, D. W. (1965): The effect of valence and coordination in *K* series diagram and non-diagram lines of magnesium, aluminum and silicon, *Advan. X-Ray Anal.*, **8**: 371–383.

BEEBY, J. (1964): Electronic structure of alloys, *Phys. Rev.*, **A135**: 130–143.

BETHE, H. A., and SALPETER, E. E. (1957): Quantum mechanics of one- and two-electron systems, *Handb. Physik*, **35**: 88–436.

BLANDIN, A. P. (1967): Theoretical considerations of the Hume-Rothery rules, in J. STRINGER and R. I. JAFFE (eds.), *Phase stability in metals and alloys*, McGraw-Hill Book Company, New York.

BLOKHIN, M. A. (1957): *The physics of x-rays*, 2d ed. (State Publishing House of Technical-Theoretical Literature, Moscow), English translation by U.S. Atomic Energy Commission, document AEC-tr-4502.

―――― and SACHENKO, V. P. (1960): Concerning the shape of energy bands in solids, *Bull. Acad. Sci. USSR, Phys. Ser. (English Transl.)*, **24**: 410–418.

BONNELLE, C. (1968): Distributions *d* et *f* de quelques Métaux et Composes obtenues par Spectroscopie X, in D. J. FABIAN (ed.), *Soft x-ray band spectra and the electronic structure of metals and materials*, pp. 163–172.

BREWER, L. (1967): Viewpoints of stability of metallic structures, in *Phase stability in metals and alloys* [see Blandin (1967)].

CARDONA, M. (1970): Optical properties and the electronic density of states, *J. Res. Natl. Bur. Std.*, **A74**.

CATTERALL, J. A., and TROTTER, J. (1958): Soft x-ray emission spectrum from liquid aluminum, *Phil. Mag.*, **8**: 897–899.

CAUCHOIS, Y. (1948): *Les spectres de rayons x et la structure electronique de la matière*, Gauthier-Villars, Paris.

CRISP, R. S. (1960): Soft x-ray emission from potassium metal in the 40–1000 Å range, *Phil. Mag.*, **5**: 1161–1169.

––––– and WILLIAMS, S. E. (1960): The *K*-emission spectrum of metallic lithium, *Phil. Mag.*, **5**: 525–527.

CUTHILL, J. R., MCALISTER, A. J., WILLIAMS, M. L., and DOBBYN, R. C. (1968): Soft x-ray spectra for nickel and nickel alloys and comparison with the theoretical densities-of-states, in D. J. FABIAN (ed.), *Soft x-ray band spectra and the electronic structures of metals and materials*, pp. 151–162.

DEKKER, A. J. (1962): *Solid state physics*, The Macmillan Company, New York.

DIMOND, R. K. (1970): "Soft x-ray spectra from aluminum-magnesium," Ph.D. thesis, University of Western Australia.

FABIAN, D. J. (ed.) (1968): *Soft x-ray band spectra and the electronic structures of metals and materials*, Academic Press, Inc., London.

FANO, U., and COOPER, J. W. (1968): Spectral distribution of atomic oscillator strengths, *Rev. Mod. Phys.*, **40**: 441–507.

FISCHER, D. W., and BAUN, W. L. (1965): $K\beta$ x-ray emission from solid and liquid aluminum, *Phys. Rev.*, **A138**: 1047–1049.

FONG, C. Y., and COHEN, M. L. (1970): Energy-band structure of copper by the empirical pseudo-potential method, *Phys. Rev. Lett.*, **24**: 306–309.

FRIEDEL, J. (1954): Electronic structure of primary solid solutions in metals, *Advan. Phys.*, **3**: 446–507.

HARRISON, W. A. (1966): *Pseudo-potential in the theory of metals*, W. A. Benjamin, Inc., New York.

HERMANN, F., and SKILLMAN, S. (1963): *Atomic structure calculations*, W. A. Benjamin, Inc., New York.

HOLLIDAY, J. E. (1968): X-ray emission bands and bonding for transition metals, solutions, and compounds, in D. J. FABIAN (ed.), *Soft x-ray band spectra and the electronic structure of metals and materials*, pp. 101–132.

KINGSTON, R. H. (1951): Spectroscopy of the solid state: Potassium and calcium, *Phys. Rev.*, **84**: 944–949.

LIEFELD, R. J. (1968): Soft x-ray emission spectra at threshold excitation, in D. J. FABIAN (ed.), *Soft x-ray band spectra and the electronic structure of metals and materials*, pp. 133–150.

LYTLE, F. W. (1966): Determination of interatomic distances from x-ray absorption fine structure, *Adv. X-Ray Anal.*, **9**: 398–409.

MAHAN, G. D. (1967): Excitons in metals: Infinite hole mass, *Phys. Rev.*, **163**: 612–617.

MATTHEISS, L. F. (1964): Energy bands for the iron transition series, *Phys. Rev.*, **A134**: 970–973.

MESSIAH, A. (1964): *Quantum mechanics*, North-Holland Publishing Company, Amsterdam.

MIZUNO, Y., and ICHIKAWA, J. (1968): *J. Phys. Soc. Japan*, **25**: 627–632.

NEMNONOV, S. A. (1968): The band structure from the electron-energy spectra of transition metals and their alloys (in Russian), in *Certain questions on the electronic structure of transition metals*, Scientific Council for Problems in Solid State Physics, Institute of Metal-physics, Academy of Science, Kiev.

NEMOSHKALENKO, V. V. (1968): X-ray spectra and the electronic structure of elements in the first and second transition series (in Russian), in *Certain questions on the electronic structure of transition metals* [see Nemnonov (1968)].

—— (ed.) (1969): *X-ray spectra and the electronic structure of materials* (in Russian), Institute of Metal-physics, Academy of Science, Kiev.

PARRATT, L. G. (1959): Electronic band structure of solids by x-ray spectroscopy, *Rev. Mod. Phys.*, **31**: 616–645.

PICK, R. M. (1967): A self-consistent OPW calculation of pair-interaction functions in some normal metals, in *Phase stability in metals and alloys* [see Blandin (1967)].

PINES, D. (1955): Electron interactions in metals, *Solid State Phys.*, **1**: 367–450.

PIRENNE, J., and LONGE, P. (1964): Double electron transitions to the soft x-ray emission bands of metals, *Physica*, **30**: 277–292.

ROOKE, G. A. (1968): Interpretation of aluminum x-ray band spectra. I. Intensity distribution, *J. Phys. Chem.*, **1**: 767–775. II. Determination of effective potentials from experimental $L_{II,III}$ emission spectra, *Ibid.*, 776–783.

SAGAWA, T. (1968): Soft x-ray emission and absorption spectra of light metals, alloys and alkali halides, in D. J. FABIAN (ed.), *Soft x-ray band spectra and the electronic structure of metals and materials*, pp. 29–44.

SCHIFF, L. I. (1955): *Quantum mechanics*, 2d ed., McGraw-Hill Book Company, New York.

SEITZ, F. (1940): *The modern theory of solids*, McGraw-Hill Book Company, New York.

SKINNER, H. W. B. (1940): Soft x-ray spectroscopy of solids, *Phil. Trans. Roy. Soc. London*, **A239**: 95–134.

SOVEN, P. (1969): Contribution to the theory of disordered alloys, *Phys. Rev.*, **178**: 1136–1144.

TINKHAM, M. (1964): *Group theory and quantum mechanics*, McGraw-Hill Book Company, New York.

TOMBOULIAN, D. H. (1957): The experimental methods of soft x-ray spectroscopy, *Handb. Physik*, **30**: 246–304.

WATSON, L. M., DIMOND, R. K., and FABIAN, D. J. (1968): Soft x-ray emission spectra of magnesium and beryllium, in D. J. FABIAN (ed.), *Soft x-ray band spectra and the electronic structure of metals and materials*, pp. 15–58.

——, NEMOSHKALENKO, V. V., KAPOOR, K. S., KRIVITSKII, V. P., GORSKII, V. V., and NIKOLAEV, L. I. (1972): Soft x-ray spectra from aluminum-transition-metal alloys (in Russian), *Bull. Acad. Sci. USSR*, **36**: 415–419.

WILSON, A. H. (1936): Optical phenomena, chap. 4, in *Theory of metals*, 1st ed., Cambridge University Press, London.

YAKOWITZ, H., and CUTHILL, J. R. (1962): Annotated bibliography on soft x-ray spectroscopy, *Natl. Bur. Std. Monograph*, 52.

MANY-BODY EFFECTS

Lars Hedin

I.   INTRODUCTION

We choose to distinguish in x-ray processes between three "types" of electrons, namely, core, valence, and free electrons. Core electrons are tightly bound and hardly change their wavefunctions in different chemical surroundings. Free electrons are so highly excited that, similarly to the core electrons, they can be regarded as distinguishable in a classical sense from the other electrons in the system. The valence electrons are the remaining ones including, e.g., d electrons and conduction electrons. The valence electrons are by far the most interesting to study, and in x-ray spectroscopy the core electrons and the free electrons are mostly regarded only as probes for testing the properties of the valence-electron system.

The gross features of most x-ray spectra can be interpreted in terms of a one-electron theory, but many-body effects can sometimes be quite pronounced. They can give rise to extra structure, change line shapes, and cause less spectacular but no less important smooth changes in the intensity curves.

Many-body effects in x-ray spectra can be viewed both as a nuisance obscuring the study of important one-electron features in the valence-electron system, and, of course, also as a tool to discover and understand many-body interactions. The former viewpoint asks for quantitatively reliable, possibly parametrized, ways of separating out the many-body effects, while the latter primarily asks for a qualitative understanding. We will try here to take up both these viewpoints.

Interesting qualitative discussions of many-body effects have been given in the two "classical" articles by Skinner (1940) and by Parratt (1959). Since then a highly sophisticated theory has been developed, which has served to clarify many of the questions posed by Skinner and Parratt. The original literature is often rather technical, but most basic results can be explained in simple physical terms. We will here avoid the use of the full many-body machinery and keep to explanations in terms of well-known quantum-mechanical concepts. For a recent review of more technical aspects pertinent to x-ray spectra, we refer to a survey article by Hedin and Lundqvist (1969).

The understanding we have today of many-body effects is still mostly qualitative. This is really not surprising since even the quantitative mapping of one-electron properties is meager. While we have a wealth of energy-band results, we have many fewer good density-of-states curves and very little oscillator-strength data. Full-band calculations including the oscillator-strength part are crucial both to the mapping of the one-electron picture and as a basis for a quantitative evaluation of the many-body effects. We emphasize that this type of work, which pushes for a more detailed understanding, urgently requires more attention.

In comparing theory with experiment we will often refer to soft x-ray spectra, that is, spectra which involve the outermost core electrons. There is a large amount of good experimental data also from hard x-ray spectra, which involve the deep core levels. These levels are, however, much broader, and we have a very difficult deconvolution problem to cope with, a problem where, however, more progress very well could be made.

In Section II we give the "general theory" by discussing a sample of fairly well understood and clear-cut many-body effects, developing the necessary equations as we proceed. These relations then serve as a reference for later sections and as a steppingstone for understanding the often more sophisticated problems taken up in the later sections.

In Section III we discuss the many-body effects in x-ray photoemission and in x-ray absorption and emission that occur in simple metals. The theories for these effects are strongly interwoven in the literature, but we will try here instead to describe them by a stepwise increase in complication.

In Section IV we take up various other interesting effects like isochromats, inelastic x-ray scattering, and appearance spectroscopy, and in Section V we review aspects connected with the band structure. In Section VI we give a summary of important points, unsolved problems, and possible future developments.

Little is said about insulators besides the more general comments in Section II. This reflects partly the author's smaller involvement in these problems, and partly a comparatively less well developed theory.

II. BASIC MANY-BODY MECHANISMS

A. Some basic relations

We here collect and comment on a few textbook results from quantum mechanics. The emission or absorption intensity I, measured as the number of photons per second, is given by the "Golden Rule" expression

$$I(\omega) = C\omega \sum_f |\langle \Psi_f | T | \Psi_i \rangle|^2 \, \delta(\omega - E_f + E_i) \qquad (5\text{-}1)$$

T is the operator which couples the electrons to the radiation field

$$T = \sum_i \mathbf{n} e^{i\mathbf{k} \cdot \mathbf{r}_i} \mathbf{p}_i \qquad (5\text{-}2)$$

$\mathbf{r}_i$ and $\mathbf{p}_i$ are the position and momentum operators of electron i, and $\mathbf{n}$ is a unit vector for the polarization of the radiation field, which has wavelength $\lambda = 2\pi/k$. The exponential can be set equal to unity (the *dipole approximation*) for soft but not, in general, for hard x-ray spectra. The factor C in (5-1) is a constant in emission and describes the intensity of the incident radiation field in absorption. The initial and final states Ψ_i and Ψ_f describe the electrons and phonons with all their many-body interactions. The phonons are usually of little importance except for some edge effects, and will mostly be neglected here. When necessary, an averaging should be made over initial states. In the dipole approximation, we can transform equations (5-1) and (5-2) to actually include the dipole moment

$$I(\omega) = C\omega^3 \sum_f \left| \left\langle \Psi_f \left| \sum_i \mathbf{n} \mathbf{r}_i \right| \Psi_i \right\rangle \right|^2 \delta(\omega - E_f + E_i) \qquad (5\text{-}3)$$

In comparison with experiment it should be noted that there are many types of intensity curves. Often the results are given as the number of counts in a photomultiplier-per-wavelength unit $dN/d\lambda$. Such data should be multiplied by ω^{-2} before they are compared to $I(\omega)$ in (5-1) since

$$I(\omega) \sim \frac{dN}{d\omega} = \frac{dN}{d\lambda} \frac{d\lambda}{d\omega} \sim \frac{1}{\omega^2} \frac{dN}{d\lambda} \qquad (5\text{-}4)$$

We should also note that while photomultipliers give counts closely proportional to the number of photons, the blackening of a photographic film requires more extensive corrections for the wavelength dependence. A correct bookkeeping for the power of ω is important to assess the relative intensity of satellites in soft x-ray spectra, and to some extent also for the proper shapes of the main bands.

In the simplest approximation, the wavefunctions Ψ_i and Ψ_f are taken as Slater determinants. We then have the result of the one-electron approximation

$$I(\omega) = C\omega \sum_{fi} g_i |p_{fi}|^2 \delta(\omega - \varepsilon_f + \varepsilon_i) \qquad (5\text{-}5)$$

where p_{fi} is the matrix element

$$p_{fi} = \int \psi_f{}^*(\mathbf{r})(\mathbf{p} \cdot \mathbf{n})\psi_i(\mathbf{r}) \, d\mathbf{r} \qquad (5\text{-}6)$$

between the one-electron wavefunctions involved, the ε are the one-electron energies, and g_i is the probability for the initial state i being occupied. The one-electron approximation plays a basic role as a first step in most cases.

B.   Lifetimes and decay mechanisms

1.   *Qualitative discussion* If we compare x-ray spectroscopy with UV atomic spectroscopy, we have an important difference in the lifetimes or energy spreads involved. The relation between lifetime Δt (in sec) and energy spread ΔE (in eV) is

$$\Delta t = \frac{\hbar}{\Delta E} \cong \frac{6.5 \times 10^{-16}}{\Delta E} \qquad (5\text{-}7)$$

While a typical excited state in UV spectroscopy may have a ΔE of 10^{-8} eV, in x-ray spectroscopy we have ΔE's ranging between 10^{-2} and 10^2 eV. There are two decay mechanisms in operation, Auger decay and radiative decay. Auger decay usually contributes 0.01 to 0.1 eV to a core-electron width and substantially more to a valence-electron width. The radiative width is, according to classical theory (Jackson, 1962, Chap. 17),

$$\Delta E_{rad} = \frac{2}{3} \frac{\alpha}{mc^2} E^2 \cong 0.952 \times 10^{-8} E^2 \qquad (5\text{-}8)$$

The correct radiative width is often some 10 to 20% smaller than the classical result. The Auger width thus dominates for energies smaller than 1 keV, or using the relation between energy (in eV) and wavelength (in A),

$$\lambda = \frac{hc}{E} \cong \frac{12,400}{E} \qquad (5\text{-}9)$$

for wavelengths larger than ≈ 10 A, that is, in the soft x-ray region. A recent calculation of radiative widths has been made by Scofield (1969) and of radiative and Auger widths by Kobayasi and Morita (1970).

The most long-lived x-ray states ($\Delta E \approx 0.01$ eV) have a lifetime of $\approx 10^{-13}$ sec. This is the same time as for one typical lattice vibration and much longer than the relaxation time of the conduction electrons. The latter time is not well known, but it should not be much longer than the time of one plasma vibration or the passage time through a Wigner-Seitz cell of an electron on the Fermi surface, both these times being about 10^{-16} sec for a typical metal.

The longest lifetimes are found for the outermost core levels in the light elements up to about potassium. Starting with the transition metals, the Auger widths seem to be considerably larger than for the light elements. Thus the width of the outermost core level is estimated as 0.01 to 0.02 eV for Na, Mg, Al, and K (Skinner, 1940), while for Ni the estimate is 0.2 eV (Cuthill et al., 1967).

We have so far talked about the widths only of the core holes, but not of the conduction electrons. In a good metal, at low temperatures and close to the Fermi surface, these widths can be exceedingly small and comparable to optical widths

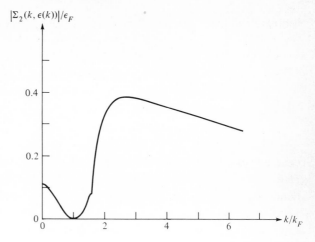

FIGURE 5-1
HWHM for the quasi-particle peak of an electron gas at the density of aluminum ($r_s = 2$), as approximated by the imaginary part of the self-energy Σ_2. Σ_2 is expressed in units of the Fermi energy ε_F (for $r_s = 2$, $\varepsilon_F = 12.5$ eV). The calculation was made by B. I. Lundqvist (1969a) using the Lindhard dielectric function.

(10^{-8} eV). For states in the bottom of the Fermi sea on the other hand, Auger-type processes are strong and give widths in the range of 0.5 to 2 eV. Estimates for an electron gas (Hedin, 1965a) are shown in Table 5-1.

The parameter r_s gives the electron density ρ according to the relation

$$\frac{4\pi(r_s a_0)^3}{3} = \frac{1}{\rho} \qquad (5\text{-}10)$$

Metallic densities correspond to r_s equals 2 to 5; for example, the r_s values for Al, Li, and Na are 2.07, 3.26, and 4.00, respectively. The width of the $k = 0$ state is, according to the table, about 15 percent of the bandwidth ε_F.

Electron states above the Fermi sea similarly become more and more broadened the further away they are from the Fermi level. In Fig. 5-1 we show electron-gas

Table 5-1   WIDTHS IN eV FOR STATES AT THE BOTTOM OF THE FERMI SEA

	$r_s = 2$	$r_s = 3$	$r_s = 4$	$r_s = 5$
Width $\Delta\varepsilon$	1.7	0.8	0.5	0.3
Fermi energy ε_F	12.6	5.6	3.1	2.0

estimates obtained by B. I. Lundqvist (1969a) of the half-widths HWHM[1] = Γ in units of the Fermi energy ε_F. In Table 5-2 we give the maximum value of FWHM = 2Γ and the corresponding energy above the Fermi level for different r_s values. Note that the estimates in Fig. 5-1 do not include contributions from decay processes through excitation of core electrons. The FWHM curves for actual metals should thus lie above those in the figure, when the energy lies above threshold for core excitation (Baer et al., 1970a).

The widths of the one-electron type of quasi-particle states that we have discussed here will influence the x-ray spectra. This will not, however, occur in a simple fashion since the actual spectra depend on the excitation process and the measuring conditions, as we will discuss later. Thus observed widths may come out to be either larger or smaller than we would expect from a simple quasi-particle picture. Extra broadening may be due to transitions that involve not only two quasi-particles but also, e.g., plasmons, coupled quasi-particles and plasmons, and cascades of quasi-particles, while we have narrowing when an electron leaves the solid, like in x-ray photoemission, and thus the idealization of an infinitely large solid breaks down. We can also have narrowing through the process of self-absorption (Subsection II F).

2.   *Mathematical formulation* Suppose that at time $t = 0$, the system we study is in a nonstationary state described by the wavefunction Φ. We expand Φ in terms of the stationary states Ψ_n:

$$\Phi = \sum_n \alpha_n \Psi_n \qquad (5\text{-}11)$$

The energy distribution of the initial state clearly is[2]

$$D(E) = \sum_n |\alpha_n|^2 \, \delta(E - E_n) \qquad (5\text{-}12)$$

where E_n is the energy of the state Ψ_n. As time passes the state will change (put $\hbar = 1$)

$$\Phi(t) = \sum_n \alpha_n \Psi_n e^{-iE_n t} \qquad (5\text{-}13)$$

Table 5-2   MAXIMUM VALUE OF FWHM AND THE CORRESPONDING ENERGY (IN eV) FOR AN ELECTRON GAS FOR DIFFERENT r_s VALUES

	$r_s=2$	$r_s=3$	$r_s=4$	$r_s=5$
Maximum of FWHM	10	7	6	5
Energy above Fermi level	90	50	30	20

[1] "HWHM" means half-width at half-maximum.
[2] Note that if we choose $\Phi = T\Psi_i$, then $D(\omega + E_i)$ is identical to $I(\omega)/\omega$ in (5-1) except for a constant.

and have a smaller and smaller overlap $S(t)$ with the initial state

$$S(t) = \langle \Phi(0) \mid \Phi(t) \rangle = \sum_n |\alpha_n|^2 e^{-iE_n t} \qquad (5\text{-}14)$$

We immediately see that the overlap S and the distribution D are simply related

$$S(t) = \int_{-\infty}^{\infty} D(E) e^{-iEt} \, dE \qquad (5\text{-}15)$$

$$D(E) = \frac{1}{2\pi} \int_{-\infty}^{\infty} S(t) e^{iEt} \, dt \qquad (5\text{-}16)$$

The decay rate of the initial state is approximately given by the "Golden Rule" expression

$$w = -\frac{d}{dt} |S(t)|_{t=0}^2 = 2\pi \sum_f |\langle \Psi_f | V | \Psi_i \rangle|^2 \, \delta(E_f - E_i) \qquad (5\text{-}17)$$

where $\Psi_i = \Phi$ is an eigenfunction of H_0, the true Hamiltonian being $H_0 + V$. In many cases of radiative decay and Auger decay, D is a Lorentzian and the decay is exponential. Thus if D is given by

$$D(E) = \frac{1}{\pi} \frac{\Gamma}{(E - E_i)^2 + \Gamma^2} \qquad (5\text{-}18)$$

then S according to (5-15) becomes

$$S(t) = e^{-iE_i t} e^{-\Gamma |t|} \qquad (5\text{-}19)$$

and the decay rate for $t = 0$ is

$$w = -\frac{d}{dt} |S(t)|_{t=0}^2 = 2\Gamma \qquad (5\text{-}20)$$

We note that a straightforward perturbation calculation gives for the coefficients α_n in (5-11)

$$\alpha_n = \frac{-V_{ni}}{E_i - E_n} \qquad (5\text{-}21)$$

and from equations (5-12), (5-17), and (5-20) we have

$$D(E) = \frac{1}{\pi} \frac{\Gamma}{(E - E_i)^2} \qquad (5\text{-}22)$$

Such a calculation clearly cannot describe the central portion of the line; perturbation theory breaks down when the energy denominators $E_i - E_n$ in (5-21) become too small. We thus see that the actual calculation of line shapes is a nontrivial undertaking.

3. Competing decay mechanisms We consider the case when there are two groups of final states in (5-17) which contribute to the "Golden Rule" expression. This is the case, for example, when Auger decay (1) and radiative decay (2) occur simultaneously. The perturbation V then consists of two parts: $V_1 \approx 1/r_{12}$ and $V_2 \approx p$. These two parts couple the initial state to different types of final states, and we have

$$w = w_1 + w_2 \qquad (5\text{-}23)$$

where, to lowest order,

$$w_1 \approx |\langle \Psi_f^{(1)} | V_1 | \Psi_i \rangle|^2 \qquad (5\text{-}24)$$

and similarly for w_2.

The total decay rate w is thus the sum of the rates w_1 and w_2 that we would have if only one of the processes could occur. In (5-22) the width Γ is now given by the sum

$$\Gamma = \tfrac{1}{2}(w_1 + w_2) \qquad (5\text{-}25)$$

and we expect in simple cases to have a Lorentzian energy distribution and an exponential decay as given by equations (5-18) to (5-20). The decay time $\tau = 1/\Gamma$ is given by

$$\frac{1}{\tau} = \frac{1}{\tau_1} + \frac{1}{\tau_2} \qquad (5\text{-}26)$$

and we see that if one decay process is much faster than the other, say $\tau_1 \ll \tau_2$, then τ_1 determines the observed rate, $\tau \cong \tau_1$. In particular, we cannot expect to have two time constants involved: an expression like

$$|S(t)|^2 = \alpha e^{-w_1 t} + \beta e^{-w_2 t} \qquad (5\text{-}27)$$

is correct at $t = 0$ if $\alpha + \beta = 1$, but yields

$$w = -\frac{d}{dt}|S(t)|^2_{t=0} = \alpha w_1 + \beta w_2 \qquad (5\text{-}28)$$

which is different from $w_1 + w_2$. For example, in soft x-ray emission the (fast) Auger decay is thus very important by setting the time scale for the emission process; the time scale determines, for example, the amount of high-energy satellites (Subsection II F).

C.   Fano-type resonances in atoms

In the late fifties Lassettre and Silverman found some interesting structure in the inelastic scattering of electrons in a helium gas. The structure occurred as an energy loss at 60 eV, far above the ionization limit of 24.6 eV, and corresponded to the autoionizing level $2s2p\ ^1P$. The shape of the resonance was not Lorentzian, but of the unsymmetrical Breit-Wigner or Fano type. Similar effects were later found in far UV photoabsorption by Madden and Codling (1963). In a now classical paper Fano (1961) gave the theory for the resonance. This helium resonance is comparatively simple to discuss while it still brings out very important many-body features, and we will give a brief recapitulation of Fano's work supplemented by a few additional comments. A fuller account of these problems is given by Fano and Cooper (1968).

Fano gave the exact solution of the line-shape problem in an important special case. He considered the case of one discrete state φ (e.g., $2s2p$) embedded in a continuum ψ_E corresponding to a Hamiltonian H defined by

$$
\begin{array}{ll}
\langle \varphi | H | \varphi \rangle = E_0 & \langle \varphi \, | \, \varphi \rangle = 1 \\
\langle \psi_E | H | \varphi \rangle = V_E & \langle \psi_E \, | \, \varphi \rangle = 0 \\
\langle \psi_E | H | \psi_{E'} \rangle = E \, \delta(E - E') & \langle \psi_E \, | \, \psi_{E'} \rangle = \delta(E - E')
\end{array} \tag{5-29}
$$

The exact solutions to the Schrödinger equation $H\Psi_E = E\Psi_E$ were written

$$
\Psi_E = a\varphi + \int b_{E'} \psi_{E'} \, dE' \tag{5-30}
$$

which gives the secular equations

$$
E_0 a + \int V_{E'}^{*} b_{E'} \, dE' = Ea \tag{5-31}
$$

$$
V_{E'} a + E' b_{E'} = E b_{E'} \tag{5-32}
$$

We solve (5-32) for $b_{E'}$:

$$
b_{E'} = \left[\frac{P}{E - E'} + Z(E)\delta(E - E') \right] V_{E'} a \tag{5-33}
$$

Here P stands for the principal part and $Z(E)$ is an undetermined constant. Inserting b_E in (5-31) gives

$$
E = E_0 + F(E) + Z(E)|V_E|^2 \tag{5-34}
$$

where

$$
F(E) = P \int \frac{|V_{E'}|^2}{E - E'} \, dE' \tag{5-35}
$$

Note that $Z(E)$ turns out to be a real quantity. In a rigorous theory we thus cannot compact equation (5-33) into $b_{E'} = V_{E'} a/(E - E' + i\delta)$; nevertheless this "solution" has some significance as we shall see shortly.

The model problem is now solved except for the normalization of Ψ_E. The evaluation of $\langle \Psi_E | \Psi_{E'} \rangle$, which should give $\delta(E - E')$, is somewhat intricate mathematically due to the singular functions involved, but the result is simple

$$
|V_E a|^2 (\pi^2 + Z(E)^2) = 1 \tag{5-36}
$$

We next want to utilize the solution of the model problem. First we consider the case when the initial state Φ, discussed in Subsection II B2, is just the discrete state $\Phi = \varphi$. The energy distribution $D(E)$ defined in (5-12) then is

$$
D(E) = \int |\langle \Phi \, | \, \Psi_{E'} \rangle|^2 \delta(E - E') \, dE' = |\langle \Phi \, | \, \Psi_E \rangle|^2 = a(E)^2 \tag{5-37}
$$

From equations (5-34) and (5-36) we obtain

$$D(E) = \frac{1}{\pi} \frac{\pi|V_E|^2}{[E - E_0 - F(E)]^2 + (\pi|V_E|^2)^2} \tag{5-38}$$

If V_E and $F(E)$ have a slow variation in energy when we pass the resonance, $D(E)$ gives a Lorentzian resonance peak. This simple result can be obtained heuristically by solving equations (5-31) and (5-32) without caring more about the singularity than inserting an infinitesimal imaginary part, thus obtaining

$$E = E_0 + \int \frac{|V_{E'}|^2}{E - E' + i\delta} dE' = E_0 + F(E) - i\pi|V_E|^2 \tag{5-39}$$

If we (completely incorrectly) use the complex energy E to describe the time dependence of Φ

$$\Phi(t) = e^{-iEt}\Phi(0) \tag{5-40}$$

we obtain the $S(t)$ in (5-19) and the Lorentzian energy distribution in (5-38).

Let us next study photoabsorption. The absorption probability is given according to (5-1) by

$$I(E) = C|\langle \Psi_E|T|\Psi_G \rangle|^2 \tag{5-41}$$

where Ψ_G is the ground-state wavefunction and

$$T = \sum_i \mathbf{np}_i \tag{5-42}$$

If we define an initial state $\Phi = T\Psi_G$, we see from (5-37) that $I(E)$ equals $D(E)$ apart from a constant. Writing

$$\Phi = T\Psi_G = \alpha\varphi + \int \beta(E')\psi_{E'} \, dE' \tag{5-43}$$

we obtain after a simple calculation

$$I(E) = C|\beta(E)|^2 \frac{(\varepsilon + q)^2}{\varepsilon^2 + 1} \tag{5-44}$$

where

$$\varepsilon = \frac{E - E_0 - F(E)}{\pi|V_E|^2}$$

$$q = \frac{\alpha + P \int \frac{\beta(E')V_{E'}}{E - E'} dE'}{\pi\beta(E)V_E} \tag{5-45}$$

Equation (5-44) gives the Breit-Wigner or Fano spectral shape depicted in Fig. 5-2.

The experimental photoabsorption results for He were very well described by (5-44) with parameter values $q = -2.80$ and $\Gamma = \pi|V_E|^2 = 0.019$ eV. These values

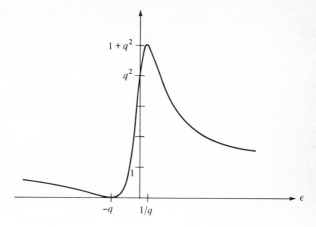

FIGURE 5-2
Breit-Wigner or Fano spectral line $(\varepsilon + q)^2/(\varepsilon^2 + 1)$ drawn for $q = 2$.

also agree very well with a priori calculations using the Fano truncation in (5-29) for the full Hamiltonian. To describe resonances in more complicated systems than He, a systematic treatment by diagram techniques could be employed advantageously (Wendin, 1970).

The theory for inelastic electron scattering is quite similar to the one for photo-absorption. The scattering probability is, in the Born approximation,

$$w \approx |\langle \Psi_E | T | \Psi_G \rangle|^2 \qquad (5\text{-}46)$$

where the transition operator T is

$$T = \frac{4\pi e^2}{q^2} \sum_i e^{i\mathbf{q} \cdot \mathbf{r}_i} \qquad (5\text{-}47)$$

$\mathbf{q}$ being the momentum transfer of the scattered electron. The operator T here is quite different from the momentum operator in (5-42), and thus the "initial states" $T\Psi_G$ are different. This difference does not affect the position of the resonance line or its width Γ, but only the parameter q. Thus for the $2s2p$ resonance in He with 500-eV energy of the incident electrons, we have $q = 1.8$, while for photoabsorption $q = -2.80$.

We note that the transition operator T depends on the energy of the incident electrons and the angle of observation. For high enough incident energies we can only have small momentum transfers so that $\exp(i\mathbf{q} \cdot \mathbf{r}) \cong 1 + i\mathbf{q} \cdot \mathbf{r}$. Since the 1 does not contribute, T becomes proportional to the dipole operator of optical absorption. To sum up, spectral lines need not be Lorentzian, and their shapes depend on how they are excited.

D. Shake-up effects

1. *The neon charge spectrum and photoelectron distribution* Consider x-ray absorption in a gas of neon atoms when the radiation energy is considerably higher than necessary to excite the K electrons. According to the one-electron theory, we would expect the radiation to give a certain distribution of singly ionized Ne atoms, with holes in the $K(1s)$, $L_I(2s)$, or $L_{II,III}(2p_{1/2},2p_{3/2})$ shells. The K holes would decay preferentially by Auger transitions (the $1s$ energy is 866.9 eV), leaving doubly ionized atoms, while the L_I holes would decay by a radiative transition since no Auger transition is energetically allowed. The result after some 10^{-13} sec would be a distribution of singly and doubly ionized atoms with holes in the $L_{II,III}$ shell. In an experiment by Carlson and Krause (1965) using 1.5-keV x rays the distribution of ions was analyzed 10^{-6} sec after the irradiation. They found, as could be expected, that the doubly ionized atoms were most abundant, but they also found a considerable number of more highly charged ions. See Table 5-3. The presence of the higher charges was explained by "shake-up" (or "shake-off") processes. The picture is then the following. The high-energy photon (1.5 keV) kicks out one electron, which leaves so quickly that the remaining electrons do not have time to readjust, and are thus left in a nonrelaxed state. This nonrelaxed state Φ has certain overlaps with the stationary states Ψ_n, and we thus obtain, according to the sudden approximation, probabilities $|\langle\Psi_n\,|\,\Phi\rangle|^2$ to end up in states Ψ_n, which may be multiply ionized (Wolfsberg and Perlman, 1955). As an example, take $\Phi = |1s,2s^2,2p^6\rangle$ and $\Psi_n = |1s,2s^2,2p^5,3p\rangle$, and evaluate Φ and Ψ_n in the Hartree-Fock approximation. The one-electron functions in Φ should be calculated from the configuration $1s^22s^22p^6$ (the initial state), while those of Ψ_n come from the configuration $1s2s^22p^53p$. The two sets of orbitals are then different, and $\langle\Psi_n\,|\,\Phi\rangle \approx \langle 3p\,|\,2p\rangle$ and thus nonzero. Extensive such calculations in the Hartree-Fock approximation have been made by Krause et al. (1968), and by Aberg (1967), with quite reasonable results.

More detailed results are obtained if also the energy of the photoelectron is registered. The experiment is then called *x-ray photoemission* (XPS). In Fig. 5-3 we show results[1] obtained by Siegbahn et al. (1969) for the $1s$ photoelectron from a neon

Table 5-3 CHARGE SPECTRUM OF Ne IRRADIATED BY 1.5-keV X RAYS

Charge	1	2	3	4	5
%	5.7	70.2	20.8	3.0	0.28

[1] In the figure, structure irrelevant to the present discussion has been smoothed out. The experimental data are found in Fig. 4.2 of Siegbahn et al. (1969).

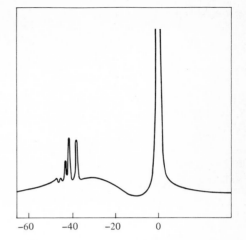

FIGURE 5-3
X-ray photoemission result from the $1s$ electron in Ne. The satellite structure at about 40 eV is due to $1s2s^22p^5\ np$ configurations. Structure irrelevant to the present discussion has been omitted. [*After Siegbahn et al.* (1969).]

atom. The dominating line comes from those $1s$ electrons which leave a relaxed ion behind them (configuration $1s2s^22p^6$) while the satellite structure to the left comes from $2s$ electrons of some 40-eV lower energy that have left the ion in excited states corresponding to configurations $1s2s^22p^5\ np$. Such states were called *valence-electron-configuration* (VEC) states by Parratt (1959). Another closely related aspect shows up as double-electron excitation processes in absorption spectra (Schnopper, 1963).

2.   *Threshold effects studied by Auger spectroscopy*   Use of the sudden approximation clearly breaks down when the excitation energy approaches threshold for the creation of multiple holes. This can conveniently be studied in Auger spectra (Mehlhorn and Stalherm, 1968). An ejected K electron can leave the atom with either a K hole or, by shake-up, a KL hole. The Auger electrons ejected in K-LL and KL-LLL processes have slightly different energies. A measurement of the relative intensity of the Auger lines should give the probability for the shake-up process. Results for Ne of this probability, obtained with electron bombardment by Carlson et al. (1970), are shown in Fig. 5-4. The energy E_e of the impact electron is measured in units of the binding energy of the $2p$ electron ($E_0 = 50$ eV), starting the energy scale at E_i, the binding energy of the $1s$ electron. Double excitation starts as soon as energy conservation allows, at $(E_e - E_i)/E_0 = 1$. The probability rises steeply, and within about $10E_0$ it reaches the same order of magnitude as the asymptotic value given by the shake-up theory. Note also the interesting threshold structure in Fig. 5-4 which stretches over some $50E_0$. The limits where suddenness applies have been discussed by Sachenko and Demekhin (1966), Sachenko and Burtsev (1968), and by Åberg (1969). A recent discussion for the atomic case has been given by Meldner and Perez (1971).

FIGURE 5-4
The probability for creating, by electron bombardment, a KL hole in Ne, as compared to the probability for creating a K hole. The energy argument is $(E_e - E_i)/E_0$, where E_e is the energy of the bombarding electron, E_i the threshold energy for creating a K hole, and $E_i + E_0$ that for creating a KL hole. [*After Carlson et al.* (1970).]

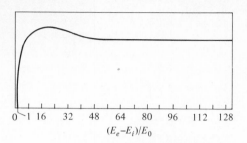

$(E_e - E_i)/E_0$

In the Auger experiments just discussed, electrons were used to excite the atoms. It has been checked by Graeffe et al. (1969), and by Krause et al. (1970), that the same asymptotic limit for shake-up probabilities is reached whether x rays or electrons are used for excitation. For energies where the asymptotic limit is not yet reached, there is, however, no equivalence of "direct" (by electrons) and "indirect" (by photons) excitations. This is in accordance with the results in Subsection II C, where we showed that the spectral shapes, in principle, depend on the excitation probe. Such dependences are well established experimentally for electrons as compared to photons in solids (Cauchois, 1968). If protons or other heavy particles are used, the probe dependence is even more marked, and for shake-up the whole pattern is strongly modified (Edwards and Rudd, 1968).

3. *Schnopper's electron bombardment (EB) and K-capture (KC) study* An interesting experiment where shake-up effects seem to be absent has been made by Schnopper (1967). He compared x-ray emission lines, obtained (1) after electron bombardment (EB) of Mn ($Z = 25$), using electron energies of about 15 keV, and (2) after K capture (KC) in Fe^{55} ($Z = 26$). The bombardment energy is sufficiently high so that we should expect to produce a $1s$ hole in Mn with no relaxation of the outer electrons. With the K capture, on the other hand, we produce the $1s$ hole in Mn practically without perturbing the outer electrons. These cases of no relaxation and complete relaxation gave the same $K\alpha_{12}(2p \rightarrow 1s)$ and $K\beta_{13}(3p \rightarrow 1s)$ spectra within experimental accuracy. This fact at first seems unexpected, and so we must take a closer look at the situation.

There is no reason to expect that EB does not give shake-up effects as usual. These effects are strongest in the outer shells. What happens in the outer shells, however, does not affect the $K\alpha$ and $K\beta$ spectra since both of the inner levels involved are affected in the same way by the outer electrons. We can also have shake-up effects on the inner-core electrons, but these are usually weak and give intensity in another spectral region than the one under observation anyhow. What might be affected is the $K\beta'$ structure. The origin of this structure is not agreed upon except to say that it comes from the influence of the valence electrons on the $3p \rightarrow 1s$ transition. Parratt (1959) has argued that it comes from shake-up, while Tsutsumi and Nakamori (1968) have attributed it to spin splitting of the core states from interactions with the d electrons (cf Subsection V C). In any case, the lack of difference between the EB and KC spectra can be taken as an indication that the lifetime of the core hole ($\approx 10^{-15}$ sec in this case) is long enough for the conduction electrons to relax before the transition takes place. This possibility of relaxation before emission is typical of solids (cf Subsection II F).

4. *"Radiative Auger" effect of Aberg et al.*   We have discussed shake-up effects in x-ray absorption; they have their counterpart also in x-ray emission. In an investigation of the satellite structure of the $K\alpha_{12}$ line ($2p \rightarrow 1s$) in some second-row elements, Aberg and Utriainen (1969) found a weak structure about 200 eV below the main line having about 0.5 percent of the intensity. This was interpreted as the excitation of one $2p$ electron to the unoccupied states when another fell down into the $1s$ hole. No detailed calculations have been made so far, but the effect is of the right order of magnitude to be explained in the same way as a shake-up effect. It would then be proportional to a basic dipole matrix element times an overlap integral, e.g.,

$$w \sim |\langle 3p' \mid 2p \rangle \langle 2p | \mathbf{np} | 1s \rangle|^2 \qquad (5\text{-}48)$$

This gives a probability that is $|\langle 3p' \mid 2p \rangle|^2$ smaller than for the main line, or a total probability for shake-up of $\approx 1 - \langle 2p' \mid 2p \rangle^{12}$. This total probability has been estimated by Aberg (1967) as 2 or 3 percent which differs considerably from the previously mentioned experimental result of 0.5 percent; however, both the experimental and theoretical numbers are quite crude. Aberg and Utriainen termed their effect "radiative Auger" or "atomic internal Compton," for obvious reasons.

5.   X-ray photoemission   If we are looking at an XPS spectrum and thus *know* that one electron has a large velocity, we can write the final state in relation (5-1) (in second quantization)[1]

$$|\Psi_f\rangle = a_\kappa^{+}|N - 1, s\rangle \qquad (5\text{-}49)$$

where κ is the state of the high-velocity electron and $|N - 1, s\rangle$ a state of the other $N - 1$ electrons, some of which may be leaving the atom (but then with low velocity compared to electron κ which has most of the energy). Writing the transition operator T as

$$T = \sum_{mn} P_{mn} a_m^{+} a_n \qquad (5\text{-}50)$$

---

[1] The development here follows closely the one given by Hedin (1967) and Hedin and Lundqvist (1969).

where m and n refer to the one-electron states associated with the ground state $\Psi_i = |N\rangle$, we have for the photoelectron distribution $I(\varepsilon)$

$$I(\varepsilon) = C\omega \sum_{\kappa s} \left| \left\langle N - 1, s \left| a_\kappa \sum_{mn} P_{mn} a_m{}^+ a_n \right| N \right\rangle \right|^2 \delta(\omega - \varepsilon_\kappa + \varepsilon_s)\delta(\varepsilon_\kappa - \varepsilon) \qquad (5\text{-}51)$$

where

$$\varepsilon_s = E(N) - E(N - 1, s) \qquad (5\text{-}52)$$

If the state κ has high enough energy, it is outside the cloud of one-electron states that form the correlated ground state $|N\rangle$, and thus only $m = \kappa$ contributes to the sum in (5-51). We have

$$I(\varepsilon) \sim \sum_s \left| \left\langle N - 1, s \left| \sum_n P_{\kappa n} a_n \right| N \right\rangle \right|^2 \delta(\omega - \varepsilon + \varepsilon_s) \qquad (5\text{-}53)$$

At this point we reach contact with the more intuitive arguments put forward earlier. The state $a_n|N\rangle$ is the unrelaxed state Φ (think of a Hartree-Fock approximation), while $|N - 1, s\rangle$ is a stationary state Ψ. Only the overlap between these states enters weighted by the oscillator strengths.

To make contact with the Green's-function technique, we note that (5-53) can be written

$$I(\varepsilon) \approx \sum_{mn} P_{\kappa m}^* P_{\kappa n} A_{mn}(\varepsilon - \omega) \qquad (5\text{-}54)$$

where $A_{mn}(\omega)$ is the spectral function associated with the one-electron Green's function (Hedin and Lundqvist, 1969, pp. 26 and 158). X-ray spectra and optical spectra usually require knowledge of at least a two-particle Green's function for their proper description. These higher order Green's functions are more difficult to calculate and to describe.

The descriptions we have of shake-up do not give the dynamical details of the process. We thus only have a qualitative picture of an incoming photon that kicks out a primary photoelectron, then some secondaries may appear, and finally we have radiative decay or Auger decay. We do not know, however, how long a time it takes after the primary electron is emitted until the secondaries appear and until the decay takes place, and we do not know the energy distribution of the secondaries. This information should be possible to obtain from theory and also from experiment by coincidence measurements.

E.   Configuration interaction effects

The shake-up effects can be viewed as a simple type of configuration interaction effects, allowing themselves to be described by making two Hartree-Fock (H-F) calculations, one for the initial and one for the final state. We will now give a few examples of excitation effects which probably cannot be described as shake-up but are easily understood in terms of configuration mixing.

Cooper and La Villa (1970) found satellites in the $3s \rightarrow 2p$ emission spectra of Ar and KCl. These satellites corresponded energetically to final states of configuration $3s^2 3p^4 nd\ ^2S$ as compared to the configuration $3s3p^6\ ^2S$ of the main line. Configuration interaction strongly mixes these two configurations (Kjöllerström et al., 1965) and thus supplies an explanation of the rather strong (≈ 10 percent) satellites, while an interpretation in terms of two H-F calculations probably does not work. Cooper and La Villa termed their effect "semi-Auger."

Wertheim and Rosencwaig (1971) found an analogous effect in the x-ray photoemission spectra of a series of alkali halides. Thus in RbCl, for example, the Rb ($4s$) line had a strong satellite about 10 eV below the main line which was explained by the configuration mixing between $4s4p^6\ ^2S$, $4s^2 4p^2 5s\ ^2S$, and $4s^2 4p^4 4d\ ^2S$. Again the satellite is quite strong and cannot easily be explained by "shake-up" effects.

F. High-energy satellites in x-ray emission; incomplete relaxation; self-absorption

In Subsections II C and II D we discussed the importance of how an initial state was prepared. With high enough energy of the excitation radiation (e.g., electrons or photons) we found that the initial state could be multiply ionized either through shake-up or through Auger transitions involving a lower lying core level. In the latter case we speak of Coster-Kronig transitions (Richtmyer, 1937). The doubly- (or multiply-) ionized states give rise to high-energy satellites in the x-ray emission. The excitation radiation can also create a cloud of low-energy electron-hole excitations in metals, again either through shake-up or through an Auger process involving closely spaced core levels (e.g., L_{II} and L_{III} levels). If the core lifetime is short enough so that the excitation cloud does not have time to decay or diffuse away, this cloud gives a broadening of the emission edge. We may then speak of *incomplete relaxation*. Even for light metals where the relaxation should be very nearly complete, there is a little "foot" at the bottom of the Fermi edge, as noted already by Skinner (1940). For transition metals with their shorter core lifetimes, the effect is more pronounced. In the observed spectra, however, edge broadening tends to be suppressed by self-absorption. The higher the energy of the excitation radiation, the stronger the self-absorption, since the radiation then penetrates deeper into the solid and the emitted light thus has a longer path to travel before it leaves the solid. The incomplete relaxation is best studied with radiation not too far from threshold. If the radiation energy gets very close to threshold, there is no high-energy structure, since there is then not sufficient energy to excite a lower core level which can Auger-decay and since the shake-up probability rapidly tends to zero; however, with large radiation energy the high-energy structure is self-absorbed away. Self-absorption has been studied by Liefeld and others (Liefeld, 1968; Chopra, 1970) and by Fischer (1970) for transition metals, and by Dimond (1967) for second-row alloys. Self-absorption can largely be avoided by using a large incident-electron angle and a 90° x-ray take-off angle (Fischer, 1970; Holliday, 1970).

The importance of self-absorption has been stressed by Parratt (1959), but seems to have been generally recognized only lately. Earlier emission measurements, particularly of edge widths, should thus be taken with some caution.

A full theoretical treatment of incomplete relaxation has not yet been carried out. In the simplest version of the theory we would have a transport problem describing how the initially created excitation cloud is leaving the core hole during the emission process, while in a full treatment we should have one dynamical process rather than a set of independent events. A recent contribution to the problem of dynamic perturbations in metals has been made by Müller-Hartmann, Ramakrishnan, and Toulouse (1971).

G.   Phonon effects

Phonon effects on optical spectra are often important, particularly in semiconductors and insulators. A well-known case is the cascade involving some 20 to 30 phonons which accompanies the optical transition in an *F* center (Brown, 1967). The theory for these phonon effects is well mapped out in papers by Huang and Rhys (1950), Lax (1952), Hopfield (1962), and others. This theory can to a large extent be taken over for the electron-hole cascades, and plasmon emissions associated with x-ray transitions in metals. It also applies to phonon effects on line shapes and edges in the x-ray case. These effects are nonnegligible for metals as well as insulators; however, since the resolution is worse in x-ray spectra than in optical spectra, they are of less importance relatively.

A comprehensive discussion of the phonon effects has been given by Lax (1952), and we find it expedient for the later sections to recapitulate the essential features of his long and detailed paper. In its simplest or semiclassical form, the Franck-Condon principle describes optical (or x-ray) transitions, assuming that the nuclei are standing still. The energy necessary for the transition depends on the instantaneous positions of the nuclei; the energy spread is obtained by averaging over the nuclear positions in the initial state (see Fig. 5-5). In x-ray transitions the Coulomb potential from the valence electrons at the position of a core electron varies as the nuclei vibrate; this change in the core-electron energy is the main source of phonon-induced energy spreads in x-ray transitions.

The semiclassical case has been discussed by Parratt (1959) who cites an equation derived by Overhauser. McAlister (1969) recently applied the Overhauser relation to

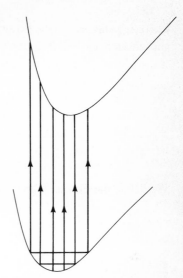

FIGURE 5-5
The semiclassical form of the Franck-Condon principle: The energy spread of an electronic transition is obtained by averaging over the nuclear configurations using the probability distribution of the lattice vibrations in the initial state.

a discussion of the soft x-ray emission edge in lithium. The Overhauser relation, however, seems to overestimate the phonon effects substantially, as we shall discuss later (cf also Bergersen et al., 1971), but more conservative estimates still indicate that tens of phonons may be created, causing broadenings of the order of 0.1 eV. This effect is small but may be observable at low temperatures if the core level is narrow (e.g., like in Al).

1. *The dynamical Franck-Condon principle* In a quantum-mechanical treatment of the Franck-Condon principle we start from the basic expression for the absorption

$$I(\omega) = \sum_{if} g_i |\langle \Psi_f | T | \Psi_i \rangle|^2 \delta(\omega - E_f + E_i) \qquad (5\text{-}55)$$

The g_i are statistical weights for the initial states Ψ_i, and the treatment of phonons is made in the Born-Oppenheimer approximation with a product of the electron ψ and phonon ϕ wavefunctions

$$|\Psi_i\rangle = |\psi_i\rangle |\phi_n\rangle$$
$$|\Psi_f\rangle = |\psi_f\rangle |\phi'_m\rangle \qquad (5\text{-}56)$$

We have put a prime on the final-state phonon wavefunction to indicate that it belongs to a different interatomic potential than in the initial state. We make the usual "Condon approximation" of neglecting the dependence on nuclear coordinates in the transition-matrix element $\langle \psi_f | T | \psi_i \rangle = T_{fi}$ and obtain

$$I(\omega) = \sum_{fmn} g_n |T_{fi}|^2 |\langle \phi'_m | \phi_n \rangle|^2 \delta(\omega - \varepsilon_f + \varepsilon_i - \omega'_m + \omega_n) \qquad (5\text{-}57)$$

where ω'_m and ω_n are the energies associated with the phonons. Introducing electron and phonon spectral functions, we have

$$I(\omega) = \int I^e(\omega - \omega')D(\omega')\, d\omega'$$

$$I^e(\omega) = \sum_f |T_{fi}|^2 \delta(\omega - \varepsilon_f + \varepsilon_i) \qquad (5\text{-}58)$$

$$D(\omega) = \sum_{mn} g_n |\langle \phi'_m \mid \phi_n \rangle|^2 \delta(\omega - \omega'_m + \omega_n)$$

We define the Fourier transform $S(t)$ of $D(\omega)$ by

$$S(t) \equiv \int D(\omega)e^{-i\omega t}\, d\omega = \sum_{mn} g_n |\langle \phi'_m \mid \phi_n \rangle|^2 e^{i(\omega_n - \omega'_m)t} \qquad (5\text{-}59)$$

and the transforms $I(t)$ and $I^e(t)$ of $I(\omega)$ and $I^e(\omega)$ in a similar way. We then have $I(t) = I^e(t)S(t)$. In particular we can obtain the behavior of $I(\omega)$, for small ω, from that of $I(t)$ for large t by well-established methods of Fourier analysis (e.g., see Lighthill, 1958).

To evaluate $S(t)$ we introduce the operator

$$U(t) = e^{iH_0 t}e^{-i(H_0 + \Delta V)t} \qquad (5\text{-}60)$$

where H_0 and $H_0 + \Delta V$ are the Hamiltonians for the phonons in the initial and final states. We then have, by standard methods of many-body theory (e.g., see Schultz, 1964),

$$S(t) = \sum_n g_n \langle \phi_n | U(t) | \phi_n \rangle$$

$$U(t) = T \exp\left[-i \int_0^t \Delta V(t')\, dt' \right] \qquad (5\text{-}61)$$

where T is the time-ordering operator and

$$\Delta V(t) = e^{iH_0 t}\, \Delta V e^{-iH_0 t} \qquad (5\text{-}62)$$

Keeping to linear terms in the lattice displacements, we can write ΔV

$$\Delta V = \sum_v (B_v a_v + B_v^* a_v^+) \qquad (5\text{-}63)$$

where a_v^+ and a_v are the phonon creation and annihilation operators of H_0

$$H_0 = \sum_v \omega_v (a_v^+ a_v + \tfrac{1}{2}) \qquad (5\text{-}64)$$

The time-dependent operator $\Delta V(t)$ becomes

$$\Delta V(t) = \sum_v (B_v e^{-i\omega_v t} a_v + B_v^* e^{i\omega_v t} a_v^+) \qquad (5\text{-}65)$$

The operators in the exponent of $U(t)$ with different v values commute, and it is sufficient to study one v value. Since each B_v is of order $N^{-1/2}$, we need to develop the exponent only up to the first nontrivial term (we leave out the index v for simplicity). Omitting the algebra, we have

$$S(t) = \sum_n g_n \left\langle \phi_n \left| T \exp\left[-i \int_0^t (Be^{-i\omega t'}a + B^*e^{i\omega t'}a^+)\, dt' \right] \right| \phi_n \right\rangle$$

$$= 1 + i \frac{|B|^2}{\omega} t + \frac{|B|^2}{\omega^2} \{(e^{-i\omega t} - 1)[n(\omega) + 1]$$

$$+ (e^{i\omega t} - 1)n(\omega)\} + \cdots \qquad (5\text{-}66)$$

Here n is the occupation number; thus

$$n(\omega) = (e^{\beta\omega} - 1)^{-1} \qquad \beta = \frac{1}{kT} \qquad (5\text{-}67)$$

Since $S(t)$ is the product of a large number ($\approx N$) of terms, each of the form $1 + C/N$, we have exactly in the $N \to \infty$ limit

$$S(t) = \exp\left(\int g(\omega)\{i\omega t + (e^{-i\omega t} - 1)[n(\omega) + 1] + (e^{i\omega t} - 1)n(\omega)\}\, d\omega \right) \qquad (5\text{-}68)$$

where the phonon distribution $g(\omega)$ is defined from

$$\sum_v \left|\frac{B_v}{\omega_v}\right|^2 = \int g(\omega)\, d\omega \qquad (5\text{-}69)$$

The basic equation (5-68) has been derived for the $T = 0$ case by Langreth (1970) without any assumption that the B's were infinitesimal. A heuristic and much shorter derivation of (5-68) has been given by Hopfield (1969).

2. Moments; connection with the semiclassical case Knowing the "phonon spectrum" $g(\omega)$, $S(t)$ can be calculated and hence also $D(\omega)$. At high temperatures $D(\omega)$ resembles a Gaussian (Lax, 1952); at low temperatures it may have a complicated shape (Bergersen, McMullen, and Carbotte, 1971). The low moments of D are simply related to g; thus

$$\int D(\omega)\, d\omega = S(t = 0) = 1$$

$$\int \omega D(\omega)\, d\omega = iS'(t = 0) = 0$$

$$\int \omega^2 D(\omega)\, d\omega = i^2 S''(t = 0) = \int [2n(\omega) + 1]\omega^2 g(\omega)\, d\omega \qquad (5\text{-}70)$$

$$\int \omega^3 D(\omega)\, d\omega = i^3 S'''(t = 0) = \int \omega^3 g(\omega)\, d\omega$$

From the limit of $S(t)$ for large t we see from (5-68) that $D(\omega)$ has a δ-function contribution, the "no-phonon" peak

$$D(\omega) = e^{-n}\delta(\omega - E_0) + \cdots \qquad (5\text{-}71)$$

where E_0 is the energy shift,

$$E_0 = -\int \omega g(\omega)\, d\omega \qquad (5\text{-}72)$$

and n is the number of emitted phonons,

$$n = \int [2n(\omega) + 1] g(\omega)\, d\omega \qquad (5\text{-}73)$$

These results are easily checked by integrating $D(\omega)$ over an infinitesimal interval including $\omega = E_0$. Lax also showed that the important second moment in (5-70) is the same as obtained in the semiclassical version of the Franck-Condon principle. Assuming a ΔV linear in the displacements as in (5-63), the semiclassical theory gives a perfectly Gaussian shape for $D(\omega)$.

3.   *Change of interatomic potential in an x-ray transition* So far the results are quite general, and we will use them also in later sections. Here we will apply them to discuss the phonon contributions to the shapes of core states. The energy necessary to take out a core electron from an ion core embedded in a solid above the energy required in a free ion core can be written (Hedin and Johansson, 1969)

$$\Delta E = V_H + \tfrac{1}{2}V_P \qquad (5\text{-}74)$$

Here V_H is the Coulomb potential from all the valence electrons at the position of the ion core evaluated before the core electron has been taken out, and V_P is the "polarization potential," that is, the extra Coulomb potential coming from the polarization of the valence electrons by the core hole. Both V_H and V_P depend on the atomic configuration. In a linear theory, where the distortion of the valence-electron charge density is the sum of the distortions created by each ion, V_P and the contribution to V_H coming from the ion "i" that has a core hole are constants. The change ΔV in the interatomic potential thus comes from the change in V_H at the ion "i" caused by changes in positions of ions "j" ($j \neq i$). Hence we need the Coulomb potentials only outside the ion cores and can use a pseudopotential formulation without bothering about the difference between pseudo-wavefunctions and true wavefunctions (neglecting the small "orthogonalization-hole" effect). Taking the potential from the bare pseudo-ions as

$$V_0(\mathbf{r}) = \sum_j w_0(\mathbf{r} - \mathbf{R}_j) \qquad (5\text{-}75)$$

we have for the Fourier transform of the crystal potential $V(\mathbf{r})$

$$V(\mathbf{q}) = \int e^{i\mathbf{q}\cdot\mathbf{r}} V(\mathbf{r})\, d\mathbf{r} = \sum_j e^{i\mathbf{q}\cdot\mathbf{R}_j} w(q) \qquad (5\text{-}76)$$

where

$$w(q) = \frac{w_0(q)}{\varepsilon(q)} \qquad (5\text{-}77)$$

is the bare pseudopotential $w_0(q)$ screened by the dielectric function $\varepsilon(q)$ appropriate to changes in charge density (Hedin and Lundqvist, 1971). The quantity ΔV in (5-63) that we need is, apart from a constant,

$$\Delta V = V(\mathbf{R}_i) = \sum_j e^{i\mathbf{q}\cdot(\mathbf{R}_j - \mathbf{R}_i)} w(q) \frac{d\mathbf{q}}{(2\pi)^3} \qquad (5\text{-}78)$$

The term containing the nuclear positions can be expanded as

$$\sum_j e^{i\mathbf{q}\cdot(\mathbf{R}_j - \mathbf{R}_i)} = \sum_j e^{i\mathbf{q}\cdot(\mathbf{R}_j{}^0 - \mathbf{R}_i{}^0)}[1 + i\mathbf{q}\cdot(\mathbf{u}_j - \mathbf{u}_i) + \cdots] \qquad (5\text{-}79)$$

Omitting the constant terms and expanding $\mathbf{u}_i$ in phonon operators, we have

$$\Delta V = \sum_j \int e^{i\mathbf{q}\cdot(\mathbf{R}_j{}^0 - \mathbf{R}_i{}^0)} i\mathbf{q}\cdot(\mathbf{u}_j - \mathbf{u}_i) w(q) \frac{d\mathbf{q}}{(2\pi)^3} = \sum_{\mathbf{k}\lambda}^{BZ} (B_{\mathbf{k}\lambda} a_{\mathbf{k}\lambda} + B_{\mathbf{k}\lambda}^* a_{\mathbf{k}\lambda}{}^+) \qquad (5\text{-}80)$$

with

$$B_{\mathbf{k}\lambda} = \frac{i}{\Omega_0} \sum_{\mathbf{K}} \left(\frac{\hbar}{2NM\omega_{\mathbf{k}\lambda}}\right)^{1/2} \varepsilon_{\mathbf{k}\lambda}[(\mathbf{K}+\mathbf{k})w(\mathbf{K}+\mathbf{k}) - \mathbf{K}w(\mathbf{K})] \qquad (5\text{-}81)$$

where $\mathbf{K}$ is a reciprocal-lattice vector, Ω_0 the volume of a unit cell $(\Omega_0 = \Omega/N)$, M the mass of the ion (taking a monoatomic crystal), and $\varepsilon_{\mathbf{k}\lambda}$ the polarization vector. From equations (5-63) and (5-69) we have, expressing the $\omega_{\mathbf{k}\lambda}$ in units of the Debye frequency ω_D and momenta in units of the Fermi momentum,

$$g(\omega) = \frac{3Z}{8\pi} \frac{m}{M} \left(\frac{\varepsilon_F}{\hbar\omega_D}\right)^3 \frac{1}{\omega^3}$$

$$\times \sum_\lambda \int \left|\sum_{\mathbf{K}} \varepsilon_{\mathbf{k}\lambda}\left[(\mathbf{k}+\mathbf{K})\frac{w(\mathbf{k}+\mathbf{K})}{\varepsilon_F\Omega_0} - \mathbf{K}\frac{w(\mathbf{K})}{\varepsilon_F\Omega_0}\right]\right|^2 \frac{dS}{|\nabla\omega_{\mathbf{k}\lambda}|} \qquad (5\text{-}82)$$

where Z is the charge of the ion. Since $\varepsilon_F/\hbar\omega_D$ is of the order $(M/m)^{1/2}$, the numerical factor in g is of the same order and thus fairly large. Keeping only the $\mathbf{K} = 0$ term in the sum and neglecting the transverse phonons, we have

$$g(\omega) = \frac{3Z}{2} \frac{m}{M} \left(\frac{\varepsilon_F}{\hbar\omega_D}\right)^3 \frac{k^4}{\omega^3|\nabla\omega_{\mathbf{k}}|} \left(\frac{w(\mathbf{k})}{\varepsilon_F\Omega_0}\right)^2 \qquad (5\text{-}83)$$

We note that $g(\omega) \approx \omega$ for small ω.

4. **Estimates of phonon effects** In a crude approximation we can write $g(\omega) = \alpha\omega \cdot \theta(\omega_D - \omega)$, choosing, e.g., $\alpha = g(\omega_D)/\omega_D$. We then have for the key quantities at $T = 0$:

$$n = \int g(\omega)\, d\omega = \tfrac{1}{2}\alpha\omega_D{}^2$$

$$E_0 = -\int \omega g(\omega)\, d\omega = -\tfrac{1}{3}\omega_D{}^3 = -n\omega_D \qquad (5\text{-}84)$$

$$\Delta E^2 = \int \omega^2 g(\omega)\, d\omega = \tfrac{1}{4}\alpha\omega_D{}^4 = \tfrac{1}{2}n\omega_D{}^2$$

For a Gaussian shape the half-width at half-maximum is related to ΔE by

$$\text{HWHM} = (2 \ln 2)^{1/2}\, \Delta E \cong 1.12\, \Delta E \qquad (5\text{-}85)$$

Results from crude estimates based on the above formulas, using the Shaw (1968) pseudopotential form factor $(\sin qR)/(qR)$ and the Hedin and Lundqvist (1971) dielectric function, are given in Table 5-4, together with results obtained by Bergersen, McMullen, and Carbotte (1971). In the Bergersen, McMullen, and Carbotte paper no pseudopotential form factor was used, but the calculation was made with full details of the phonon spectrum. They also transformed the sum over $\mathbf{K}$ in (5-82) to a sum over $\mathbf{R}$, which was carried to convergence. The two sets of results in the table agree well enough considering all uncertainties involved; in particular they predict much smaller values than by the original Overhauser relation. The type of approximations used in our estimates and in Bergersen's indicates that all results in Table 5-4 are somewhat large.

We note that the strength of the "zero-phonon" line is very small and that the widths are barely large enough to be observable at low temperatures. By comparison, the contribution from thermal excitations in the Fermi sea to the edge width (from

Table 5-4   PHONON EFFECTS IN SIMPLE METALS (ENERGIES IN eV)

Metal	Present estimates			Bergersen et al.		
	n	E_0	HWHM	n	E_0	HWHM
Li	7	−0.17	0.08	2.8	−0.041	0.025
Na	6	−0.06	0.03	5.3	−0.036	0.019
Al	60	−1.3	0.21			

5 to 95%) is

$$5.89kT \cong 5 \times 10^{-4}T \quad \text{eV} \quad (5\text{-}86)$$

The value of HWHM for Al given in Table 5-4 seems about a factor of 4 larger than the results by the DESY group (Kunz, in a private communication, estimates a value not larger than 0.06 eV). There is thus still a large discrepancy between theory and experiment.

We finally comment on the lifetime questions. In molecular spectroscopy there is often a difference between absorption and emission frequencies, the latter being smaller due to relaxation through phonon transitions. In x-ray spectroscopy of simple metals there does not seem to be any such differences. Thus in the case of Al the absorption and emission edges coincide to within 0.02 eV or better. According to Table 5-4, we could have a phonon relaxation energy of -1.3 eV (E_0) for Al. This number is probably too large. If we assume that HWHM is 0.06 eV instead of 0.21 eV (cf, our previous discussion of HWHM), then by equation (5-84) n is down by a factor $(0.21/9.06)^2 = 12$, which gives $E_0 = -0.11$ eV. Also this value is considerably larger than the accuracy 0.02 eV with which absorption and emission edges coincide, which indicates that there is no phonon relaxation going on. This is consistent with the lifetime estimate of about 10^{-13} sec, which barely allows one lattice vibration. The often-used phrase "complete relaxation" thus refers only to the electrons and not to the phonons.

III. X-RAY PHOTOEMISSION AND SOFT X-RAY SPECTRA IN SIMPLE METALS

A. Basic features

Before starting the discussion of many-body effects, we will shortly summarize some basic features. X-ray photoemission (XPS), or electron spectroscopy for chemical analysis (ESCA) as it is often called by the main users of the method, cf introduction to Chap. 8, has during the last decade developed into a precise and sophisticated tool for the investigation of the properties of valence electrons. The most important data from the photoelectron spectra are the positions of the core levels; small "chemical" shifts in these positions signal changes in the charge distribution and polarizability of the valence electrons. The line shapes and the satellite structures give other important information. This latter information is most clear-cut in spectra from gases, where the effects of secondary scattering can easily be estimated by changing the gas pressure. The other main drawback, besides secondary scattering, of the XPS method is the comparatively large width (about 1 eV) of the excitation radiation. XPS spectra have been obtained not only from the core electrons but also from the valence electrons; these map directly the occupied part of the band structure. The XPS method is described in detail by Hagström and Fadley in Chap. 8.

Soft x-ray spectra (SXS) are traditionally used for obtaining information on the unoccupied (absorption) and occupied (emission) parts of the band structure. For metals, the SXS spectra, like the XPS spectra, have important many-body modifications giving rise to edge singularities and to satellite structures. In emission the satellites come out clearly below the main band, but in absorption they fall on top of the one-body structure, and it is very difficult to assess experimentally how large they are.

SXS spectra are often directly associated with the one-electron density of states. Disregarding edge effects and satellites, there are, however, still two factors which disrupt this simple interpretation. One is the oscillator strength, and the other is the perturbing effect of the core hole. It is now well established (Subsection V A) that the oscillator strengths are about as important as the density of states itself, and it seems likely that the core hole perturbs the spectra appreciably. SXS spectra are described in detail by Cuthill in Chap. 3 and Rooke in Chap. 4 (emission) and by Azároff and Pease in Chap. 6 (absorption).

The many-body effects are mainly concerned with cascades of electron-hole pairs and phonons, with plasmon emission and with the coupling between quasi-particles and plasmons. We also have the high-energy satellites discussed in Subsection II F and the Auger tailing. The latter is due to the finite lifetime of the quasi-particles and to shake-up of electron-hole pairs. To help visualize these effects, sketches of the main features are given in Fig. 5-6.

The theories for XPS and SXS spectra are strongly interwoven. There is, however, a stepwise increase of complication in going from XPS to SXS in absorption and to SXS in emission. In our discussion we will tend to follow these steps rather than the historical development of the subject. To aid the reader who has had some contact during the years with the development of the theory, we will begin, however, by giving a short historical sketch.

B.   Review of contributions to the theory of edge and satellite effects

The possibility of plasmon creation in soft x-ray emission was suggested by Ferrell (1956) in a review paper on positron annihilation. The proposed mechanism for the plasmon creation was the shock on the valence-electron system when the core hole disappeared. The effect was observed by Rooke (1963) and found to be much weaker than predicted by Ferrell. It was then realized that the actual shock was produced by the difference in the potentials from the core electron and the jumping valence electron, rather than from the core electron alone. This refined explanation was outlined in a treatment based on the Bohm-Pines theory by Brouers (1964; 1967) and

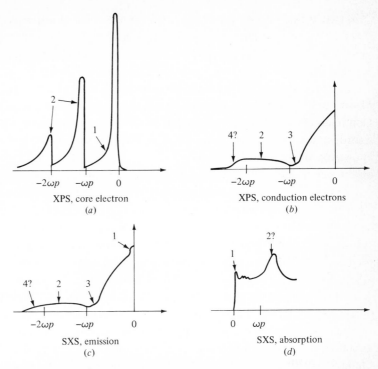

FIGURE 5-6

Sketch of many-body effects in spectra from metals. (*a*) XPS from a core electron; (*b*) XPS from conduction electrons; (*c*) SXS in emission; (*d*) SXS in absorption. (1) Effect due to cascade of electron-hole pairs; (2) plasmon satellite; (3) Auger tail; (4) plasmaron edge due to coupling between quasi-particles and plasmons.

in a semiclassical description by Ferrell (1965*a* and *b*). Elaborate many-body calculations of the effect have been made by Longe and Glick (1969) and others.

The development of the theory for electron-hole cascades started in the spring of 1967 with independent contributions by Mahan and Anderson, who were both inspired by private communications from Hopfield. Mahan discussed excitons in metals and found indications of a logarithmic divergence at the threshold (1967*a*); shortly afterward (1967*b*) he changed "logarithmic" to "power law." Anderson (1967*a*) showed that the ground state of a Fermi gas with a localized impurity potential was orthogonal to a gas without impurity; he sharpened this discussion mathematically slightly later (1967*b*). Mahan and Anderson's work clearly demonstrated the importance of electron-hole excitations of low energy, and they both suggested that the effect should be seen in SXS. We mention in passing an interesting difference brought out by Mahan and Anderson's work between spectral and ground-state properties, namely that the latter vary completely smoothly when an impurity potential is introduced (Kohn and Majumdar, 1965), while the former change character.

The work by Mahan and Anderson gave rise to intensive activity which has resulted in dozens of papers on these problems. The essential features of the SXS problem were clarified in independent contributions by Mizuno and Ishikawa (1968), Morita and Watabe (1968), Bergersen and Brouers (1969), and Nozières and de Dominicis (1969). The last of these papers is the longest and the most complete and rigorous; the first is a short but still quite comprehensive note: it was written without knowledge of the Mahan-Anderson work. The second and third papers concentrate on the important problem of obtaining a description for the whole main band, while the other two papers were concerned only with the exponent of the singular term.

The XPS case was first discussed by B. I. Lundqvist (1969b) who, from the consideration of a low-order diagram, concluded that the quasi-particle δ-function peak should be softened into a skew line and give away a substantial part of its oscillator strength to a satellite structure. The basic physics of both the SXS and XPS problems was analyzed in a review article by Hedin and S. Lundqvist (1969). The implications of the Nozières and de Dominicis work for the XPS case was clearly recognized by Doniach and Sunjic (1970) who discussed at length the possibilities of verifying the skew line shape experimentally. The treatment of the XPS satellite structure given by B. I. Lundqvist (1969b) was extended in an important way by Langreth (1970), and the quantitative consequences of Langreth's results were mapped out by Hedin, Lundqvist, and Lundqvist (1970).

The possibility of a coupling between plasmons and quasi-particles was discussed by Hedin, Lundqvist, and Lundqvist (1967) and in more detail by B. I. Lundqvist (1967a; 1967b; 1968; 1969a; 1969b). The consequences on SXS in emission were discussed by Hedin (1967; 1968). While the plasmon emission proposed by Ferrell (1956) should manifest itself as a high-energy edge of the plasmon satellite structure, the coupled plasmon-electron aggregate, the "plasmaron," should give rise to a low-energy edge (Fig. 5-6).

When Ferrell (1956) discussed the effect of the core hole in SXS, he made an analogy with positron annihilation and predicted a large enhancement in the SXS emission intensity. Harrison (1968), however, pointed out that the structure of the core, which gives an effective pseudopotential that is much weaker than the bare Coulomb potential of a positron, should make the SXS intensity enhancement much smaller than that in positron-annihilation rates. This conclusion was also argued by Hedin (1968) from a somewhat different viewpoint and supported by explicit calculations by Hedin and Sjöström (1971) and by Brouers, Longe, and Bergersen (1970). The strong many-body effects on the energy position of XPS peaks and SXS edges were pointed out and estimated by Hedin (1965b; 1967) and their origin explained by Hedin and Johansson (1969).

C.   X-ray photoemission spectra

X-ray photoemission (XPS) was discussed in Subsection II D 5, where we derived the basic relation (5-53) for the intensity of the photoelectrons

$$I(\varepsilon) \approx \sum_s \left| \left\langle N - 1, s \left| \sum_n P_{\kappa n} a_n \right| N \right\rangle \right|^2 \delta(\omega - \varepsilon + \varepsilon_s) \qquad (5\text{-}87)$$

Here $\varepsilon_s = E_N - E_{N-1,s}$, and κ is defined by $\varepsilon_\kappa = \varepsilon$. XPS was also discussed in Subsection II D 1, with specific application to Ne gas.

1. *Spectra from the valence electrons*   The basic equation (5-87) can be written in terms of the spectral function A [cf (5-54)]

$$I(\varepsilon) \sim \sum_{mn} P_{\kappa m}^* P_{\kappa n} A_{mn}(\varepsilon - \omega) \qquad (5\text{-}88)$$

Using the vector label $\mathbf{k}$ for the one-electron Bloch states, we have, approximating $A_{\mathbf{kk'}}$ by its diagonal part,

$$I(\varepsilon) \sim \sum_{\mathbf{k}} |P_{\kappa \mathbf{k}}|^2 A(\mathbf{k}, \varepsilon - \omega) \qquad (5\text{-}89)$$

In the one-electron approximation $A(\mathbf{k},\omega)$ is just a δ function, $A(\mathbf{k},\omega) = \delta(\omega - E_{\mathbf{k}})$, while for interacting electrons A can have considerable structure (see, e.g., Hedin and Lundqvist, 1969, p. 92). If the matrix elements $P_{\kappa \mathbf{k}}$ were constant, $I(\varepsilon)$ would give the density of states

$$N(\omega) = \sum_{\mathbf{k}} A(\mathbf{k},\omega) \qquad (5\text{-}90)$$

The density-of-states curve is also expected to show strong deviations from the parabolic form, when electron interactions are considered and look something like the sketch in Fig. 5-6 for the "XPS conduction electrons" case.

The matrix elements are not constant, but they could hardly vary so much that the strong satellite indicated by the theoretical results for the density of states would fail to come through in the experimental spectra. Thus careful XPS measurements in the tail region of the valence-electron spectra of simple metals should be very interesting. The interpretation of the experimental data is, however, obscured by the broadening effects of the incident radiation and secondary scattering and also by the more or less unknown background intensity.

2. *Spectra from the core electrons*   In core-electron spectra, the final states $|N - 1, s\rangle$ in (5-87) have one core electron missing. The operator a_n must thus destroy a core electron in the initial state $|N\rangle$, and we have to a good approximation

$$I(\varepsilon) \approx |P_{\kappa c}|^2 \sum_s |\langle N_v^*, s | N_v \rangle|^2 \delta(\varepsilon - \varepsilon_0 - E_s) \qquad (5\text{-}91)$$

where $|N_v\rangle$ is the ground state of the valence electrons and $|N_v^*, s\rangle$ is an excited state with energy E_s calculated in the presence of the perturbing potential from the core hole. The matrix element $P_{\kappa c}$ varies only weakly with $\varepsilon(= \varepsilon_\kappa)$ in the energy range of interest, and the photoelectron distribution $I(\varepsilon)$ thus maps an energy distribution like the $D(E)$ discussed in Subsection II B 2 [cf equations (5-11) and (5-12)]. Such energy distributions were discussed explicitly in Subsection II G on phonons [cf (5-57)], and much of that discussion can be taken over here.

The problem of the singular edge has been solved rigorously by Nozières and de Dominicis (1969). We will, however, not recapitulate their work here but content ourselves with two simplified derivations, one based on Anderson's theorem and the other on neglect of correlations between low-energy electron-hole pairs. We will then discuss the satellite structure, following closely the same lines of derivation as for the edge (and the phonon case). These discussions of the edge singularity and the satellite structure start from different model Hamiltonians. To obtain a unified treatment we introduce the diagram technique and derive the value of the coupling constant in the model Hamiltonians.

a. The skew quasi-particle line The Anderson theorem states that the overlap between the ground states of a uniform electron gas containing N electrons and a gas with a localized impurity potential is [1]

$$\langle N^* \mid N \rangle = N^{-\delta^2/(3\pi^2)} \qquad (5\text{-}92)$$

where δ is the $l = 0$ phase shift[2] for the impurity potential.

From this result we can obtain the singular limiting value of $I(\varepsilon)$ for small ε by a heuristic argument. Consider (5-14) with $|\Phi(t)\rangle$ taken as the state of the valence-electron system at time t after the photoelectron has been kicked out of the ion core. We may picture $|\Phi(t)\rangle$ as having a completely relaxed spherical region around the core hole, with one shell of incomplete relaxation and one of no relaxation outside. If we assume that the radius of the inner sphere is vt and neglect the intermediate region, we have

$$S(t) = \langle \Phi(0) \mid \Phi(t) \rangle = \langle N_2|\langle N_1 \mid N_1^* \rangle|N_2 \rangle = \langle N_1 \mid N_1^* \rangle \qquad (5\text{-}93)$$

where $|N_1\rangle$ and $|N_2\rangle$ are the state vectors for the inner and outer regions. Since by assumption $N_1 \approx (vt)^3$, we have from Anderson's theorem

$$S(t) \approx t^{-\delta^2/\pi^2} \qquad (5\text{-}94)$$

and hence from the asymptotic properties of Fourier transforms (see, e.g., Lighthill,

---

[1] The extra factor of 2 in the exponent as compared to Anderson's (1967b) formula is a spin factor.

[2] The contributions from higher l values are neglected.

1958, p. 43)

$$\lim_{E \to 0} D(E) \approx \left(\frac{1}{E}\right)^{1 - \delta^2/\pi^2} \qquad (5\text{-}95)$$

which gives the limiting behavior of $I(\varepsilon)$ in (5-91). This very crude argument is not sufficient to produce the correct exponent $1 - 2\delta^2/\pi^2$, but it may help to clarify the connection between Anderson's theorem and the spectral properties. In reality we cannot expect a sharp boundary between relaxed and unrelaxed regions which propagates with velocity v; the detailed description of the relaxation process remains an unsolved and very interesting problem.

The threshold singularity can also be derived in a straightforward and more stringent way by using the same technique as for the phonon case in Subsection II G (Hopfield, 1969). Consider the model Hamiltonian used by Nozières and de Dominicis (1969)

$$H = \sum_k \varepsilon_k a_k^+ a_k + \sum_{kk'} V_{kk'} a_k^+ a_{k'} b b^+ + E_0 b^+ b \qquad (5\text{-}96)$$

Here a_k and b are annihilation operators for a conduction electron and a core electron, while $V_{kk'}$ is the scattering potential from the core hole. $|N_v\rangle$ in (5-91) is the ground state of the first term in (5-96), and is thus a filled Fermi sea of plane-wave states. The $|N_v^*, s\rangle$ in (5-91) are excited states of the Hamiltonian

$$H^* = \sum_k \varepsilon_k a_k^+ a_k + \sum_{kk'} V_{kk'} a_k^+ a_{k'} \qquad (5\text{-}97)$$

In the spirit of the Tomonaga (1950) paper we treat the electron-hole pairs $a_k^+ a_{k'}$ as independent Boson-type operators. Neglecting certain commutators, we can then write H^* in (5-97)

$$H^* = \sum_v \omega_v c_v^+ c_v + \sum_v (B_v c_v + B_v^* c_v^+) \qquad (5\text{-}98)$$

with the index v labeling an electron-hole pair $(\mathbf{k}, \mathbf{k'})$ where $k > k_F$ and $k' < k_F$, and with $\omega_v = \varepsilon_k - \varepsilon_{k'}$, $c_v^+ = a_k^+ a_{k'}$ and $B_v = V_{kk'}$. We then have exactly the same problem as in the phonon case with $H = H_0 + \Delta V$, H_0 and ΔV being given by (5-64) and (5-63). The distribution function $g(\omega)$ of (5-69) is determined by

$$\int g(\omega) \, d\omega = \sum_k^{\text{unocc}} \sum_{k'}^{\text{occ}} \left| \frac{V_{kk'}}{\varepsilon_k - \varepsilon_{k'}} \right|^2$$

$$= 2|V_0 N_F|^2 \int d\omega \int_0^\infty d\varepsilon_1 \int_{-\infty}^0 d\varepsilon_2 \frac{\delta(\omega - \varepsilon_1 + \varepsilon_2)}{(\varepsilon_1 - \varepsilon_2)^2} \qquad (5\text{-}99)$$

and thus becomes

$$g(\omega) = \frac{2|V_0 N_F|^2}{\omega} \theta(\omega) \qquad (5\text{-}100)$$

Here N_F is the density of s states at the Fermi surface, V_0 is the matrix element between two s states, and 2 is a spin factor. In the Born approximation we have $V_0 N_F = -\delta/\pi$, where δ is the $l = 0$ phase shift.

From (5-72) we see that the energy shift E_0 is well defined, while (5-73) shows that the number of emitted electron-hole pairs diverges, as it should according to Anderson's theorem. As a comparison to the singular behavior of $g(\omega) \approx 1/\omega$ for electron-hole pairs, we found in the phonon case that $g(\omega) \approx \omega$.

To obtain the threshold behavior of $I(\varepsilon) = D(\varepsilon_0 - \varepsilon)$, we need the small ω behavior of $D(\omega)$, which follows from the large t behavior of $S(t)$. We have from (5-68), disregarding the E_0 factor and introducing a cutoff energy ω_0,

$$S(t) = \exp\left[\int_0^{\omega_0} \frac{A}{\omega}(e^{-i\omega t} - 1)\,d\omega\right] \qquad (5\text{-}101)$$

where $A = 2|V_0 N_F|^2$. We rewrite the integral as

$$\int_0^{\omega_0 t} \frac{e^{-ix} - 1}{x}\,dx = \int_0^1 \frac{e^{-ix} - 1}{x}\,dx + \int_1^{\omega_0 t} \frac{e^{-ix}}{x} - \int_1^{\omega_0 t} \frac{dx}{x} \qquad (5\text{-}102)$$

and have for $S(t)$ in the limit of large t

$$S(t) = \exp\left[A(\text{const} - \log \omega_0 t)\right] \sim (\omega_0 t)^{-A} \qquad (5\text{-}103)$$

which leads to the correct power-law singularity for $D(\omega)$

$$\lim_{\omega \to 0} D(\omega) \approx \omega^{1-A} \qquad (5\text{-}104)$$

We note that the choice for the cutoff energy ω_0 is immaterial for the limiting behavior of $D(\omega)$, which thus is determined by electron-hole excitations of very low energy.

b.   *The satellite structure*   The satellite structure of the core spectra is due to plasmon excitation. We can discuss this structure in terms of a Hamiltonian (Lundqvist, 1969b; Langreth, 1970)

$$H = \sum_q \omega_q c_q^+ c_q + \sum_q g_q(c_q + c_{-q}^+)bb^+ + E_0 b^+ b \qquad (5\text{-}105)$$

where in comparison with equations (5-96) and (5-98) we have replaced the individual electron-hole operators $c_v^+ = a_k^+ a_{k'}$ by the coherent superposition

$$c_q^+ = \sum_p \alpha_p a_{p+q}^+ a_p \qquad (5\text{-}106)$$

which forms the creation operator for a plasmon (Sawada et al., 1957). Like the low-energy electron-hole operators the plasmon operators are assumed to obey Boson commutation relations. Their energy spectrum however starts with a constant value, the plasmon energy of order 10 eV, and increases quadratically in the momentum, while the electron-hole energies start linearly from zero, $\varepsilon_{\mathbf{k}+\mathbf{q}} - \varepsilon_{\mathbf{k}} = [1/(2m)] \times$ $(\mathbf{kq} + \mathbf{q}^2)$. It is a reasonable approximation to choose the dispersion

$$\omega_q = \omega_p + \frac{1}{2m}q^2 \qquad (5\text{-}107)$$

and the coupling constant

$$g_q^{\;2} = \frac{1}{\Omega}\frac{4\pi e^2}{q^2}\frac{\omega_p^{\;2}}{2\omega_q} \qquad (5\text{-}108)$$

as we will discuss later. We obtain the "plasmon-distribution function" $g(\omega)$ from (5-69)

$$\int g(\omega)\,d\omega = \sum_q \left|\frac{g_q}{\omega_q}\right|^2 = \frac{e^2}{\pi}\int \frac{\omega_p^{\;2}\,d\omega_q}{\omega_q^{\;3}(d\omega_q/dq)} \qquad (5\text{-}109)$$

thus

$$g(\omega) = \frac{e^2}{\pi}\left(\frac{m}{2}\right)^{1/2}\frac{\omega_p^{\;2}\theta(\omega-\omega_p)}{\omega^3(\omega-\omega_p)^{1/2}} \qquad (5\text{-}110)$$

We note that $g(\omega)$ is zero for $\omega < \omega_p$ and starts with an integrable singularity. The number n of emitted plasmons, (5-73), is thus finite, and we have

$$S(t) = e^{-n}\exp\left[\int_{\omega_p}^{\infty} g(\omega)e^{-i\omega t}\,d\omega\right]e^{-iE_0 t} \qquad (5\text{-}111)$$

which can be expanded as [since $g(\omega)$ is integrable]

$$S(t) = e^{-n}e^{-iE_0 t}\left[1 + \int g(\omega)e^{-i\omega t}\,d\omega \right.$$
$$\left. + \tfrac{1}{2}\int g(\omega)g(\omega')e^{-i(\omega+\omega')t}\,d\omega\,d\omega' + \cdots\right] \qquad (5\text{-}112)$$

Disregarding the energy shift E_0, we obtain for $D(\omega)$

$$D(\omega) = e^{-n}\left[\delta(\omega) + g(\omega) + \tfrac{1}{2}\int g(\omega')g(\omega-\omega')\,d\omega' + \cdots\right] \qquad (5\text{-}113)$$

The spectrum thus consists of a δ-function peak (the quasi-particle), a satellite $g(\omega)$ starting with a $\omega^{-1/2}$ singularity, at an energy ω_p, a second satellite starting with a constant value at $2\omega_p$, and further contributions starting at $3\omega_p$, $4\omega_p$, etc. The latter carry only a few percent of the total oscillator strength. The spectrum is illustrated in Fig. 5-6 for XPS, core electron. Note that the energy is measured in different directions for I and D, $I(\varepsilon) = D(\varepsilon_0 - \varepsilon)$.

c. Unified discussion of the skew quasi-particle line and the satellite The discussions of the main peak and the satellite structure can be given more systematically if we

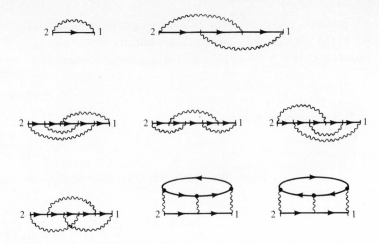

FIGURE 5-7
Diagrams representing the expansion of the self-energy Σ. The arrows represent Green's functions, the wiggly lines screened interactions.

use diagrams. From (5-89) we see that the basic quantity involved is the one-electron spectral function A or equivalently the one-electron Green's function G. The latter can be represented by diagrams as shown in Fig. 5-7, where the arrows represent Green's functions and the wiggly lines screened interactions. The contribution to A from the first diagram was calculated by B. I. Lundqvist (1969b), neglecting the finite extension of the core-electron wavefunction, using the zero-order Green's function, and taking the Lindhard dielectric function in the screened interaction. The result is shown by the dashed and dotted curve in Fig. 5-8. The effect of the finite size of the core-electron wavefunction was estimated to be very small. The spectrum was also very insensitive to the detailed choice of dielectric function. In particular the simple choice

$$\frac{1}{\varepsilon(q,\omega)} = 1 + \frac{\omega_p{}^2}{\omega^2 - \omega_q{}^2} \qquad (5\text{-}114)$$

which does not contain any electron-hole excitations, and thus gives a δ-function quasi-particle peak, left the satellite structure practically unchanged in shape and strength as compared to choosing the Lindhard dielectric function.

The result obtained with the model Hamiltonian in (5-105), using the coupling constant in (5-108), is precisely the same as obtained by summing all diagrams consisting of one core-hole line dressed in all possible ways with interactions that are screened with the dielectric function in (5-114). The result is shown as the full-drawn curve in Fig. 5-8. Comparison with the dotted curve, obtained with the first diagram

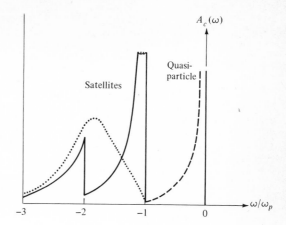

FIGURE 5-8
Theoretical results for the XPS core-electron spectrum of an idealized metal with the density of sodium ($r_s = 4$). The full-drawn curve gives the exact solution for the model Hamiltonian in (5-105), while the dotted curve is the result when only the first diagram in Fig. 5-7 is kept. The dashed line indicates the result from the first diagram when the screening includes electron-hole pairs and not only plasmons.

only, does not look very encouraging. Even if the shape of the satellite comes out poorly, however, at least the position in energy and the strength come out quite well.

The shape of the quasi-particle peak obtained with the first diagram using the Lindhard dielectric function (dashed line in Fig. 5-8) is singular as it should be; however, the form is $(\omega \ln^2 \omega)^{-1}$ rather than the correct $(1/\omega)^{1-A}$ result. The singularity is due to the infinite number of electron-hole pairs contained in one screened interaction. The model Hamiltonian in (5-96), however, gives rise to a much wider class of particle-hole contributions, and there is no reason to expect the first diagram to describe the shape of the quasi-particle line better than it describes the satellite.

To sum up, the first diagram in Fig. 5-7 gives only a very rough idea of the core-electron spectrum. The strength and the position of the structure seem right, and a skew line is predicted; however, the detailed shape of the spectrum seems completely unreliable.

d. Comparison with experiment The experimental information on the shape of the core spectra is not very detailed due to broadenings from the excitation radiation and from the energy losses of the outgoing photoelectron. Results for aluminum obtained by P. O. Hedén (1970, unpublished) are shown in Fig. 5-9. The relative strength of the satellite is about 40% as compared to about 35% obtained from the theory discussed above (Hedin, 1971, unpublished). The excess strength obtained experimentally is easily accounted for by the secondary scattering. The predicted singular satellite structure and the skew edge shape seem, however, well hidden by the experimental broadenings. Improved experimental conditions with better vacuum, better surfaces, and grazing incidence of the excitation radiation (to

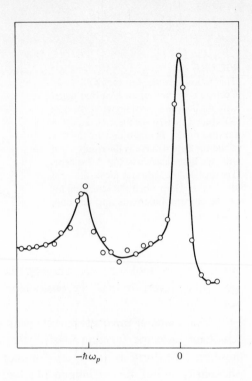

FIGURE 5-9
X-ray photoemission result from a $2p$ electron in aluminum metal obtained by P. O. Hedén. The skew line shape and satellite are partly due to intrinsic effects and partly to inelastic scattering of the outgoing photoelectron.

minimize secondary scattering) could, however, throw more light on these questions.

Also the positions of the quasi-particle peaks are affected by many-body effects. The factors determining these positions were discussed in Subsection II G 3. The agreement between theory and experiment is good for simple metals (Hedin, 1965b; Hedin and Lundqvist, 1969; Hedin and Johansson, 1969). For a further discussion of core-level positions we refer to Chap. 8 by Hagström and Fadley. Core-level positions in transition metals will be discussed in Subsection V C.

D.   X-ray emission and absorption spectra

1.   *Formulation of the problem* The basic theory for XPS developed in Subsection II D 5 can easily be adapted to SXS, if in emission we assume that the core-hole lifetime is long enough to allow complete relaxation (Subsection II F). Thus we write the initial and final states in absorption and emission as

$$
\begin{aligned}
\text{absorption} && |\Psi_i\rangle &= |N_v\rangle|N_c\rangle \\
&& |\Psi_f\rangle &= a_c|N_v + 1^*, s\rangle|N_c\rangle \\
\text{emission} && |\Psi_i\rangle &= a_c|N_v^*\rangle|N_c\rangle \\
&& |\Psi_f\rangle &= |N_v - 1, s\rangle|N_c\rangle
\end{aligned}
\qquad (5\text{-}115)
$$

where v and c stand for valence and core, and the star serves to remind that a core-hole potential is present. The core part in the problem again drops out to a good approximation, and the transition probability in (5-1) becomes, for absorption,

$$I(\omega) \approx \omega \sum_f \left| \left\langle \Psi_f \left| \sum_{mn} P_{mn} a_m^+ a_n \right| \Psi_i \right\rangle \right| \delta(\omega - E_f + E_i)$$

$$= \omega \sum_s \left| \left\langle N_v + 1^*, s \left| \sum_m P_{mc} a_m^+ \right| N_v \right\rangle \right|^2 \delta(\omega - \omega_0 - E_s) \qquad (5\text{-}116)$$

where E_s is the excitation energy of the state $|N_v + 1^*, s\rangle$, and for emission

$$I(\omega) \sim \omega \sum_s \left| \left\langle N_v - 1, s \left| \sum_n P_{cn} a_n \right| N_v^* \right\rangle \right|^2 \delta(\omega - \omega_0 + E_s) \qquad (5\text{-}117)$$

where E_s now is the excitation energy of the state $|N_v - 1, s\rangle$. The threshold energy ω_0 is the same in emission and absorption, namely the energy to take out a core electron and place it on top of a completely relaxed valence-electron sea (Hedin, 1967).

2. *The edge singularities* If we compare equations (5-91) for XPS and (5-116) and (5-117) for SXS, we see that they are the same if in SXS we define "initial states" by $\sum_m P_{mc} a_m^+ |N_v\rangle$ and $\sum_n P_{cn} a_n |N_v^*\rangle$; we can forget about the difference between $|N_v\rangle$ and $|N_v \pm 1\rangle$ since we are discussing metals. Like in XPS, we thus expect singular edges in SXS. In XPS we have a $1/E$ singularity with exponent $1 - 2(\delta/\pi)^2$. The relevant phase shift in SXS comes from the difference between the potential from the core hole and the fictitious potential that gives the "initial state." The wavefunction of the "initial state" in configuration space is (Hedin and Lundqvist, 1969, p. 66) the antisymmetrized product of (in the absorption case)

$$\varphi(\mathbf{r}_1)\Psi_{N_v}(\mathbf{r}_2, \mathbf{r}_3, \ldots, \mathbf{r}_{N_v+1}) \qquad \text{where } \varphi(\mathbf{r}) = \sum_m P_{mc}\psi_m(\mathbf{r}) = P\psi_c(\mathbf{r})$$

Here ψ_m and ψ_c are the one-electron wavefunctions for valence and core electrons. Since ψ_c is a strongly localized function, so is $\varphi = P\psi_c$, and the "initial state" thus has one bound electron. This bound electron has a definite spin (that of ψ_c), and the potential corresponding to the "initial state" has a phase shift of π for, say, s electrons (if we consider L spectra) of spin up neglecting the small d component of φ. The exponent of $1/E$ thus becomes

$$1 - \left(1 - \frac{\delta}{\pi}\right)^2 - \left(\frac{\delta}{\pi}\right)^2 = 2\left(\frac{\delta}{\pi}\right) - 2\left(\frac{\delta}{\pi}\right)^2 \qquad (5\text{-}118)$$

where δ is the phase shift of the core-hole potential (we consider only s waves). The preceding qualitative discussion is based on the work by Schotte and Schotte (1969a; 1969b) and by Hopfield (1969).

The rigorous solution to the edge problem for the Hamiltonian in (5-96) was developed by Nozières and de Dominicis (1969). They obtained the following expression for the intensity at the edge

$$\sum_{l=0}^{\infty} W_l \left(\frac{1}{E}\right)^{\alpha_l} \qquad (5\text{-}119)$$

where

$$\alpha_l = 2\left(\frac{\delta_l}{\pi}\right) - 2 \sum_{l=0}^{\infty} (2l + 1) \left(\frac{\delta_l}{\pi}\right)^2 \qquad (5\text{-}120)$$

When the core state has p symmetry, only W_0 and W_2 are different from zero, while for an s-type core state only W_1 contributes. For simple metals there are reasons to expect that δ_0 is the largest phase shift. Friedel's sum rule then tells us that $\delta_0 \cong \pi/2$, and from (5-120) we can expect that $\alpha_0 \cong \frac{1}{2} > 0$ while the other α's are negative. Thus the L spectra should show a singularity, while the K spectra should be rounded (since only α_1, which presumably is negative, contributes). These conclusions are the same for emission spectra.

The Li K spectrum has a rounded edge as opposed to the sharp edges of the other simple light metals (Na, Mg, Al, K). The rounding can be explained at least partially by the many-body effect. This was first proposed by Mizuno and Ishikawa (1968), and the argument was developed with more rigor by Ausman and Glick (1969). The effect also may be due partially to resonance scattering of the conduction electrons by the core hole (Allotey, 1967).

3. **The main band in soft x-ray emission**   The experimental results for soft x-ray emission spectra of the most free-electron-like metals, sodium and potassium, indicate that the edge singularity is rather weak and that it affects the spectrum only close to the edge. Theoretical calculations with the one-electron approximation (Subsection V A)

$$I_0(\omega) \sim \omega \sum_k |P_{ck}|^2 \delta(\omega - \varepsilon_k + \varepsilon_c) \qquad (5\text{-}121)$$

reproduce the experimental results to a reasonable degree in the cases that have been investigated (Li, Na, K, Rb, Cs, Al, Cu, Si, Ge). With these facts in mind, we may attempt the simplest possible extension of the one-electron theory, namely to include only one virtual electron-hole excitation. Thus

$$|N_v^*\rangle = \left(\alpha + \sum_{pq} \alpha_q{}^p a_p{}^+ a_q\right) |N_v\rangle \qquad (5\text{-}122)$$

The correct wavefunction is built from an infinite number of electron-hole pairs as shown by Anderson's theorem; however, (5-122) may be a good enough approximation away from the edge. If we were to include an infinite number of electron-hole pairs and thus account correctly for the singular edge, we might proceed by solving the Nozières and de Dominicis (1969) equations. These equations are quite difficult to solve, however; so far the only numerical results are estimates of the exponent in the singular term.

The result for the intensity $I(\omega)$ in (5-117), when we use the approximation in (5-122), is (Hedin, 1967)

$$I(\omega) \sim \omega \sum_{k}^{\text{occ}} \left| \alpha P_{ck} + \sum_{q} \alpha_k{}^q P_{cq} \right|^2 \delta(\omega - \varepsilon_k + \varepsilon_c)$$

$$+ \frac{\omega}{2} \sum_{k}^{\text{unocc}} \sum_{k_1 k_2}^{\text{occ}} |\alpha_{k_2}{}^k P_{ck_1} - \alpha_{k_1}{}^k P_{ck_2}|^2 \delta(\omega - \varepsilon_{k_1} - \varepsilon_{k_2} + \varepsilon_k + \varepsilon_c) \qquad (5\text{-}123)$$

The first term (I_1) is the same as the one-electron result I_0 in (5-121), except that P_{ck} has been replaced by an effective matrix element, involving states above the Fermi surface. The second term (I_2) gives rise to multiple excitations. Estimates of I_2 by variational calculations and using constant matrix elements (Hedin, 1967; Hedin and Sjöström, 1971) indicate that it is much smaller than I_1; however, it contributes in the tailing region where I_1 is small or zero. Its contribution to the tail appears to be comparable to, but definitely smaller than, the contribution from the lifetime of the quasi-particles discussed by Landsberg (1949), and by Blokhin and Sachenko (1960). The multiple excitation term (I_2) was first seriously discussed by Pirenne and Longe (1964); they, however, left out the second (exchange) matrix elements in (5-123) and obtained a contribution far too large.

Estimates of the main term $I_1(\omega)$ can be discussed in terms of an enhancement factor $\gamma(\omega)$

$$I_1(\omega) = \gamma(\omega) I_0(\omega) \qquad (5\text{-}124)$$

Calculations using constant matrix elements (Hedin and Sjöström, 1971; Brouers, Longe, and Bergersen, 1970) indicate that γ is quite sensitive to the choice of potential for the core hole. If a Coulomb potential is used, there is a large enhancement (factors of 5 or 10), much as for positron annihilation. If, on the other hand, we take account of the internal structure of the ion core by using a pseudopotential, the enhancement goes down to much smaller values (25 or 50%). In all cases γ turned out to be almost linear in ω, and to increase by about 50% from the bottom to the top of the Fermi sea.

Absolute intensities cannot presently be measured, but results for the enhancement are still of interest. Thus a γ value of, say, 1.25 at the Fermi surface shows that [cf (5-123)] the states above the Fermi surface contribute significantly to the effective matrix element in I_1.

We have discussed the one-particle-hole approximation where the coefficients α_q^p were evaluated from the variational principle. Another approach was used by Bergersen and Brouers (1969) who made a perturbation calculation. Their result can be written $I = I_0 + I_1 + I_2 + \cdots$, where I_0 is the one-electron approximation

$(\approx \sqrt{\omega})$, I_1 contains a log ω term, I_2 a log^2 ω term, etc. To improve their result Bergersen and Brouers write the intensity in the form

$$I(\omega) = I_0(\omega) \exp \frac{I_1(\omega)}{I_0(\omega)} \qquad (5\text{-}125)$$

which agrees to lowest order with the series expansion and gives a power-law singularity at the edge. Estimates using an approximation of a screened Coulomb potential gave results which looked more like an asymmetric core-electron peak than a conduction band. Later calculations based on pseudopotentials (Brouers, Longe, and Bergersen, 1970; Bergersen, Brouers and Longe, 1971) gave more reasonable but qualitative results. These results indicate the need of a nonperturbative approach to the very important problem of accounting quantitatively for the whole main band, including the singular edge.

4. *Satellites* The plasmon satellite in soft x-ray emission comes from the shock on the valence electrons when the core hole is filled and has thus the same origin as in the XPS case. The effect is weaker, however, since it is only the difference between the potential from the core hole and from the jumping electron that acts, and not the full core-hole potential as in the XPS case. A clear-cut case, where we can see that only the difference enters, is that of an x-ray line; here the relevant potential that shakes the conduction electrons no doubt is the difference between the potential of the two core holes involved. A jumping conduction electron cannot leave a small well-defined hole like a core electron does, but it should leave a hole that is localized to some extent. Ferrell (1965a; 1965b) has discussed this situation in a crude but physically plausible way. He calculated the probability for plasmon creation when a charged spherical shell (the conduction electron hole) expanded with a certain velocity v. The result was averaged for velocities ranging from zero to the Fermi velocity. The resulting probability for plasmon creation was much smaller than if only the potential from the core hole were considered, and it was in rough agreement with experiment.

Another approach to this problem was taken by Brouers (1964; 1967) who made use of the Pines (1955) method of canonical transformations. Brouers found that the virtual plasmon fields of the core hole and the conduction electron to a large extent canceled each other, again giving a result in rough accord with experiment. The problem has also been approached by straightforward many-body diagram expansions (Glick, Longe, and Bose, 1968; Brouers and Longe, 1968); the results are, however, plagued by difficulties with divergences and with large and almost canceling contributions. Other, mainly formal, many-body approaches have been attempted by Rystephanick and Carbotte (1968) and by Heaney and Rystephanick (1970).

A nonperturbative analysis has been given by Hedin (1968) who arrived at the expression

$$I(\omega) \sim \omega \sum_{k} |P_{ck}^{\text{eff}}(\omega)|^2 A(k,\omega) \qquad (5\text{-}126)$$

Here $A(k,\omega)$ is the one-electron spectral function, and an effective matrix element appears

$$P_{ck}^{\text{eff}}(\omega) = \alpha P_{ck} + P_{ck}^{(+)} + (\omega + \varepsilon_c - \varepsilon_k^{\text{H-F}})P_{ck}^{(-)} \qquad (5\text{-}127)$$

where the $P_{ck}^{(+)}$ involve the coefficients α_k^q of (5-122) and couple to Bloch states above (+) and below (−) the Fermi level. For ω in the main band, the coefficient of the $P_{ck}^{(-)}$ term is small, and we have essentially the result of the first term in (5-123). In the tailing region the $P_{ck}^{(-)}$ terms pick up in importance due to the large dispersion of the H-F energy $\varepsilon_k^{\text{H-F}}$ and thus mimic to some extent the multiple-excitation contribution in (5-123). The quasi-particle lifetime enters the picture through the width of the quasi-particle peak in $A(k,\omega)$ and is the most important source of the Auger tail.

In the satellite region the coefficient of the $P_{ck}^{(-)}$ term changes sign, and we have a partial cancellation with the first two terms. Disregarding the detailed variation of the matrix elements, we can thus expect, for a simple metal like sodium, to obtain essentially the density of states (5-90) of an electron gas (see Fig. 5-10) with the satellite scaled down, however. Thus if the electron-gas results are correct, we should expect the plasmaron edge (see Fig. 5-6) to show up at the low-energy end of the emission spectrum, but possibly with a quite small intensity due to the cancellation effects.

There is actually some slight experimental evidence of a plasmaron edge (Cuthill et al., 1968). The evidence is scanty, however, and despite careful investigations by other researchers, no more indications of an edge have appeared. We thus have strong doubts if the edge actually exists, particularly since inclusion of diagrams beyond the first-order one give strongly fluctuating results in the satellite region just as in the core-electron case (Hedin, unpublished).

There also have been suggestions that plasmon satellite structures should occur in absorption spectra (Ferrell, 1956). Thus Leder, Mendlowitz, and Marton (1956) found a sequence of peaks in absorption spectra that correlate with the creation of one or several volume plasmons, sometimes accompanied by surface plasmons. Lundqvist (1969c) has also suggested such interpretations by correlating peaks with the displaced plasmon energies obtained from his calculations, and Brown et al. (1970) have found plasma structure in alkali halide spectra. These detailed correlations must, however, still be regarded as tentative.

Disregarding the fine structure, there is one strong and dominant peak in the spectra of the simple metals Na, Mg, and Al which does not seem to have any band structure or atomic oscillator-strength explanation. It has been suggested (Hedin,

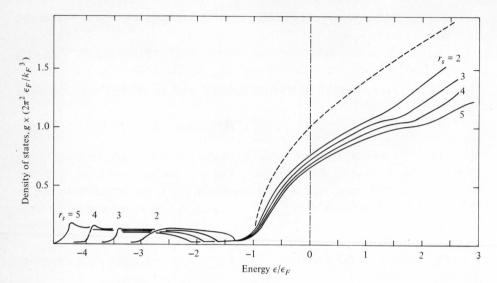

FIGURE 5-10
Density of states (5-90) for an electron gas obtained by Lundqvist (1968). The dashed curve is the Hartree approximation, and the vertical broken line indicates the Fermi level.

Lundqvist, and Lundqvist, 1970) that this peak could be due to some collective electron-gas effect. No detailed investigations have been made so far, but we point out that, for absorption, the jumping electron being above the Fermi surface can be expected to be less localized than in the emission case, and thus the cancellation of the core-hole potential should be less complete and the plasmon effect stronger, like in the XPS case.

IV.   VARIOUS TYPES OF EXPERIMENTS

We will give a short discussion of a few experiments which so far have not been subject to any detailed theoretical analysis but which could yield important information on many-body mechanisms.

A.   The isochromat method

A high-energy electron produced in an x-ray tube makes a one-quantum jump to a lower unoccupied state (in the anode), emitting a photon. The photon is observed in a spectrometer set on a fixed energy (hence the name "isochromat"), and the unoccupied states are scanned by varying the energy of the incident electron (the anode potential). The experimental results from isochromat measurements are discussed in Chap. 6, Section V, by Azároff and Pease.

In the isochromat method there is no problem either with the widths of inner-core levels or with the perturbing effects of core-hole potentials. The energy spreads involved are essentially the Maxwellian ones picked up by the electrons when in the cathode (typically 0.5 eV in the present-day arrangements). To develop a theory, we again apply the approximation of considering the high-energy electron entering the anode as free, and we write the initial and final states as

$$|\Psi_i\rangle = a_\kappa^+|N\rangle$$
$$|\Psi_f\rangle = |N + 1, s\rangle \tag{5-128}$$

The transition probability then becomes

$$w \approx \sum_s |\langle N + 1, s| \sum_m P_{m\kappa} a_m^+ |N\rangle|^2 \delta(\omega - E_{N-1,s} + E_N + \varepsilon_\kappa)$$

$$= \sum_{kk'} P_{k'\kappa}^* P_{k\kappa} A_{kk'}(\omega + \varepsilon_\kappa) \cong \sum_k |P_{k\kappa}|^2 A(k, \omega + \varepsilon_\kappa) \tag{5-129}$$

We thus have a direct measure of the one-electron spectral function just as in the XPS case, but this time for the $(N + 1)$ part, while in the XPS case we obtained the $(N - 1)$ part. Estimates of the spectral function of an electron gas (Lundqvist, 1968; 1969c) predict that there should be a broad plasmon peak in front of the ordinary quasi-particle peak and displaced somewhat more than the plasmon energy. The observation of this intrinsic effect will, however, just as in the XPS case, be mixed up with plasma losses and other inelastic losses suffered by the incoming electron.

Due to elastic scattering and the Maxwellian energy spread of the incident electrons, there will be an averaging over the k values of the incident electrons, which makes the experiments give the density of states $N(\omega)$ in (5-90) rather than the more detailed information about $A(k\omega)$ that (5-129) might lead us to expect. This averaging is particularly effective when a hot cathode and a polycrystalline anode are used, which so far has been the case.

B. Inelastic x-ray scattering

The spectrum of x rays scattered at a certain angle, in general, has many contributions; it is only at very high energies compared to the binding energies that Compton scattering dominates. For more moderate energies we also have a Rayleigh peak (elastic scattering) and different Raman-type peaks (inelastic scattering). The Raman peaks may come from plasmon excitation and from core-electron excitation. The plasmon peak was predicted by Pines (1963) but has only recently been observed experimentally (Priftis, Theodossiou, and Alexopoulos, 1968; Tanokura, Hirota, and Suzuki, 1969). The information obtained in addition to the plasmon energy is the dispersion and the width of the excitation, that is, the same information as from energy losses of fast electrons (Raether, 1965). X-ray scattering from plasmons is valuable as an independent check on the energy-loss method, and could possibly be improved both experimentally and theoretically. X-ray Raman scattering from core electrons in metals has recently been discussed by Doniach et al. (1971). This scattering should start with a singular edge, much as for x-ray emission and absorption. For x-ray scattering, however, the nature of the singularity should depend on the scattering angle. Thus in the case of Li, the edge structure should change from suppression, like in a Li *K* spectrum, to enhancement, like in a Na *L* spectrum, when the scattering angle increases. This change comes from the dependence of the matrix element on the scattering angle. The difficulty in comparing theory and experiment again comes from the resolution; the edge structures are a few tenths of an electronvolt wide, while the x radiation usually has a much larger width.

C.   X-ray appearance potential spectroscopy

In recent measurements (Park, Houston, and Schreiner, 1970; Houston and Park, 1971) the total soft x-ray fluorescence yield as a function of anode potential was observed right at the appearance potential of characteristic x rays. The derivative of the photomultiplier current rises abruptly at threshold and may display quite a detailed structure (cf fig. 4, Houston and Park, 1971). Since the spectrometer used is nondispersive, the method is experimentally very simple; yet it gives quite accurate and detailed results. Little theoretical analysis of the effect has been carried out so far, but since it occurs at threshold, there may be quite interesting many-body effects involved. Recently, Nilsson and Kansky (1972; 1973) have observed appearance spectra in simple metals, and the theory has been developed by Langreth (1971), Chang and Langreth (1972), and Laramore (1971; 1972).

V.   BAND-STRUCTURE ASPECTS

A.   Calculations of oscillator strengths for x-ray emission bands

While calculations of oscillator strengths for optical transitions in free atoms are comparatively simple, they are quite involved in solids since we need Bloch functions for both the initial and final states and have a **k**-space integration with complicated boundary conditions. For x-ray emission we have an intermediate situation since one of the states involved, the core state, is always the same and since the **k** integration runs only over occupied states. Such calculations have now been carried out in a number of cases: for lithium by McAlister (1969), for copper by Goodings and Harris (1969), for Si and Ge by Klima (1970), for the alkali metals by McMullen (1970), and for aluminum by Smrcka (1971). A comparison between theory and experiment for Cu has been made by Dobbyn et al. (1970).

Most of the important test cases for the simple one-electron approximation have thus been covered; alkalies, valence semiconductors, transition metals (Cu), and polyvalent metals (Al). These calculations clearly demonstrate that the oscillator strengths are just as important as the density of states. The calculations represent an important step forward; particularly the results for Si, Ge, and Cu demonstrate the importance of the underlying one-electron picture. The results also make it very clear that there are important aspects left out in the one-electron approximation. Thus, for instance, the lithium spectrum still seems to remain an unsolved problem. The band-structure calculations give a sharp Fermi edge, and it is doubtful if the combined effects of phonon broadening (McAlister, 1969) and the inverted Fermi edge singularity (Mizuno and Ishikawa, 1968) are sufficient to explain the very broad edge observed experimentally. Allotey (1967) has suggested that some resonance scattering particular to lithium should be involved. While this seems a reasonable explanation, it must still be regarded as a tentative one.

If we attempt a guess at the order in which we should add improvements to the simple one-electron picture, we would suggest, first, the replacement of the matrix element P_{ck} by P_{ck}^{eff} which includes states above the Fermi surface; second, the inclusion of multiple excitations to get the tails correctly; and third, a correct matching of the singular edge structure to the rest of the band. The precise form of the crystal potential may also be important, particularly for scaling the Brillouin-zone structures correctly in energy.

B. Alloy spectra

Problems with alloys and disordered systems form a vital and a very important area of solid-state research. Soft x-ray spectra which view the valence electrons from the localized core sites have contributed important information in this area, and there is still much work which could be done here. One large problem for the further development is to be able to measure the small intensity from solute atoms present in low concentrations.

The many-body effects in x-ray spectra from metals are present also in alloys. For alloys, however, the interest is concentrated on the one-electron problem, but one electron in a disordered random lattice. While this problem does not involve many-electron interactions, it does involve interactions with the many centers of the random lattice and utilizes for its treatment many-body techniques. The problems of alloys form a classical field, where early discussions can be found in the book by

Mott and Jones (1936). Among later contributions we would like to mention those of Friedel (1952), Varley (1954), and the recent development with the coherent potential started by Soven (1967) and Velicky, Kirkpatrick, and Ehrenreich (1968).

We will not try in any way to cover this rich and rapidly expanding field; we will give only a few simple aspects. For a fuller review of x-ray spectra in alloys we refer to Fabian (1970). To start the discussion, we note that x-ray emission in a pure metal may be looked upon as a process involving an impurity atom; the emitting atom with its core hole certainly is quite distinct and gives a different contribution to the crystal potential than the rest of the atoms. If this impurity potential is not strong enough to bind one extra electron, its effect is essentially to replace the matrix element P_{ck} by an effective matrix element, and the spectrum can still, to a first approximation, be discussed in terms of the pure "solvent." A good way to test the perturbing effect of the core hole is to study the high-energy satellites which come from doubly-ionized cores. A classical study here is the paper by Catterall and Trotter (1958); they found no perceptible change between the main band and the high-energy satellite in the rather extreme case of lithium where the satellite comes from transitions to an ion core without any core electrons at all! Other satellite studies have given similar results. There is no well-documented case of a core-hole potential producing a bound state in a metal or an alloy.

If we next consider real alloys, we would like to point at a few facts. First, the spectra from the different constituents in the alloy seem to be more similar to the spectra of pure metals of the respective constituents than to a picture for a common shared-valence band. The typical case here is magnesium-aluminum alloys which have been carefully investigated over a wide range of concentrations (Appleton and Curry, 1965; Dimond, 1970). Also alloys of transition metals show this behavior (Curry, 1968). The latter case can sometimes be readily explained from explicit band calculations. Thus Arlinghaus (1967) found for β brass (CuZn) that the d bands were split in two groups, one where the wavefunctions were centered on the Cu atoms and one where they were centered on the Zn atoms. This is what could be expected for the rather narrow d bands involved, but also for simple metals the situation seems to be rather similar. A calculation by Jacobs (1969) illustrates this to some extent. Jacobs studied a model for AlMg where the magnesium atoms had zero potential and the aluminum atoms had a square-well potential. The atoms with the deeper potential (Al) gave the broader band; the states in the bottom of the energy band had their wavefunctions heavily localized on the Al atoms. The model, however, did not elucidate the critical question, namely, that the width of the Al band in the alloy should be the same as for pure Al; instead the Al bandwidth in the alloy agreed with the virtual-crystal approximation. Very interesting from the theoretical viewpoint is the problem of very dilute alloys. What happens when the distance between the solute atoms becomes so large that there eventually is no overlap between their valence electrons? Should we always expect localized states to appear?

Another interesting feature of alloys is the possibility of cross transitions—an electron preferentially centered on one atom makes a transition to a core level on another atom. This is essentially the inverse effect of the one just discussed. The stability of the bandwidths during alloying depends on preferential localization for the itinerant conduction electrons, while the cross transitions depend on overlap between electrons preferentially localized on different atoms. The cross transitions may not always be describable from the simple overlap picture; hybridization effects may sometimes be important (Marshall et al., 1969). The importance of cross transitions theoretically and experimentally has recently been discussed at lengths by Fischer (1970).

The problem of edge broadening in alloys has also been discussed. Experiments by Catterall and Trotter (1962) and by Gale, Catterall, and Trotter (1969) seem to show that when an ordered alloy becomes disordered, there is an increase in the edge breadth. They found correlations between this increase and the shorter lifetime of the conduction electrons due to increased scattering on disordering. The problem from a many-body standpoint is now why an increase in edge breadth should occur. The shorter lifetime is clearly of importance for a property like the electrical conductivity, while the sharp edge in x-ray emission does not seem to depend on having well-defined k values, only on having a sharp Fermi level, whatever classifications are used for the states involved. A probable explanation of the broadening is that it comes from the fluctuating Coulomb fields of the neighbors. These fields, thinking of screened pseudo-potentials, are certainly quite small, but they may be sufficiently large to explain the increases in edge widths.

C. Spectra from transition-metal and rare-earth-metal atoms

Investigations of magnetic solids form another very important area of solid-state research. As for alloys, we can only touch upon this large area and indicate a few important points. We have earlier cited investigations of copper (Goodings and Harris, 1969; Dobbyn et al., 1970), which showed quite good agreement between theory and experiment. The valence band of paramagnetic Ni has also been carefully measured in SXS emission and discussed by Cuthill et al. (1967). An interesting example of the importance of the oscillator strengths rather than the density of states is provided by results for the $K\beta_5$ spectra for transition metals obtained by Adelson and Austin (1969). The $K\beta_5$ spectra are due to $3d \rightarrow 1s$ transitions made possible by hybridization with the sp conduction band. Thus the intensities depend on the amount of nonlocalization. The measured intensities give the inverse of the Slater-Pauling magnetic moment curve, in good agreement with the above considerations.

Very interesting results on transition and rare-earth metals have been obtained with XPS. Thus extensive mappings of d bands have been presented by Fadley and Shirley (1970a) and by Baer et al. (1970b) and of f bands by Hedén, Löfgren, and

Hagström (1971). The *d* bands have also been studied with core electrons as probes. Multiplet splittings of the core levels were found due to interactions with the magnetic moment of the *d* electrons (Fadley et al., 1969; Fadley and Shirley, 1970*b*). For iron, the splitting was found to be the same below and above the Curie temperature. Also for Ni the (UV) photoemission spectra have recently been found to remain unchanged when passing the Curie temperature (Pierce and Spicer, 1970). These results, no doubt, are very important for our understanding of magnetism.

The XPS results for core splittings may be closely connected to the $K\beta'$ structure observed in SXS emission. This structure is only found in magnetic materials. Tsutsumi and Nakamori (1968) have interpreted this structure as due to Hund's rule coupling between the core electrons and the *d* electrons. Recent experimental results and some theoretical considerations are discussed by Ekstig et al. (1970).

VI.   CONCLUDING REMARKS

X-ray spectroscopy is a time-honored subject; in the twenties it contributed importantly to the foundations of atomic physics and in the thirties to our understanding of metals. After the pioneering work by Mott and Skinner on metals there followed a period of criticism and deeper penetration. In particular the Parratt and Friedel schools pointed at many uncertainties in the earlier experimental results as well as in their interpretation and offered some new, more comprehensive theories. Their critical comments concerned points like self-absorption, distortions by the core-hole potential, surface effects, and dependence on the energy of the excitation radiation. We have tried in this chapter to show that a number of the more serious of these difficulties now seem to have been brought under control. We have also given the recently discovered edge and satellite effects their due attention.

In our discussions we have attempted to bring out the common features and similarities in the different effects. To that end we have emphasized three important basic points, the roles of the core-hole potential and of the matrix elements, and the concept of shake-up. The two first points become quite important in a one-electron theory. Thus the core-hole potential enhances the main-band emission intensity and contributes to the tailing as discussed in Subsection III D 3. The enhancement is strong but, owing to pseudopotential effects, probably not enormous, as in the positron-annihilation case. Matrix elements have a large influence; this is discussed by Rooke in Chap. 4, and here it is taken up in Section V. In pure metals the matrix elements can largely magnify some of the Brillouin-zone structures and suppress others rather completely, as well as causing more smoothly varying distortions. In alloys they can amplify one portion of the energy band and make the observed spectra look more like the density of states of one constituent than of the alloy. The explanation for this is that itinerant electrons can be localized in the sense that they have a much larger amplitude on one type of atom than on another. If we take the opposite viewpoint and think in terms of localized electrons, then we can see their delocalization through cross transitions, where the electron on one atom jumps to a core hole on another.

The core-hole potential also gives rise to a set of many-body effects, which can be discussed in terms of the concept "shake-up" (Subsection II D). The name "shake-up" comes from the picture of the electron system suffering a shock or shake-up when an electron (usually a core electron) is suddenly created or removed. Quantum mechanically, we describe this as the sudden approximation, and calculationwise we need the overlap between a state where the valence electrons are kept immobile during the change in the core-hole occupation and a state where they have relaxed. Shake-up produces finite probabilities for the valence electrons to end up in excited configurations not allowed in the usual one-electron picture. Examples are the multi-electron-hole configurations, which lead to the celebrated edge singularities in metals, and electron-plasmon configurations, which give rise to plasmon satellites in the spectra. These satellites are quite strong in the case of x-ray photoemission while in x-ray emission they are weaker, since the shake-up then comes from the difference between the potentials from the core hole and from the hole left by the jumping-valence electron. In x-ray absorption we also have a difference potential and, since in this case the valence electron is less localized, the plasmon effect may possibly be stronger. The satellite is superposed on ordinary one-electron-band features, however, and hard to distinguish for that reason.

Another point where our understanding has increased during recent years concerns the relation between spectral shapes and excitation conditions. Thus, in Subsection II C, we discuss Fano-type resonances in atoms where the large difference in spectra produced by UV absorption and by inelastic electron scattering are well understood. Another clear-cut case is the dependence of the spectral shape on the scattering angle for inelastic x-ray scattering, discussed in Subsection IV B. We also have effects on spectral shapes in x-ray emission like the competition between incompletely relaxed shake-up structure and self-absorption discussed in Subsection II F.

We now turn our attention from achievements of the past and try to say something about the future. First, it seems quite probable that x-ray spectroscopy, spurred by the advances in our basic understanding, will continue to serve an increasingly important role for chemical analysis, and the x-ray photoemission method especially is likely to expand greatly in importance. Investigations of alloys and magnetic materials have recently given some very interesting results; it seems likely that research in these directions will expand. Improvements in resolution, surface treatment, and data interpretation could largely aid such an expansion. The fresh possibilities provided by work with synchrotron radiation (Chap. 7) do not seem to be exhausted.

Looking at specific many-body effects, we find that there is still a large discrepancy between the amount of detail which the theoreticians expect should be found and the resolution that the experimentalists can produce. For further progress, there is need of more detailed theoretical work which can give, for example, not only the exponent of a singular term but also its strength. The dynamics of plasmon creation and the problems of incomplete relaxation also remain unsolved—a challenge to theoreticians as well as experimentalists.

ACKNOWLEDGMENTS

I would like to thank Stig Lundqvist for a clarifying discussion of phonon cascades, Carl Olof Almbladh for valuable comments on lifetime problems, and W. F. Brinkman for a discussion of alloy edges. I am also indebted to T. Arai, B. Bergersen, S. Doniach, D. Fabian, S. Hagström, N. March, L. Parratt, G. Rooke, M. Stott, and others for discussions and comments.

REFERENCES

ABERG, T. (1967): Theory of x-ray satellites, *Phys. Rev.*, **156**: 35–41.

——— (1969): Multiple excitation of a many-electron system by photon and electron impact in the sudden approximation, *Ann. Acad. Sci. Fennicae*, **Ser. A. VI.** *Physica*, no. 308, 1–46.

——— and UTRIAINEN, J. (1969): Evidence for a "radiative Auger effect" in x-ray photon emission, *Phys. Rev. Lett.*, **22**: 1346–1348.

ADELSON, E., and AUSTIN, A. E. (1969): Dependence of x-ray transitions from the conduction band upon non-localized 3d-electrons, *Solid State Comm.*, **7**: 1819–1820.

ALLOTEY, F. K. (1967): Effect of electron-hole scattering resonance on x-ray emission spectrum, *Phys. Rev.*, **157**: 467–479.

ANDERSON, P. W. (1967a): Infrared catastrophe in Fermi gases with local scattering potentials, *Phys. Rev. Lett.*, **18**: 1049–1051.

——— (1967b): Ground state of a magnetic impurity in a metal, *Phys. Rev.*, **164**: 352–359.

APPLETON, A., and CURRY, C. (1965): Soft x-ray emission spectra of non-dilute aluminum-magnesium alloys, *Phil. Mag.*, **12**: 245–252.

ARLINGHAUS, F. J. (1967): Energy bands in ordered beta-brass, *Phys. Rev.*, **157**: 491–499.

AUSMAN, G. A., and GLICK, A. J. (1969): Threshold behavior of the soft x-ray spectra in metals, *Phys. Rev.*, **183**: 687–691.

BAER, Y., HEDÉN, P. F., HEDMAN, J., KLASSON, M., and NORDLING, C. (1970a): Determination of the electron escape depth in gold by means of ESCA, *Solid State Comm.*, **8**: 1479–1481.

BAER, Y., HEDÉN, P. F., HEDMAN. J., KLASSON. M., and SIEGBAHN, K. (1970b): Band structure of transition metals studied by ESCA, *Physica Scripta*, 1: 55–65.

BERGERSEN, B., and BROUERS, F. (1969): The soft x-ray spectra of metals near the emission edge, *J. Solid State Phys.*, 2: 651–660.

———, ———, and LONGE, P. (1971): Influence of correlations and of the core hole on metal x-ray spectra, *J. Phys.*, F1: 945–959.

———, MCMULLEN, T., and CARBOTTE, J. P. (1971): Effect of lattice relaxation on the soft x-ray spectra of metals, *Can. J. Phys.*, 49: 3155–3165.

BLOKHIN, M. A., and SACHENKO, V. P. (1960): Concerning the shape of energy bands in solids, *Bull. Acad. Sci. USSR, Phys. Ser. (English Transl.)*, 24: 410–418.

BROUERS, F. (1964): Theoretical intensity estimation of plasmon satellite bands in soft x-ray emission spectra, *Phys. Lett.*, 11: 297–298.

——— (1967): Plasmon satellites of soft x-ray spectra, *Phys. Stat. Sol.*, 22: 213–221.

——— and LONGE, P. (1968): A new perturbative interpretation of the satellite plasmon emission band, in D. J. Fabian (ed.), *Soft x-ray band spectra and the electronic structures of metals and materials*, Academic Press, Inc., New York.

———, ———, and BERGERSEN, B. (1970): The effect of the core hole on the shape of soft x-ray spectra in metals, *Solid State Comm.*, 8: 1423–1426.

BROWN, F. C. (1967): *The physics of solids*, W. A. Benjamin, Inc., New York.

———, GÄHWILLER, C., KUNZ, A. B., and LIPARI, N. (1970): Soft x-ray spectra of the lithium halides and their interpretation, *Phys. Rev. Lett.*, 25: 927–930.

CARLSON, T. A., and KRAUSE, M. O. (1965): Electron shake-off resulting from K-shell ionization in neon measured as a function of photoelectron velocity, *Phys. Rev.*, A140: 1057–1064.

———, MODDEMAN, W. E., and KRAUSE, M. O. (1970): Electron shake-off in neon and argon as a function of energy of the impact electron, *Phys. Rev.*, A1: 1406–1410.

CATTERALL, J. A., and TROTTER, J. (1958): The interpretation of soft x-ray emission spectra, *Phil. Mag.*, 3: 1424–1431.

——— and ——— (1962): The broadening of soft x-ray emission edges in metals and alloys, *Phil. Mag.*, 7: 671–676.

CAUCHOIS, Y. (1968): Sur les spectres x des metaux: Quelques commentaires et exemples, in D. J. Fabian (ed.), *Soft x-ray band spectra and the electronic structures of metals and materials*, Academic Press, Inc., London.

CHANG, J. J., and LANGRETH, D. C. (1972): Deep hole excitations in solids. I. Fast-electron-plasmon effects, *Phys. Rev.*, B5: 3512–3522.

CHOPRA, D. (1970): Ni L self-absorption spectrum, *Phys. Rev.*, A1: 230–235.

COOPER, J. W., and LA VILLA, R. E. (1970): Semi-Auger processes in L_{23} emission in Ar and KCl, *Phys. Rev. Lett.*, 25: 1745–1748.

CURRY, C. (1968): Soft x-ray emission spectra of alloys and problems in their interpretation, in D. J. Fabian (ed.), *Soft x-ray band spectra and the electronic structures of metals and materials*, Academic Press, Inc., London.

CUTHILL, J. R., DOBBYN, R. C., MCALISTER, A. J., and WILLIAMS, M. L. (1968): Search for plasmaron structure in the soft x-ray L_{23} emission spectrum of Al, *Phys. Rev.*, 174: 515–517.

CUTHILL. J. R., MCALISTER, A. J., WILLIAMS, M. L., and WATSON, R. E. (1967): Density of states of Ni: Soft x-ray spectrum and comparison with photoemission and ion neutralization studies, *Phys. Rev.*, **164**: 1006–1017.

DIMOND, R. K. (1967): Self-absorption in soft x-ray spectra of alloys, *Phil. Mag.*, **15**: 631–634.

—— (1970): "Soft x-ray spectra from aluminum-magnesium alloys," Ph.D. thesis, University of Western Australia.

DOBBYN, R. C., WILLIAMS, M. L., CUTHILL, J. R., and MCALISTER, A. J. (1970): Occupied band structure of Cu: Soft x-ray spectrum and comparison with other deep-band-probe studies, *Phys. Rev.*, **B2**: 1563–1575.

DONIACH, S., PLATZMAN, D. M., and YUE, J. T. (1971): X-ray Raman scattering in metals, *Phys. Rev.*, **B4**: 3345–3350.

—— and SUNJIĆ, M. (1970): Many-electron singularity in x-ray photoemission and x-ray line spectra from metals, *J. Phys.*, **C3**: 285–291.

EDWARDS, A. K., and RUDD, M. E. (1968): Excitation of auto-ionizing levels in neon by ion impact, *Phys. Rev.*, **170**: 140–144.

EKSTIG, B., KÄLLNE, E., NORELAND, E., and MANNE, R. (1970): Electron interaction in transition metal x-ray emission spectra, *Physica Scripta*, **2**: 38–44.

FABIAN, D. (1970): Soft x-ray emission and electronic structure of alloys, *Mater. Res. Bull.*, **5**: 591–606.

FADLEY, C. S., and SHIRLEY, D. A. (1970a): Electronic densities of states from x-ray photo-electron spectroscopy, *J. Res. Natl. Bur. Std.*, **A74**: 543–558.

—— and —— (1970b): Multiplet splitting of metal-atom electron binding energies, *Phys. Rev.*, **A2**: 1109–1120.

——, ——, FREEMAN, A. J., BAGUS, P. S., and MALLOW, J. V. (1969): Multiplet splitting of core-electron binding energies in transition-metal ions, *Phys. Rev. Lett.*, **23**: 1397–1401.

FANO, U. (1961): Effects of configuration interaction on intensities and phase shifts, *Phys. Rev.*, **124**: 1866–1878.

—— and COOPER, J. W. (1968): Spectral distribution of atomic oscillator strengths, *Rev. Mod. Phys.*, **40**: 441–507.

FERRELL, R. A. (1956): Theory of positron annihilation in solids, *Rev. Mod. Phys.*, **28**: 308–337.

—— (1965a): Plasmon production in x-ray emission of metals, *Bull. Am. Phys. Soc.*, **10**: 1218.

—— (1965b): Plasmon excitation in x-ray emission, *Techn. Rept.*, 485, July 1965, University of Maryland.

FISCHER, D. W. (1970): Chemical bonding and valence state—nonmetals, *Advan. X-Ray Anal.*, **13**: 159–181.

FRIEDEL, J. (1952): X-ray transition probabilities with special reference to K absorption in lithium, *Phil. Mag.*, **43**: 1115–1139.

GALE, B., CATTERALL, J. A., and TROTTER, J. (1969): Soft x-ray L_{23} emission edge breadth in ordered and disordered Mg_3Cd, *Phil. Mag.*, **20**: 79–87.

GLICK, A. J., LONGE, P., and BOSE, S. M. (1968): The effect of electron interaction on soft x-ray emission spectra of metals, in D. J. FABIAN (ed.), *Soft x-ray band spectra and the electronic structures of metals and materials*, Academic Press, Inc., New York.

GOODINGS, D. A., and HARRIS, R. (1969): Calculations of the x-ray emission bands of copper using augmented plane wave Bloch functions, *J. Phys.*, **C2**: 1808–1816.

GRAEFFE, G., SIIVOLA, J., UTRIAINEN, J., LINKOAHO, M., and ABERG, T. (1969): X-ray $K\alpha$ satellite spectra in primary and secondary excitation, *Phys. Lett.*, **A29**: 464–465.

HARRISON, W. A. (1968): Electronic structure and soft x-ray spectra, in D. J. Fabian (ed.), *Soft x-ray band spectra and the electronic structures of metals and materials*, Academic Press, Inc., London.

HEANEY, W. J., and RYSTEPHANICK, R. G. (1970): Tailing of the soft x-ray emission spectrum in metals, *Phys. Lett.*, **A31**: 221–222.

HEDÉN, P. O., LÖFGREN, H., and HAGSTRÖM, S. B. M. (1971): $4f$ electronic states in the metals Nd, Sm, Dy and Er studied by x-ray photoemission, *Phys. Rev. Lett.*, **26**: 432–434.

HEDIN, L. (1965a): New method for calculating the one-particle Green's function with application to the electron-gas problem, *Phys. Rev.*, **A139**: 796–823.

——— (1965b): Effect of electron correlation on band structure of solids, *Arkiv Fysik*, **30**: 231–258.

——— (1967): Many-body effects in soft x-ray emission in metals, *Solid State Comm.*, **5**: 451–454.

——— (1968): Many-body effects in the soft x-ray emission from metals, in D. J. Fabian (ed.), *Soft x-ray band spectra and the electronic structures of metals and materials*, Academic Press, Inc., London.

——— and JOHANSSON, A. (1969): Polarization corrections to core levels, *J. Phys.*, **B2**: 1336–1346.

——— and LUNDQVIST, B. I. (1971): Explicit local exchange-correlation potentials, *J. Phys.*, **C4**: 2064–2083.

——— and LUNDQVIST, S. (1969): Effects of electron-electron and electron-phonon interactions on the one-electron states of solids, in F. Seitz, D. Turnbull, and H. Ehrenreich (eds.), *Solid State Physics*, vol. 23, pp. 1–181, Academic Press, Inc., New York.

———, LUNDQVIST, B. I., and LUNDQVIST, S. (1967): New structure in the single-particle spectrum of an electron gas, *Solid State Comm.*, **5**: 237–239.

———, ———, and ——— (1970): Beyond the one-electron approximation: Density of states for interacting electrons, *J. Res. Natl. Bur. Std.*, **A74**: 417–431.

——— and SJÖSTRÖM, R. (1971): Effect of the core hole on soft x-ray emission in metals, *Proceedings of International Conference on Electronic Density of States*, Natl. Bur. Std. Publ. 323.

HOLLIDAY, J. E. (1970): Soft x-ray valence state effects in conductors, *Advan. X-Ray Anal.*, **13**: 136–157.

HOPFIELD, J. J. (1962): Electron-phonon coupling in impurity states, *Proc. Intern. Conf. Semicond. Phys., Exeter*, pp. 75–80, Institute of Physics and Physics Society, London.

——— (1969): Infrared divergencies, x-ray edges and all that, *Comm. Solid State Phys.*, **2**: 40–49.

HOUSTON, J. E., and PARK, R. L. (1971): Anomalous fine structure in the soft x-ray appearance potentials of non-metals, *J. Vac. Sci. Technol.*, **8**: 91–93.

HUANG, K., and RHYS, A. (1950): Theory of light absorption and nonradiative transitions in F-centres, *Proc. Roy. Soc. (London), Ser. A*, **204**: 406–423.

JACOBS, R. L. (1969): The soft x-ray spectra of concentrated binary alloys, *Phys. Lett.*, **A30**: 523–524.

KJÖLLERSTRÖM, B., MÖLLER, N. H., and SVENSSON, H. (1965): Configuration interaction in Ar II, *Arkiv Fysik*, **29**: 167–173.

KLIMA, J. (1970): Calculation of the soft x-ray emission spectra of silicon and germanium, *J. Phys.*, **C3**: 70–85.

KOBAYASI, T., and MORITA, A. (1970): Theoretical investigation of the x-ray level widths of light metals, *J. Phys. Soc. Japan*, **28**: 457–466.

KOHN, W., and MAJUMDAR, C. (1965): Continuity between bound and unbound states in a Fermi gas, *Phys. Rev.*, **A138**: 1617–1620.

KRAUSE, M. O., CARLSON, T. A., and DISMUKES, R. D. (1968): Double electron ejection in the photoabsorption process, *Phys. Rev.*, **170**: 37–47.

——, STEVIE, F. A., LEWIS, L. J., CARLSON, T. A., and MODDEMAN, W. E. (1970): Multiple excitation of neon by photon and electron impact, *Phys. Lett.*, **A31**: 81–82.

KUNZ, C., HAENSEL, R., KEITEL, G., SCHREIBER, P., and SONNTAG, B. (1971): Photoabsorption measurement of Li, Be, Na, Mg, and Al in the vicinity of K and L_{23} edges, *Proceedings of International Conference on Electronic Density of States*, Natl. Bur. Std. Publ. 323.

LANDSBERG, P. T. (1949): A contribution to the theory of soft x-ray emission bands of sodium, *Proc. Phys. Soc. (London), Ser. A*, **62**: 806–816.

LANGRETH, D. C. (1970): Singularities in the x-ray spectra of metals, *Phys. Rev.*, **B1**: 471–477.

—— (1971): Born-Oppenheimer principle in reverse: Electrons, photons and plasmons in solids—singularities in their spectra, *Phys. Rev. Lett.*, **26**: 1229–1233.

LARAMORE, G. E. (1971): Threshold singularities in appearance-potential spectroscopy, *Phys. Rev. Lett.*, **27**: 1050–1053.

—— (1972): Plasmon emission in appearance-potential spectroscopy, *Solid State Comm.*, **10**: 85–89.

LAX, M. (1952): The Franck-Condon principle and its application to crystals, *J. Chem. Phys.*, **20**: 1752–1760.

LEDER, L. B., MENDLOWITZ, H., and MARTON, L. (1956): Comparison of the characteristic energy losses of electrons with the fine structure in the x-ray absorption spectra, *Phys. Rev.*, **101**: 1460–1467.

LIEFELD, R. J. (1968): Soft x-ray emission spectra at threshold excitation, in D. J. Fabian (ed.), *Soft x-ray band spectra and the electronic structures of metals and materials*, Academic Press, Inc., London.

LIGHTHILL, M. J. (1958): *Fourier analysis and generalized functions*, Cambridge University Press, London.

LONGE, P., and GLICK, A. J. (1969): Electron-interaction effects on the soft x-ray emission spectrum of metals. I. Formalism and first-order theory, *Phys. Rev.*, **177**: 526–539.

LUNDQVIST, B. I. (1967a): Single-particle spectrum of the degenerate electron gas. I. The structure of the spectral weight function, *Phys. Kondens. Materie*, **6**: 193–205.

—— (1967b): Single-particle spectrum of the degenerate electron gas. II. Numerical results for electrons coupled to plasmons, *Phys. Kondens. Materie*, **6**: 206–217.

LUNDQVIST, B. I., (1968): Single-particle spectrum of the degenerate electron gas. III. Numerical results in the random phase approximation, *Phys. Kondens. Materie*, **7**: 117–123.

——— (1969a): Some numerical results on quasi-particle properties in the electron gas, *Phys. Stat. Sol.*, **32**: 273–280.

——— (1969b): Characteristic structure in core electron spectra of metals due to the electron-plasmon coupling, *Phys. Kondens. Materie*, **9**: 236–248.

——— (1969c): "Spectra of interacting electrons in metals," Thesis, Chalmers University of Technology, Gothenburg.

MADDEN, R. P., and CODLING, K. (1963): New autoionizing atomic energy levels in He, Ne, and Ar, *Phys. Rev. Lett.*, **10**: 516–518.

MAHAN, G. D. (1967a): Excitons in metals, *Phys. Rev. Lett.*, **18**: 448–450.

——— (1967b): Excitons in metals: Infinite hole mass, *Phys. Rev.*, **163**: 612–617.

MARSHALL, C. A. W., WATSON, L. M., LINDSAY, G. M., ROOKE, G. A., and FABIAN, D. J. (1969): Interpretation of soft x-ray emission spectra of aluminum-silver alloys, *Phys. Lett.*, **A28**: 579–580.

MCALISTER, A. J. (1969): Calculation of the soft x-ray *K*-emission and absorption spectra of metallic Li, *Phys. Rev.*, **186**: 595–599.

MCMULLEN, T. (1970): A calculation of the soft x-ray emission spectra of the alkali metals, *J. Phys.*, **C3**: 2178–2185.

MEHLHORN, W., and STALHERM, D. (1968): Die Auger-Spektren der L_2- und L_3-Schale von Argon, *Z. Physik*, **217**: 294–303.

MELDNER, H. W., and PEREZ, J. D. (1971): Observability of rearrangement energies and relaxation times, *Phys. Rev.*, **A4**: 1388–1396.

MIZUNO, Y., and ISHIKAWA, K. (1968): Anomalies in edges of soft x-ray emission spectra of metals, *J. Phys. Soc. Japan*, **25**: 627–628.

MORITA, A., and WATABE, M. (1968): Theory of soft x-ray emission spectra of light metals, *J. Phys. Soc. Japan*, **25**: 1060–1068.

MOTT, N. F., and JONES, H. (1936): *The theory of the properties of metals and alloys*, Oxford University Press, London.

MÜLLER-HARTMANN, E., RAMAKRISHNAN, T. V., and TOULOUSE, G. (1971): Localized dynamic perturbations in metals, *Phys. Rev.*, **B3**: 1102–1119.

NILLSON, P. O., and KANSKI, J. (1972): The appearance potential spectrum of aluminum, *Phys. Lett.*, **A41**: 217–218.

——— and ——— (1973): Appearance potential spectroscopy of simple metals, *Surface Science*, in press.

NOZIÈRES, P., and DE DOMINICIS, C. T. (1969): Singularities in the x-ray absorption and emission of metals. III. One-body theory exact solution, *Phys. Rev.*, **178**: 1097–1107.

PARK, R. L., HOUSTON, J. E., and SCHREINER, D. G. (1970): A soft x-ray appearance potential spectrometer for the analysis of solid surfaces, *Rev. Sci. Instr.*, **41**: 1810–1812.

PARRATT, L. G. (1959): Electronic band structure of solids by x-ray spectroscopy, *Rev. Mod. Phys.*, **31**: 616–645.

PIERCE, D. T., and SPICER, W. E. (1970): Photoemission studies of ferromagnetic and paramagnetic nickel, *Phys. Rev. Lett.*, **25**: 581–584.

PINES, D. (1955): Electron interaction in metals, in F. SEITZ, and D. TURNBULL (eds.), *Solid State Physics*, vol. 1, pp. 367–450, Academic Press, Inc., New York.

──── (1963): *Elementary excitations in solids*, pp. 204–207, W. A. Benjamin, Inc., New York.

PIRENNE, J., and LONGE, P. (1964): Contribution of the double electron transitions to the soft x-ray emission bands of metals, *Physica*, **30**: 277–292.

PRIFTIS, G., THEODOSSIOU, A., and ALEXOPOULOS, K. (1968): Plasmon observation in x-ray scattering, *Phys. Lett.*, **A27**: 577–579.

RAETHER, H. (1965): Solid state excitations by electrons, in *Springer tracts in modern physics*, **38**: 84–157.

RICHTMYER, F. K. (1937): The multiple ionization of inner electron shells of atoms, *Rev. Mod. Phys.*, **9**: 391–402.

ROOKE, G. A. (1963): Plasmon satellites of soft x-ray emission spectra, *Phys. Lett.*, **3**: 234–236.

RYSTEPHANICK, R. G., and CARBOTTE, J. P. (1968): Soft x-ray emission in metals, *Phys. Rev.*, **166**: 607–615.

SACHENKO, V. P., and BURTSEV, E. V. (1968): Probability for multiple ionization of atoms under photon excitation, *Bull. Acad. Sci. USSR, Phys. Ser. (English Transl.)*, **31**: 980–984.

──── and DEMEKHIN, V. F. (1966): Satellites of x-ray spectra, *Soviet Phys. JETP (English Transl.)*, **22**: 532–535.

SAWADA, K., BRUECKNER, K. A., FUKUDA, N., and BROUT, R. (1957): Correlation energy of an electron gas at high density: Plasma oscillations, *Phys. Rev.*, **108**: 507–514.

SCHNOPPER, H. W. (1963): Multiple excitation and ionization of inner atomic shells by x-rays, *Phys. Rev.*, **131**: 2558–2560.

──── (1967): Atomic readjustment to an inner-shell vacancy: Manganese K x-ray emission spectra from an Fe^{55} K-capture source and from the bulk metal, *Phys. Rev.*, **154**: 118–123.

SCHOTTE, K. D., and SCHOTTE, U. (1969a): Tomonaga's model and the threshold singularity of x-ray spectra of metals, *Phys. Rev.*, **182**: 479–482.

──── and ──── (1969b): Threshold behavior of the x-ray spectra of light metals, *Phys. Rev.*, **185**: 509–517.

SCHULTZ, T. D. (1964): *Quantum field theory and the many-body problem*, Gordon & Breach, New York and London.

SCOFIELD, J. H. (1969): Radiative decay rates of vacancies in the K and L shells, *Phys. Rev.*, **179**: 9–16.

SHAW, R. W. (1968): Optimum form of a modified Heine-Abarenkov model potential for the theory of simple metals, *Phys. Rev.*, **174**: 769–781.

SIEGBAHN, K., et al. (1969): *ESCA applied to free molecules*, North-Holland Publishing Company, Amsterdam.

SKINNER, H. W. B. (1940): The soft x-ray spectroscopy of solids, *Phil. Trans. Roy. Soc. London*, **A239**: 95–134.

SMRCKA, L. (1970): Energy band structure of aluminum by the augmented plane wave method, *Czech. J. Phys.*, **B20**: 291–300.

──── (1971): Calculation of soft x-ray emission spectra of aluminum by the APW method, *Czech. J. Phys.*, **B21**: 683–692.

SOVEN, P. (1967): Coherent-potential model of substitutional disordered alloys, *Phys. Rev.*, **156**: 809–813.

TANOKURA, A., HIROTA, N., and SUZUKI, T. (1969): X-ray plasmon scattering, *J. Phys. Soc. Japan*, **27**: 515.

TOMONAGA, S. (1950): Remarks on Bloch's method of sound waves applied to many-fermion problems, *Progr. Theoret. Phys. (Kyoto)*, **5**: 544–569.

TSUTSUMI, K., and NAKAMORI, H. (1968): X-ray *K* emission spectra of chromium in various chromium compounds, *J. Phys. Soc. Japan*, **25**: 1418–1423.

VARLEY, J. H. O. (1954): The calculation of heats of formation of binary alloys, *Phil. Mag.*, **45**: 887–916.

VELICKY, B., KIRKPATRICK, S., and EHRENREICH, H. (1968): Single-site approximations in the electron theory of simple binary alloys, *Phys. Rev.*, **175**: 747–766.

WENDIN, G. (1970): Atomic resonances from a many-body point of view: I, *J. Phys.*, **B3**: 455–465.

WERTHEIM, G. K., and ROSENCWAIG, A. (1971): Configuration interaction in the x-ray photoelectron spectra of alkali halides, *Phys. Rev. Lett.*, **26**: 1179–1182.

WOLFSBERG, M., and PERLMAN, M. L. (1955): Multiple electron excitation in Auger processes, *Phys. Rev.*, **99**: 1833–1835.

6

X-RAY ABSORPTION SPECTRA

Leonid V. Azároff and Douglas M. Pease

I. INTRODUCTION

When an x-ray beam passes through a medium, the intensity of the transmitted beam I is attenuated logarithmically. According to the classical absorption equation, if the incident beam of intensity I_0 has traveled a distance x, then the absorption coefficient

$$\mu_l = -\frac{1}{x} \ln \frac{I}{I_0} \qquad (6\text{-}1)$$

It follows from (6-1) that the dimensions of μ_l are reciprocal centimeters so that it represents the attenuation of the beam per unit length traveled and is called the *linear absorption coefficient*. It turns out to depend on the energy (wavelength) of the x rays, the atomic numbers of the constituent atoms, and on their state of aggregation.

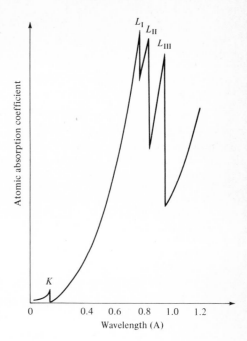

FIGURE 6-1
Variation of absorption coefficient of
lead as a function of x-ray wavelength.
(*After Azároff*, 1968.)

Calculation of the linear absorption coefficient is complicated by this last fact so that
it is more convenient to use a *mass* absorption coefficient

$$\mu_m = \frac{\mu_l}{\rho} \qquad (6\text{-}2)$$

where ρ is the density of the absorbing medium and μ_m is the absorption cross section
of the mass unit.

In the analysis of the dependence of μ on wavelength and atomic number, it is
most convenient to define an *atomic* absorption coefficient

$$\mu_a = \frac{A}{N_0} \mu_m \qquad (6\text{-}3)$$

where A is the atomic weight of the element and N_0 is Avogadro's number. A plot
of μ_a against wavelength, for any element, has the general appearance of the curve
shown in Fig. 6-1. The absorption increases with increasing wavelength (decreasing
energy of incident x rays) until an abrupt discontinuity occurs. After this discontinuity, called an *absorption edge*, the absorption increases with wavelength until
new discontinuities are encountered. An empirical study of the relation between the
atomic absorption coefficient and the x-ray wavelength and the atomic number of the
absorbing atoms shows that curves like that in Fig. 6-1 can be described by

$$\mu_a = CZ^m\lambda^n + \sigma_a(Z,\lambda) \qquad (6\text{-}4)$$

where the coefficient C takes on different values on each side of an absorption edge and the "best" values for the two exponents are $m = 4$ and $n = 3$. The second term in (6-4) is called the *atomic scattering cross section* σ_a and represents the intensity lost due to the scattering in directions other than that of the incident beam. For wavelengths that are large compared to electron-electron distances in atoms ($\lambda > 0.5$ A), σ_a is nearly independent of wavelength and increases in proportion to Z. Since the first term in (6-4) increases with both Z and λ much more rapidly, the second term may be neglected by comparison. At shorter wavelengths, the λ dependence becomes more complex but, because of severe experimental difficulties in measuring the scattering part of (6-4), an accurate dependence has not been established. Since the second term is small (typically, $\sigma/\rho \sim 0.10$ to 0.20 cm^2/g), it is normally neglected in the discussion of absorption coefficients. This is an acceptable practice for all but the lightest elements, for which the total absorption coefficient may have commensurate magnitudes. (For carbon, $\mu_m < 4.0$ cm^2/g for $\lambda < 1.5$ A.)

The first term in (6-4) is of primary interest, not only because it is relatively much larger but also because it represents the wavelength and Z dependencies of the photoelectric absorption process for x rays. As can be seen in Fig. 6-1, as the wavelength of the transmitted x rays increases, the photoelectric absorption increases $\propto \lambda^3$ until a critical wavelength is reached. For wavelengths longer than that at the absorption edge, the incident x rays have insufficient energy to knock out a particular kind of bound electron in the atom, and the photoelectric ejection of such electrons no longer contributes to the absorption process. The absorption edges in Fig. 6-1 are called the K, L_I, L_{II}, and L_{III} edges, respectively, according to the inner electron whose binding energy equals the energy at the absorption edge. Following the photoelectric ejection of an inner electron, the excited atom typically emits an x-ray photon, as an outer electron falls into the newly created hole, so that the first term in (6-4) is called the *fluorescence term*.

The height of the discontinuity at an absorption edge turns out to be proportional to the ratio of r of the wavelengths of the succeeding edge to that of the edge being considered $[r(K) \sim \lambda_{L_I}/\lambda_K]$. Although this relationship to the electron energy levels is of obvious interest, only approximately obeyed empirical relations have been found. Such relationships, similar to (6-4), are of practical utility primarily when it is necessary to deduce values of absorption coefficients that, for various reasons, cannot be measured with better accuracy.

When the horizontal wavelength scale in Fig. 6-1 is expanded, the saw-toothed character of an absorption edge is replaced by a fine structure that extends for hundreds of electronvolts on the high-energy (low-wavelength) side of the main edge. A fine structure is also apparent in the main edge itself, and, occasionally, some structure has been found to occur on the low-energy side of the edge as well. The first to explain the structure appearing at an absorption edge, Kossel (1920) pointed out that the ejected photoelectron can undergo transitions to a series of unoccupied quantum states in an atom provided that such transitions are allowed by quantum-mechanical selection rules. The fine structure appearing at and near the observed absorption edge (within a few tens eV), therefore, reflects the distribution of unoccupied states for the absorbing atom. As is well known, the proximity of other atoms modifies such quantum states and their relative densities, and so the absorption-edge fine structure can be utilized in their study, as discussed in detail below.

In molecular aggregates, particularly in solids, the fine structure has been found to extend for hundreds of eV from the edge. This was first explained by Kronig (1931) for crystalline absorbers by suggesting that the fine structure was caused by a grouping of the energy levels possible in a solid into allowed and forbidden energy bands. Thus it is usual to interpret the near-in or Kossel structure in terms of atomic levels (as modified by the environment) and the extended Kronig structure in terms of the energy levels produced by the state of aggregation.

It is clear that the detailed fine structure reflects the ability of a photoexcited electron to undergo a transition to an allowed and unoccupied quantum state. Thus the magnitudes of the observed fine-structure undulations should be proportional to the density of allowed and unoccupied states $N(E)$, and the probability that a transition to a state having this energy E will take place $T(E)$:

$$\mu(E) \propto N(E)T(E) \qquad (6\text{-}5)$$

where it is usual to express the transition probability by a matrix element like

$$T(E) \propto \left| \int \psi_i \frac{\partial}{\partial_x} \psi_f \, d\tau \right|^2 \qquad (6\text{-}6)$$

in which ψ_i and ψ_f are, respectively, the wavefunctions of the initial and final quantum states occupied by the electron. In the simplest case (6-6) reduces to the familiar dipole selection rules, it being generally believed that quadrupole transitions make a negligibly small contribution to absorption spectra. In most cases of practical interest, however, the calculation of the transition probabilities in (6-6) requires an explicit knowledge of the two wavefunctions. The correct wavefunction representation for the final state of the ejected electron is further complicated by the uncertainty regarding what effect the presence of the hole has on the systems. It is clear that the presence of a hole must affect the energies of the other electrons, but whether it is a simple perturbation of the original states or whether a new set of states must be invoked is not clear. [Another way of looking at this is to compare the lifetime of the absorption process to the relaxation time of the perturbed system, as suggested but not developed by Blokhin (1969).] In view of this difficulty, most of the interpretations of absorption-spectra fine structures have been speculative in nature. It should be noted that interactions with plasmons also should be possible, but, to date, relatively scant evidence has been presented to demonstrate how such interactions actually affect the observed spectra.

Following a brief consideration of some of the experimental aspects that are unique to absorption spectroscopy, the absorption spectra of various atoms will be presented. The near-in spectra have been grouped according to the state of aggregation in which the absorbing atom occurs, beginning with isolated atoms in monatomic gases and concluding with crystalline solids. The extended fine structure occurring primarily in crystalline materials is considered in a separate section, and finally, isochromats are considered. Isochromats are recorded by measuring an emission process when the short-wavelength limit of the continuous spectrum is at an absorption edge, so that their fine structure bears a closer analogy to absorption spectra than to the characteristic emission spectra discussed elsewhere in this book. In selecting the absorption spectra for explicit discussion below, from among the very large number of published spectra, special attention was paid to the accuracy of the measurements, the systematics of the investigations, and the soundness of the speculative aspects of the proposed explanations of observed fine-structure details. Even so, only representative spectra can be considered in detail within the scope of the present chapter. It is important to stress here that the limitations imposed by the absence of a straightforward process for evaluating the transition probabilities (6-6) require experimental interpretations to be made on a relative basis by comparing details in the fine structure that change due to the variation of some single parameter, such as temperature or composition. Except where approximate calculations of the relevant factors have been attempted, comparisons of the absorption spectra of the same atomic species in radically different environments are without firm foundation.

II. EXPERIMENTAL CONSIDERATIONS

A. Instrumentation

The basic considerations underlying the selection of spectrometers for absorption spectroscopy are quite similar to those already discussed for emission spectroscopy. The resolving power and sensitivity of an instrument in any spectral range clearly does not depend on whether emission or absorption spectra are to be measured. Provided that only qualitative comparisons of fine-structure details recorded under identical conditions on the same instrument are to be made, it does not matter whether photographic film or counters are used. The same applies to angular (energy) location of one prominent feature relative to another. When quantitative estimates of the fine structure are sought, however, counters are clearly preferable. This is so not

only because counters provide numerical data directly, eliminating the need for subsequent photometry of the film, but also because the density of a film viewed with
visible light bears a logarithmic relation to the transmitted intensity. When this is
combined with the variation in the absorption characteristics of the photographic
emulsion as a function of wavelength, a quantitative evaluation of a photometer
tracing becomes most difficult. Admittedly, the gas in, say, a proportional counter
also exhibits a spectral variation. In fact, near an absorption edge, variation in the
counter's response can be used to determine the fine structure at the absorption edge
of a gas (Schnopper, 1964). Because it is recommended practice to measure both I
and I_0 at each point, however, when a counter is used, such variations automatically
cancel out once the ratio of the two intensities is calculated. This brings out the second
difference between films and counters. A film records the transmitted-beam intensity,
not the absorption coefficient. For qualitative comparisons this does not matter, but
for quantitative measurements it does.

B.   Specimen characteristics

Because of the high absorption by all materials of x rays having energies close to
absorption edges, the chief experimental problem in absorption spectroscopy is the
preparation of homogeneous samples having the requisite thickness, typically of the
order of a few microns. Ideally, nothing but the substance itself should be placed
in the beam. This can be accomplished by reducing thin sheets by successive rolling
and annealing (stress relieving) or by evaporation of films onto a substrate that is subsequently removed. In our laboratory, sputtering of brittle alloys onto a halite substrate has proved feasible, but only in a limited number of cases. When the material
is not ductile, a fine powder is prepared either by grinding or by chemical precipitation,
and the powder is mixed with some binder such as parlodian and coated onto a glass
plate. After drying, the film can be removed by flotation in water and further reduced
in thickness by rolling the plastic sheet. (Soluble compounds also can be absorbed
on a very thin paper blotter.) Alternatively, the fine powder can be coated onto a
substrate which, however, must be transparent in the x-ray region to be studied.
Generally, this is the least satisfactory method of sample preparation because the
specimen must be extra-thin to compensate for the absorption by the substrate.
Perel (1970) has suggested an analytical function that corrects for nonuniformities
in such a sample and allows estimation of the effective sample thickness.

The fact that the thickness of an absorber affects the details of the fine structure
being recorded has been appreciated for some time. Several theoretical and experimental explorations of this effect have been made, but no apparent consensus seems
to have emerged. First it should be noted that when photographic recording is
employed, usually in a bent-crystal spectrograph, then it is the transmitted-beam

intensity not the absorption coefficient that is being recorded. Counter detectors, whether in single-crystal or two-crystal spectrometers, usually measure I and I_0 at each point so that the quantity being determined is μx or the absorption coefficient μ, within the accuracy with which the thickness x is known. Despite the difference, the detail visible in the fine structure is similarly affected in both cases, and two factors are important here: the "window" function of the spectrometer, i.e., the instrumental broadening, and the thickness of the absorber.

If only the thickness effect is considered, it appears that different criteria exist depending on whether the structure at the edge itself or the extended fine structure is to be examined (Blokhin, 1957; Sandström, 1957). If I_1 and μ_1 are the intensity and linear absorption coefficient on the low-energy side of the absorption edge, and I_2 and μ_2 these values on the high-energy side, then Sandström (1930) first suggested that the optimum thickness can be determined by setting $d(I_1 - I_2)/dx = 0$, which gives

$$x_0 = \frac{\ln (\mu_2/\mu_1)}{\mu_2 - \mu_1} \qquad (6\text{-}7)$$

When the extended fine structure is considered, Kurylenko (1939) argued that the difference that should be maximized is $\Delta I_2 = -\mu_2 x e^{-\mu_2 x}\, dx$. This leads to an optimum thickness

$$x_0 = \frac{1}{\mu_2} \qquad (6\text{-}8)$$

A comparison of (6-7) and (6-8) shows that the thickness in (6-8) is smaller. In the most recent experimental test of the thickness in photographic recording, Krishnan and Nigam (1967) recorded the extended fine structure of three nickel foils, 4, 6, and 8 microns thick, in a relatively low-resolution, bent-crystal spectrometer. They found that the detailed structure within the first 20 eV was more pronounced in the thinner sample, that from 20 to 100 eV it remained unchanged, and that above 100 eV it was enhanced with increasing thickness. Unfortunately, their thickest foil had a thickness very close to that given by (6-8) so that the implication that contrast in the extended structure increases with thickness should not be extrapolated from their results.

When spectrometers employing counters are used, then an additional criterion is imposed by counting statistics. Assuming that the statistical precision of all experimental points is the same (fixed-count method) and negligible, then Parratt, Hempstead, and Jossem (1957) have suggested as a rule of thumb that the thickness should be such that $I/I_0 \simeq 0.10$ which leads to

$$x_0 = \frac{-\ln 0.10}{\mu} \qquad (6\text{-}9)$$

The main thrust of their paper, however, was to point out that the thickness of the sample modifies the transmitted beam throughout the entire spectral range transmitted at each spectrometer setting. Clearly this affects the accuracy (absolute value) of the measurements. Unfortunately, it also affects the relative values of two adjacent points because the overlap due to the tails diminishes the contrast between them. In order to minimize this effect, the sample should be made as thin as possible, consistent with retaining sample homogeneity (Klems, Das, and Azároff, 1963). Incidentally, it is assumed in this discussion that overlapping spectra from second-order and higher order diffraction by the analyzing crystal have been eliminated by suitably adjusting the excitation voltage on the x-ray tube.

From his analysis of the origins and magnitude of statistical errors in x-ray absorption measurements, Nordfors (1960) concluded that it is more dangerous to choose a specimen thickness that is too thin than one that is too thick. This is based on an analysis of the dependence of the standard deviation in the absorption measurement on the product μx, which tends rapidly to infinity as $\mu x \rightarrow 0$. The optimum measurement of I and I_0, as well as the influence of background, are also considered by Nordfors.

C. Specimen placement

It is generally believed that the absorber can be placed anywhere in the x-ray beam path provided that a single-crystal spectrometer is employed. Actually it is better to place the absorber ahead of the crystal so that any fluorescent radiation emitted by the absorber does not reach the detector. Such placement is particularly desirable in a two-crystal spectrometer whose increased dispersion also minimizes the transmission of all radiation scattered by the sample in other than the forward-beam direction. Placement of the absorber ahead of the first crystal in a two-crystal spectrometer is also recommended whenever the specimen characteristics cause a significant amount of small-angle scattering components of the incident beam. The importance of beam divergence and other factors in this case has been discussed by Parratt et al. (1959) who also compared the ratios of I/I_0 for several kinds of absorbers placed ahead of the first crystal, between the crystals, and following the second crystal.

III.   ABSORPTION EDGES

A.   Introduction

As first suggested by Kossel (1920), a rapid rise in the absorption of incident x rays should take place as soon as the energy of the x-ray photons equals the work done in "promoting" a bound electron from an inner level to the first empty energy state of the absorbing atom. Since this rise actually takes place over a range of several electron-volts, a question arises regarding the energy value of the "true" edge. (Historically, the term "absorption *edge*" of an x-ray absorption spectrum has been used to denote the inflection point in the initial rise of the absorption curve. It is probably better usage to denote the position of an edge as the ionization threshold of the absorbing atom, a terminology that is more consistent with that employed in atomic physics.) Richtmyer, Barnes, and Ramberg (1934) considered an energy continuum of discrete states having equally sharp energies and equal transition probabilities. This model gives rise to a series of absorption lines having equal half-widths (widths at half-maximum) Γ_E and an arctangent energy dependence for the absorption coefficient

$$\mu(E) = C\left(\frac{1}{2} + \frac{1}{\pi}\tan^{-1}\frac{E - E_0}{\Gamma_E/2}\right) \tag{6-10}$$

where C is a constant and E_0 is the energy difference between the initial and first allowed empty energy level.

The maximum value of the absorption coefficient should occur at $E = \infty$ at which value $\mu(\infty) = \mu_{max} = C$, as can be seen by direct substitution in (6-10). Note also that when $E = E_0$, then $\mu(E_0) = \frac{1}{2}\mu(\infty)$, so that the midpoint (inflection) in the arctangent curve marks the energy value of the first allowed empty state. Since the final states were assumed to be quite sharp by Richtmyer, Barnes, and Ramberg, the actual widths of the absorption lines must be due to the width of the initial state. This width can be determined by substituting $C = \mu_{max}$ in (6-10) and by considering the energy values $E_{1/4}$ and $E_{3/4}$ at which the absorption coefficient equals, respectively, one-quarter and three-quarters of its maximum value. It follows directly from (6-10) that

$$\mu(E_{1/4}) = \mu_{max}\left(\frac{1}{2} + \frac{1}{\pi}\tan^{-1}\frac{E_{1/4} - E_0}{\Gamma_{E/2}}\right) = \frac{\mu_{max}}{4} \tag{6-11}$$

and

$$\mu(E_{3/4}) = \mu_{max}\left(\frac{1}{2} + \frac{1}{\pi}\tan^{-1}\frac{E_{3/4} - E_0}{\Gamma_{E/2}}\right) = \frac{3\mu_{max}}{4} \tag{6-12}$$

so that

$$E_{3/4} - E_{1/4} = \Gamma_E \tag{6-13}$$

from which the half-width of the initial state can be determined. Unfortunately, the validity of the assumptions underlying the derivation of (6-10) is uncertain, so that the only part of the above conclusions that finds general acceptance is that the inflection in the initial absorption rise (displaying very nearly an arctangent dependence in most cases) marks the energy value of the onset of allowed energy levels for the ejected inner electron (Nordfors, 1960).

There is another way to determine the energy value corresponding to the first allowed transition for an inner electron. After setting the spectrometer to detect a strong line of the characteristic (emission) spectrum, the voltage on the x-ray tube is gradually increased until the emission line is first detected (see Section V, Iso-chromats). Since this is the energy at which the incident electrons are first able to create an inner vacancy, it must correspond to an absorption edge of the target atoms. Such measurements were carried out by Nilsson (1953) for several elements in the first long period, and his values agree, within experimental error, with those obtained from the inflection points in the initial absorption rise of the same elements.[1] Apart from the intrinsic interest in determining the binding energies of inner electrons, the exact location of the inflection point is necessary to compare any shifts in the onset of absorption due to changes in the absorbing atom's environment.

The Kossel model of single-electron transitions to unoccupied states has been applied to the interpretation of the absorption-edge structure of isolated atoms (inert gases) as well as to molecules and solids, in which case use is made of band-model calculations, including the possible existence of quasi-stationary bound states such as exciton states. Among possible many-electron mechanisms, several attempts have been made to explain observed fine structures in terms of two-electron excitations and plasma excitations. In his review paper, Parratt (1959) suggested another many-electron mechanism in which the energy levels of the outer valence states are perturbed by the inner electron vacancy during its finite lifetime. This model has been utilized in explaining Cr emission spectra from chromium halides by Bergwoll and Nigavekar (1968), but has been applied to absorption spectra analyses in qualitative ways only.

B.   Atoms in gases

For the first careful analysis of the absorption spectrum of an inert gas, Parratt (1939) chose the K absorption edge of argon, reproduced in Fig. 6-2. In the free atoms, the final states should be the very sharp atomic states which are broadened to about 0.58 eV in argon because this is the finite width of the inner K level. Thus only the first two individual absorption lines can be resolved in Fig. 6-2; the others overlap and merge with the continuous absorption curve which marks the series limit of the atomic levels. In calculating the atomic states, Parratt assumed that "so far as the outer levels are concerned, the argon atom with a missing electron behaves approximately as a potassium atom." This approximation—that the inner electron from an atom of atomic number Z undergoes transitions to the optical levels associated with an atom having number $Z + 1$—is central to the analysis of atomic absorption in monoatomic and polyatomic gases as well as in solutions.[2]

---

[1] It should be noted that a comparable experiment could be carried out using x-ray instead of electron excitation by irradiating a specimen with a carefully mono-chromated x-ray beam whose energy is being constantly changed, say, by changing the angle of a crystal monochromator. The energy at which a strong fluorescence line first appears then would mark the onset of the absorption transitions.

[2] This approximation was examined by Mitchell (1965) for neon. He found that the radial wavefunctions of sodium and of neon in the K state are similar, but cautioned that this approximation may not hold for very light atoms.

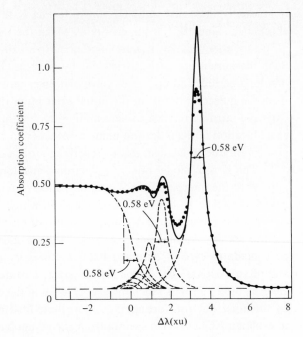

FIGURE 6-2
Absorption spectrum of argon. The solid line represents the absorption spectrum (experimental points) corrected for instrumental broadening effects. (*After Parratt*, 1939.)

Parratt further assumed that dipole selection rules governed the transition possibilities so that the first allowed transition in argon is $1s \rightarrow 4p$, followed by transitions to the subsequent np states whose energy separation is the same as that of the optical levels in potassium. The continuous absorption curve, having an arctangent shape for its initial rise, can then be combined with the discrete series of absorption lines shown in Fig. 6-2 to reconstruct the experimentally observed absorption curve of argon. Watanabe (1965b) used this one-electron model to analyze the argon K edge by making a seven-parameter fit to the experimental curve in which the resonance absorption lines have a Lorentzian shape, a consequence of the transition to a state having a single K hole and no additional excitation. A concurrent analysis making use of self-consistent-field wavefunctions was carried out by Bagus (1965), including relaxation effects. He gets good agreement with the experimentally determined value of the argon K ionization limit and attributes this agreement, at least in part, to the predominance of the one-electron contributions to the Hartree-Fock operator over the two-electron contributions. The inclusion of dipole transitions between Hartree-Fock states in the analysis has made possible a semiquantitative interpretation of the argon

K absorption spectrum that is more complete than that of any other substance. Two questions that remain are a quantitative explanation of the high continuum slope just after the main edge and the relative importance of the K hole and electron correlation in the double excitation responsible for the extended fine structure in the argon spectrum discussed below. Some aspects of such two-electron ejection in inert gases have been discussed by Carlson (1967).

In a different approach, Vainshtein and Narbutt (1945; 1950) (cf Vainshtein, 1950) assumed hydrogenic wavefunctions for the initial and final states and calculated the square of the dipole transition matrix between an initial $1s$ and final np state. Keeping in mind that the Coulomb field acting on a $1s$ electron is much stronger than that acting on a highly screened np state, they found the area of the absorption line Θ_n for the nth level:

$$\Theta_n \propto \frac{n^2 - 1}{n^5} \qquad (6\text{-}14)$$

Making use of (6-14), they then calculated the relative intensities of the various absorption lines in an analysis that otherwise was identical to the one carried out by Parratt.

Nadzhakov and Barinskii (1960) modified the above procedure by deriving the ratio of the heights of the nth absorption line τ_n to that of the continuous absorption curve τ_∞, using the same assumptions. This serves to bring out an explicit dependence on the total charge of the absorbing atom η (equal to unit positive charge in the K shell plus an effective charge η' at the periphery of the atom) and an effective quantum number n (not necessarily an integer) that increases by unit for successive absorption lines (exciton levels). The ratio is

$$\frac{\tau_n}{\tau_\infty} = \frac{4\eta^2}{\pi \Gamma_n} \frac{n^2 - 1}{n^5} \qquad (6\text{-}15)$$

where Γ_n is the half-width of the nth line. The parameters η and n are determined by curve fitting, as demonstrated by Barinskii and Nadzhakov (1960) for a number of gas atoms and molecules, including Cl_2 and $GeBr_4$ whose absorption edges can be reconstructed from a single Rydberg series obeying (6-15). The authors suggest that in certain cases it may be necessary to consider molecular (as distinct from atomic) levels or Stark splitting of the exciton levels by the electric field due to other atoms present. It is not clear, however, whether the analysis of molecular spectra in terms of a single Rydberg series is a valid procedure to follow.

Barinskii and Malynkov (1962) specifically considered the possible influence of the Stark effect on exciton levels by analyzing the sulfur K edges in SO_2, $SOCl_2$, SO_2Cl_2, and H_2S. The absorption spectra for the first three gases are identical, implying that molecular levels need not be considered since the absorption spectra can be reconstructed by a single Rydberg series of unsplit lines. The peripheral charge

$\eta' = \eta - 1$ calculated in the curve-fitting process turns out to be zero for these three gases, leading the authors to conclude that the atoms in the molecules are co-valently bonded. This, in turn, accounts for the absence of the Stark effect. For H_2S, however, a single series of unsplit levels did not fit the absorption curve. Beginning with an approximate "fit," the authors obtained a charge for the sulfur atom and, using first-order perturbation theory, calculated the level splitting and claimed good fit with the experimental curve; however, this agreement was not illustrated. Barinskii and Malynkov (1963) have questioned the validity of their own perturbation calculations. They suggested that, in addition to possible Stark splittings, it is necessary to take account of hybridized empty levels having admixed p and d symmetry so that $1s \rightarrow nd$ transitions are allowed by the selection rules. This enables them to explain the presence of "additional" peaks in the germanium K edge in $GeBr_4$.[1] The analysis is predicated, however, on a very limited number of experimental data points.

Subsequently, LaVilla and Deslattes (1966) remeasured the absorption edge of H_2S which differed somewhat in appearance from the above-reported curve. It is also shifted from the sulfur edge in SF_6, a fact which is qualitatively adduced as showing differences in binding energies. The SF_6 absorption spectrum has been reexamined by Best (1967) who pointed out that the active-electron approximation, in which only one electron changes quantum numbers, is not applicable to SF_6 even though it gives an adequate description for most molecular spectra. Best makes use of the Hartree-Fock approximation to expand the dipole transition moment for x-ray absorption and shows that the slopes near the absorption edge can be different from those resulting from the active-electron approximation. He also suggests that the SF_6 spectrum may differ from that of other molecules because the d-electron binding is different in SF_6 and also the ionic bonding between the parent and the residue molecular species.

The L spectra of gases offer additional interpretative difficulties since the electrons from L_{II} and L_{III} levels can undergo transitions to nd or ns states. In addition, the different L levels may have small energy separations. Barinskii (1961) recorded the L_{III} spectra of krypton and xenon and used the L_{III} edge of argon previously measured by Prins (1934) to test a modification of the theory presented above. It turns out that the argon curve can be resolved into ns and nd absorption peaks plus the corresponding arctangent curves at the series limits. The krypton and xenon curves, however, can be fitted only by τ_∞ values that differ considerably from calculated values for both the s and d states. Lukirskii and Zimkina (1963) show $L_{II,III}$ absorption edges that display fine structure not recorded by Prins and suggest that the L_{II} and L_{III} absorption edges of argon probably merge. The L absorption spectrum

---

[1] It is also typical for molecular gases to exhibit a fine structure extending for several hundred eV from the initial rise. The discussion of such structures is postponed till Section IV.

of xenon also has been recorded by Watanabe (1965*a*) who used the one-electron model to determine the oscillator strengths for the resonance absorptions. Wuilleumier (1966) shows the $L_{II,III}$ edge of krypton with a fine structure not recorded by Barinskii and explains it qualitatively in terms of simultaneous transitions of two *p* electrons. A table of such double excitations and their energies is presented.

Multiple excitation also was proposed by Schnopper (1963) to explain a new resonance absorption structure he observed in the *K* absorption spectrum of argon (see Fig. 1-3). Using a newly built two-crystal spectrometer, Schnopper explored the spectrum up to 50 eV above the first maximum and discovered a new maximum at about 22 eV which was followed by an apparent continuum. Using a bent-crystal spectrometer and photographic recording, Bonnelle and Wuilleumier (1963) observed not only a maximum at 21.8 eV but also one at 33.8 eV above the first line in argon. They attribute both of these maxima to double ionizations $K \rightarrow M_{II,III}$ and $K \rightarrow M_I$, respectively.

Lukirskii, Zimkina, and Brytov (1964) have investigated *M* absorption edges of krypton and the *N* edges of xenon. For the xenon N_{IV} spectrum, they propose that two of the four maxima recorded can be explained by single excitations and the other two by double excitations. The other spectra are also discussed semiquantitatively by trying to fit individual absorption lines having Lorentzian shapes plus an arctangent curve at the limit to the observed curves. The absorption edges of gases also have been studied using electron-synchrotron radiation and are discussed by Madden in Chapter 7 of this book.

C. Molecules and ions in solutions

It is possible to use a solution containing dissolved ions or molecules as an absorber in a manner quite similar to that used for solid absorbers. The first systematic attempt to compare the fine structures of ions in different solutions was carried out by Beeman and Bearden (1942) who measured the *K* absorption spectra of Ni^{++}, Cu^{++}, and Zn^{++} in aqueous solutions. Typically, they found two absorption maxima whose energies, as compared to the inflection point in the initial rise, agreed with what would be expected from calculated optical terms for the transitions $1s \rightarrow 4p$ for the first maximum and unresolved $1s \rightarrow np$ ($n = 5, 6, 7, \ldots$) for the second maximum. The peaks are considerably broader than the expected width of the excited *K* state, and the extra broadening was attributed to interactions with the unfilled $3d$ shells and Stark splitting, possibly caused by the surrounding molecules. In addition to the above ions, Hanson and Beeman (1949) examined the absorption edges of Mn^{++} in solutions and found that the $1s \rightarrow 4p$ transition occurred at 14 eV as compared to about 17 eV for both nickel and copper.

Vainshtein and Kavetsky (1953) undertook an examination of the role that the solvent plays in the absorption process. Comparison of the fine structures of divalent copper and zinc ions in aqueous and several nonaqueous solutions shows that the main features of the fine structures are similar, but detailed differences do occur. The width and height of some of the maxima change with solvent, and a slight displacement of the second maximum relative to the first one is observed. Vainshtein and Kavetsky find a relation between these changes and the dipole moments of the solvent molecules. This relationship is not exact, and since details of the experimental measurements are not presented, it is hard to judge whether some of the changes may be due to an effect analogous to the "thickness" effect discussed above. Also the possibility of complex ion formation cannot be ignored, as demonstrated strikingly by the identity of their curves for aqueous solutions of Zn^{++} derived from $ZnCl_2$ and from $Zn(NO_3)_2$ as opposed to marked differences in the Zn edges when the two salts are dissolved in methyl alcohol.

Nadzhakov and Barinskii (1960) have demonstrated that their analysis of the K absorption spectra of argon can be applied to ions in solutions as well. Using (6-15), they calculated the exciton lines for Zn^{++} ions in solution and, by adjusting the parameters η and n, obtained a good fit to the experimental curve. Narbutt and Izraeleva (1963) reexamine the fine structure of the zinc-ion edge in an attempt to evaluate the relative breadths of the exciton lines, which they believe to exceed the combined instrumental and K-level widths by 2 to 6 eV. Comparing their curve to the Z^{++} absorption edges in crystalline hydrates (Cotton and Hanson, 1958), they find sufficient agreement to assume that the zinc ion in solution is also octahedrally coordinated by water molecules. This makes Stark splitting of the exciton lines impossible unless the octahedra are distorted. Despite calculational attempts to estimate such distortions, the authors fail to obtain a quantitative explanation of the observed half-widths.

Analyses similar to those described above have been applied to ions and molecules in solids. Their detailed discussion is postponed until Subsection III D 4, but it is of interest here to juxtapose the considerations that enter when the absorbing atom is covalently bonded to surrounding atoms. Mitchell and Beeman (1952) tried to infer the actual bond types formed, i.e., whether planar dsp^2 or tetrahedral sp^3 hybridized orbitals provided a better fit to the fine structures of Cr, Mn, Fe, and Ni atoms in various compounds. They were able to establish an empirical correlation which, however, was later questioned by one of the authors. Mitchell (1962) reexamined the details of the fine structures of three planar nickel compounds and concluded that the first absorption maximum probably represented a transition to an empty $4p$ orbital in nickel, but that the second maximum was most likely due to plasma oscillations. This proposal is supported by an order-of-magnitude calculation of the plasmon energy. Although Mitchell's analysis clearly throws doubt on the validity of inferring a bonding scheme from the presence of $1s \rightarrow 4p$ transitions in nickel compounds, his calculations of plasmon energies involve too many approximations to allow an assessment of the validity of considering plasmon transitions. Best (1966) applied molecular orbital theory to calculations of available orbitals for transitions allowed in MnO_4^-, CrO_4^{--}, and VO_4^{3-} ions, while Seka and Hanson (1969) made similar calculations for 43 transition-metal complexes. A qualitative correlation between the observed fine structures and the calculated transitions is found to exist, indicating that x-ray absorption spectra provide a useful additional tool for verifying stereochemical deductions.

D. Atoms in solids

1. *Pure metals* The absorption spectra of metals can be explained in terms of a one-electron model by assuming that absorption is proportional to the product of the density of allowed but unoccupied states and the probability of transition to these states (6-5). This approach was first put on a sound footing by the analysis of the K edges of Fe, Co, Ni, Cu, Zn, Ga, and Ge by Beeman and Friedman (1939), who made use of the first band calculations carried out for metals by Krutter (1935) and Slater (1936). Such qualitative interpretations have been successful for all but the lightest elements. Ausman and Glick (1969) have shown that, when many-electron interactions are taken into account, then the absorption-edge structure of Li and Na can be explained satisfactorily. They do this by calculating phase shifts for the electrons scattered by the inner vacancy, making use of the Friedel sum rule and a simplified model proposed by Nozières and de Dominicis (1969). [Schotte and Schotte (1969) have presented an alternative derivation of this result.] Since many-body effects have been discussed in Chap. 5 by Hedin, they are not further considered here. It is pertinent to note here, however, that the possible role of plasmons in x-ray absorption processes has been considered by Ferrell (1956), who suggested that the absorption edge might be reproduced in shape and shifted toward shorter wavelengths by an amount equal to the plasmon energy. Sobelman and Feinberg (1958), who considered the region well to the high-energy side of the initial absorption edge, and Schmidt (1961b), who considered the initial rise of the "main" edge, found that the probability of x-ray absorption with simultaneous plasmon formation is of the same order of magnitude as transitions without plasmon formation. Nozières and Pines (1959) point out that the x-ray transition usually does not involve major charge-distribution distortions, so that the transition probabilities are small. They further point out that plasmons would produce a satellite band edge which should be masked by the Kronig structure. The proposed correlations between this extended fine structure and characteristic electron-energy losses are further cited in Section IV.

In their original analysis, Beeman and Friedman (1939) suggested that relation (6-5) was applicable to the absorption-edge structure of metals. They furthermore assumed that the transition probabilities were either constant or very slowly varying

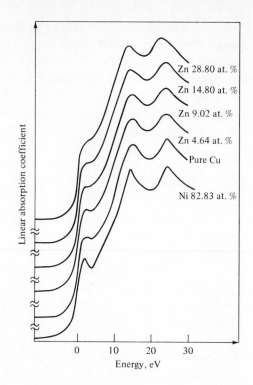

FIGURE 6-3
Copper *K*-edge fine structure in several copper-zinc solid solutions. (*After Yeh and Azároff*, 1965.)

functions of energy in the regions studied, so that the absorption edge should resemble the density-of-states curves of the metal. They made explicit use of the density-of-states curve of copper calculated by Rudberg and Slater (1936) and found a very satisfactory qualitative correlation to exist. The validity of this correlation was re-emphasized by Burdick (1963) who found that the absorption edge of copper also agreed with his recalculation of the energy-band structure of copper. The absorption edge of copper (lower curve in Fig. 6-3) shows an initial rise corresponding to the onset of empty states in the 4*s* band having admixed 4*p* character. Although the density of states remains virtually constant according to Burdick, about 4 eV above the inflection in the initial rise (Fermi energy), the symmetry of the empty states assumes a predominantly *s* character, so that the transition probabilities decline at this energy, causing a decline in the absorption. As the energy increases further, however, the states assume pure 4*p* character, and the absorption coefficient increases even more rapidly than the rise in the density-of-states curve's maximum at the energy of the 4*p* band. The next maximum visible in Fig. 6-3 is believed to be due to states having predominantly 5*p* character, although it is probably overlapped by higher bands as well. Similar qualitative correlations have been proposed for the other metals neighboring copper in the periodic table.

Rooke (1968) has warned that the transition probability in (6-5) may have a strong dependence on the **k** vector, i.e., the energy in different parts of a band. Despite the fact that a correlation of the absorption coefficient directly to the density of the final one-electron states may be an oversimplification, a qualitative correlation is nevertheless possible and permits the interpretation of the absorption spectra of pure metals as well as of solid-solution alloys. Irkhin (1961) and Shepeleva and Irkhin (1961) have tried direct calculation of the transition probabilities for Ni and for Fe, respectively, making use of hybridized *d-s-p* wavefunctions in the dipole approximation, and they claim fair agreement with experimental curves.

Borovskii and Batyrev (1960) present photographically recorded spectra of the elements Ti through Cu recorded in a bent-crystal spectrograph. After correcting the data for somewhat pronounced instrumental effects, they reveal a fine structure not observed by other investigators and relate it to plasmons. Borovksii and Schmidt (1960) examine the *K* absorption edge of iron as a function of temperature up to 950°C and relate it to characteristic electron losses also measured by them. The Fe *K* edge in this case was measured in a two-crystal spectrometer and after correction for instrumental effects looks more like the photographically recorded spectrum cited above than like the spectrum reported by other investigators. Again they present arguments against explaining the fine structure in terms of density-of-states curves and in favor of x-ray-induced plasma oscillations. Using the same curve, Singh (1962) demonstrates, however, that the maxima corresponding to the fine structure reportedly repeated by the plasmon absorption are in precisely the same relative positions as the first peaks in the extended Kronig structure. Thus he supports the conclusions reached previously by Nozières and Pines. It is interesting to note that Borovskii and Schmidt paid no attention to possible shifts of the edge or of any maxima in the iron curves except to note a change in the fine structure above the temperature for the transition of α-Fe to γ-Fe. Trapeznikov (1956) examined the *K* edges of Fe and Ni above and below their magnetic Curie points and found in the position of the inflection in the initial rise no change larger than experimental errors. Hagarman (1962) did find a systematic shift of the inflection point (Fermi energy) with temperature in Ni, but no departures from this trend at temperatures above the magnetic Curie point. These results contravene predictions made by Sokolov (1956) regarding anomalous changes to be expected in the density of states near the Curie point in ferromagnetic crystals.

Data on soft x-ray absorption in metals published prior to 1956 have been reviewed by Tomboulian (1957). Since then, significantly different results have been obtained with synchrotron radiation, and these are described in Chap. 7 in this book. The *K* spectra of light metals and the *L* and *M* spectra of heavier metals illustrate the relative difficulty in distinguishing exactly which regions represent the Kossel structure and which the extended fine structure. The problem of overlapping absorption lines for transitions from *L* or *M* levels to higher empty states, encountered in the discussion of inert gases, is considerably aggravated in metals because of the extensive band overlapping that normally occurs. Thus most of the discussions center on the ability of the extended-fine-structure theories (Section IV) to explain the observed spectra. In a very careful experimental study of the three L edges of silver, Nordfors (1961) measured the structure extending for about 70 eV from each edge but was unable to explain the observed maxima by the Kronig theory. Nordfors and Noreland (1961) similarly measured the L edges of the elements Cd-Te but again could not find a simple explanation for the observed fine structures.

2. *Solid-solution alloys* Following the successful interpretation of the K absorption spectra of Ni, Cu, and Zn, Beeman and coworkers examined the solid-solution alloys formed by these three metals: Cu-Ni (Friedman and Beeman, 1940), Ni-Zn (Bearden and Beeman, 1940), and Cu-Zn (Bearden and Friedman, 1940). They noted that the absorption edges of the two constituents in the same alloy were dissimilar and resembled quite closely the fine structure evinced, respectively, by the pure metals. Changes in the composition of the alloys (atomic ratios) did produce visible changes in the Zn edges in Cu-Zn and Ni-Zn alloys and in the Cu edge in Cu-Zn. No visible changes were reported for either metal in Cu-Ni. Like several later investigators who studied solid-solution alloys, Beeman et al. expected to find that the constituents of an alloy would display fine structures that resembled each other, i.e., that they would reflect the common density-of-states curve of the alloy. The fact that the absorption edges of the constituent atoms in an alloy resemble the respective atomic (elemental) curves suggests that the final state for the ejected electron probably cannot be described by the usual Bloch functions used to describe stationary states for a metal (alloy). Instead, the final states should be treated as localized states that are characteristic of the absorbing atom's environment and how it is affected by alloying.

Pursuing this line of reasoning and paying special attention to small but systematically regular changes in the fine structure of the absorption edges, Azároff and Das (1964) reexamined the absorption spectra of Cu-Ni solid solutions and detected a progressive decline in the first maximum of the Cu K edge, as successively more copper was added to nickel. This was interpreted to indicate a progressive decline in the unfilled $3d$ states (actually d-s-p hybrids) at Ni atoms with an attendant increase in the empty $4s$ (actually s-p hybrid) states of Cu. Subsequently Yeh and Azároff (1965; 1967) reexamined the absorption spectra of Cu-Zn and Ni-Zn alloys and found corresponding changes taking place in their fine structures. As an example, the decline of the first maximum in the Cu K edge is illustrated by the series of curves reproduced in Fig. 6-3. By comparison, it should be noted that, although the Zn absorption edge in these alloys resembles that of pure zinc, as the zinc concentration increases, a new maximum appears in the initial portion of the zinc curve and rises with an increasing percent of zinc present. Since, as shown in Fig. 6-3, the corresponding maximum in the copper edge shows a decline, this is interpreted to indicate a rise

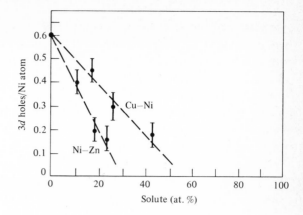

FIGURE 6-4
Decline in 3d hole density at nickel atoms in Cu-Ni and Ni-Zn solid solutions.
(*After Azároff*, 1967*b*; *Yeh and Azároff*, 1967.)

in the number of empty 4*s*-4*p* states at zinc sites and an accompanying decline in such empty states at copper-atom sites. Azároff (1967*b*) suggested how the observed changes could be related on a semiquantitative basis by measuring the areas under different portions of an absorption curve. By transposing the observed area changes into numbers of empty 3*d* states at nickel atoms and by prorating the areas to those of pure nickel, it proved possible to obtain the dependencies on atomic composition illustrated in Fig. 6-4.

The contention that x-ray *K* absorption (and, correspondingly, emission) spectra reflect the distribution of quantum states in the absorbing atom's vicinity is based in part on the short lifetime of the excited state. That the relative distribution in energy of such states is affected by the atom's environment is attested to by the changes produced, for example, by alloying. To put this another way, band-model calculations have been useful in demonstrating that the inflection point in the initial absorption rise corresponds to the Fermi energy (crossover of emission and absorption edges) and in allowing the calculation of the symmetries of various quantum states. The creation of an inner vacancy perturbs the system in different ways (Friedel, 1954, 1958; Parratt, 1959) causing localized states to appear that are characteristic of the absorbing atom and of the energy bands (crystal structure). One would expect, therefore, that two atoms having very similar electron structures, say, two neighboring transition elements, and placed in identical environments (crystal structures) should have similar x-ray absorption spectra. To test this premise, Azároff (1967*a*) compared the fine structures at the *K* edges of Fe, Co, and Ni in face-centered cubic (fcc) and in body-centered cubic (bcc) solid solutions and in their equiatomic alloys with aluminum (CsCl structure). As expected, the three curves for each structure type are quite similar, the chief difference being a progressive increase in the first absorption maximum of the atoms in the sequence Ni → Co → Fe, reflecting the increase in empty $3d$ states at the absorbing atom. The curves for the absorbing atoms are characteristically different, however, when their fine structures in the three different alloy types are compared. Nemnonov and Kolobova (1967) similarly found that the K edge of Ti and Fe resembled each other in a bcc solid solution and in the equiatomic alloy, even though they differed markedly in the pure metals. (Fe is bcc, while Ti has a hexagonal structure.)

Other studies of alloys have been limited, largely, to reports of certain spectral features. For example, Lucasson-Lemasson (1958) studied the K and L absorption spectra of Cu with Al, Ni, and Zn and noted shifts in the initial rise of some of the edges. Later Lucasson (1960) reexamined these curves as well as those of Zn, Ga, and Ge in a more extensive review paper. She concluded that the structure at the initial absorption rise was related to band considerations like those discussed previously by Friedel (1954) and by Parratt (1959) and that the extended fine structure was related to characteristic electron-energy losses. As another example, Bally and Benes (1959) reported some shifts in the inflection points in the K edges of Fe and Ni for several Fe-Ni solid solutions. Donahue and Azároff (1967) tried to relate progressive changes in both bcc and fcc solid solutions of Fe-Ni alloys to $3d$ hole densities and found fairly good agreement with deductions from magnetic measurements for the body-centered alloys. Bally and Müller (1959) examined the temperature behavior of Fe-Ni solid solutions and noted changes in the fine structure, but detected no systematics, possibly because of experimental difficulties of measurement. Mandé (1960) reviewed his measurements of the K and L edges of palladium and the L and M edges of gold in the pure metals and in several solid-solution alloys. Displacements in some extended-fine-structure maxima were noted and related to characteristic electron losses. Sorokina and Nemnonov (1967) focused attention on the sharp peak that appears in the initial absorption rise at the Pd L_{III} edge. (It is frequently called a *white line* because of its appearance on a photographically recorded spectrum.) According to band-model considerations, this absorption line should correspond to the transition $2p → 4d$, so that alloying with silver should cause a gradual occupation of the empty states (as evidenced also by a declining electron specific heat and magnetic susceptibility) and a commensurate decline in the intensity of the white line. Sorokina and Nemnonov found that the peak persisted, however, regardless of the amount of silver present. Excluding the possibility that the band-model considerations are fallacious, since they appear to explain the other alloy properties, the authors conclude that possibly localized states are responsible for the observed absorption line.

The fact that x-ray absorption spectra reflect the absorbing atom's environment suggests the possibility of studying the influence of changes in the crystal structure—for example, those due to order-disorder transitions in alloys. That such correlations are indeed possible has been demonstrated for dilute solutions of aluminum in nickel by Das and Azároff (1965) and of Al in iron by Murty and Azároff (1967). Because only the transition-metal spectra were examined, however, the interpretations assigned are more speculative than those of the other investigations described in this section. This points up an important aspect of such studies. In the absence of a more rigorous theory, the qualitative or semiquantitative interpretations must be based on comparisons of changes produced by varying a single parameter, say composition, while keeping all others, such as crystal structure, essentially constant. Only then is it "reasonable" to assume that the transition probabilities for two different alloys (at the same energy value) remain virtually unchanged. Moreover, the spectra of both components of a binary alloy should display complementary changes in their fine structures. Finally, when possible, two or more kinds of edges should be examined to see whether the postulated transitions are being obeyed consistently.

3. Intermetallic compounds   When dealing with specific compounds, as distinct from solid solutions, it becomes necessary to interpret the fine structure of an absorption edge in terms of what is known about the quantum states (band model) of that compound. In the absence of such information, speculations about the appearance or disappearance of maxima or other details in the fine structure are no better than the intuition of the speculator. An example of one of the more restrained "hand-waving" types of arguments can be found in the interpretation of the K spectra of some copper and nickel sulfides by Ovsyarnikova and Borovskii (1960) in which the authors compare changes observed in the emission and absorption spectra of both components of each compound. Despite the apparent reasonableness of their arguments, more reliably founded correlations must await a detailed knowledge of the energy-band structures. Men'shikov (1963) used an approach similar to that of Beeman and Friedman (1939) to interpret the K spectrum of chromium carbides. Although clearly speculative in nature, the supporting evidence presented lends credence to the interpretations proposed by Men'shikov in this case.

Many studies of intermetallic compounds have been limited to measurements of the position of the absorption edge. Typically, the inflection point in the initial absorption rise is located for both components in a compound and compared to its energy value in the pure elements. From the magnitudes of the apparent shifts and from the directions of these shifts (whether to higher or lower energies), inferences are made regarding possible bonding models or charge distributions among the constituents. Because of their practical importance, semiconducting compounds have received particular attention. Since it is not instructive to consider all the measurements that have been made, the so-called III-V compounds (sphalerite structure) are discussed by way of example.

Gusatinskii and Nemnonov (1964a; 1964b) measured K and L_{III} absorption spectra for the compounds InSb, AlSb, InP, GaP, and AlP and, assuming that they were tetrahedrally bonded via sp^3 hybridized orbitals, explained the observed absorption spectra in terms of transitions to empty antibonding states. Kantelhardt and Waidelich (1969) measured the K absorption edges of Ga and As in GaAs, GaSb, and InAs and interpreted the observed shifts as indicative of the ionic (charge-transfer) type of bonding rather than the covalent bonding assumed by Gusatinskii and Nemnonov. Without supporting evidence from actual energy-level calculations, it is hard to see how unequivocal conclusions can be drawn regarding the true causes of the observed absorption-edge shifts. In the case of InSb, Kantelhardt and Waidelich interpret Gusatinskii and Nemnonov's data to indicate the absence of any charge transfer (purely covalent bonding) despite evidence from x-ray diffraction and electrical conductivity measurements to the contrary (Attard and Azároff, 1963). Such arbitrary interpretations of absorption-edge shifts are particularly suspect because energy-band calculations are suggestive of partly covalent and partly ionic bonding in these compounds.

4. *Transition-metal compounds* One of the first systematic investigations of transition-metal oxides was carried out by Coster and Kiestra (1948) who used photographic recording and collodion foils of powder approximately 200 microns thick. This thickness was estimated to equal about 8 to 10 microns of metal, and the foil was moved at right angles to the x-ray beam during the exposure in order to minimize sample inhomogeneities. The absorption curves obtained in this manner for pure Mn and Fe powders resembled closely those measured earlier by Beeman and Friedman, while the MnO_2 curve was quite similar to one obtained later by Hanson and Beeman (1949) using a two-crystal spectrometer. Coster and Kiestra found that the metal absorption edges in the oxides resembled those in the pure metal and ascribed the main maximum in both to $1s \rightarrow 4p$ transitions. Due to the more nearly atomic character of the electron wavefunctions in the oxides, they explained the diminished initial rise at the absorption edge in the oxide as being due to a decreased admixture of p symmetry to the empty s-d states.

Hanson and Beeman (1949) compared the K edges of manganese in the metal to those in ferromanganese (81 percent Mn) and several oxides and other compounds. Like Coster (1924), they observed a sharp absorption maximum at the low-energy side of the edge in $KMnO_4$ (called a *white line* because of its appearance in photographically recorded spectrum) and a lesser one in K_2MnO_4, MnO_2, and MnS. They suggested that a low-lying level is formed through the hybridization of orbitals in these covalently bonded compounds, using analyses of the bonds previously proposed by Pauling. In a subsequent investigation of covalently bonded Cr, Mn, Fe, and Ni compounds, Mitchell and Beeman (1952) attempted to correlate the bonding type to the presence (or absence) of such a white line. This analysis, in terms of four planar dsp^2 bonds, for tetrahedral sd^3 or sp^3 bonds, and eight octahedral d^2sp^3

bonds, with unfilled low-lying $4p$ orbitals responsible for the sd^3 hybrid, appeared to meet with success. Subsequently, Mitchell (1962) questioned some of these assignations and suggested that plasmon formation maybe was necessary to explain the complete observed fine structure.

Kauer (1956), Böke (1957*a*; 1957*b*), and Fiedler (1963) proposed different interpretations for some of the series of transition-metal compounds that they examined. For example, Böke assigned the central chromium ion in the carbonyl $Cr(CO)_6$ a charge of -6 due to electrons transferred from the six surrounding CO groups. Since this requires all empty $4d$, $4s$, and $4p$ states to be filled when electron-pair bonds are formed with the six CO groups, he proposed that the next unfilled states having p symmetry were the $5p$ states. In support of this contention, Böke argued that the $5p$ level in the negative chromium ion lay at a sufficiently lower energy than the $4p$ level of the positive ion to explain the energy position of the white line. Similar analyses involving molecular orbital theory and charge transfers are used to explain the fine structures of most of the complexes studied. There is some doubt, however, about the validity of the "effective" charges assigned to the metal atoms.

The effect of crystal-field splitting of energy levels on the absorption-edge fine structure was examined by Cotton and Ballhausen (1956) and Cotton and Hanson (1956; 1958). For covalently bonded metal ions, Cotton and Hanson found molecular orbital theory adequate to explain absorption-edge fine structures, but they suggested that Stark splitting of the metal orbitals is necessary for ion-ion or ion-dipole binding. Thus Cotton and Ballhausen found in their theoretical analyses that the occurrence of splitting depended on the coordination symmetry, in agreement with most of, but not all, the spectra that Cotton and his coworkers examined.

In his discussion of the K absorption spectra of the metal atoms in MnO_4^-, CrO_4^{--}, and VO_4^{3-} complexes, Best (1966) assumed that the electron ejected from the K shell felt the negative charge of the radical as a whole, so that the positive charge due to the hole was neutralized and the onset of the main absorption rise was unobscured by excitons. Best also suggested that it was possible to assign ionization potentials to electrons occupying certain molecular orbitals by considering the absorption spectra together with the emission spectra. The ionization energy sought is given by the difference between the energies of the main absorption edge and the K emission line for the appropriate valence level. Use is made of the Ritz combination principle although uncertainties in locating the continuum may obscure matters, as was pointed out by Deslattes (1969). Best also observed a sharp low-energy peak in the absorption spectrum of VO_4^{3-} similar to that found in the manganate and chromate and attributed it to an unoccupied $4p$ orbital. Transitions to empty $4p$ orbitals were also proposed by Cotton and Hanson (1957) to explain the white lines. More recently, Seka and Hanson (1969) ascribed the white line in tetrahedral complexes to tetrahedral energy levels derived by a more sophisticated molecular orbital theory.

The presence of the white line in the absorption spectra of chromate and vanadate radicals was utilized in an ingenious way by Lytle (1967). The chemical formula of the compound $CrVO_4$ sometimes is written $VCrO_4$. Because of the close proximity of their atomic scattering factors, x-ray diffraction studies have been unable to locate correctly the vanadium and chromium atoms in the structure. By comparing the Cr K edges in $NiCrO_4$ and the V K edges in Na_3VO_4 and $CrVO_4$, Lytle was able to show that the Cr edge in $CrVO_4$ did not exhibit a white line whereas the V edge did, and he concluded that the compound contains vanadate radicals.

The appearance of a sharp white line in vanadium pentoxide has been observed by Vainshtein et al. (1961) in their study of several vanadium compounds. Refractory compounds of titanium and vanadium also have received considerable attention. For example, Chirkov, Blokhin, and Vainshtein (1967) made use of the band calculations of Ern and Switendick (1965) in explaining the absorption spectra of three titanium compounds. The procedures used were similar to those developed by Barinskii for complexes in solutions (Subsection III C), although Vainshtein and his collaborators disagreed with Barinskii regarding formalisms employed (cf Barinskii, 1962; 1967).

Fischer and Baun (1968) have studied the $L_{II,III}$ absorption spectra of titanium and several of its compounds by recording the self-absorption of the $L_{II,III}$ emission spectra excited by an electron beam inclined at various angles to the target. By forming a ratio between the more heavily absorbed spectra to that of the least-absorbed spectrum, an effective absorption curve is obtained which compares well with absorption curves obtained in the usual manner. They found that their spectra correlated better with the band calculations of Williams and Lye (1965), who assumed a charge transfer of 1.3 electrons from carbon to titanium, than with the calculations of Ern and Switendick (1965), who assumed neutral atoms. The relative shifts of the absorption edges of TiN and TiC led Fischer and Baun to conclude that titanium has an ionic character in both compounds.

A change in the binding energy of the inner electrons produces a shift in the energy of an absorption edge. Several investigators have made use of this effect in adducing different bond types. Glen and Dodd (1968) have shown that the Mn K edge shifts to higher energies as the manganese-oxidation state increases. They warned, however, that the coordination of an ion also affects the energy. Somewhat more involved arguments were employed by Böke (1957a) to deduce the effective charges of the absorbing metal atom in various complexes. Agarwal and Verma (1968) observed a shift to higher energies of the As K edge with increasing oxidation state in As_2O_3 and As_2O_5. Many other examples of such shifts can be cited.

Other interpretations for the observed energy shifts also have been proposed. Karal'nik (1957) argued that the outer electrons screen the inner electrons from the nucleus. Removal of an outer electron, therefore, should cause all the electron levels to undergo shifts. The detailed qualitative arguments, however, appear to be speculative at this time. Gianturco and Coulson (1968) used Hartree self-consistent-field procedures to estimate the consequences of very rapid removal of an electron with insufficient time for relaxation to the case where relaxation takes place. Subsequent calculations of the energy shift expected as a function of ionization degree showed very small differences for the two models employed. A more definitive discussion of the adiabatic approximation, wherein the electron travels slowly through the system, and the sudden approximation, wherein multiple-hole states can be formed by electron shake-off, was presented by Schnopper (1967) who compared the emission spectra produced by electron bombardment and by K capture and found them quite similar.

5. *Alkali halides* The K absorption spectra of potassium halides were first measured by Nuttall (1928), using a plane-crystal spectrometer and subsequently by numerous other investigators employing two-crystal spectrometers (Trischka, 1945) and bent-crystal spectrometers (cf Sugiura, 1961). Although the gross features of these spectra resemble each other, considerable differences are present in the fine structures reported by different investigators. Sugiura (1961) noted that the position of the inflection point in the initial absorption rise is the same for KF, KCl, KBr, and KI, within experimental error. As early as 1949, Cauchois and Mott observed a sharp peak on the low-energy side of the absorption edge (white line) of certain insulators and suggested that such maxima are caused by transitions to exciton states. They proposed that the ejected electron leaves a positive hole in the K shell such that a hydrogenic potential $V(r) = -e^2/(\kappa r)$, where κ is the high-frequency dielectric constant and e and r have their usual meanings, is added to the potential of the relaxed crystal. The bound electron has a series of discrete exciton levels in the band gap, having the base of the conduction band as a series limit.

Parratt and Jossem (1957) used exciton states to interpret the absorption peaks in the KCl spectrum and assigned the bottom of the conduction band to a position in the vicinity of the second absorption maximum (called peak B). A calculation of the separation of the first maximum (peak A) from the conduction band was carried out along the lines outlined by Cauchois and Mott. Assuming that the effective reduced mass of the hole and electron equals the mass of an electron and that the dielectric constant has the high-frequency value $\kappa = 2.13$, Parratt and Jossem obtained a value of 3 eV for the binding energy of the lowest exciton state. The authors pointed out, however, that the limited extent of the exciton wavefunction probably causes the effective dielectric constant to have a smaller value than κ, which, in turn, should increase the calculated binding energy. The observed separation between the A and B peaks is 3.9 eV for chlorine and 3.5 eV for potassium, which agrees approximately with the values deduced by assigning the bottom of the conduction band to peak B. Parratt and Jossem warned, however, that the simple hydrogenic calculation should not be taken too seriously.

The placement of the conduction-band bottom at the second absorption maximum (peak B) is made by analogy to the generally accepted interpretation of the ultraviolet absorption spectrum of KCl (cf Mott and Gurney, 1957) in which the second absorption shoulder is associated with photoconductivity and corresponds to the base of the conduction band. Parratt and Jossem assigned the chlorine B peak to a position slightly above the base of the conduction band because of a "symmetry-exchange" effect. In a crystal, band states can have different symmetry near sites which are not translationally equivalent (Bell, 1954; Mazalov, Blokhin, and Vainshtein, 1967). The bottom of the conduction band in KCl is believed to be the chlorine $4s$ band. Because of symmetry exchange, the potassium electron may go to the bottom of this band by a dipole transition. The chlorine $1s$ electron, however, has to go to higher states in the band to find a state having some p admixture.

Muto and Okuno (1956) calculated the binding energy of the KCl x-ray excitons using a formalism developed by Wannier and Slater. Assuming an effective mass of $0.8m$ and a dielectric constant of 1.5, they calculated the energy position of an x-ray exciton around a potassium ion to be 3.39 eV from the conduction-band bottom, as compared to 3.2 eV measured by Parratt and Jossem. Using the same values then gives 2.5 eV for the x-ray exciton around a chlorine atom as compared to 4.13 eV measured by Parratt and Jossem. This led Muto and Okuno to suggest that the band bottom in the chlorine absorption spectrum might be better located in the valley between peaks A and B. Parratt and Jossem (1957) examined these calculations in the light of their experimental findings and suggested several other views of x-ray-excited states, including the possibility that the series limit of such states may well be above the bottom of the continuum. This enabled them to reach certain conclusions regarding the width of the valence band in KCl. Parratt and Jossem (1955) also questioned the "atomic" approach of Vainshtein, Narbutt, and Barinskii (1952) because they fail to take into account solid-state effects.

A different approach to interpreting the absorption spectra of KCl was taken by Mazalov, Blokhin, and Vainshtein (1967) who attributed the maxima and minima in the fine structure to Van Hoven singularities in the density of conduction-band states. Making use of calculations for optical transitions, they attempted to relate the electron energies corresponding to observed absorption maxima and minima to the energy values of conduction states in different parts of the Brillouin zone. In another study, Narbutt and Smirnova (1963) have proposed that lattice defects induced in KCl by doping it with AgCl enable the formation of F centers during the x irradiation of the crystal which leads to the formation of weak absorption peaks on the low-energy side of the absorption edge.

The *L* spectra of alkali halides have been studied by Lukirskii and Zimkina (1964) who examined the $L_{II,III}$ absorption spectra of potassium and chlorine in KCl and attribute the first five maxima in the chlorine curve to transitions to exciton states, whereas the subsequent general rise in the absorption curves is attributed to the L_{II} and L_{III} absorption edges. The potassium spectra are more poorly resolved. In a subsequent study of several alkali halides, Zimkina and Lukirskii (1965*a*) placed the bottom of the conduction band in the middle of the general rise in the absorption curve (following a pronounced fine structure of peaks); since only one such rise appears to predominate in the spectra of LiCl, NaCl, RbCl, and CsCl, they conclude that all the fine structure is due to absorption by the L_{III} level of chlorine. No explanation for the lack of L_{II} absorption was presented. In still another paper, Zimkina and Lukirskii (1965*b*) compared the absorption spectra of the alkali halides to the corresponding inert gases and concluded that crystal-lattice effects play a minimal role in the solids.

The sodium $L_{II,III}$ absorption spectra were investigated in several alkali halides by Nakai and Sagawa (1969). They observed sharp absorption bands, which they attributed to excitons, including a doublet, which they explained by spin-orbit splitting and an electron-hole exchange interaction. The other peaks were attributed to excitons, and transitions to the conduction band. An attempt to relate band calculations in alkali halides to their soft x-ray spectra as well as other properties has been made by Kunz (1970). Using his local-orbital theory and the Hartree-Fock approximation, Kunz calculated one-electron energy bands and with the aid of empirical pseudopotentials calculated a density of states in the tight-binding approximation. This approach holds out considerable promise in the interpretation of x-ray spectra.

IV.   EXTENDED FINE STRUCTURE

A.   Introduction

The fine structure extending for several hundred electronvolts on the high-energy side of the initial rise was first examined theoretically by Kronig (1931; 1932*a*; 1932*b*) and therefore bears his name. The first two papers relate the inflection points between maxima and minima in crystalline absorbers to the groupings of final quantum states for the ejected electron according to allowed and forbidden energies (band model). The third paper considers a diatomic molecule and develops a theory for polyatomic gases based on the scattering of electron waves by nearest neighbor atoms. In essence, the electron, initially in a ground state ψ_0, undergoes a transition to a continuum state described by the wavefunction $\psi(\mathbf{k})$ and the absorption coefficient as a function of the wave vector $\mathbf{k}$:

$$\mu(\mathbf{k}) \propto \rho(\mathbf{k})|\psi_0 V \psi(\mathbf{k})|^2 h\nu \qquad (6\text{-}16)$$

where $\rho(\mathbf{k})$ is the density of allowed continuum states, V is the perturbation potential, h is Planck's constant, and ν is the frequency of the incident radiation. The initial wavefunction is assumed to be localized within a volume whose cross section is much smaller than the x-ray wavelength, and since only K absorption has been dealt with explicitly, it is assumed that $\psi_0 = \psi_K$ and is known so that the theories concern themselves with $\rho(\mathbf{k})$ and $\psi(\mathbf{k})$ in (6-16). For crystalline solids, the density of allowed states is considered to be of paramount importance, whereas for gas molecules, the transition probabilities are assumed to play a dominant role.

Since the initial state in (6-16) has s symmetry, the final state must have p-type symmetry. Kronig assumed that the density of such states was sufficiently high that, for almost every allowed $\mathbf{k}$ value, an appropriate final state exists. The predominant role of $\rho(\mathbf{k})$ in crystalline absorbers is supported by comparing x-ray absorption spectra to the characteristic losses suffered by electrons transmitted through thin films or reflected from crystalline solids. Although the initial states (valence band) are different in the electron case, it is reasonable to expect that the final states are the same. The correspondence between the extended-fine-structure K absorption in aluminum and characteristic electron losses in the same metal was first pointed out by Cauchois (1952). Since then, many others have described such relationships, and a judicious comparison for several metals is given by Leder, Mendlowitz, and Marton (1956).

Following Kronig's pioneering papers, several other investigators have proposed theoretical formulations purporting to yield better agreement between theory and experiment. As shown in a review of most of them (Azároff, 1963), they can be classified into two groups. In one, plane waves (Bloch functions) are used to describe the ejected electron, and the energies of allowed final states $\rho(\mathbf{k})$ in a crystal are calculated. Since this presupposes a three-dimensionally ordered crystal (with or without local defects), such theories have been named *long-range-order* (*lro*) *theories*. In the other treatment, the final states are considered to be those following a scattering of the electron by neighboring atoms, so that they have been named *short-range-order* (*sro*) *theories*. The various theories proposed so far, including more recent modifications, are most successful in predicting the relative energies at which maxima (minima) in the absorption curves occur and much less successful in reproducing actual spectral curves. In no case, however, is agreement with experiment complete enough to demonstrate clear superiority for any of the theories.

B. Long-range-order theories

Representing the ejected electron by a Bloch-type plane-wave solution of the time-independent Schrödinger equation, Kronig (1932a) found that discontinuities in the allowed energy values occurred whenever

$$E = \frac{h^2}{8m} \frac{(\alpha^2 + \beta^2 + \gamma^2)}{a^2 \cos^2 \varphi} \qquad (6\text{-}17)$$

where α, β, and γ are integers, a is the periodic spacing length, and φ is the angle between the ejected electron's propagation direction and the normal to a "reflecting" plane in the crystal, while h and m have their usual meaning. For an unpolarized x-ray beam (or a polycrystalline absorber), it is necessary to integrate over all possible propagation directions, which reduces (6-17) to

$$E = \frac{h^2}{8m} \frac{(\alpha^2 + \beta^2 + \gamma^2)}{a^2} \qquad (6\text{-}18)$$

The three integers in (6-18) can be likened to the Miller indices of crystallographic planes and the absorption process to that of electron diffraction by certain critical planes, viz., those forming the Brillouin-zone boundaries. Since the Brillouin-zone boundaries separate the quantum states having allowed energy values from those in the forbidden region, Kronig argued that these boundary energies should correspond to the inflection points between adjacent absorption maxima (allowed energies) and minima (forbidden energies). Using (6-18), he calculated the energies for several elemental metals and found that they were much too closely spaced for experimental resolution. He grouped them into regions of higher density and assumed that their midpoints corresponded to the inflections in actual absorption curves. It is most important to keep this in mind because the actual groupings, and consequently the energies corresponding to the inflections, were selected with a certain degree of arbitrariness, the actual possible solutions of (6-18) being quite large in view of the large number of possible values that $\alpha^2 + \beta^2 + \gamma^2$ can assume. It is quite possible that more detailed band-model calculations would provide a more satisfactory explanation of the extended fine structure.

In the first of many papers, Hayasi (1949a) assumed that the ejected electron was not like a nearly free electron traveling through a crystal, but instead that it would be reflected back to the parent atom by crystallographic planes formed by its nearest neighbor atoms at right angles to possible propagation directions. This has the effect of setting up a localized standing-wave pattern which Hayasi called a *quasi-stationary state*. The amplitude of the reflected waves, and hence their relative importance, is determined by some power of the number i of atoms coordinating the absorbing atom and lying in the reflecting planes and inversely proportional to some power of the distance r to these i atoms. With the aid of the Bragg equation, it is then possible to obtain the following expression for a cubic metal crystal

$$E_{hkl} = 150 \frac{(h^2 + k^2 + l^2)}{4a^2} \quad \text{eV} \quad (6\text{-}19)$$

where h, k, and l are the Miller indices of the reflecting planes and a (in A) is the length of the cubic cell edge.

The energy values predicted by (6-19) are supposed to be those corresponding to maxima in the absorption curve; in fact, Hayasi (1949b) tried to reconstruct actual absorption curves by placing Gaussian-shaped maxima onto a monotonically varying background, with somewhat limited success. The relative "displacement" of the values given by the parallel equations (6-18) and (6-19) cannot be tested experimentally because of the difficulties in locating the true absorption "edge." In an attempt to give a sounder basis for his quasi-stationary states, Hayasi (1960) considered a one-dimensional Schrödinger equation and a periodically varying potential containing a "hole" which gives rise to a set of discrete localized (quasi-stationary) states lying in each of the forbidden-energy bands of a perfect crystal. Their energy levels are given by (6-19), and the absorption spectrum can be considered to consist of two additive parts: A continuous monotonic absorption curve due to transitions to the allowed energy states (continuum) in a perfect crystal and absorption maxima produced by transitions to the quasi-stationary states. Since the three-dimensional Schrödinger equation can be separated into one-dimensional equations, according to components whose product forms the three-dimensional solution (wavefunction), Hayasi used linear combinations of the eigenstates for each such possible solution to determine the energy values of possible quasi-stationary states. By considering suitable combinations to describe the final states he was thus able to obtain matches for the maxima not only in the K but also in the $L_{II,III}$ absorption spectrum of aluminum and, subsequently, in copper (Hayasi, Fujimori, and Suzuki, 1962). Although the Hayasi theory directly considers only the atoms surrounding the absorbing atoms, its formalisms require the presence of a triply-periodic crystal so that it is correctly classified as one of the lro theories. Like the Kronig theory, its apparent ability to match the relative energy values of maxima in an absorption spectrum depends on "judicious" selection of values predicted by (6-19) or comparable relations for L spectra.

Although not realized by its author, the formalism underlying still another lro theory suggested by Lytle (1966) bears a certain similarity to the Hayasi theory. The chief difference is that Lytle makes use of solutions of the Schrödinger equation for a spherical "hole" which leads to possible energy values given by

$$E = \frac{h^2}{8mr^2} Q \quad (6\text{-}20)$$

where r now is the radius of the spherical hole and Q takes on the values of the zero roots of half-order Bessel functions forming the radial parts of the solutions. The more restricted number of possible values that Q can have (for p levels, the values are, roughly, 2, 6, 12, 20, 30, etc.) makes it impossible for Lytle to match much more than half of the maxima observed in the extended structure of copper, a most popular foil for testing various theories (Azároff, 1963). This did not deter Mandé and Joshi (1968), who included Q values corresponding to d-type levels, when necessary, to obtain a more complete fit. Literally, equation (6-20) comes from considering a spherical hole inside which the electron "sees" a zero potential while outside of it the potential is infinite. To provide additional degrees of freedom in determining Q values, Chivate et al. (1968) considered solutions given by Schiff (1949) for the case when the potential outside the hole is large, but finite. This allowed them to match 3 to 5 maxima observed to occur within 150 eV of the absorption edges of several manganese oxides. Despite attempts to impart a physical meaning to such calculations, it is doubtful that they shed any real light on the processes actually responsible for the observed fine structure, a point, incidentally, openly admitted by Lytle (1966).

C.   Short-range-order theories

In his third paper, Kronig (1932b) considered the case of an isolated molecule, specifically, diatomic Cl_2. For this case he argued that the density of allowed states $\rho(\mathbf{k})$ in (6-16) should vary monotonically so that the absorption process is determined by the transition probabilities and the ejected electron can be described by a plane wave which is scattered by the surrounding atoms whose potentials (scattering amplitudes) enter into the expression for the absorption coefficient. An extension of this treatment by Hartree, Kronig, and Petersen (1934) enabled them to account for the extended fine structure recorded for $GeCl_4$ by Coster and Klamer (1934), by expressing Kronig's equations in terms of phase shifts in the scattered wave produced by the four chlorine atoms. Subsequently, Petersen (1936) reexamined this analysis, simplifying the equations somewhat, and introduced an additional phase term expressing the influence of the absorbing atom (germanium). A remeasurement of the K absorption spectrum by Shaw (1946) disclosed a somewhat different structure, but a recalculation of it by Corson (1946), using Petersen's equations, reaffirmed the applicability of this approach. The K spectrum was remeasured by Stephenson (1947) and by Glaser (1951), and the L spectrum was measured by Mott (1966) who claimed that the theory developed by Petersen satisfactorily explained the extended structure.

The first to apply the above formalism developed for finite molecules to calculations of the extended fine structure of solids (infinite molecules), Kostarev (1941) argued that it was not the irregularities in the density of states but in the transition probabilities that accounted for the observed fine structure. He expressed this as a ratio of the transition probabilities P_{KF} (from an initial K state to a final F state) in the solid to that in the free atom

$$\chi(E) = \frac{|P_{KF}|^2 \text{ solid}}{|P_{KF}|^2 \text{ free atom}} \qquad (6\text{-}21)$$

He then made use of the equations developed by Hartree, Kronig, and Petersen to calculate the final-state wavefunctions by assuming a Wigner-Seitz-model potential distribution in metals and of the Wentzel-Kramers-Brillouin (WKB) method to calculate the phase shifts caused by the atoms surrounding the absorbing atom. In a later paper, Kostarev (1949) developed a compact expression (in atomic units)

$$\chi(E) = \left\{ 1 + \frac{2}{k} \sum_{i=1}^{\infty} z_i \int_0^{\infty} \overline{U_i(r)} \sin 2[kr + \delta(k)] \, dr \right\}^{-1} \qquad (6\text{-}22)$$

where r is the radial distance from the absorbing atom, k is the wave-vector magnitude, and $\delta(k)$ the scattering phase for the ejected electron wave; z_i is the number of atoms in the ith shell about the absorbing atom, and $U_i(r)$ is the potential energy of the electron in the field of an atom in the ith shell, averaged over all values of r. He calculated and tabulated separately the energy values of the maxima and minima predicted by (6-22) for all the odd i shells surrounding an atom in copper. (The contributions from even i shells were considered to be smaller than the experimental errors.) Although the exact energy values calculated for each i differed, using his own measured K absorption curve for copper, Kostarev was able to claim reasonable agreement between theory and experiment.

Shiraiwa, Ishimura, and Sawada (1958) assumed that the density-of-states distribution in metals could be likened to a nearly free-electron curve so that

$$\mu(E) \propto E^{1/2} P(E) \qquad (6\text{-}23)$$

and the transition probabilities govern the appearance of a fine structure in the absorption spectrum. To determine the appropriate wavefunction for the final state in (6-16), they considered the elastic and inelastic scattering of the ejected-electron wave by the surrounding atoms. The wavefunction is written as a sum

$$\psi(k) = \psi_i + \sum_s \psi_s + \sum_{s,t} \psi_{st} + \sum_{s,t,n} \psi_{stn} + \cdots \qquad (6\text{-}24)$$

where ψ_i is the emitted wave, ψ_s is the emitted wave scattered by the sth atom, ψ_{st} is this wave rescattered by the tth atom, and so forth. Each wavefunction is multiplied by an attenuation factor which falls off with distance, and the waves are combined in the vicinity of the absorbing atom, yielding an expression that has a mathematical form similar to that obtained by Kronig for the diatomic molecule. The scattering amplitudes are calculated by assuming a square-well potential and by using the Born approximation. A fairly complete set of calculations along with examples of absorption (and emission) spectra are presented by Sawada et al. (1959). The result of these calculations is an actual absorption curve whose maxima show a fairly good agreement with experimental curves for Cu, Ni, and Fe, but whose actual shape differs, even when as many as fifteen nearest neighbor shells are considered. (The attenuation factor is an adjustable parameter and is determined by the best fit of the calculated curves.) Subsequently, Shiraiwa (1960) used Hartree self-consistent-field calculations to determine phase shifts of the scattered waves, and obtained a better fit with experimental curves for Cu and Ti. The agreement with the absorption spectra of TiO_2 was not as good, possibly because the density-of-states distribution assumed in (6-23) is clearly not applicable to an insulator. Alternatively, the nearly-free-electron approximation implicit in all these theories may be appropriate only in metal absorbers. Vishnoi and Agarwal (1966) have carried out a similar calculation for copper, assuming zero attenuation, and going out to seven nearest neighbor shells.

Sayers, Lytle, and Stern (1970) have modified the above analyses by assuming muffin-tin type of potential for each atom which leads to spherical wavefunctions (Hankel functions) whose phase shifts due to scattering are analyzed in the presence of the absorbing (ionized) atom. Except for some calculations by Kozlenkov (see below), not since Petersen (1936) had any of the theories taken the absorbing atom's influence directly into account. A Hartree self-consistent field is used to determine the potential of the absorbing atom and the WKB method to determine the phase shift. The surrounding atoms are replaced by point atoms (delta functions), and multiple scattering, as distinct from backscattering, is neglected. The agreement between experimental curves for Cu, F, and Ge and calculated curves is comparable to that of the other theories. As more and more surrounding shells are included in the calculations, increasingly more "fine structure" becomes evident in the curves. Partly this can be controlled by altering the phase-shift magnitudes and partly it can be washed out by assuming an instrumental broadening function (Subsection II B).

Kozlenkov (1961) compared the theoretical treatments of Petersen (1936), Kostarev (1949), and Shiraiwa, Ishimura, and Sawada (1958), and argued that the sro theories were more general than the lro theories of Kronig and, presumably, therefore more valid. [Although it is true that sro theories can be applied to finite molecules and to infinite ones (crystals), it does not necessarily follow, as asserted by some authors, that the appearance of an extended fine structure in the absorption spectra of gases and solids proves that sro theories are the only correct ones possible.] Finding that the basic equations obtained by the different investigators bear close similarities, Kozlenkov demonstrated that the assumption of a simple square-well potential for an atom permits the calculation of the phases with an accuracy no worse than that obtained in more rigorous calculations. This allowed him to come up with a considerably simplified expression for the absorption coefficient (largely controlled by transition probabilities)

$$\mu(k) \propto - \sum_i \frac{s_i}{r_i^2} \sin \left[2kr_i + 2\delta(k) \right] \qquad (6\text{-}25)$$

where the subscript i denotes the ith shell, containing s_i atoms, at a distance r_i from the absorbing atom. Using phases δ tabulated by Mott and Massey (1933), Kozlenkov obtained a satisfactory agreement with experimental curves for Cu, Fe, Ti, and Zn. Kozlenkov also showed that the spacing between maxima and minima is inversely proportional to a^2, as originally proposed by Kronig in equation (6-18). This follows from the $1/r_i^2$ dependence in (6-25) which allows calculation of interatomic distances, as independently noted by Lytle (1966).

In a subsequent paper, Kozlenkov (1963) took into account the overlap of averaged potentials of neighboring atoms, leading to an average lattice potential U_0. Using Hartree potentials, he calculated the locations of the most intense maxima in the absorption spectra for Cu, Ca, Al, Be, K, and Li. In a sequel paper, Batyrev (1963) attempted to deduce the scattering phases of Cu and Zn in beta and eta brasses by curve-fitting calculations based on Kozlenkov's equations. Although interesting, the physical reality of such calculations is not clear. Kozlenkov (1964) further considered the averaged Hartree potentials and noted that they contained a barrier position as well as the portion extending below U_0. Taking this barrier into account, Kozlenkov claimed, considerably improved the agreement of theory with experiment. He also considered the possible role of the K vacancy and compared three alternatives:

1 Disregarding the hole and treating the crystal as continuously periodic
2 Treating the hole as adequately screened outside its polyhedron (Wigner-Seitz cell)
3 Treating the hole as unscreened

Kozlenkov claimed that he got equally good results with assumption *1* or *2* but not with *3*. Thus he concluded that, in metals, the hole should be either disregarded or assumed to be screened. More recently, Anikin, Borovskii, and Kozlenkov (1967) have extended these calculations to the L spectra of silver.

Brümmer, Dräger, and Starke (1970) used the Hayasi, Shiraiwa, and Kozlenkov theories and different potential models to calculate the extended fine structures for bcc chromium, hexagonal CrH, and cubic (diamond-type) germanium. They found that the sro theories reproduced the fine structure of the experimental curves qualitatively and some of the energy values of the maxima quantitatively. They concluded that the theories must be improved by more detailed considerations of interactions between the photoelectron and surrounding atoms and conduction electrons in order to reproduce the experimental fine structure in all details. This is another way of stating that more detailed band calculations are necessary to explain correctly the extended fine structure. A similar conclusion was reached by Perel and Deslattes (1970) who compared the experimental curves for three spinel-type oxides with predictions of sro and lro theories. They definitely rejected the applicability of the Kronig theory, but were somewhat less conclusive about the sro approach. These authors overlooked, however, the fact that the Kronig theory explicitly and the sro theories implicitly assumed that the photoelectron behaves like a nearly free electron in the crystals. Until the detailed band structure of the absorber is incorporated in the theoretical models, therefore, it is premature to draw any conclusions regarding the most appropriate calculational approach to employ.

D. Comparison of theories

1. Polarization anisotropy All the theories proposed to explain the extended fine structure of x-ray absorption spectra include some kind of averaging operation leading to essentially spherical symmetry about the absorbing atom. When a plane-polarized x-ray beam is incident on an anisotropic (single-crystal) absorber, then such averaging should not be necessary, and a more critical test of competing theories becomes possible. Izraeleva (1966) examined theoretically the effect that this averaging should have on the sro theories of Kostarev and Kozlenkov, and concluded that the transition probabilities depend only on the mean potential over the surface of a sphere surrounding the absorbing atom so that the process of averaging used in them is valid for polycrystalline absorbers (and gases). She did obtain a separate term in the expression which should show up in the spectrum of a single-crystal absorber.

Stephenson (1933) pointed out that the use of a plane-polarized x-ray beam should cause a shift in the fine-structure maxima of a single-crystal absorber according to Kronig's equation (6-17). He failed to observe such a relative shift for different orientations of a germanium single crystal as an absorber with unpolarized radiation (Doran and Stephenson, 1957) but recorded clearly discernible shifts when using a 7 percent polarized beam (El-Hussaini and Stephenson, 1958). These measurements were repeated in his laboratory by Singh (1961) who confirmed the peak shifts and reported a few new maxima not previously observed. Alexander et al. (1963) remeasured the absorption spectra, using 90 percent polarized radiation, and observed intensity changes in the absorption maxima but no shifts in their positions. The absence of shifts is explained by them by theoretical arguments that the direction of the photoejected electron is independent of the polarization direction so that no anisotropies should be expected in cubic crystals.

Single-crystal copper foils were examined in different orientations by Boster and Edwards (1962), using 82.4 percent polarized x rays, without significant changes observed in either the peak positions or magnitudes. A similarly negative result was obtained by Fujimoto (1965), using 60 percent polarized radiation on copper single-crystal foils epitaxially grown parallel to different planes. Weber (1962) examined single-crystal foils of orthorhombic gallium after orienting a 20 percent polarized beam parallel to three different crystallographic directions. When the beam was parallel to the c axis, a considerable decline in the intensity of the fine structure was observed as compared to either the a or b axes. Subsequently, Weber (1967) observed a similar change in peak heights but not in their energy positions, when examining a cadmium single crystal parallel and perpendicular to its c axis at room temperature and at 83°K. Alexander et al. (1965) found similar intensity changes in gallium, when comparing polarization directions parallel to a and to c, and also reported observing shifts in the peak positions. Perel (1966), one of his collaborators, generalized the sro theory of Shiraiwa, Ishimura, and Sawada (1958) to include polarization effects, and his calculated curves for polarization directions parallel to a and c do show the reported difference in peak intensity and peak locations, although a comparison with the shifts reported by Alexander et al. is not presented.

In concluding his paper, Perel states that the sro methods of Kostarev and Kozlenkov cannot be applied to this case. Actually, Kostarev (1965a) had already developed a modification of his theory taking account of polarization anisotropy and found that the intensity of absorption maxima should decrease as a function of direction in a single-crystal absorber but their positions should remain unchanged. Kostarev (1965b) applied this theory to Weber's data on gallium and claimed good qualitative agreement. Subsequently, Kostarev (1967) applied his theory to the absorption spectra obtained from germanium crystals by Alexander et al. (1963). In order to obtain a satisfactory agreement, he emphasized the relative importance of different coordination shells in determining the fine structure in different parts of the spectrum.[1]

Izraeleva (1969) used a totally different approach in which she considered what effect the electric-field tensor has on the transmission of a plane-polarized beam. Thus she expected, and obtained, a change in the intensity of the entire spectrum as a function of crystallographic symmetry. She cited the experimental curves of Brümmer and Dräger (1966) on $FeCO_3$ and of Schnopper (1966) on $KClO_3$ as examples and claimed agreement with earlier measurements on gallium as well.

2. *Temperature effects* As early as 1931, Hanawalt observed that the contrast between maxima and minima in the iron K absorption spectrum lessened with increasing temperature. Coster and Veldkamp (1931) seized on this to attribute the

---

[1] A shift of the energy at which absorption first sets in with polarization direction also has been observed by Brümmer and Dräger (1966) in the K spectra of Fe_2O_3 and $FeCO_3$ single crystals and by Schnopper (1966) in $KClO_3$.

more "washed out" structure of zinc as compared to copper to the former's lower melting point. By now it has been clearly demonstrated by many investigators that the contrast of the fine structure can be enhanced considerably by lowering the temperature of the absorber. Experimentally, this has the very real advantage of improving the precision with which the energy of maxima can be located and of extending the discernible structure to higher energies.

A theoretical treatment of the temperature dependence of the fine structure was first presented by Schmidt (1961a), who began with the Kozlenkov sro theory developed primarily for pure metals. Beginning with equation (6-25) Schmidt assumed that the ejected electron sees only the instantaneous atomic distribution because he estimated the electron's lifetime to be 1000 times shorter than the period of thermal vibration. The sum over coordination spheres in (6-25) thus can be replaced by a sum over individual atomic positions, and the atomic displacements are included in r_i. Expanding the sine term in (6-25) and treating the displacements as small and governed by the Debye specific-heat theory, he finally obtained an expression for a time-averaged absorption coefficient exactly like (6-25) except that the sine term now is multiplied by

$$\frac{S_i}{r_i} \exp\left\{ -\frac{7.7 k^2 h^2 r_i^2}{\kappa m \Theta V^{2/3}} \left[F\left(\frac{T}{\Theta}\right) + \frac{1}{2} \right] \right\} \qquad (6\text{-}26)$$

where κ is Boltzmann's constant, m the atomic mass, Θ the Debye temperature, V the atomic volume, $F(T/\Theta)$ is an integral expression taking account of the phonon spectrum, and the $\frac{1}{2}$ represents the zero-point energy.

It is evident from (6-26) that some "washing out" of the fine structure occurs even at the absolute zero; for example, a peak lying 147 eV above the initial rise in the copper K absorption spectrum has only 63 percent of the magnitude it would have if the atoms were completely at rest. It is also clear that the maxima at higher energies are affected more by increasing the temperature than are the peaks closer to the initial rise. Similarly, the higher frequency contribution for larger values of i, that is, from the more distant atoms, declines more rapidly than the contribution from the first coordination shell. This is probably the chief reason why relatively little improvement in agreement between experiment and sro theories is obtained by including contributions from even more distant shells.

A comparison with data published for iron by Trapeznikov and Nemnonov (1956) shows a fair agreement up to 600°C and a sharp divergence above that temperature. Schmidt (1963) remedied this by modifying his theory to include higher energy phonons (shorter phonon wavelengths) which leads to a modification of the functional form of $F(T/\Theta)$. The result is a better agreement between peak heights predicted by (6-26) and observed peaks in iron, over a temperature range of 73 to 1073°K. A fair but by no means exact agreement with Schmidt's theory was found by Boster and Edwards (1968) for their measurements on niobium and copper at 300 7,7, and

4.2°K. As expected from (6-26), the element having the larger Debye temperature (niobium) shows a more pronounced increase in the fine structure with temperature in alloys and in compounds (cf Sharkin, 1964). Because the theory developed by Schmidt applies only to pure metals, the changes observed can be discussed in qualitative terms only.[1]

V.   ISOCHROMATS

A.   Atomic constant determination

Consider a spectrometer set to record the $K\alpha_1$ line of some element constituting the target of an x-ray tube. As the voltage applied to the tube is slowly raised from just below to just above the excitation voltage for the K spectrum, the intensity of the $K\alpha_1$ line starts to rise and then displays the dependence on tube voltage shown in Fig. 6-5. The curves in Fig. 6-5 are called *isochromats* because they represent the intensity variation at a single wavelength. The tube voltage at which the K spectrum first appears, i.e., the K excitation voltage V_K, determines the maximum energy eV_K that an electron striking the tube target can give up. This must equal the x-ray energy at the K absorption edge, so that

$$eV_K = h\,\frac{c}{\lambda_K} = h\nu_K \qquad (6\text{-}27)$$

Since V_K and ν_K are determined experimentally, (6-27) enables the determination of h/e from which the "x-ray value" of Planck's constant can be calculated.

It should be noted that it is possible to determine the value of h/e also by substituting the minimum energy of the continuous spectrum and the maximum voltage applied to the x-ray tube in (6-27). The use of the short-wavelength limit was first proposed by Duane and Hunt (1915), and essentially two different experimental procedures can be employed. In one, the continuous spectrum produced at a fixed voltage is scanned by a spectrometer to establish the short-wavelength or Duane-Hunt limit. In the other, the spectrometer is set to transmit x rays of a fixed wavelength, and the voltage on the tube is increased stepwise, as discussed above.

There are a number of experimental difficulties in such measurements, as reviewed by Stephenson (1957) and by Cohen and Du Mond (1957). Among them

---

[1] Jope (1969) has proposed that the fractional amplitude increase in the fine structure, on approaching the absolute zero, should be about 3 times greater for a correction appropriate to the lro theory than that for the sro theory developed by Schmidt. In the absence of a more rigorous theoretical justification or a clear-cut experimental example, this contention cannot be verified.

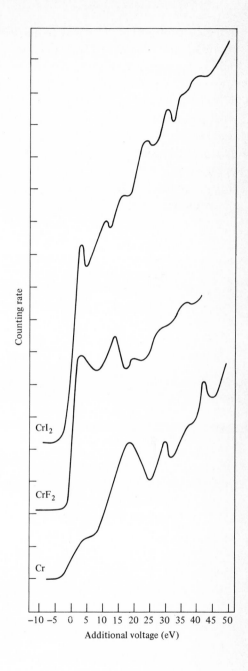

FIGURE 6-5
Isochromats of Cr in the pure metal and
in two chromium halides. (*Courtesy of
P. Ohlin.*)

is the question of how to make an allowance for the work function of the electrons emitted by the cathode. Ohlin (1943) questioned the need for such a correction and indicated a better self-consistency in his measurements if the correction were omitted. Subsequent workers, however, have demonstrated clearly that such corrections are necessary. In fact, in their 1957 review article, Cohen and Du Mond state that the most important remaining source of systematic errors lay in the determination of the true isochromat excitation voltage in view of the finite instrumental resolving power. The recommended process is to determine the maximum in the second derivative of the isochromat curve, although Bearden and Thomsen (1957) have pointed out that the spectrometer window effect (Subsection II A) makes its precise location doubtful.

Another source of difficulty is the fine structure appearing near the initial rise in the isochromat curve. Although a more extensive consideration of this fine structure is delayed until the next section, it should be noted here that Ulmer (1959) analyzed various solid-state effects and concluded that the correct value for the excitation voltage is given by locating the point on the isochromat lying at one-half the maximum value of the initial rise. The way to eliminate such fine-structure effects was shown by Spijkerman and Bearden (1964) who built a high-power, mercury-vapor-target x-ray tube and measured an isochromat at the short-wavelength limit of the continuous spectrum emitted by this tube. The absence of the fine structure normally present when solid targets are employed and a very careful technique enabled Spijkerman and Bearden to obtain an x-ray value for $h/e = (1.37949 \pm 30 \text{ ppm}) \times 10^{-17}$ erg-sec/esu, which differs from the best non-x-ray value by only 2 ppm (see also Chap. 2).

B.   Fine structure

A fine structure, not unlike that present near an x-ray absorption edge, can be seen to occur in the isochromats of solid targets in Fig. 6-5. It was first observed by Ohlin in his doctoral dissertation (Ohlin, 1941; 1952) when he recorded isochromats, using relatively low excitation potentials. The experimental problems attendant to isochromat measurements have been considered in detail by Nilsson (1953), who discusses target contamination, tube-voltage stability, and other experimental factors. It should be noted that Nilsson measured the voltage variation near the K edge, similarly to the method preceding equation (6-27) in this text. In this case, the radiation monitored is the $K\alpha$ line and corresponds to transitions from the L to the K shell after a K electron has been knocked out so that the isochromat produced reflects, among other factors, the probabilities that a K-excitation collision occurs. In the usual isochromat measurements, the radiation emitted at the Duane-Hunt limit is recorded, and this involves direct transitions of the incident cathode electrons to unoccupied states in the target.

The first to propose an explanation for the Ohlin structure in terms of solid-state effects, Nijboer (1946) considered the allowed and forbidden energy regions for crystals similarly to the earlier discussions of the fine structure in x-ray absorption. Stephenson (1957) drew attention to some apparent experimental inconsistencies with Nijboer's predictions, and Lublow (1960) suggested that the fine structure is caused by transition radiation from electrons crossing the surface potential of the target. Subsequent investigators have concluded, however, that solid-state effects are primarily responsible for the fine structure in an isochromat, so that a close analogy to x-ray absorption spectra is believed to exist.

Johansson (1960) recorded the isochromats of W, Ni, Cr, and Cr_2O_3 and observed a "foot" structure in the initial rise for the last three targets not unlike that observed in their absorption edges (cf Fig. 6-3). For the case of nickel and chromium, he interpreted this as arising from a radiative transition of the incident electrons to the empty $3d$-$4s$ states in the metals. The chromium-oxide curve is shifted to slightly higher voltages, but also exhibits a maximum in the initial rise which is explained by transitions to empty $3d$ states in a schematic band model of the insulator. Bergwoll and Tyagi (1965) recorded the isochromats of several transition metals and of yttrium oxide. They attribute the first maximum in the initial rise ("foot" structure) to transitions to empty d states and draw attention to the existence of apparent "echoes" of the first two maxima, a phenomenon which is consistent with plasmon theory. In a subsequent study of the rare-earth elements Gd through Lu, Bergwoll (1966) pointed out that "primary" electrons, whose kinetic energy is converted into radiation in one single process, give rise to the initial maximum in the isochromat curves and reflect the density of states at the Fermi level. The "secondary" electrons, those undergoing energy losses characteristic of the surface or volume states, produce radiation lying at regular intervals from the first peak and are also related to the density of empty states. Their displacement in the isochromat curve produces the "echo effect" mentioned above. Because an inner level is not involved in this process, selection rules do not appear to be involved.

The transition probabilities are considered by Ulmer (1969) in a most interesting juxtaposition of x-ray and isochromat spectroscopic methods. He suggests that the incident electrons are nearly free in the target so that the transition probabilities for these electrons are independent of the symmetry considerations attendant to inner-electron transitions and can therefore be assumed constant. The validity of these considerations is demonstrated by the isochromats of a series of Rh-Pd and Pd-Ag alloys. By suitably superimposing the isochromat curves, Ulmer shows that a progressive shift of the initial rise to higher voltages corresponds to an expected shift in the Fermi level in the density-of-states curve (rigid-band model). At the same time, the height of the first maximum declines, corresponding to a decline in the unoccupied

$4d$ states in the alloys. Similar interpretations based on the rigid-band model had been proposed by Claus and Ulmer (1963; 1965) to explain the isochromats of Ta, Nb, W, and Mo and of Ir, Rh, Pt, and Pd, and by Merz and Ulmer (1968a; 1968b) for several body-centered cubic and hexagonal closest-packed transition metals. Some of these models are in sharp disagreement with those derived from electron specific-heat measurements, but Bergwoll and also Claus and Ulmer propose that the discrepancies may come from electron-phonon interaction contributions to the specific heat. In this connection it should be noted that others also have questioned the direct relationship between specific-heat data and density-of-states curves.

A more direct analogy to the extended fine structure in x-ray absorption spectra has been suggested by Auleyten and Lidén (1962) and by Fujimoto (1965) who attempted to explain their isochromat curves in terms of the Hayasi theory (Subsection IV B). Fujimoto compared an isochromat curve of copper measured at 8 kV to the K absorption edge of a copper foil and suggested that the absorption and isochromat maxima occur at nearly the same values calculated for the quasi-stationary states. If valid, then such comparisons extend the same uncertainties, already discussed for the case of absorption fine structures, to the fine structure evident in isochromats.

VI. ACKNOWLEDGMENTS

The authors are indebted to Dr. P. Best for reading and criticizing this manuscript and for drawing their attention to several relevant papers.

REFERENCES

AGARWAL, B. K., and VERMA, L. P. (1968): Chemical effects in x-ray K absorption spectra, *J. Phys. Chem.*, Sec. 2, **1**: 208–209.

ALEXANDER, E., FELLER, S., FRAENKEL, B. S., and PEREL, J. (1965): The fine structure of the extended K absorption edge of a Ga single crystal, *Nuovo Cimento*, **35**: 311–312.

———, FRAENKEL, B. S., PEREL, J., and RABINOVITCH, K. (1963): Orientation dependence of K-absorption extended fine structure of a single crystal of germanium, *Phys. Rev.*, **132**: 1554–1559.

ANIKIN, A. V., BOROVSKII, I. B., and KOZLENKOV, A. I. (1967): Investigation of the fine structure of the K and L absorption spectra of metallic silver, *Bull. Acad. Sci. USSR, Phys. Ser. (English Transl.)*, **31**: 1032–1038.

ATTARD, A., and AZÁROFF, L. V. (1963): Electron distribution in indium antimonide, *J. Appl. Phys.*, **34**: 774–776.

AULEYTEN, A., and LIDÉN, B. (1962): Die Struktur der Bremsstrahlungs-Isochromaten von Silicium und Germanium, *Arkiv Fysik*, **23**: 41–48.

AUSMAN, G. A., and GLICK, A. J. (1969): Threshold behavior of the soft x-ray spectra in metals, *Phys. Rev.*, **183**: 687–691.

AZÁROFF, L. V. (1963): Theory of extended fine structure of x-ray absorption edges, *Rev. Mod. Phys.*, **35**: 1012–1022.

———— (1967a): Characterization of metal atoms in alloys by soft x-ray absorption spectroscopy, *Mater. Res. Bull.*, **2**: 137–143.

———— (1967b): X-ray *K* absorption edges of alloys. I. Correlation to band models, *J. Appl. Phys.*, **38**: 2809–2812.

———— (1968): *Elements of x-ray crystallography*, McGraw-Hill Book Company, New York.

———— and DAS, B. N. (1964): X-ray *K* absorption spectra of Cu-Ni alloys, *Phys. Rev.*, **A134**: 747–751.

BAGUS, P. S. (1965): Self-consistent-field wave functions for hole states of some Ne-like and Ar-like ions, *Phys. Rev.*, **A139**: 619–634.

BALLY, D., and BENES, L. (1959): La structure fine des discontinuités d'absorption des rayons X du nickel et du fers dans les alliages Ni-Fe, *Compt. Rend.*, **248**: 2327–2329.

———— and MÜLLER, L. (1959): La structure fine des discontinuités *K* d'absorption du nickel et du fer dans les alliages Ni-Fe, aux bases temperatures, *Compt. Rend.*, **249**: 1099–1101.

BARINSKII, R. L. (1961): Experimental investigation and calculation of the *L* spectra of noble gases, *Bull. Acad. Sci. USSR, Phys. Ser. (English Transl.)*, **25**: 958–964.

———— (1962): On the paper by E. E. Vainshtein and Yu. F. Kopelev, "X-ray spectral analysis of aromatic complexes of the transition elements," *J. Struct. Chem. (USSR) (English Transl.)*, **3**: 442–445.

———— (1967): Study of the chemical bond by x-ray spectra, *J. Struct. Chem. (USSR) (English Transl.)*, **8**: 805–816.

———— and MALYNKOV, B. A. (1962): Interpretation of the *K* absorption spectra of sulfur in molecules and crystals in terms of the Stark effect in excitons, *Bull. Acad. Sci. USSR, Phys. Ser. (English Transl.)*, **26**: 412–418.

———— and ———— (1963): The Stark effect and hybridization in *K* absorption spectra, *Bull. Acad. Sci. USSR, Phys. Ser. (English Transl.)*, **27**: 360–367.

———— and NADZHAKOV, E. G. (1960): Calculation of the charge of atoms in molecules from the *K* absorption spectra, *Bull. Acad. Sci. USSR, Phys. Ser. (English Transl.)*, **24**: 419–425.

BATYREV, V. A. (1963): Determination of the asymptotic scattering phases in some metals and alloys from the fine structure of the x-ray absorption spectra, *Bull. Acad. Sci. USSR, Phys. Ser. (English Transl.)*, **27**: 389–391.

BEARDEN, J. A., and BEEMAN, W. W. (1940): The *K* absorption edges and $K\beta_{2,5}$ emission lines of two Zn-Ni alloys, *Phys. Rev.*, **58**: 396–399.

———— and FRIEDMAN, H. (1940): The x-ray $K\beta_{2,5}$-emission lines and *K*-absorption limits of Cu-Zn alloys, *Phys. Rev.*, **58**: 387–395.

———— and THOMSEN, J. S. (1957): A survey of atomic constants, *Nuovo Cimiento Suppl.*, **5**: 267, 360.

BEEMAN, W. W., and BEARDEN, J. A. (1942): The K-absorption edges of metal ions in aqueous solution, *Phys. Rev.*, **61**: 455–458.

——— and FRIEDMAN, H. (1939): The x-ray K absorption edges of the elements Fe(26) to Ge(32), *Phys. Rev.*, **56**: 392–405.

BELL, D. (1954): Group theory and crystal lattices, *Rev. Mod. Phys.*, **26**: 311–320.

BERGWOLL, S. (1966): X-ray isochromats of some rare earth elements, *Z. Physik*, **193**: 13–22.

——— and NIGAVEKAR, A. S. (1968): Experimental evidence for the Parratt x-ray excitation theory, *Phys. Rev.*, **175**: 33–35.

——— and TYAGI, R. K. (1965): An investigation of the Ohlin structure near the quantum limit of the continuous x-ray spectrum, *Arkiv Fysik*, **29**: 439–454.

BEST, P. (1966): Electronic structure of the MnO_4^-, CrO_4^{2-}, and VO_4^{3-} ions from the metal K x-ray spectra, *J. Chem. Phys.*, **44**: 3248–3253.

——— (1967): X-ray absorption in molecules; SF_6, *J. Chem. Phys.*, **47**: 4002–4006.

BLOKHIN, M. A. (1957): *The physics of x-rays*, 2d ed. (State Publishing House of Technical-Theoretical Literature, Moscow), English translation by U.S. Atomic Energy Commission, document AEC-tr-4502.

——— (1969): Possibilities and problems in x-ray spectroscopy (in Russian), *Proc. Kiev Conf.*, **I**: 6–12.

BÖKE, K. (1957a): Die K-Kanten-Feinstruktur von Komplexen der Übergangselemente Chrom bis Zink (Teil I), *Z. Physik. Chem.*, *Neue Folge* **10**: 45–58.

——— (1957b): Die K-Kanten-Feinstruktur von Komplexen der Übergangselemente Chrom bis Zink (Teil II), *Z. Physik. Chem.*, *Neue Folge* **10**: 59–82.

BONNELLE, C., and WUILLEUMIER, F. (1963): Mise en évidence d'excitations doubles dans des couches atomiques internes sous l'action des rayons X, *Compt. Rend.*, **256**: 5106–5109.

BOROVSKII, I. B., and BATYREV, V. A. (1960): Nature of the fine structure of the main K absorption edge of elements of the iron group, *Bull. Acad. Sci. USSR, Phys. Ser.* (*English Transl.*), **24**: 449–450.

——— and SCHMIDT, V. V. (1960): Investigation of the correlation between the fine structure of x-ray absorption spectra and the characteristic electron energy losses, *Bull. Acad. Sci. USSR, Phys. Ser.* (*English Transl.*), **24**: 438–441.

BOSTER, T. A., and EDWARDS, J. E. (1962): Kronig-structure variations in the x-ray K-absorption spectra of a Cu single crystal, *J. Chem. Phys.*, **36**: 3031–3033.

——— and ——— (1968): X-ray K-absorption fine structure of niobium and copper at cryogenic temperatures, *Phys. Rev.*, **170**: 12–16.

BRÜMMER, O., and DRÄGER, G. (1966): Investigations of the orientation dependence of the x-ray K-absorption spectra of Ge, Fe, Fe_2O_3, and $FeCO_3$ single crystals using linearly polarized x-radiation, *Phys. Stat. Solid.*, **14**: K175–K179.

———, ———, and STARKE, W. (1970): Zur Deutung der Kantenfernen Feinstruktur der Röntgen-K-Absorbtionsspektren, *Ann. Physik*, **24**: 200–217.

BURDICK, G. A. (1963): Energy band structure of copper, *Phys. Rev.*, **124**: 138–150.

CARLSON, T. A. (1967): Double electron ejection resulting from photo-ionization in the outermost shell of He, Ne, and Ar, and its relationship to electron correlation, *Phys. Rev.*, **156**: 142–149.

CAUCHOIS, Y. (1952): Étude comparée du spectre K d'absorption de l'aluminium métallique, *Acta Cryst.*, **5**: 351–356.

—— and MOTT, N. F. (1949): The interpretation of x-ray absorption spectra of solids, *Phil. Mag.*, **40**: 1260–1269.

CHIRKOV, V. I., BLOKHIN, M. A., and VAINSHTEIN, E. E. (1967): Study of x-ray K spectra of titanium in its nitride and carbide, *Soviet Phys.-Solid State (English Transl.)*, **9**: 873–877.

CHIVATE, P., DAMLE, P. S., JOSHI, N. V., and MANDÉ, C. (1968): A model for extended fine structure in x-ray absorption spectra, *J. Phys. Chem.*, sec. 2, **1**: 1171–1175.

CLAUS, H., and ULMER, K. (1963): Untersuchung zur Energiebänderstruktur von Ta, b, W and Mo mit Isochromatenmessungen, *Z. Physik*, **173**: 462–475.

—— and —— (1965): Untersuchungen der Zustandsdichte und der charackteristischen Energieveluste von Ir, Rh, Pt und Pd mit Isochromatenmessungen, *Z. Physik*, **185**: 139–154.

COHEN, E. R., and DU MOND, J. W. M. (1957): The fundamental constants of atomic physics, in S. Flügge (ed.), *Handb. Physik*, **35**: 1–87.

CORSON, E. M. (1946): The theory of the fine structure of x-ray absorption limits in polyatomic molecules, *Phys. Rev.*, **10**: 645–652.

COSTER, D. (1924): Über die Absorptionsspektren im Röntgengebiet, *Z. Physik*, **25**: 83–98.

—— and KIESTRA, S. (1948): On the empty electron bands of lowest energy of the transition metals Mn and Fe and their oxides, *Physica*, **15**: 175–188.

—— and KLAMER, G. (1934): Experimental determination of the fine structure of x-ray absorption edges of the polyatomic vapors $GeCl_4$ and $AsCl_3$, *Physica*, **1**: 889–894.

—— and VELDKAMP, J. (1931): Bestimmung des Absorptionskoeffizienten für Röntgenstrahlen in der Nähe der K-Absorptionskante der Elemente Cu und Zu, *Z. Physik*, **70**: 306–316.

COTTON, A., and BALLHAUSEN, C. J. (1956): Soft x-ray absorption edges of metal ions in complexes. I. Theoretical considerations, *J. Chem. Phys.*, **25**: 617–619.

—— and HANSON, H. P. (1956): Soft x-ray absorption edges of metal ions in complexes. II. Cu K edge in some cupric complexes, *J. Chem. Phys.*, **25**: 619–623.

—— and —— (1957): K x-ray absorption edges of Cu, Mn, Fe, Co, Ni ions in complexes, *J. Chem. Phys.*, **26**: 1758–1759.

—— and —— (1958): Soft x-ray absorption edges of metal ions in complexes. II. Zinc (II) complexes, *J. Chem. Phys.*, **28**: 83–87.

DAS, B. N., and AZÁROFF, L. V. (1965): X-ray K absorption spectra of Ni-Al alloys, *Acta Met.*, **13**: 827–833.

DESLATTES, R. D. (1969): Relative energy measurements in the K series of argon, *Phys. Rev.*, **186**: 1–4.

DONAHUE, R. J., and AZÁROFF, L. V. (1967): X-ray K absorption edges of alloys. II. Nickel-cobalt and nickel-iron solid solutions, *J. Appl. Phys.*, **38**: 2813–2817.

DORAN, D. G., and STEPHENSON, S. T. (1957): K x-ray absorption spectrum of a single crystal of germanium, *Phys. Rev.*, **105**: 1156–1157.

DUANE, W., and HUNT, F. L. (1915): On x-ray wavelengths, *Phys. Rev.*, **6**: 166–171.

EL-HUSSAINI, J. M., and STEPHENSON, S. T. (1958): Single-crystal orientation effects in K x-ray absorption spectra of Ge, *Phys. Rev.*, **109**: 51–54.

ERN, V., and SWITENDICK, A. C. (1965): Electronic band structure of TiC, TiN, and TiO, *Phys. Rev.*, **A137**(6): 1927–1936.

FERRELL, R. A. (1956): Theory of positron annihilation in solids, *Rev. Mod. Phys.*, **28**(3): 308–337.

FIEDLER, G. (1963): Die Feinstruktur der Röntgen-K-Absorptionskante von Kovalenten Carbonyl- und Cyano-Komplex-verbindungen des Mangams und Chroms, *Z. Physik. Chem., Neue Folge* **37**: 79–98.

FISCHER, D. W., and BAUN, W. L. (1968): Band structure and the titanium $L_{II,III}$ x-ray emission and absorption spectra from pure metal, oxides, nitride, carbide, and boride, *J. Appl. Phys.*, **39**: 4757–4776.

FRIEDEL, J. (1954): Electronic structure of primary solid solutions in metals, *Advan. Phys.*, **3**: 446–507.

———— (1958): Sur la Structure Électronique des Métaux et Alliages de Transition et des Métaux Lourds, *J. Phys. Radium*, **19**: 573–581.

FRIEDMAN, H., and BEEMAN, W. W. (1940): Copper and nickel x-ray $K\beta_2$ and $K\beta_5$ emission lines and K-absorption limits in Cu-Ni alloys, *Phys. Rev.*, **58**: 400–406.

FUJIMOTO, Y. (1965a): Investigation of the K absorption spectra of copper single crystals by polarized x-rays, *Sci. Rept. Tohoku Univ., First Ser.*, **49**: 28–31.

———— (1965b): Isochromat structure of copper at 8 kV, *Sci. Rept. Tohoku Univ., First Ser.*, **49**: 32–36.

GLASER, H. (1951): The absolute absorption coefficient of germanium and the fine structure in the K edge of some of its compounds, *Phys. Rev.*, **82**: 616–621.

GIANTURCO, F. A., and COULSON, C. A. (1968): Inner-electron binding energy and chemical bonding in sulphur, *Molecular Physics*, **14**: 223–232.

GLEN, G. L., and DODD, C. G. (1968): Use of molecular orbital theory to interpret x-ray K-absorption spectral data, *J. Appl. Phys.*, **39**: 5372–5377.

GUSATINSKII, A. N., and NEMNONOV, S. A. (1964a): Investigation of the L_{III} absorption spectra and the K absorption spectra of phosphorous in binary semiconductor compounds of the $A_{III}B_V$ type, *Bull. Acad. Sci. USSR, Phys. Ser.* (*English Transl.*), **28**: 828–838.

———— and ———— (1964b): Nature of the fine structure of the L_{III} x-ray absorption edge of indium in semiconducting compounds, *Phys. Metals Metallog.*, **17**(3): 42–48.

HAGARMAN, V. (1962): "Ferromagnetism and its influence on x-ray absorption spectra," Doctoral dissertation, University of Texas.

HANAWALT, J. D. (1931): Der Einfluss der Temperatur auf die K-Absorption des Eisens, *Z. Physik*, **70**: 293–305.

HANSON, H., and BEEMAN, W. W. (1949): The Mn K absorption edge in manganese metal and manganese compounds, *Phys. Rev.*, **76**: 118–121.

HARTREE, D. R., KRONIG, R. DE L., and PETERSEN, H. (1934): A theoretical calculation of the fine structure for the K-absorption band of Ge in GeCl$_4$, *Physica*, **1**: 895–924.

HAYASI, T. (1949a): Zur Theorie der Feinstruktur des Röntgen K-Absorptionsspektrums. I. Die Wellenlängen der Absorptionsmaxima in K-Absorptionsspektrum vom Metalle —Nickel und Kupfer, *Sci. Rept. Tohoku Univ., First Ser.*, **33**: 123–132.

HAYASI, T. (1949*b*): Zur Theorie der Feinstruktur des Röntgen *K*-Absorptionsspektrums. II. Der Einflus der Thermischen Schwingungen des Kristallgitters auf die Feinstruktur, *Sci. Rept. Tohoku Univ., First Ser.*, **33**: 183–194.

—— (1960): The *K*- and $L_{\text{II,III}}$-absorption spectrum of aluminium, *Sci. Rept. Tohoku Univ., First Ser.*, **44**: 87–94.

——, FUJIMORI, K., and SUZUKI, M. (1962): Fine structure in $L_{\text{II,III}}$ absorption spectrum of copper, *Sci. Rept. Tohoku Univ., First Ser.*, **46**: 144–148.

IRKHIN, YU. P. (1961): Theory of x-ray *K*-absorption spectra of transition metals, *Fiz. Metal. i Metalloved.*, **11**(1): 10–19.

IZRAELEVA, L. K. (1966): Theory of short-wave *K*-absorption regions of single crystals and polycrystalline aggregates, *Soviet Phys. "Doklady" (English Transl.)*, **11**: 506–508.

—— (1969): Absorption of polarized and unpolarized x-radiation in single crystals (in Russian), in *X-ray spectra electronic struct., Proc. Intern. Symp. Kiev (1968)*, **2**: 211–221.

JOHANSSON, P. (1960): On the structure near the short wavelength limit of the continuous x-ray spectrum, *Arkiv Fysik*, **18**: 329–338.

JOPE, J. A. (1969): Temperature dependence of x-ray absorption fine structure in metals, *J. Phys. Sci.*, **2**: 1817–1821.

KANTELHARDT, D., and WAIDELICH, W. (1969): Röntgen-Absorptionskanten von III-V Halbleitern, *Z. Angew. Phys.*, **26**: 239–245.

KARAL'NIK, S. M. (1957): External screening and the fine structure of x-ray spectra, *Soviet Akad. Sci. Bull., Phys. Ser.*, **21**: 1432–1439.

KAUER, F. (1956): Die Feinstruktur der Röntgen-*K*-Absorptionskante von Kovolenten Komplexverbindungen des Eisens, Kobalts, und Nickels, *Z. Physik. Chem., Neue Folge* **6**: 105–117.

KLEMS, G., DAS, B., and AZÁROFF, L. V. (1963): Single-crystal spectrometer for x-ray absorption spectroscopy, *Develop. Appl. Spectry.*, in J. R. FERRARO and J. S. ZIOMEK (eds.), Plenum Press, New York, **2**: 275–284.

KOSSEL, W. (1920): Zum Bau der Röntgenspektren, *Z. Physik*, **1**: 119.

KOSTAREV, A. I. (1941): Theory of the fine structure of x-ray absorption spectra in solids (in Russian), *Zh. Eksperim. i Teor. Fiz.*, **11**: 60–73.

—— (1949): Illucidation of the "super-fine" structure of x-ray absorption spectra in solids (in Russian), *Zh. Eksperim. i Teor. Fiz.*, **19**: 413–420.

—— (1965*a*): Theory of the x-ray *K* absorption spectrum anisotropy of single crystals, *Phys. Metals Metallogr.*, **19**: 801–808.

—— (1965*b*): Fine structure of the x-ray *K* absorption spectrum of a gallium single crystal, *Phys. Metals Metallogr.*, **20**: 26–32.

—— (1967): On the *K* absorption spectrum of polarized x-rays in a germanium single crystal, *Opt. Spectry. (USSR) (English Transl.)*, **22**: 163–165.

KOZLENKOV, A. I. (1961): Theory of the fine structure of x-ray absorption spectra, *Bull. Acad. Sci. USSR, Phys. Ser. (English Transl.)*, **25**: 968–987.

—— (1963): Calculation of the fine structure of the x-ray absorption spectra of metals using Hartree fields, *Bull. Acad. Sci. USSR, Phys. Ser. (English Transl.)*, **27**: 373–388.

—— (1964): Calculation of the fine structure of the *K* absorption spectrum of copper, *Bull. Acad. Sci. USSR, Phys. Ser. (English Transl.)*, **28**: 794–805.

KRISHNAN, T. V., and NIGAM, A. N. (1967): Thickness effect in the x-ray K absorption of nickel, *Proc. Indian Acad. Sci.*, **65**: 45–48.

KRONIG, R. DE L. (1931): Zue Theorie der Feinstruktur in den Röntgenabsorptionspektren, *Z. Physik*, **70**: 317.

——— (1932a): Zur Theorie der Feinstruktur in den Röntgenabsorptionspektren: II, *Z. Physik*, **75**: 191–210.

——— (1932b): Zur Theorie der Feinstruktur in den Röntgenabsorptionspektren: III, *Z. Physik*, **75**: 468–475.

KRUTTER, H. M. (1935): Energy bands in copper, *Phys. Rev.*, **48**: 664–671.

KUNZ, A. B. (1970): Energy bands and the optical properties of LiCl, *Phys. Rev. Abstr.*, **1**: 21.

KURYLENKO, C. (1939): "Franges en voisinage de la discontinuité K des rayons X," Ph.D. dissertation, University of Paris.

LAVILLA, R. E., and DESLATTES, R. D. (1966): K-absorption fine structure of sulfur in gaseous SF_6, *J. Chem. Phys.*, **44**: 4399–4400.

LEDER, L. B., MENDLOWITZ, H., and MARTON, L. (1956): Comparison of the characteristic energy losses of electrons with the fine structure of the x-ray absorption spectra, *Phys. Rev.*, **101**: 1460–1467.

LUBLOW, D. (1960): Ph.D. dissertation, Universität von Köln.

LUCASSON, A. (1960): Contribution à l'étude du zinc, du gallium, du germanium et d'alliages de cuivre par spectrographie X, *Ann. Phys. (Paris)*, **5**: 509–565.

LUCASSON-LEMASSON, A. (1958): Spectres L et K d'absorption du cuivre de divers alliages, *Compt. Rend.*, **246**: 94–97.

LUKIRSKII, A. P., and ZIMKINA, T. M. (1963): Fine structure of the L_{II}-L_{III} absorption edge of argon, *Bull. Acad. Sci. USSR, Phys. Ser. (English Transl.)*, **27**: 333–338.

——— and ——— (1964): $L_{II,III}$ absorption spectra of potassium and chlorine in KCl, *Bull. Acad. Sci. USSR, Phys. Ser. (English Transl.)*, **28**: 674–680.

———, ———, and BRYTOV, I. A. (1964): Fine structure of the M absorption edges of krypton and the N edges of xenon, *Bull. Acad. Sci. USSR, Phys. Ser. (English Transl.)*, **28**: 681–688.

LYTLE, F. W. (1966): Determination of interatomic distances from x-ray absorption fine structure, *Advan. X-Ray Anal.*, **9**: 398–409.

——— (1967): Determination of metal atom positions in $CrVO_4$ by x-ray absorption spectroscopy, *Acta Cryst.*, **22**: 321.

MANDÉ, C. (1960): Contribution à l'étude de l'or du palladium et de leurs alliages par spectrographie X, *Ann. Phys. (Paris)*, **5**: 1560–1614.

——— and JOSHI, N. V. (1968): Study of the fine structure associated with the K absorption discontinuity of copper, *Indian J. Pure Appl. Phys.*, **6**: 371–374.

MAZALOV, L. N., BLOKHIN, S. M., and VAINSHTEIN, E. E. (1967): Anomalies in the influence of the band structure on the x-ray spectra of solids, *Soviet Phys.-Solid State*, **8**: 1926–1931.

MEN'SHIKOV, A. Z. (1963): Problems of interpreting the x-ray emission and absorption spectra of transition metals, *Phys. Metals Metallogr.*, **15**: 29–33.

MERZ, H., and ULMER, K. (1968a): Zur Zustandsdichte der Übergangsmetalle. I. Isochromaten-spektroskopische Untersuchungen und kubish-raumzentrierten Übergangsmetallen, *Z. Physik*, **210**: 92–110.

—— and —— (1968b): Zur Zustandsdichte der Übergangsmetalle. II. Isochromaten-spektroskopie hexagonaler III A- und IV A-Metalle, *Z. Physik*, **212**: 435–448.

MITCHELL, G. R. (1962): X-Ray *K*-absorption edges of three-planar nickel compounds, *J. Chem. Phys.*, **37**: 216–219.

—— (1965): Approximations for the interpretations of x-ray *K* absorption spectra, *Develop. Appl. Spectry.*, **4**: 109–117.

—— and BEEMAN, W. W. (1952): The x-ray *K* absorption edges of covalently bonded Cr, Mr, Fe, and Ni, *J. Chem. Phys.*, **20**: 1298–1301.

MOTT, D. L. (1966): Spectra of x-ray absorption in the *K* shell of Si in gaseous $SiCl_4$ and in the *L* shell of Ge in gaseous $GeCl_4$, *Phys. Rev.*, **144**: 94–96.

MOTT, N. F., and GURNEY, R. W. (1957): *Electronic processes in ionic crystals*, 2d ed., Oxford University Press, London.

—— and MASSEY, H. S. W. (1933): *Theory of atomic collisions*, Clarendon Press, Oxford.

MURTY, N. H., and AZÁROFF, L. V. (1967): X-ray *K* absorption spectra of iron-aluminum alloys, *Acta Met.*, **15**: 1655–1659.

MUTO, T., and OKUNO, H. (1956): Electronic structure of the exciton, *J. Phys. Soc. Japan*, **11**: 633–644.

NADZHAKOV, E. G., and BARINSKII, R. L. (1960): A new procedure for calculating x-ray *K* absorption spectra, *Soviet Phys. "Doklady" (English Transl.)*, **4**: 1319–1323.

NAKAI, S., and SAGAWA, T. (1969): Na^+ $L_{2,3}$ absorption spectra of sodium halides, *J. Phys. Soc. Japan*, **26**: 1427–1434.

NARBUTT, K. I., and IZRAELEVA, L. K. (1963): Structure of the *K* absorption spectrum of Zn^{2+}, *Bull. Acad. Sci. USSR, Phys. Ser. (English Transl.)*, **27**: 356–359.

—— and SMIRNOVA, I. S. (1963): *K* absorption spectrum and the conduction band of alkalide halide crystals, *Bull. Acad. Sci. USSR, Phys. Ser. (English Transl.)*, **27**: 349–356.

NEMNONOV, S. A., and KOLOBOVA, K. M. (1967): X-ray spectral analysis of the energy spectrum of Ti-Fe alloys, *Phys. Metals Metallogr.*, **23**: 66–71.

NIJBOER, B. R. A. (1946): On the intensity-distribution of the continuous x-ray spectrum near its short-wavelength limit, *Physica*, **12**: 461–465.

NILSSON, A. (1953): A precision determination of the *K* excitation potential of some 3*d* elements, *Arkiv Fysik*, **6**: 513–592.

NORDFORS, B. (1960): The statistical error in x-ray absorption measurements, *Arkiv Fysik*, **18**: 37–47.

—— (1961): The x-ray *L* absorption spectrum of silver: An investigation with a bent crystal and proportional counter, *Arkiv Fysik*, **19**: 259–288.

—— and NORELAND, E. (1961): The x-ray *L*-absorption spectra of 48 Cd-52 Te, *Arkiv Fysik*, **20**: 1–23.

NOZIÈRES, P., and DE DOMINICIS, C. T. (1969): Singularities in the x-ray absorption and emission of metals. III. One-body theory exact solution, *Phys. Rev.*, **178**: 1097–1107.

NOZIÈRES, P., and PINES, D. (1959): Electron interactions in solids, *Phys. Rev.*, 113: 1254–1267.

NUTTALL, J. M. (1928): The *K* absorption edges of potassium and chlorine in various compounds, *Phys. Rev.*, 31: 742–747.

OHLIN, P. (1941): Ph.D. dissertation, Uppsala Universitets Arsskrift.

——— (1943): Short wave-length limit of the continuous x-ray spectrum and determinations of *h/e*, *Nature*, 152: 329–330.

——— (1952): A spectrometer for an investigation of angular dependence of the structure in the continuous x-ray spectrum, *Arkiv Fysik*, 4: 387–390.

OVSYARNIKOVA, I. A., and BOROVSKII, I. B. (1960): Investigation of the fine structure of the *K* absorption spectra of some sulfides, *Bull. Acad. Sci. USSR, Phys. Ser. (English Transl.)*, 24: 444–448.

PARRATT, L. G. (1939): X-ray resonance absorption lines in the argon *K* spectrum, *Phys. Rev.*, 56: 295–297.

——— (1959): Electronic band structure of solids by x-ray spectroscopy, *Rev. Mod. Phys.*, 31(3): 616.

———, HEMPSTEAD, C. F., and JOSSEM, E. L. (1957): "Thickness effect" in absorption spectra near absorption edges, *Phys. Rev.*, 105: 1228–1232.

——— and JOSSEM, E. L. (1955): X-ray spectroscopy of the solid state, *Phys. Rev.*, 97: 916–926.

——— and ——— (1957): X-ray excited states ("excitons") and width of valence band in KCl, *J. Phys. Chem. Solids*, 2: 67–71.

———, PORTEUS, J. O., SCHNOPPER, H. W., and WATANABE, T. (1959): X-ray absorption coefficients and geometrical collimation of the beam, *Rev. Sci. Instr.*, 30: 344–347.

PEREL, J. (1970): The determination of x-ray absorption coefficients and the thickness of nonuniform thin samples, *J. Phys.*, E3: 268–270.

——— and DESLATTES, R. D. (1970): Extended fine structure in x-ray absorption spectra of certain perovskites, *Phys. Rev.*, B2: 1317–1323.

PETERSEN, H. (1936): Zur Theorie der Röntgenabsorptionen molekularen Gase: III, *Z. Physik*, 98: 569–575.

PRINS, J. (1934): Transitions to optical levels in the argon *L* x-ray absorption spectrum, *Nature*, 133: 795–796.

RICHTMYER, F. K., BARNES, S. W., and RAMBERG, E. (1934): The widths of the *L*-series lines and of the energy levels of Au(79), *Phys. Rev.*, 46: 843.

ROOKE, G. A. (1968): Interpretation of aluminum x-ray band spectra. I. Intensity distribution, *Proc. Phys. Soc.*, 2(2): 767–775.

RUDBERG, E., and SLATER, J. C. (1936): Theory of inelastic scattering of electrons from solids, *Phys. Rev.*, 50: 150–158.

SANDSTRÖM, A. (1930): Röntgenspektroskopische Messungen der *L* Absorption der Elemente 74 Wolfram bis 92 Uran, *Z. Physik*, 65: 632.

——— (1957): Experimental methods of x-ray spectroscopy: Ordinary wavelength, in S. FLÜGGE (ed.), *Handb. Physik*, 30: 78–245.

SAWADA, M., TSUTSUMI, K., SHIRAIWA, T., ISHIMURA, T., and OBASHI, M. (1959): Some contributions to the x-ray spectroscopy of solid state, *Ann. Rept. Sci. Works, Fac. Sci., Osaka Univ.*, 7: 1–87.

SAYERS, D. E., LYTLE, F. W., and STERN, E. A. (1970): Point scattering theory of x-ray K absorption fine structure, *Advan. X-Ray Anal.*, **13**: 248–271.

SCHIFF, L. F. (1949): *Quantum mechanics*, McGraw-Hill Book Company, New York.

SCHMIDT, V. V. (1961*a*): Contribution to the theory of the temperature dependence of the fine structure of x-ray absorption spectra, *Bull. Acad. Sci. USSR, Phys. Ser. (English Transl.)*, **25**: 988–993.

——— (1961*b*): On the effect of interelectron interaction on the fine structure of x-ray spectra, *J. Exp. Theor. Phys. (English Transl.)*, **12**: 886–890.

——— (1963): Contribution to the theory of the temperature dependence of the fine structure of x-ray absorption spectra. II. Case of high temperatures, *Bull. Acad. Sci. USSR, Phys. Ser. (English Transl.)*, **27**: 392–397.

SCHNOPPER, H. W. (1963): Multiple excitation and ionization of inner atomic shells by x-rays, *Phys. Rev.*, **131**: 2558–2560.

——— (1964): Argon K absorption edge by the counter response method, *Phys. Rev.*, **A133**: 627–628.

——— (1966): in *Röntgenspektren und chemische Bindung* (Proc. 1965 Intern. Conf. Leipzig), 303–305.

——— (1967): Atomic readjustment to an inner-shell vacancy: Manganese K x-ray emission spectra from an Fe^{55} K-capture source and from the bulk metal, *Phys. Rev.*, **154**: 118–123.

SCHOTTE, K. D., and SCHOTTE, U. (1969): Threshold behavior of the x-ray spectra of light metals, *Phys. Rev.*, **185**: 509–517.

SEKA, W., and HANSON, H. P. (1969): Molecular orbital interpretation of x-ray absorption edges, *J. Chem. Phys.*, **50**: 344–350.

SHARKIN, O. P. (1964): I. Concerning the influence of temperature on the fine structure of the K absorption spectra of potassium halides, *Bull. Acad. Sci. USSR, Phys. Ser. (English Transl.)*, **28**: 814–820.

SHAW, C. H. (1946): Secondary structure in the K absorption limit of germanium tetrachloride, *Phys. Rev.*, **70**: 643–645.

SHEPELEVA, I. M., and IRKHIN, YU. P. (1961): K absorption spectra in iron, *Fiz. Metal. i Metalloved.*, **11**: 313–314.

SHIRAIWA, T. (1960): II. The theory of the fine structure of the x-ray absorption spectrum, *J. Phys. Soc. Japan*, **15**: 240–250.

———, ISHIMURA, T., and SAWADA, M. (1958): The theory of the fine structure of the x-ray absorption spectrum, *J. Phys. Soc. Japan*, **13**: 847–859.

SINGH, J. N. (1961): Fine structure of the K x-ray absorption edge of germanium, *Phys. Rev.*, **123**: 1724–1729.

——— (1962): Fine structure of the x-ray K-absorption edge of Fe, *Physica*, **28**: 131–132.

SLATER, J. C. (1936): The ferromagnetism of nickel, *Phys. Rev.*, **49**: 537–545.

SOBELMAN, A., and FEINBERG, E. (1958): Some optical effects of plasma oscillations in a solid, *J. Exp. Theor. Phys. (English Transl.)*, **7**: 339–344.

SOKOLOV, A. V. (1956): On the absorption and emission of x-rays by ferromagnetic metals, *Bull. Acad. Sci. USSR, Phys. Ser. (English Transl.)*, **20**: 103–107.

SOROKINA, M. F., and NEMNONOV, S. A. (1967): Concerning the white line in the L_{III} absorption spectrum of palladium, *Bull. Acad. Sci. USSR, Phys. Ser. (English Transl.)*, **31**: 1039–1042.

SPIJKERMAN, J. J., and BEARDEN, J. A. (1964): Precision evaluation of the high-frequency limit of the continuous x-ray spectrum using a gas target x-ray tube, *Phys. Rev.*, **A134**: 871–876.

STEPHENSON, S. T. (1933): Fine structure in the K x-ray absorption spectrum of bromine, *Phys. Rev.*, **44**: 349–352.

—— (1947): X-ray absorption structure of $GeCl_4$ and $AsCl_3$, *Phys. Rev.*, **71**: 84–87.

—— (1957): The continuous x-ray spectrum, in S. FLÜGGE (ed.), *Handb. Physik*, **30**: 337–370.

SUGIURA, C. (1961): X-ray K absorption spectra of potassium in KF, KCl, KBr, and KI, *Sci. Rept. Tohoku Univ., First Ser.*, **45**: 248–259.

TOMBOULIAN, D. H. (1957): The experimental methods of soft x-ray spectroscopy and the valence band spectra of the light elements, in S. FLÜGGE (ed.), *Handb. Physik*, **30**: 246–304.

TRAPEZNIKOV, V. A. (1956): Position of the initial K absorption edge of iron and nickel for pure metals in the ferromagnetic and paramagnetic state (in Russian), *Fiz. Metal. i Metalloved.*, **3**: 561–562.

—— and NEMNONOV, S. A. (1956): Investigation of bond strengths in Fe-Mo solid solutions from their x-ray absorption fine structures (in Russian), *Fiz. Metal. i Metalloved.*, **3**: 314–320.

TRISCHKA, J. W. (1945): Structure in the x-ray K absorption edges of solid potassium chloride, *Phys. Rev.*, **67**: 318–320.

ULMER, K. (1969): Solid state isochromat spectroscopy, in *X-ray spectra and electronic structure of matter, Proc. Intern. Symp. Kiev (1968)*, **2**: 79–89.

VAINSHTEIN, E. E. (1950): X-ray spectra of atoms in molecules of chemical compounds and in alloys (in Russian), *Akad. Nauk SSSR, Moscow*.

——, BARINSKII, R. L., and NARBUTT, K. I. (1955): Certain aspects of the x-ray K-absorption edge structure of atoms in alkali-metal halides (in Russian), *Dokl. Akad. Nauk SSSR*, **105**: 1196–1199.

—— and KAVETSKY, V. S. (1953): X-ray absorption spectra of Ni, Cu, and Zn ions in aqueous and nonaqueous solutions (in Russian), *Dokl. Akad. Nauk SSSR*, **91**: 775–778.

—— and NARBUTT, K. I. (1945): Structure of the K-edge of x-ray absorption spectra of atoms in the molecules of a gas (in Russian), *Izv. Akad. Nauk SSSR*, **1**: 71–73.

—— and —— (1950): II. Structure of the K-edge of atomic x-ray absorption spectra in gaseous molecules (in Russian), *Izv. Akad. Nauk SSSR*, **4**: 344–349.

——, ——, and BARINSKII, R. L. (1952): Structure of the x-ray absorption edge of atoms in polar crystals and its relation to ultraviolet absorption (in Russian), *Dokl. Akad. Nauk SSSR*, **82**: 701–704.

——, ZHURAKOVSKII, E. A., NESHPOR, V. S., and SAMSONOV, G. V. (1961): Fine structure of x-ray K-spectral absorption and the Hall effect in vanadium silicides, *Soviet Phys. "Doklady" (English Transl.)*, **5**: 996–998.

VISHNOI, A. N., and AGARWAL, B. K. (1966): Theory of the extended x-ray absorption fine structure, *Proc. Phys. Soc. (London)*, **89**: 799–804.

WATANABE, T. (1965*a*): Measurement of the *L* absorption spectra of xenon, *Phys. Rev.*, **A137**: 1380–1382.

—— (1965*b*): Theoretical fitting of the argon *K* absorption spectrum on a one-electron model, *Phys. Rev.*, **A139**: 1747–1751.

WEBER, W. M. (1962): Anisotropy of the *K* absorption in gallium single crystals, *Physica*, **28**: 689–694.

—— (1967): Anisotropic *K* absorption in a cadmium single crystal, *Phys. Lett.*, **A25**: 590–591.

WILLIAMS, W. S., and LYE, R. G. (1965): U.S. Air Force Tech. Doc. Rept. ML-TDR-64-25, part 2.

WUILLEUMIER, F. (1966): Mise en évidence d'excitations doubles dans le spectre d'absorption *L* du krypton gazeux, *Compt. Rend.*, **265**: 450–452.

YEH, H. C., and AZÁROFF, L. V. (1965): Electronic structure of copper-zinc solid solutions deduced from their x-ray *K* absorption spectra, *Appl. Phys. Lett.*, **6**: 207–208.

—— and —— (1967): X-ray *K* absorption edges of alloys. III. Copper-zinc and nickel-zinc systems, *J. Appl. Phys.*, **38**: 4034–4038.

ZIMKINA, T. M., and LUKIRSKII, A. P. (1965): Fine structure of the $L_{II,III}$ absorption spectra of Cl in the compounds LiCl, NaCl, KCl, RbCl, and CsCl, *Soviet Phys.-Solid State (English Transl.)*, **7**: 1175–1179.

7

SYNCHROTRON RADIATION AND APPLICATIONS

Robert P. Madden

I. INTRODUCTION

In the 1960s, there were written a number of comprehensive articles and reviews dealing with the subject of synchrotron radiation. These are readily available and will be referred to below. For this reason the subject will not be re-reviewed in depth here. However, a modern text on x-ray spectroscopy would be incomplete without dealing to some extent with this important source of radiation. In this chapter we will attempt to achieve an appropriate depth of understanding of synchrotron radiation for the non-specialist and indicate some of the more important applications of this modern tool.

Historically, electron accelerators were developed for the study of the inter-action of high-energy electrons with matter and as sources of gamma rays for photo-nuclear and high-energy photoionization studies. The gamma rays are produced when the electrons are stopped by solid targets (Brehmsstrahlung). Today cyclic accelerators have become important as sources of soft x-ray radiation for experiments in atomic, molecular, and solid-state physics. This radiation, most often referred to as *synchrotron radiation*, is produced when the direction of motion of the charged particles is altered. Such an angular displacement of the velocity constitutes an acceleration which classically must be accompanied by the radiation of electro-magnetic energy. In a synchrotron, such an acceleration is required to keep the charged particles in a closed orbit. This "radial" acceleration results from the inter-action of the charged particle with the magnetic field; hence the radiation is occa-sionally referred to as *magnetic Brehmsstrahlung*. Because of the small electron/proton mass ratio, this radiation is most significant for electron accelerators, and henceforth only electron synchrotron radiation will be considered.

The energy which is radiated by this mechanism must be resupplied from the accelerating system of the synchrotron if the electrons are to maintain a stable orbit. This energy loss becomes increasingly important as higher electron energies are attained, and it was this fact which led to the first real interest in synchrotron radiation in the 1940s. In a relatively short period of time, the full theory of synchrotron radiation was developed, and the first experimental observations were made. The real potential of this radiation source for physical experiments, however, was not realized until 1956, when Professor Tomboulian and his coworkers at Cornell Univer-sity conducted an experiment on the 320-MeV electron synchrotron. The published account of their experiment clearly delineated the advantages of the synchrotron as a soft x-ray source. From this time on many groups of spectroscopists and solid-state physicists expressed interest in utilizing these machines. It was not until 1961, how-ever, that development of an electron synchrotron facility for spectroscopy was initiated at the National Bureau of Standards. Synchrotron spectroscopy quickly gained momentum in the 1960s, with solid-state physicists utilizing synchrotrons in Italy, Japan, and West Germany and a storage ring in the United States.

The applications of synchrotron radiation are many and diverse. In addition to its continuous spectral coverage, the radiation is highly polarized and is generated in a clean high-vacuum environment. All these properties have been exploited to open new avenues of research in the soft x-ray region. As predicted by Tomboulian and Hartman in 1956, the accurate calculability of the spectral and angular distribution of the radiation has been utilized to make the synchrotron an absolute flux standard for the far ultraviolet. Today there is no questioning the usefulness of this source in extending all the usual measurements of physics into the previously difficult soft x-ray region of the spectrum. It is, in this author's opinion, the most significant experimental development in soft x-ray physics of the 1960s.

The following sections are devoted to a short review of the development of the theory of synchrotron radiation, including simple equations which allow approximate calculations of its more interesting properties, a review of the development of the technology of synchrotron spectroscopy, and illustrations of the scientific achievements obtained with this new tool.

II. THEORETICAL DEVELOPMENT

The classical relativistic theory of synchrotron radiation was developed rather completely during the late 1940s and early 1950s, with refinements on the quantum effects extending into the 1960s. However, attempts to calculate the radiation from centripetally accelerated charges date back to the nineteenth century. Today, a number of reviews of synchrotron radiation exist which include equations describing its more important properties (Tomboulian and Hartman, 1956; Godwin, 1969; Sokolov and Temov, 1968). For a more detailed treatment of the classical theory one should return to Schwinger's (1946; 1949) original work. A recent very comprehensive review, particularly of the Russian contributions, which includes the combined effects of quantum-mechanical fluctuations and radiation damping on electron orbital motion has been written by Sokolov and Temov (1968).

In this section we will consider the most fundamental concepts, avoid the detailed mathematical development, and present the more important equations of practical interest.

A. Lienard's equation for the total radiated power

A nonrelativistic expression for the rate at which energy is radiated by a charge undergoing acceleration was given by Larmor toward the end of the nineteenth century. The familiar expression for the instantaneous power radiated is

$$P = \frac{2}{3} \frac{e^2}{c^3} a^2 \qquad (7\text{-}1)$$

where e is the electric charge, c is the velocity of light, and a is the instantaneous acceleration. The relativistic implication is that the accelerating charge is instantaneously at rest when the equation applies. Since

$$\mathbf{a} = \frac{d\mathbf{v}}{dt} = \frac{1}{m}\frac{d\mathbf{p}}{dt}$$

where $\mathbf{v}$ is the velocity, $\mathbf{p}$ the momentum, and m the rest mass of the charged particle, the Larmor relation may also be written in the convenient form

$$P = \frac{2}{3}\frac{e^2}{m^2c^3}\left(\frac{d\mathbf{p}}{dt}\cdot\frac{d\mathbf{p}}{dt}\right) \qquad (7\text{-}2)$$

The electrons which radiate usable synchrotron radiation have very high energy; hence we need to determine the relativistically invariant form of this expression. Since energy and time transform in the same way, the instantaneous power [left-hand side of equation (7-2)] is already an invariant. The proper Lorentz-invariant form of the right-hand side is obtained by substituting for $\mathbf{p}$ the momentum-energy four-vector $\mathbf{p}$ and taking the derivative with respect to the Lorentz-invariant proper time:

$$P = \frac{2}{3}\frac{e^2}{m^2c^3}\left(\frac{d\mathbf{p}}{d\tau}\cdot\frac{d\mathbf{p}}{d\tau}\right) \qquad (7\text{-}3)$$

The components of $\mathbf{p}$ consist of the spatial components, $\mathbf{p}_i$, for $i = 1, 2, 3$, which are related to the components of the ordinary momentum by $\mathbf{p}_i = \gamma p_i$, where $\gamma = (1 - v^2/c^2)^{-1/2}$, and $\mathbf{p}_4 = iE/c$, where E is the total energy of the particle, $E = \gamma mc^2$. The element of proper time $d\tau$ is related to the element of ordinary time dt by $\gamma\, d\tau = dt$.

If we evaluate the four-vector scalar product in (7-3), we obtain

$$P = \frac{2}{3}\frac{e^2}{m^2c^3}\left[\left(\frac{d(\gamma\mathbf{p})}{d\tau}\right)^2 - \frac{1}{c^2}\left(\frac{dE}{d\tau}\right)^2\right] \qquad (7\text{-}4)$$

Equation (7-4) is a relativistically invariant form of (7-3) in which the spatial and energy terms have been separated. Further substitutions will show that (7-4) in fact reduces to (7-2) in the limit as $\beta = v/c \to 0$.

Equation (7-4), in slightly different form, was obtained by Lienard (1898). Lienard also recognized that in the case of a charge in circular motion, the rate of change of the energy is small compared to the rate of change of the momentum. Neglecting the second term in the bracket of equation (7-4), and assuming circular motion so that

$$\left(\frac{d(\gamma\mathbf{p})}{d\tau}\right)^2 = \gamma^4\left(\frac{d\mathbf{p}}{dt}\right)^2 = \gamma^4\omega^2|\mathbf{p}|^2$$

we see that equation (7-4) reduces to

$$P = \frac{2}{3}\frac{e^2}{m^2c^3}\cdot\gamma^4\omega^2|\mathbf{p}|^2$$

Replacing γ by $E/(mc^2)$ and letting $\omega = v/R$, where R is the radius of the orbit,

$$P = \frac{2}{3} \frac{e^2 c \beta^4}{R^2} \left(\frac{E}{mc^2}\right)^4 \qquad (7\text{-}5)$$

This result, obtained by Lienard, expresses the important fact that the total instantaneous power radiated by high-energy ($\beta \approx 1$) electrons in circular motion is proportional to the fourth power of the energy and to the inverse second power of the radius of the orbit. The energy radiated *per revolution* is

$$\delta E = \frac{2\pi R}{c\beta} P = \frac{4\pi}{3} \frac{e^2}{R} \beta^3 \left(\frac{E}{mc^2}\right)^4 \qquad (7\text{-}6)$$

For $\beta \approx 1$, a useful expression for estimating is

$$\delta E \text{ (keV)} = 88.5 \frac{[E \text{ (BeV)}]^4}{R \text{ (m)}} \qquad (7\text{-}7)$$

The considerations of Lienard were extended by Schott (1907; 1912) who, in attempting to develop an atomic model to explain the discrete nature of radiation from atoms, considered the radiation from an electron moving in a homogeneous magnetic field. Schott actually developed expressions describing the angular distribution of radiation from an electron moving relativistically in a circular orbit as a function of the harmonic of the orbiting frequency.

The development of electron radiation theory stopped at this point as atomic theory took a new and rewarding direction. Interest revived in the 1940s when high-energy electron accelerators were being developed. Ivanenko and Pomeranchuk (1944) pointed out that an upper limit in electron energy is reached in a cyclic accelerator when the energy input to the electron per revolution becomes equal to the energy loss per revolution given by equation (7-6). Immediately following this realization, several efforts to develop the theory of radiation from centripetally accelerated electrons were begun. The classical theory, complete in most details, was presented by Schwinger (1946; 1949) and by Ivanenko and Sokolov (1948).

B.   Schwinger's treatment

Equation (7-5) tells us only the total instantaneous power radiated by an electron undergoing centripetal acceleration. The goal of a complete theory would be to examine the nature of that radiation in detail: its wavelength distribution, angular-radiation pattern, polarization properties, and the dependence of all these characteristics on the controllable machine parameters.

Schwinger developed the radiation theory for an arbitrary electron trajectory from a consideration of the rate at which the electron does work on the electromagnetic field. He then applied his general solution to the special case of circular motion, developing solutions for the radiation in terms of harmonics of the fundamental orbiting frequency of the electron.

The classical equations derived by this method have been confirmed by quantum-mechanical calculations, which show that for electron energies $E \ll mc^2(mcR/\hbar)^{1/2}$ the classical theory of the characteristics of the radiation is correct. This limit ($>10^4$ BeV) is well beyond the energy of existing or contemplated accelerators. However, detailed calculations have shown that for existing machines fluctuations in the electron orbits can be caused by reaction of the electrons to the quantum-mechanical nature of the radiation. These perturbations on the orbit are reduced, however, by the damping effect of the radiation itself. This subject has been well treated [Sokolov and Temov (1968)] and will not be important to the remainder of this chapter.

Schwinger's results show that the most important relativistic effects are: (1) The instantaneous dipole radiation pattern expected from an accelerating charge at low energy is warped forward (even the back lobe) and condensed into the forward-moving direction of the electron as its energy increases—that is, the radiation emanates from the moving electron as the beam from the headlight of a train moving on a circular track. The low and high electron energy radiation patterns are illustrated in Fig. 7-1. (2) The radiated power shifts to higher and higher harmonics of the fundamental orbital frequency as the electron energy is increased.

The relativistic warping of the radiation into the forward direction can be quantitatively expressed rather simply. For the total power radiated the angular half-width of the distribution at half-maximum is given by

$$\delta = \frac{mc^2}{E} \qquad (7\text{-}8)$$

where mc^2, the electron rest energy, ≈ 0.51 MeV, and δ gives us a measure of the angular divergence of the "headlight" beam.

From (1), as quantified by equation (7-8), we see that an observer located in the plane of the orbit looking along a tangent toward the electron trajectory would receive a pulse of radiation each time the electron passed the tangent point moving instantaneously in the direction of the observer. The power spectrum of the radiation would be given by the Fourier components in this pulse, and since the pulses repeat at the orbital frequency of the electron, the spectrum would consist of harmonics of this frequency. As the energy of the electron increases, the angular compression of the radiation into the forward-moving direction increases with γ, decreasing the pulse width. The observed pulse is further compressed proportional to γ^2 by the Doppler

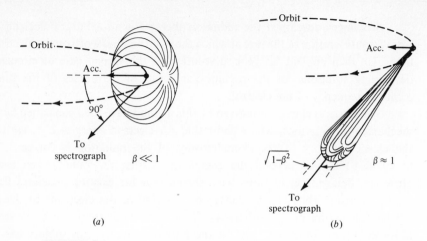

FIGURE 7-1
Angular radiation patterns for orbiting electrons of greatly different energies.
(*a*) Low energy (nonrelativistic), showing the typical dipole pattern; (*b*) high
energy (very relativistic), showing that the dipole pattern has been greatly warped
around to the forward direction of the electron. (*From Tomboulian and Hartman,*
1956.)

effect, since the source is moving toward the observer at nearly the velocity c. Since
$\gamma = (mc^2/E)$, the pulse duration goes roughly as E^{-3}, and the radiated power shifts
to higher and higher harmonics of the orbital frequency as the energy increases, which
qualitatively explains effect (2) above.

As the electron proceeds around its circular orbit, the narrow cone of radiation
it emits in the forward direction sweeps across the field of observation, causing the
time-average distribution of radiation to be completely uniform in a direction parallel
to the orbital plane. However, there is a strong angular dependence of the distribution
observed in a direction perpendicular to the orbital plane. This distribution can be
described as a function of the *azimuth angle of observation*, which is defined as the
angle between the line of observation and its projection on the orbital plane (the
projection being tangent to the orbit).

C.   Angular and polarization distribution of the instantaneous power as a function of wavelength

The angular and spectral distribution of instantaneous power radiated by a monoenergetic electron in circular motion, according to Schwinger (1949), in a convenient
form [Tomboulian and Hartman (1956)] is given by:

$$P(\psi,\lambda) = \frac{8}{3} \frac{\pi e^2 c^2}{\omega_0 \lambda^4} \left(\frac{mc^2}{E}\right)^4 (1 + x^2)^2 \left[K_{2/3}^2(\xi) + \left(\frac{x^2}{1 + x^2}\right) K_{1/3}^2(\xi) \right] \qquad (7\text{-}9)$$

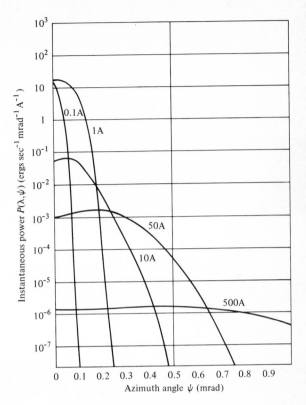

FIGURE 7-2
The angular (azimuth) distribution of the instantaneous power radiated by a 6-BeV electron moving in a 31.7-m-radius orbit. The parameter is the wavelength in A. (*From Haensel and Kunz, 1967.*)

where $P(\psi,\lambda)$ is the power in ergs per sec per rad per unit wavelength and where $x = [E/(mc^2)]\psi$ and $\xi = [2\pi R/(3\lambda)](mc^2/E)^3(1 + x^2)^{3/2}$. Here ψ is the azimuth angle of observation measured relative to the orbital plane, λ is the wavelength, E is the energy of the electron, $\omega_0 = c/R$ is the orbital angular frequency, and $K_{1/3,2/3}$ are modified Bessel functions of the second kind. Using equation (7-9), we can calculate the angular distribution for any wavelength as a function of electron energy and orbital radius. As an example Fig. 7-2 shows the angular distribution of the total instantaneous power radiated per unit wavelength by a monoenergetic electron of energy 6 BeV in a 31.7-m-radius orbit for several wavelengths. Note how small the azimuth angles are in this practical case, and that the distribution can have a "dip" at $\psi = 0$.

Equation (7-9) contains all the information regarding polarization of the radiation. The first term in the brackets gives the angular distribution of the radiation

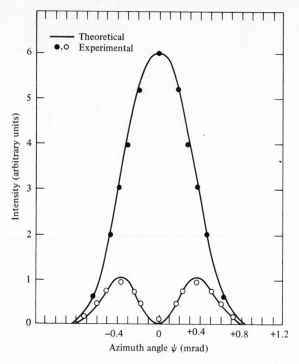

FIGURE 7-3

The angular and polarization distribution of the instantaneous power radiated at 5000 A by a 120-MeV electron moving in an 0.83-m-radius orbit. The solid lines are theoretical curves calculated from the two parts of equation (7-9) (see text). The high, single-peak curve is for the component with the electric vector parallel to the orbital plane, while the lower, double-peak curve represents the component with electric vector perpendicular to the orbital plane. The experimental points were obtained by a scanning photomultiplier-filter-polarizer system. The experimental data were normalized to the peak maximum; however, the relative magnitudes of the parallel peak to the perpendicular peaks are shown as observed experimentally. (*From Codling and Madden, 1965a.*)

with the electric vector parallel to the electron orbit, while the second term gives the angular distribution of the radiation with the electric vector perpendicular to the electron orbit. In general the radiation is elliptically polarized since the parallel and perpendicular components are 90° out of phase. Note that the second term goes to zero as $x \to 0$ ($\psi \to 0$). Thus if we observe exactly in the plane of the orbit, the radiation should be completely plane-polarized with the electric vector parallel to the orbital plane. The solid curves in Fig. 7-3 give an example of the calculated angular distribution for both polarizations at 5000 A in the case of a 120-MeV electron in an 0.83-m orbit.

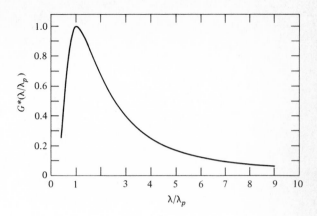

FIGURE 7-4
The universal wavelength-distribution function $G^*(\lambda/\lambda_p)$ for monoenergetic electrons as a function of reduced wavelength. The peak at $\lambda/\lambda_p = 1$ where

$$\lambda_p \, (\text{A}) = 2.35 \, \frac{R \, (\text{m})}{[E \, (\text{BeV})]^3}$$

To find the instantaneous power per unit wavelength integrated over azimuth angle at any wavelength, values from this curve can be used with equation (7-12).

D.   Wavelength distribution of total instantaneous power

The wavelength distribution of the total instantaneous power may be obtained by integrating equation (7-9) over azimuth angle ψ. The result is

$$P(\lambda) = \frac{3^{5/2}}{16\pi^2} \frac{e^2 c}{R^3} \left(\frac{E}{mc^2} \right)^7 G(y) \qquad (7\text{-}10)$$

where

$$G(y) = y^3 \int_y^\infty K_{5/3}(\eta) \, d\eta$$

with

$$y = \frac{4\pi R}{3\lambda} \left(\frac{mc^2}{E} \right)^3$$

For quick estimates, a graphical representation of the function G is useful. Changing the function slightly to express it in terms of the useful variable λ/λ_p and to normalize the peak to 1, we write $G^*(\lambda/\lambda_p)$, where λ_p is the wavelength of the peak of the distribution given by

$$\lambda_p \, (\text{A}) = 2.35 \, \frac{[R \, (\text{m})]}{[E \, (\text{BeV})]^3} \qquad (7\text{-}11)$$

The value of $G^*(\lambda/\lambda_p)$ is determined from the graph of Fig. 7-4.

Expression (7-11) tells us, as we anticipated earlier, that the instantaneous power per unit wavelength radiated into all angles will have a peak in spectral distribution

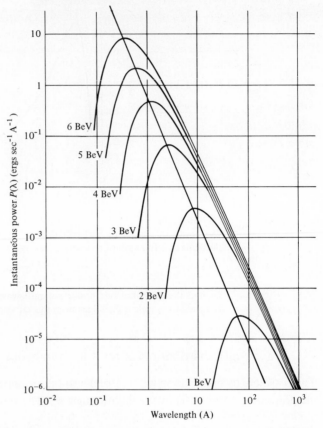

FIGURE 7-5
The wavelength distribution of the instantaneous power per unit wavelength integrated over azimuth angle for monoenergetic electrons moving in a 31.7-m-radius orbit. The parameter is the electron energy. Notice the peak shifting in accordance with equation (7-11), and that at long wavelengths the curves approach each other asymptotically. (*From Haensel and Kunz, 1967*).

at a wavelength which is inversely proportional to E^3 and proportional to R. Thus the usefulness of electron accelerators as radiation sources for a specific region of the spectrum depends critically on these two machine parameters, but especially on the electron energy.

An approximate form of equation (7-10), useful for estimating the spectral distribution of the total instantaneous power, is then given by

$$P(\lambda) = 0.90 \frac{[E \text{ (BeV)}]^7}{[R \text{ (m)}]^3} G^* (\lambda/\lambda_p) \qquad (7\text{-}12)$$

where $P(\lambda)$ has the units of erg/sec-A.

Figure 7-5 is an example of the spectral distribution of synchrotron radiation and its dependence on electron energy, for a synchrotron with a 31.7-m turning radius.

E.   Total instantaneous power

The total instantaneous power radiated into all angles and all wavelengths is obtained by integrating equation (7-10) over wavelength. If this integration is performed, the Lienard result [equation (7-5)] is obtained. This expression can be written in a very simple form for approximate calculations, namely,

$$P = 6.77 \frac{[E \text{ (BeV)}]^4}{[R \text{ (m)}]^2} \qquad (7\text{-}13)$$

where P has the units erg/sec.

F.   Practical calculations

With the above equations, the most significant properties of synchrotron radiation can be estimated quickly if the electron energy and orbital radius are known. Equation (7-13) gives the instantaneous power (erg/sec) radiated into all angles and all wavelengths. The wavelength at which the maximum of the spectral distribution occurs is given by (7-11). This value, λ_p, sets the abscissa scale for Fig. 7-4 which then gives the relative wavelength distribution of the total instantaneous power. Since $G^*(\lambda/\lambda_p)$ is normalized to 1, the relative distribution can be put on an absolute scale by evaluating the normalizing factor in equation (7-12).

Of course all the above expressions assume a single monoenergetic electron at energy E. To calculate the usable flux from a real machine, one must in addition multiply by the number n of electrons in the beam and also by the fraction of the time they will be observed. The latter factor is determined by both the duty cycle of the machine and a collection geometry factor. Let us call f the fraction of the time that usable synchrotron radiation is produced by the machine at electron energies close to E, and F the fraction of the total power radiated during this "on-time" which is accepted by the experiment. In most applications the entire vertical distribution can be accepted by the experimental system. In this case the power $N(\lambda)$ per unit wavelength at wavelength λ entering the experiment is given by

$$N(\lambda) = nfFP(\lambda) \qquad (7\text{-}14)$$

where $P(\lambda)$ is given by equation (7-10) or approximately by equation (7-12). The factor F will be evaluated for two geometrical arrangements in Subsection III B.

The detailed angular distributions require more elaborate calculations; however, the azimuth angle (measured from the orbital plane) at which the total radiated power drops to one-half is given by equation (7-8), and this is a rough estimate of the angular divergence for wavelengths in the vicinity of λ_p. Also, at higher energies, the ratio of the intensity radiated with electric vector parallel to the orbital plane to the total intensity radiated quickly approaches a value of ≈ 0.80 to 0.90 for most accelerator applications.

In certain cases it may be desirable to calculate the wavelength distribution of power averaged over the entire period of acceleration of an electron. An expression for $P(\lambda)$, averaged over a sinusoidal acceleration, has been given by Tomboulian and Hartman (1956).

In all of the above equations, the radius R is the turning radius. Many synchrotrons have straight sections, and hence an average orbital radius considerably larger than the radius in the turns.

III. OBSERVATIONAL DEVELOPMENT

A. Historical

1. *First observations*   Following the important realization (Ivanenko and Pomeranchuk, 1944) that radiative-energy losses have an important effect on the operation of electron accelerators, the first significant observations of synchrotron radiation were obtained at the General Electric Company on the 70-MeV synchrotron. This observation was reported by Elder et al. (1947); they also determined that the light was highly polarized with the electric vector parallel to the orbital plane. In a subsequent report, Elder et al. (1948) reported observing the continuum nature of the radiation with a spectrograph and showed that its spectral distribution in the visible region agreed with the theory of Schwinger.[1]

2. *First soft x-ray experiment*   An important milestone in the exploitation of synchrotron radiation occurred when Tomboulian and Hartman (1956) published the results of their experiments on the Cornell 320-MeV synchrotron. They were the first to observe the radiation at high photon energies. As a result of their careful photographic analysis with a grazing-incidence spectrograph, they were able to confirm that the spectral and angular distribution of the radiation in the spectral region

---

[1] Schwinger's theoretical calculations were communicated earlier to a number of investigators, and his essential results referred to in several papers dating back to 1945. Except for an invited paper on his calculations which was abstracted in 1946, Schwinger himself waited until 1949 to publish the details of his calculation.

80 to 300 A was in agreement with theoretical prediction. In this paper also, experimental results using the synchrotron as a source for soft x-ray spectroscopy were presented for the first time. They photographed the spectral transmission of Be and Al foils in the K and L regions, respectively. Furthermore, Tomboulian and Hartman pointed out in some detail all of the advantages of synchrotron radiation as a soft x-ray source: its continuum nature, its polarization, and its calculability which could be exploited for radiometric standards. Unfortunately these workers had access to the synchrotron for only two weeks; however, they stimulated a great deal of interest in the use of synchrotrons as sources for physical experiments. In spite of this interest, it was another five years before atomic physicists first gained access to a high-energy electron accelerator—once the exclusive domain of the high-energy physicist.

3. *Angular distributions* In the meantime, theoretical interest continued regarding the quantum-mechanical effects of the radiation, and additional observations were carried out on the polarization and angular distribution of the radiation in the visible region of the spectrum. Korolev et al. (1956) showed that the angular and spectral distributions were in fair agreement with the theory. However, the two peaks in the angular distribution for the perpendicular component (see Fig. 7-3) were not symmetrical. Their experiments utilized photographic techniques since they were also studying orbital dynamics where high-speed photography was required. Joos (1960) was the first to apply photoelectric techniques to study the angular distributions, and he also found the two peaks for the perpendicular component to be of different magnitudes.

4. *Establishment of the National Bureau of Standards facility* At this time the large-scale efforts to utilize synchrotron radiation were begun. In 1961 the National Bureau of Standards in Washington, D.C., built a tangent section into the vacuum system of their 180-MeV electron synchrotron, establishing NBS-SURF (National Bureau of Standards–Synchrotron Ultraviolet Radiation Facility). Measurements were begun to evaluate its potential as a standard source and as a source of radiation for experiments. This machine has a radius $R = 0.834$ m. Therefore, from equation (7-11), the maximum instantaneous radiated power (per unit wavelength) occurs at 335 A. The first NBS experiments utilized the continuum nature of the radiation as a background source and a 3-m grazing-incidence spectrograph to photograph absorption spectra of the noble gases between 80 and 600 A (Madden and Codling, 1963). A wealth of previously unobserved resonance structures were discovered in the photoionization continuum of these atoms due to inner-shell excitations and two-electron excitations. These discoveries further stimulated general interest in the use of synchrotron radiation.

The NBS group studied the angular and polarization distributions of the radiation from this synchrotron because of its potential application as a standard source for radiometry. Their results (Codling and Madden, 1965a) showed a very good agreement between experiment and theory for the angular and polarization distributions (see Fig. 7-3) and indicated that the two peaks in the angular distribution of the perpendicular component were mirror images of each other, reflected through the orbital plane. They also noted that the window through which the light passed was required to be quite strain-free or the polarization distributions would be distorted— an effect which may explain the earlier discrepancies.

5.   Establishment of the Frascati facility   At nearly the same time as the effort at NBS, a program was developed on the 1.15-BeV synchrotron at Frascati, Italy. This machine has a radius in the curved sections of 3.6 m; thus at full electron energy the peak of the spectral distribution of the instantaneous power (per unit wavelength) occurs at 5.5 A. In a joint effort, the Istituto di Sanita de Roma and the Laboratoire de Chimie Physique de Paris mounted first a single-crystal spectrometer and then a grazing-incidence grating spectrograph on the synchrotron, and studied thin-metal films in absorption. These new results, particularly the broad absorption structure in gold and bismuth seen in the 50- to 100-A region [Jaegle and Missoni (1966)], stimulated new theoretical efforts (Combet-Farnoux and Héno, 1967; Combet-Farnoux, 1967; Manson and Cooper, 1968) in the interpretation of soft x-ray absorption spectra.

Photoelectric measurements by the Frascati group also confirmed the agreement between theory and experiment for the angular and polarization distribution of synchrotron radiation (Missoni and Ruggiero, 1965).

6.   Establishment of the Tokyo facility   In 1962, activity was begun at the Institute for Nuclear Study in Tokyo to make available for spectroscopic studies the 750-MeV, 5-m-turning-radius electron synchrotron (updated to 1.3 MeV in 1965). A large group, designated INS-SOR (Institute for Nuclear Studies–Synchrotron Orbital Radiation), was formed combining a predominant interest in solid-state physics with other interests in gas spectroscopy and radiometry. Following an earlier report of their observations of the light (Sasaki, 1965), a large amount of data on the soft x-ray absorption of solids was published (Sagawa et al., 1966a; 1966b). In particular, they studied the metals Be, Al, Sb, Bi, and Al-Mg alloy, and the alkali halides KCl and NaCl. More recently they have also reported on gaseous absorption at photon energies above those reached by the NBS-SURF group (Nakamura et al., 1968; 1969).

7.   Establishment of the Hamburg facility   Interest in utilizing synchrotron radiation from the 6-BeV Deutsches Electronen-Synchrotron (DESY) in Hamburg, West

Germany, began in 1963. This was the highest energy machine that had yet been utilized for synchrotron radiation, and with a radius of 31.7 m in the curved sections, the instantaneous power distribution (per unit wavelength) peaks at 0.34 A, according to equation (7-11). Thus this machine provides radiation throughout the soft x-ray region, extending from 0.1 to beyond 1000 A. (DESY has now been increased to 7.5 BeV.) At DESY, the theoretical predictions were again carefully checked against observations, and excellent agreement was reported (Bathow et al., 1966). In particular, this was the first serious effort to check the wavelength distribution of synchrotron radiation, and their results showed no discrepancies with the predictions of the Schwinger theory (see also Lemke and Labs, 1967). A review of the properties of this machine, including a discussion of the experimental facility, has been published by Haensel and Kunz (1967). The first applications of this source were in the study of photoemission from aluminum near the plasma frequency (Steinmann and Skibowski, 1966), the study of metal films (Cu, Ag, Sn, Au, Bi) in absorption (Haensel et al., 1967; 1968a), and studies of the absorption spectra of alkali halides (Haensel et al., 1968b; 1968c). The experimental effort at DESY has been very strong, with many groups using the facility which has been expanded to allow many simultaneous experiments.

8. *Establishment of the Wisconsin facility* One other large-scale experimental effort utilizing synchrotron radiation was initiated in this period. In the early 1960s, a storage ring was constructed as a machine design study by an association of Midwestern universities (MURA) at Stoughton, Wisconsin. When this study was terminated, the storage ring was converted to a "synchrotron" radiation facility, beginning in about 1968. A storage ring has the advantage that the electron beam is continuously circulating, eliminating the "off" time of a synchrotron caused by the dead time in the ac cycling of the magnetic field and the time required to bring the electrons up to energy (the two effects taking typically 50 to 90 percent of the cycle time). Thus the duty factor f in equation (7-14) is unity for a storage ring. At Stoughton, several milli-amperes of current will circulate for several hours before the current has been sub-stantially reduced (due to electrons gradually scattering out of the beam via several mechanisms). Also, the only radiation hazard to the experiments and experimenters occurs when a new injection is made or an old beam is dumped. Thus the experiments can be conducted directly at the source, and the experimenters may attend to their apparatus.

The electrons in the Wisconsin storage ring have an energy of 240 MeV, with a turning radius of ~ 0.54 m. Thus, from equation (7-11), $\lambda_p \sim 92$ A, and this source has useful radiation above about 25 A. A further advantage of the storage ring, in principle, is that the circulating current can be increased by multiple injections, thereby increasing the total radiated power proportionately. However, as the current is increased, the beam lifetime is decreased—and if it is allowed to become too short, some of the advantages mentioned above will be nullified. In any event the Wisconsin storage ring is already a strong and useful source at a few milliamperes of circulating current.

The experimental program at Stoughton has grown quite rapidly. The initial interests were in solid-state physics (Fujita et al., 1969; Brown et al., 1970a; Rubloff, 1971); however, radiometric applications soon followed (Fairchild, 1970), and gas absorption spectroscopy was begun in 1971. Three experimental ports existed in 1971 and expansion was planned; thus many experiments can be run simultaneously. This facility has been operated solely for the synchrotron radiation experiments.

9.   Other facilities   A 340-MeV, 1.25-m-turning-radius synchrotron at Glasgow University, Scotland, was modified for synchrotron radiation studies and a few experiments performed in the late 1960s. Of particular interest was the accurate comparison of experiment and theory for the radiated power distribution in the 3500- to 6000-A spectral region (Key, 1970). Discrepancies for the relative wavelength distribution were found to be less than 1 percent, pointing up once again the potential of synchrotron radiation as a radiometric standard for the far-UV spectral region.

At this writing there is under construction at the Daresbury Nuclear Physics Laboratory, Cheshire, England, a synchrotron radiation laboratory (Marr and Munroe, 1971) on the 5-BeV, 20.8-m-turning-radius electron synchrotron (NINA). Tentative plans also exist to establish facilities of some nature at the Cambridge Electron Accelerator, Cambridge, Massachusetts (6 BeV, 26 m, the Stanford SPEAR storage ring (3 BeV, 12.7 m), and the 450-MeV storage ring at Orsay, France.

10.   Summary   Thus, synchrotron radiation as a source for soft x-ray physics has really come of age. The work of the 1960s has proved its importance and confirmed all aspects of the related electrodynamic theory. There is now great interest in its use, and we shall find a still larger fraction of soft x-ray physical research utilizing these sources in the future. This is quite appropriate since the higher current machines are many orders of magnitude stronger than conventional x-ray sources in the 100-to 200-A range (Parratt, 1959), and synchrotron radiation is the only truly satisfactory continuum source from 200 to 600 A. In fact, a 6-BeV machine is superior in available power to a conventional x-ray source down to below 1 A.

Table 7-1 gives a summary of the properties of the electron synchrotrons and storage rings which have been or are about to be used as soft x-ray sources. The important parameters for estimating the angular and wavelength distribution are given, although the number of electrons in the beam is only a rough approximation which requires constant updating.

B. Synchrotron instrumentation

Synchrotrons are sufficiently nonconventional light sources to have caused some evolution of standard soft x-ray instrumentation into configurations more suitable for this application. The effective source is a distribution of orbital tangent points, and from these points emanates a horizontal sheet of radiation of narrow, vertical angular divergence. The source points are fixed in space, and it is impossible to put a slit or optical element close to them.

1. Spectrometers Since the radiation is a continuum, a spectrometer is required for spectral isolation, which must be effectively coupled to the synchrotron. Due to the tangential nature of the radiation and the necessity of not perturbing the electron orbit or the magnetic field, the entrance slit of the instrument must be kept quite far from the source. (At NBS-SURF, which has only 0.83-m radius, the minimum distance is about 1.5 m.) For synchrotrons, shielding of the instrument, and indeed the entire experimental area, is a distinct advantage—in which case a further removal of the instrument from the tangent point is required. Thus, in general, instruments are located so far from the source of radiation that a light-collecting mirror is required to illuminate the spectrometer entrance slit.

The exception to this can be realized if the instrument is removed a sufficiently large distance from the source tangent point, so that the light incident on the grating is approximately parallel. A concave grating can then be used in a Wadsworth mounting (Skibowski and Steinmann, 1967), or a plane grating can be used if a telescope

Table 7-1 SUMMARY OF DATA ON MACHINES WITH ACTIVE OR PLANNED SYN-
CHROTRON RADIATION PROGRAMS

Machine	Maximum energy		Radius in curved sections, m	λ_p, A	n^*
Cambridge Synchrotron	6	BeV	26.0	0.28	6×10^{10}
Daresbury Synchrotron (NINA)	5	BeV	20.77	0.39	10^{11}
Deutsches Electronen-Synchrotron (DESY)	6	BeV†	31.7	0.34	7×10^{10}
Frascati Synchrotron	1.15	BeV	3.6	5.5	10^{10}
Glasgow Synchrotron	340	MeV	1.25	75	
National Bureau of Standards (NBS-SURF)	180	MeV	0.834	335	10^9
Orsay Storage Ring	450	MeV			
Stanford Storage Ring (SPEAR)	3	BeV	12.7	1.1	7×10^{11}
Tokyo Synchrotron	1.3	BeV	4.0	4.3	4×10^{10}
Wisconsin Storage Ring	240	MeV	0.54	92	10^9

* Estimates only. At one time probably good to a factor of 2, but require constant updating.
† Increased to 7.5 BeV.

mirror is placed after the grating to focus the monochromatic radiation on the exit slit. In either case the resolution is generally limited by the nonparallelism of the incident light, and high resolution is not achieved.

A monochromator of the latter type has been developed for use on the synchrotron at the INS-SOR facility (Miyake et al., 1969). The instrument is located far from the synchrotron, and the light is assumed sufficiently parallel, or, for higher resolution, an entrance slit can be located as near as possible to the source. This instrument is depicted in Fig. 7-6a. It consists of a plane grating followed by a spherical concave mirror which forms a monochromatic image at the fixed exit slit. The spectrum is scanned by simply rotating the grating about an axis through its pole parallel to the grating rulings. The telescope mirror remains fixed, as do the direction of illumination of the grating and the exit beam. The included angle between the incident and diffracted beams of this instrument is fixed and can be carefully selected so that the first order of the grating is highly reflected while the second order (of shorter wavelengths) occurring at the same angle is poorly reflected.

In a more elaborate version of the plane-grating case, a plane mirror can be used before the plane grating in addition to the focusing telescope mirror. This configuration, illustrated in Fig. 7-6b (Kunz, Haensel, and Sonntag, 1968), also allows fixed entrance and exit beam angles and a fixed exit slit. The instrument is scanned by rotating the grating about an axis through its pole parallel to the grating rulings, while the plane mirror is rotated and translated in proper phase relative to the grating rotation. This latter motion is designed to keep the grating at the theoretical blaze for all wavelengths. Some reduction in higher order contamination is also realized. The disadvantages of this instrument are a complex drive system and the three reflections which are inefficient at short wavelengths.

If a spectrometer cannot be removed sufficiently far from the synchrotron to assume parallel incident radiation, or if highest resolution is desired, an entrance slit must be used. In this case, a significant intensity advantage is realized by using a condenser mirror before the entrance slit *if* that mirror is used at grazing incidence (Madden et al., 1967). This arrangement is illustrated for a conventional spectrometer in Fig. 7-6c. When high resolution is desired, the entrance slit will be narrow, and the reduced image of the source tangent point on the entrance slit will overfill the width of the slit. Then, if the full vertical distribution of the radiation is intercepted by the collector mirror, it is only necessary to fill the grating with light to achieve the maximum possible intensity advantage. This condition, then, dictates the size of the collector mirror to be used. At NBS-SURF four such mirrors are used, directing the radiation to four different experimental areas. The beams all come from a single 3-in.-diameter tangent tube intercepting the synchrotron vacuum system. This is accomplished without compromising the amount of radiation being coupled into the individual experiments.

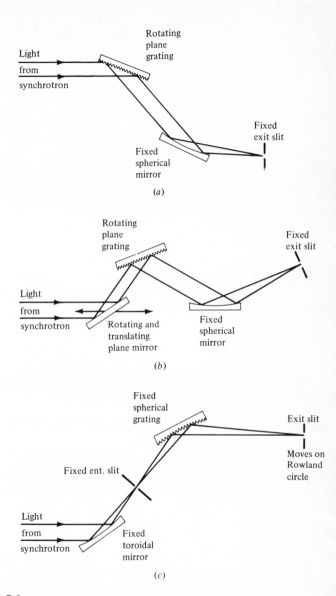

FIGURE 7-6
Diagrams of spectrometer optical systems which have been developed for use
with synchrotron sources. These are discussed in the text.

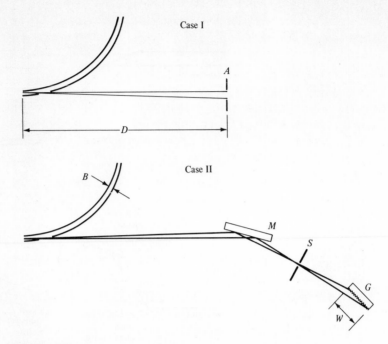

FIGURE 7-7
Two geometrical arrangements for optically coupling a spectrometer to a synchrotron radiation source. The geometrical coupling factor F is given for these two cases in the text.

2. Collection geometry factor It is of interest to consider the fraction of the radiated power which can be coupled into the optical systems mentioned above. This corresponds to evaluating the geometrical factor F appearing in equation (7-14). Two cases can be considered, as shown in Fig. 7-7. We assume that all the vertical distribution can be accepted by the experimental system. Then for case I, which represents a simple aperture of width A intercepting the radiation normal to the viewing direction,

$$F_I = \frac{A}{2\pi D} \qquad (7\text{-}15)$$

where D is the distance of the aperture from the nominal tangent point. Here A might be the width of an entrance slit of a conventional instrument used without a collector mirror, the projected width of the grating in the instrument depicted in Case I of Fig. 7-7, or the primary plane mirror in Case II of Fig. 7-7.

Case II represents the geometry of collection when a collector mirror is used to focus the synchrotron radiation onto the entrance slit of a spectrometer. If we assume that the entrance slit is narrow compared to the width of the image of the electron beam on the slit, and that the mirror is chosen to fill the grating of the spectrometer, then

$$F_{II} = \frac{S\mathscr{R}W}{2\pi BR} \qquad (7\text{-}16)$$

where S is the slit width, B the electron-beam width, R the radius of the concave grating, W the width of the grating, and $\mathscr{R}$ the effective reflectance of the collector mirror at the angle of incidence used. Notice that if the above conditions are met, the factor F_{II} does not depend on the distance of the collector mirror from the tangent point.

3. *Astigmatism* Astigmatism is generally a serious problem in grazing-incidence spectrometers. In general, a point at the entrance slit is monochromatically focused by a concave spherical grating into a long line normal to the direction of dispersion in the exit "image plane." We note that when a condenser mirror is used to focus the source tangent point onto the entrance slit, it is sufficient if the image is a line parallel to the entrance slit. Therefore, the focusing in the perpendicular plane can be adjusted to occur at a different image distance by using a mirror with different radii for the curvature in the vertical and horizontal planes. In particular, these radii can be chosen so that the source tangent point is brought to a line image parallel to the entrance slit at the entrance slit and a line image parallel to the dispersion direction in the exit "image plane." (This does not, of course, affect the focal properties of the spectrometer itself.) This focal condition will exist for only a single-image distance, hence only at one point along the Rowland circle. At this point spectrum "lines" are condensed to a very short length with a consequent advantage in detection. Particular advantage results if the spectrum is being photographed since this condensation of the spectrum will allow much shorter exposure times.

When independent focusing is desired in two orthogonal directions, the condenser mirror should have a toroidal surface. In the special case where both the vertical and horizontal images are desired at the entrance slit, as with a normal-incidence spectrometer, the mirror surface should be an ellipsoid. Unfortunately, these mirror types are difficult to manufacture and require dedicated hand-polishing to achieve high performance in the short-wavelength region. However, their usefulness in synchrotron spectroscopy cannot be overemphasized.

4. *Polarization* Since synchrotron radiation is strongly polarized with the electric vector in the plane of the orbit and mirrors have highest reflectance for the component of incident light which has the electric vector perpendicular to the plane of incidence, mirrors and gratings are used to some advantage if the planes of incidence are perpendicular to the orbital plane. This advantage is reduced at extreme grazing incidence, since the components of reflectance R_p and R_s approach equality. Thus the advantage in intensity when the plane of incidence is perpendicular to the orbital plane must be weighed against the limitation that many commercial spectrometer systems cannot function in that orientation.

A number of sophisticated solid-state experiments have now been devised which are utilizing synchrotron radiation, but in general the equipment is not specific to this source. Some of them, however, have sample and detector manipulators which can operate in either of two orientations rotated by 90° about the optic axis to take advantage of the polarization of the light.

5. *Absorption cells* Gas absorption spectroscopy requires that the light beam traverse an absorption chamber. Grazing-incidence spectrographs and mono-chromators can be constructed which themselves act as absorption cells (Madden, Ederer, and Codling, 1967), the path length of absorption being the sum of the entrance and exit arms. In such instruments the gas leak toward the synchrotron (which must be maintained at high vacuum) can be limited to the optically used portion of the entrance slit. Differential pumping can be used just outside of the entrance slit. In a monochromator a similar constraint, less easily achieved (Madden, Ederer, and Codling, 1967), can be placed on the gas leaking into the detector chamber.

For gases potentially harmful to the spectrometers, and for metal vapors, an absorption cell with thin metal, plastic, or carbon windows can be placed between the synchrotron and the spectrometers. These windows must be quite large to pass a solid angle of incident radiation which will fill the instruments used. Films 1000 A thick or less can be backed by an 80-percent-transmitting nickel mesh. In the case of the metal vapors, the cell can be heated at the center by rf coils and, for some metals, the vapor can be recirculated by the heat-pipe principle (Vidal and Cooper, 1969). A plug of buffer gas, argon or helium, at each end keeps the vapor from reaching the windows. This heat-pipe furnace has the advantage that the pressure and path length can be quite accurately determined (Ederer, Lucatorto, and Madden, 1971a).

6. *Gating* Radiation-detector electronic systems in use on synchrotrons generally use gates to keep the detectors turned off except for the most advantageous on-time. In this way noise spikes which occur when electrons are injected or dumped can be eliminated. The gate can be triggered and phased to the magnetic field. By adjusting the phase delay it is possible to accept radiation from the electrons at any time during their acceleration cycle—i.e., at any energy up to the maximum energy. One can thereby control the spectral distribution of the radiation [see equation (7-11)]. Mechanical rotating shutters have also been used.

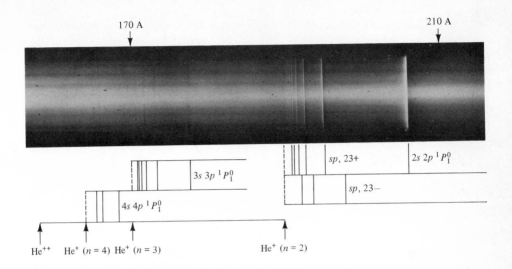

FIGURE 7-8

Absorption spectrum of helium showing the resonances in the photoionization continuum due to the existence of two-electron excitation states. The resonances are asymmetric, with white zones denoting decreased absorption and dark zones denoting increased absorption relative to the background absorption due to $1s^2 \rightarrow 1s\varepsilon p$. (*From Madden and Codling*, 1965.)

IV.   APPLICATIONS

Synchrotron spectroscopy has now developed so fully that we cannot here begin to cover all the experiments to which it has been applied. Hence we will restrict ourselves to only a few of the many interesting examples.

A.   Atomic

1.   *Autoionization in helium*   Probably the most interesting example of the application of synchrotron radiation to atomic physics was the very first experiment which was attempted. In 1963 the continuum nature of synchrotron radiation was utilized to obtain the first optical absorption spectrum of the autoionizing two-electron excitation states of helium (Madden and Codling, 1963; 1965). Helium gas was placed in a 3-m grazing-incidence spectrograph and exposed to the synchrotron radiation. A spectrogram thus obtained is reproduced in Fig. 7-8, showing Rydberg series of asymmetric resonance structures in the photoionization continuum above 60 eV.

To gain some understanding of this phenomenon, the energy-level diagram of

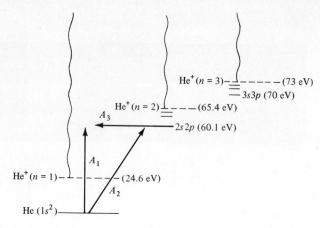

FIGURE 7-9
Schematic energy-level diagram of the two-electron excitation states in helium (not to scale).

helium shown in Fig. 7-9 is useful. We remember that the usual optically excited states of helium, the one-electron excitation states, all converge to the ionization limit at 24.6 eV. The structures we are observing in Fig. 7-8 are more than 35 eV above the ionization limit and are due to the simultaneous excitation of both electrons in helium. A strong series can be seen beginning with $2s2p$ and converging onto the first excited state of the helium ion. A higher Rydberg series can also be seen beginning with $3s3p$ and converging onto the second excited state of the ion.

  When these high-lying two-electron states are excited, the lifetime is very short, on the order of 10^{-13} to 10^{-14} sec. The decay is via autoionization, in which one of the electrons leaves the atom and the other drops back into the $n = 1$ state of He^+. The excess in energy is carried away by the ionized electron. For the two-electron states lying above the $n = 2$ state of He^+, the ion is left preferentially in the nearest lower state after autoionization.

  The resonances in Fig. 7-8 are obviously asymmetric, being comprised of adjacent dark and light zones. It must be remembered that the discrete structures shown here are actually superimposed upon a continuum of absorption due to the photoionization process $1s^2 \rightarrow 1s\varepsilon p$, beginning at 24.6 eV. Thus, the dark and light zones in Fig. 7-8 are to be interpreted as regions of absorption enhancement and absorption reduction ("windows") relative to the continuous absorption background.

  Such profiles suggest an interference effect, and indeed they can be so interpreted. In zero approximation, one can think of a matrix element A_1 associated with the photoionization transition $1s^2 \rightarrow 1s\varepsilon p$ as if the discrete structure did not exist, a matrix element A_2, associated with the discrete transition $1s^2 \rightarrow 2s2p$, ignoring the existence of the adjacent continuum, and a matrix element A_3 associated with the autoionization process. Of course we cannot long cling to the idea that the discrete state can be independent of the adjacent continuum states, and in constructing a better physical approximation to the system we find that the matrix elements A_1 and A_2 interfere with one another, giving rise to the asymmetric profiles observed. The physics of these resonance shapes has been treated in detail (Fano, 1961; Fano and Cooper, 1965), and the expression which describes the helium-resonance profile quantitatively is

$$\sigma = \sigma_B \frac{(q + \varepsilon)^2}{1 + \varepsilon^2} \qquad (7\text{-}14)$$

where σ and σ_B are the total cross section and background cross section (ignoring the resonances), respectively; ε is the photon energy in units of the half-width of the resonance measured relative to the "center" of the resonance $[\varepsilon = (E - E_0)/\frac{1}{2}\Gamma)]$; and q is the profile index, a nondimensional quantity involving the matrix elements A_1, A_2, and A_3. For the $2s2p$ resonance in Fig. 7-8, $q = -2.80$ and $\Gamma = 0.038$ eV.

Looking closely at the spectrogram of Fig. 7-8, we can see a weak second Rydberg series which converges to the limit He^+ ($n = 2$). The interpretation of this series and the explanation of the unusual notation in labeling the higher Rydberg series members is an interesting story in its own right (Burke and McVicar, 1965; Cooper, Fano, and Prats, 1963) which is left for the interested reader to pursue.

The discovery of these resonant structures in the photoionization continuum of helium precipitated a great deal of activity among theoreticians to develop the mechanics for dealing with such phenomena and to achieve quantitative agreement with the experimental observations (Burke, 1965; Smith, 1966). It also stimulated increased activity experimentally to find resonant structures in other atoms, both optically and by electron and other charged-particle scattering (Smith, 1969; Rudd and Smith, 1968).

Asymmetric resonances due to inner-shell electron excitation and to two-electron excitation were also discovered in neon (Madden and Codling, 1963; Codling, Madden, and Ederer, 1967) and the other noble gases. It should be mentioned that when q is very small, equation (7-14) allows, as a special case, a profile which has an absorption-reduction zone but not an absorption-enhancement zone. Such "window" features have in fact been found in argon (Madden, Ederer, and Codling, 1969), krypton, and xenon (Madden and Codling, 1964; Codling and Madden, 1971; Ederer, 1971).

2. *Autoionization in lithium* Lithium, the simplest atom following helium, in which two-electron excitation is possible, has also been studied (Ederer, Lucatorto, and Madden, 1970, 1971a). In this case highly excited single-electron excitation states are also possible. That is, one can look for K electron transitions of the type

$1s^2 2s \rightarrow 1s2snl$ as well as $1s^2 2s \rightarrow 1snln'l'$. Generally, one-electron excitations of the former type are one to two orders of magnitude stronger than the two-electron excitations.

To observe these high-lying transitions in lithium, a vapor furnace with aluminum end windows operating on the dynamic principles of a heat pipe was constructed at NBS-SURF (see Subsection III B 5). This furnace absorption cell was placed between the synchrotron and the spectrograph, and the lithium absorption spectrum was photographed between 50 and 70 eV (250 to 170 A). The result obtained by Ederer et al. is shown in Fig. 7-10. In this spectrum, the strongest features are the one-electron excitations $1s^2 2s \rightarrow 1s2snp$. Two series are observed, leading to the $1s2s\ ^3S$ and the $1s2s\ ^1S$ states of the ion. The measured positions of many of these absorption features are in excellent agreement with first principles Hartree-Fock calculations (Weiss, 1970). A clearly observed two-electron excitation series is also observed converging on the $1s2p\ ^1P$ state of the ion.

In this region of the spectrum, the photoionization continuum $1s^2 2s \rightarrow 1s^2 \varepsilon p$ is weak, and the resonance features have a high value of q [see equation (7-14)]; therefore, they appear simply as absorption features superimposed upon the continuous absorption. However, the various K excitation configurations interact with one another and with the $(1s2s\ ^3S)\varepsilon p$ continuum above 65 eV. Therefore the understanding of the higher lying spectra requires detailed multiconfiguration calculations, which have only just been initiated (Cooper et al., 1970). This discrete spectrum in lithium has recently been extended to still higher photon energies by replacing the aluminum furnace windows with plastic ones.

Resonances of this type can be expected for other metals. Sodium (Comerade, Garton, and Mansfield, 1971) and magnesium (Newsom, 1971) have already been studied with pulsed discharge continuum sources, and the observations on these metals have been extended (Ederer et al., 1971b) using synchrotron radiation. In fact, it is now clear that all atoms will exhibit many resonant features in the photoionization continuum. These features can be highly visible due to the interference phenomenon, even though the oscillator strength associated with the excitations may not be large.

B.   Molecular

What may be the most important application to the study of molecules, namely, photoelectron spectroscopy, is yet to get underway using synchrotron radiation. Significant measurements of the absorption cross section of molecules using synchrotron radiation were only just beginning in 1971. However, some progress has been made in the investigation of discrete structural details in the photoionization continua of molecules. In the 500-to 600-A region, discrete structures have been found which are caused by the excitation of inner-orbital electrons in the diatomic molecules

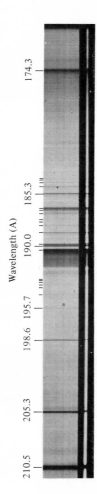

K absorption spectrum of lithium

Wavelength (A)

FIGURE 7-10

Absorption spectrum of lithium vapor in the spectral region 175 to 210 A. The discrete structures are due to transitions from the ground state to one- and two-electron excitation states involving always a *K* electron. Black denotes absorption. (*From Ederer et al., 1971b.*)

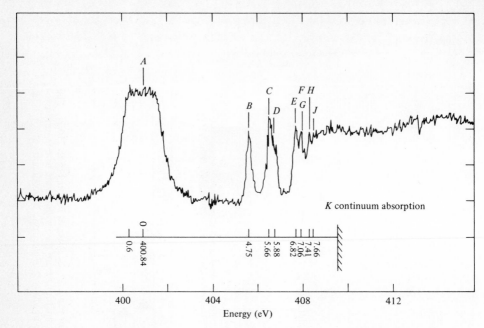

FIGURE 7-11
Densitometer trace of a photographed absorption spectrum of gaseous N_2 in the 400-eV region. The discrete structure is due to the excitation of a K electron to outer orbitals. (*From Nakamura et al.*, 1969.)

oxygen [Codling and Madden (1965*b*)], nitrogen (Codling, 1966*a*), and carbon monoxide.

The Tokyo synchrotron has been used to study molecular absorption at higher photon energies, establishing K and L ionization energies and allowing analysis of the structure below the edges due to the inner electron being excited to outer orbitals. An interesting example is the K edge of N_2 observed in absorption near 30 A by Nakamura et al. (1969). Their results are reproduced in Fig. 7-11, which is a densitometer trace of a spectrograph plate obtained with a 2-m grazing-incidence spectrograph having an optical slit width of about 0.3 A. Nitrogen gas was allowed into the spectrograph to a pressure of 1.5 torr. Oxygen, used as a filter, was also present at a pressure of 2 to 4 torr.

The ground state for N_2 has the electron configuration

$$(\sigma_g 1s)^2 (\sigma_u 1s)^2 (\sigma_g 2s)^2 (\sigma_u 2s)^2 (\Pi_u 2p)^4 (\sigma_g 2p)^2 \; ^1\Sigma_g{}^+$$

When one of the K electrons is excited to an outer-shell orbital, it is as if an electron has been added to the outer configuration. Therefore, the outer-shell orbitals are nearly the same as those of the NO molecule and should have nearly the same relative energy levels. With this assumption, the absorption features in Fig. 7-11 have been tentatively assigned as transitions to the following states:

$$A: \quad (\sigma_u 1s)^{-1}(\Pi_g 2p) \; {}^1\Pi_u$$

$$B: \quad (\sigma_u 1s)^{-1} 3s\sigma_g \; {}^1\Sigma_u{}^+$$

$$C \text{ and } D: \begin{cases} (\sigma_g 1s)^{-1} 3p\Pi_u \; {}^1\Pi_u \\ (\sigma_g 1s)^{-1} 3p\sigma_u \; {}^1\Sigma_u{}^+ \end{cases}$$

$$E, F, \text{ and } G: \begin{cases} (\sigma_u 1s)^{-1} 4s\sigma_g \; {}^1\Sigma_u{}^+ \\ (\sigma_u 1s)^{-1} 3d\Pi_g \; {}^1\Pi_u \\ (\sigma_u 1s)^{-1} 3d\sigma_g \; {}^1\Sigma_u{}^+ \end{cases}$$

$$H \text{ and } I: \quad (\sigma_g 1s)^{-1} 4p\Pi_u \; {}^1\Pi_u$$

$$\text{and/or} \quad (\sigma_g 1s)^{-1} 4p\sigma_u \; {}^1\Sigma_u{}^+$$

State A corresponds to the ground state of the NO molecule, while B corresponds to the lowest Rydberg state of NO, namely, the $A^2\Sigma^+$ state. Similarly C and D correspond to the $C^2\Pi$ and $D^2\Sigma^+$ states of NO. The energy differences in these spectra have been found to correspond well to those in NO. From the energy positions of these structures in N_2 and the ionization energy of NO, the K ionization energy for N_2 was determined to be 409.5 ± 0.1 eV.

Discrete structure has also been observed in the polyatomic molecule SF_6 in the 500-to 600-A region (Codling, 1966b). At this writing further studies of molecular absorption are underway at the Wisconsin storage ring and at DESY.

C.   Solids

Most of the experiments performed thus far with synchrotron radiation have been on solids. Transmission of thin films has been determined at room temperature and at low temperatures. Reflectance of films and bulk solids has been measured, including solid noble gases. Photoemission from films and bulk solids has been observed as a function of angle of incidence and polarization. Many metals have been studied, a few alloys, semiconductors, and, of course, the alkali halides. Most of the effort has been directed toward explaining the structure in the absorption coefficient for these various materials.

1.   *Excitons or density of states?* Valence-electron excitations, core-electron excitations to exciton levels or to the conduction band, and possible multiple excitations in the alkali halides and other ionic crystals have been investigated (Sagawa et al., 1966a; Haensel et al., 1968b, 1968c; Fujita, Gähwiller, and Brown, 1969; Brown et al., 1970a; and Rubloff et al., 1971). Haensel et al. (1969a; 1969b) have studied the

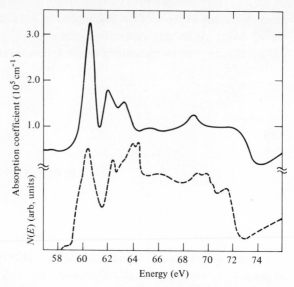

FIGURE 7-12
The upper curve shows the observed absorption coefficient for LiCl in the energy region 60 to 75 eV, where the Li$^+$ K electron is being excited. The lower dotted curve shows the theoretical density of states for the LiCl conduction band with the first peak aligned for best agreement. (*From Brown et al.*, 1970*b*.)

solid noble gases in reflection and compared the results with gaseous absorption spectra. From all this work it appears that considerable caution is required in making assignments of structure in the soft x-ray region. Often it is difficult to distinguish exciton and atomic effects from the structure due to the variation of the density of states in the conduction-band continuum. For example, the existence of two Rydberg series $[\Gamma_{\frac{3}{2},\frac{1}{2}}]$ of Wannier-exciton levels in the 3p valence-electron excitation spectrum of argon in the 11- to 14-eV spectral region has been very well established (Haensel et al., 1969*b*; Keitel, 1970). However, in spite of speculation in the literature regarding probable excitons near the Li$^+$ K edge of LiCl, new calculations (Kunz, 1969; Brown et al., 1970*b*) indicate that the observed structure can be quite satisfactorily explained as due only to the variation in the density of states in the conduction band. Figure 7-12 indicates how well the absorption spectrum of LiCl observed by Brown et al. (1970*b*) compares with the computed density of states. Apparently, the Li$^+$ K structure in LiBr can also be explained in terms of the density of states in the conduction band.

2. *Oscillator-strength effects: centrifugal barrier* The importance of considering the atomic behavior of the core ions in interpreting soft x-ray structure of solids has been

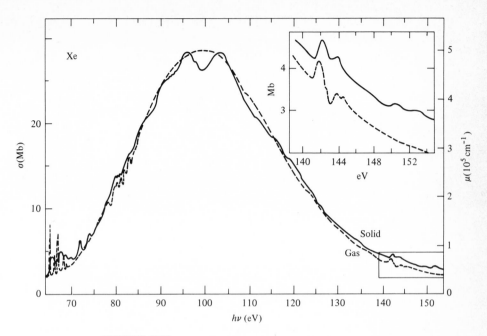

FIGURE 7-13
Photoabsorption cross section plotted vs. photon energy for solid xenon at 55°K (solid curve) and gaseous xenon (dashed curve) in the region of the $4d \to \varepsilon f$ transitions. (*From Haensel et al.*, 1969a.)

well demonstrated in the studies on metals and also in the comparison of the spectra of the noble gases in solid and gaseous form. As an example, let us consider the case of the excitation of a $4d$ electron in atoms. The elements $Z = 46$ (Pd) to $Z = 56$ (Ba) have completed $4d$ shells and open $4f$ shells; hence, at first inspection, we might expect a series of transitions $4d \to nf$ lying below the ionization energy for the $4d$ electrons. It has been experimentally shown for xenon ($Z = 54$) that this is not the case and that only transitions of the type $4d \to np$ appear below the limit (Codling and Madden, 1964). It has been shown both theoretically (Cooper, 1964) and experimentally (Ederer, 1964) that the oscillator strength for the $4d \to \varepsilon f$ transitions does not become large until the electron leaves the atom with a kinetic energy of about 20 eV. This is because the $4d$ wavefunction lies close to the nucleus and the nf wavefunctions lie far out. Thus there is very little overlap between the two, and a transition between these states has very low probability. A significant overlap occurs only for continuum f-type wavefunctions which describe states in which the electron has significant kinetic energy as it leaves. Alternatively, the effect can be described in terms of a centrifugal barrier which the electron must overcome if it has an angular momentum corresponding to $l = 3$.

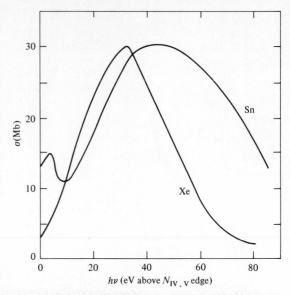

FIGURE 7-14

Photoabsorption cross section plotted vs. photon energy above the $N_{IV,V}$ edge for gaseous xenon (*from Ederer*, 1964) and for solid tin from data of Codling *et al.* (1966) in the region of the $4d \rightarrow \varepsilon f$ transitions for xenon. These cross sections are absolute, and no arbitrary normalization has been used.

Haensel et al. (1969*a*) have now shown that the overall structure of this oscillator-strength distribution in gaseous Xe is remarkably well reproduced in solid Xe. The comparison is given in Fig. 7-13. Figure 7-14 shows that solid Sn ($Z = 50$) also has been found (Codling et al., 1966) to have a structure similar to that of xenon, if the absorption spectra are plotted in energy relative to the $N_{IV,V}$ edge. Similar results suggestive of large centrifugal barriers have since been observed in other solid metals in the $Z = 46$ to 56 group, all of which have xenon-like ion cores. It has now been shown theoretically (Fano and Cooper, 1968) that these metals as free atoms should exhibit the centrifugal-barrier effect. Thus, we must conclude that this unusual distribution of oscillator strength observed in the free atoms carries over substantially unchanged to the $4d$ electron excitation to f-symmetry bands in the solid.

3. *Oscillator-strength effects in the lanthanide rare earths* The more complicated situation for the lanthanide rare earths $Z = 57$ to $Z = 71$ has been discussed by Dehmer et al. (1971) and Sugar (1971). For these elements the $4f$ shell is filling, and, significantly, this electron shell has been brought inside the potential barrier mentioned

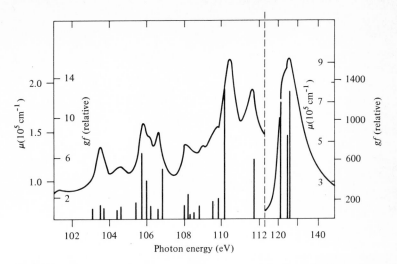

FIGURE 7-15
The measured photoabsorption cross section of cerium is plotted vs. energy as the solid curve. The calculated relative positions and line strengths of the $4d^{10}4f^N \rightarrow 4d^94f^2$ transitions are indicated by the vertical solid lines. (*From Sugar*, 1972.)

above, while electrons in higher nf shells would remain outside. Now the transition $4d^{10}4f^N \rightarrow 4d^94f^{N+1}$ becomes possible, but transitions $4d \rightarrow 5f$ and higher are prevented by the barrier. Due to a strong exchange interaction between the f electrons and the d vacancy, the excited states corresponding to $4d^94f^{N+1}$ are spread over a broad energy range, ~ 20 eV, and most of the oscillator strength is found in transitions to the higher $4f^{N+1}$ levels which lie above the ionization threshold for the $4d$ electron. Hence, they autoionize, mainly into the $4d^94f^N\varepsilon f$ continua. Consequently, the $4d$ absorption spectra of these ions is composed of a broad peak 5 to 20 eV above threshold with variable structure superimposed.

Recent developments in combining spectroscopic theory and the theory of multilevel, multichannel autoionization have resulted in the capability to calculate the expected position and strength of the $4d^{10}4f^N \rightarrow 4d^94f^{N+1}$ transitions in the free ions. Figure 7-15 shows the results of calculating the expected discrete spectrum, including intensities, for the case of triply-ionized cerium ($Z = 58$) (Dehmer et al., 1971; Sugar, 1972). The solid curve is the structure observed in absorption of solid thin films of cerium (Haensel, Rabe, and Sonntag, 1970). This excellent agreement between the atomic predictions and the observations in the solids extends also to a number of the other rare-earth metals (Zimkina et al., 1967a, 1967b; Haensel, Rabe, and Sonntag, 1970).

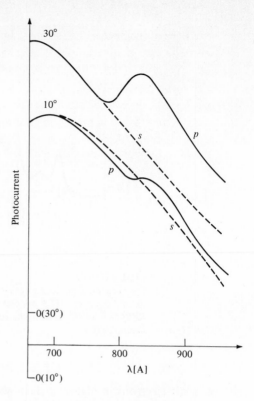

FIGURE 7-16
The photoemission current from a thin aluminum film is shown vs. wavelength in the region of the volume plasmon (827 A) for two orientations of incident plane-polarized radiation and two angles of incidence. Note the zero offsets for the different angles of incidence. (*From Steinmann and Skibowski, 1966.*)

4. *Volume plasmon decay in aluminum*   Another interesting application of synchrotron radiation in solid-state physics has been the study of photoemission from aluminum. The plasma frequency for aluminum is about 15 eV, and to excite it optically in a thin film requires a component of the electric field normal to the surface. In other words, it cannot be excited with s-polarized incident radiation, but only with p-polarized incident radiation at nonnormal incidence. If p-polarized light falls on aluminum at nonnormal incidence, an increase in absorption is expected near the plasma frequency. This energy is well above the surface-potential barrier for photoemission. The question then arises: Will the photoemission from aluminum show an equivalent increase near the plasma frequency? In general, photoemission yields are proportional to the absorption near the surface for homogeneous materials. But if the photoemission increases linearly with the increase in absorption caused by the onset of a plasmon oscillation, the implication would be that the energy involved in the collective oscillation of many electrons can, by some mechanism, be transferred to a single electron, allowing it to escape over the surface-potential barrier.

Steinmann and Skibowski (1966) capitalized on the continuum and polarization properties of synchrotron radiation to measure the photoemission yield of aluminum near the plasma frequency. Figure 7-16 shows their result for s and p polarization at two angles of incidence. The increase in photoemission for the *p* component at the plasma frequency is quite apparent, and the increase becomes greater as the angle of incidence changes from 10 to 30°. Subsequently, it was shown (Skibowski et al., 1968) that the increase could be quantitatively related to the increase in absorption just below the surface of the aluminum layer. Hence we must conclude that the volume plasmons in aluminum can decay in a manner which allows the collective energy to be transferred to a single electron, as it was already shown to be able to decay via emission of a single photon (Herickhoff et al., 1965).

D. Radiometry

1. *Source calibrations* Since the power radiated in the VUV by an orbiting electron can be accurately calculated, the total flux coming from a synchrotron at any wavelength can be determined if the number of electrons is known. Pitz (1969) devised a double-monochromator system to carry out a calibration of VUV sources. One monochromator looked at the visible radiation from the synchrotron and compared it to a tungsten standard lamp. This comparison allowed the synchrotron radiation in the visible region to be measured absolutely. Since the Schwinger theory tells us the power radiated by a single electron, it was possible from this measurement to calculate the number of electrons present in the beam. From this, the radiated power as a function of wavelength in the VUV could be calculated. Pitz's second monochromator operated in the VUV and compared the synchrotron radiation to a VUV source of unknown output. Performing these measurements simultaneously in the visible and VUV with known geometrical relationships, he was able to absolutely calibrate D_2 and Hg transfer standard sources with an accuracy of about 5 percent in the region 1650 to 2700 A.

2. *Detector calibrations* A particularly esthetic radiometry-calibration scheme is being used by Fairchild (1970) on the Wisconsin storage ring to calibrate windowed photomultipliers in the region down to 1950 A. He begins his measurement when the current in the storage ring has dropped to the order of 100 electrons in the beam. He records the signal from his photomultiplier-filter combination which has been placed in the radiation beam with a known collection geometry. When the measurement is completed, the electron beam is allowed to diminish slowly as the remaining electrons are scattered. A close observation of the detector current during this scattering process reveals that the loss of each electron is accompanied by a sudden discrete drop in the radiation received by the detector. In other words, *the radiation from a single electron is a readily measurable quantity.* By following these discrete steps down until the last electron is lost, the number of electrons in the beam at the time of the calibration can be counted exactly. Then the absolute radiated flux which had been intercepted by the photomultiplier can be calculated.

At this writing no synchrotron has yet been used for calibrations at photon energies above ~10 eV. However, there appears to be no inherent obstacle to accomplishing this goal.

REFERENCES

BATHOW, G., FREYTAG, E., and HAENSEL, R. (1966): Measurement of synchrotron radiation in the x-ray region, *J. Appl. Phys.*, **37**: 3449–3454.

BROWN, F. C., GÄHWILLER, C., FUJITA, H., KUNZ, A., SCHEIFLEY, W., and CARRERA, N. (1970*a*): Extreme-ultraviolet spectra of ionic crystals, *Phys. Rev.*, **B2**: 2126–2138.

———, ———, KUNZ, A. B., and LIPARI, N. O. (1970*b*): Soft x-ray spectra of the lithium halides and their interpretation, *Phys. Rev. Lett.*, **25**: 927–930.

BURKE, P. G. (1965): Resonance in electron scattering and photon absorption, *Advan. Phys.*, **14**: 521–567.

——— and MCVICAR, D. D. (1965): Resonances in $e^- - $ He$^+$ scattering and the photo-ionization of He, *Proc. Phys. Soc. (London)*, **86**: 989–1006.

CODLING, K. (1966*a*): Structure in the photoionization continuum of N$_2$ near 500 A, *Astrophys. J.*, **143**: 552–558.

——— (1966*b*): Structure in the photoionization continuum of SF$_6$ below 630 A, *J. Chem. Phys.*, **44**: 4401–4402.

——— and MADDEN, R. P. (1964): Optically observed inner-shell electron excitation in neutral Kr and Xe, *Phys. Rev. Lett.*, **12**: 106–108.

——— and ——— (1965*a*): Characteristics of the "synchrotron light" from the NBS 180-MeV machine, *J. Appl. Phys.*, **36**: 380–387.

——— and ——— (1965*b*): New Rydberg series in molecular oxygen near 500 A, *J. Chem. Phys.*, **42**: 3935–3938.

——— and ——— (1971): Resonances in the photoionization continuum of Kr and Xe, *Phys. Rev.*, **A4**: 2261–2263.

———, ———, and EDERER, D. L. (1967): Resonances in the photoionization continuum of Ne-I (20–150 eV), *Phys. Rev.*, **155**: 26–37.

———, ———, HUNTER, W. R., and ANGEL, D. W. (1966): Transmittance of tin films in the far ultraviolet, *J. Opt. Soc. Am.*, **56**: 189–192.

COMBET-FARNOUX, F. (1967): Theoretical study of the variation of the cross sections for photoionization of atoms in a central potential model, *Compt. Rend.*, **B264**: 1728–1731.

——— and HÉNO, Y. (1967): Calculation of the cross section for photoionization of the atoms gold and bismuth in the soft x-ray region, *Compt. Rend.*, **B264**: 138–141.

COMERADE, J. P., GARTON, W. R. S., and MANSFIELD, M. W. D. (1971): Absorption spectrum of Na-I in the vacuum ultraviolet, *Astrophys. J.*, **165**: 203–212.

COOPER, J. W. (1964): Interaction of maxima in the absorption of soft x rays, *Phys. Rev. Lett.*, **13**: 762–764.

———, CONNEELY, M. J., SMITH, K., and ORMONDE, S. (1970): Resonant structure of lithium between the 2^3S and 2^1P thresholds, *Phys. Rev. Lett.*, **25**: 1540–1543.

COOPER, J. W., FANO, U., and PRATS, F. (1963): Classification of two-electron excitation levels of helium, *Phys. Rev. Lett.*, **10**: 518–521.

DEHMER, J. L., STARACE, A. F., FANO, U., SUGAR, J., and COOPER, J. W. (1971): Raising of discrete levels into the far continuum, *Phys. Rev. Lett.*, **26**: 1521–1525.

EDERER, D. L. (1964): Photoionization of the 4d electrons in xenon, *Phys. Rev. Lett.*, **13**: 760–762.

—— (1971): Cross section profiles of resonances in the photoionization continuum of krypton and xenon (600–400 A), *Phys. Rev.*, **A4**: 2263–2270.

——, LUCATORTO, T. B., and MADDEN, R. P. (1970): Autoionization spectra of lithium, *Phys. Rev. Lett.*, **25**: 1537–1540.

——, ——, and —— (1971a): Resonances in the photoionization continuum of lithium I (55 to 70 eV), *Proc. Colloq. CNRS, Paris*, no. 196.

——, ——, and —— (1971b): Study of autoionizing spectra in metallic vapors, *Proc. Third Intern. Conf. Vac. Ultraviolet Radiation Phys. Tokyo*.

ELDER, F. R., GUREWITSCH, A. M., LANGMUIR, R. V., and POLLOCK, H. C. (1947): Radiation from electrons in a synchrotron, *Phys. Rev.*, **71**: 829–830.

——, LANGMUIR, R. V., and POLLOCK, H. C. (1948): Radiation from electrons accelerated in a synchrotron, *Phys. Rev.*, **74**: 52–56.

FAIRCHILD, T. (1970): Use of synchrotron radiation from an electron storage ring as an absolute standard of radiant flux for wavelengths from 1100–3000 A, in R. LÜST and F. LABUHN (eds.), *New techniques in space astronomy*, Proc. IAU Symposium No. 41, Garching, Aug. 1970.

FANO, U. (1961): Effects of configuration interaction on intensities and phase shifts, *Phys. Rev.*, **124**: 1866–1878.

—— and COOPER, J. W. (1965): Line profiles in the far-UV absorption spectra of the rare gases, *Phys. Rev.*, **A137**: 1364–1379.

—— and —— (1968): Spectral distribution of atomic oscillator strengths, *Rev. Mod. Phys.*, **40**: 441–507.

FUJITA, H., GÄHWILLER, C., and BROWN, F. C. (1969): Far-ultraviolet spectra due to 4d electrons in the alkali iodides, *Phys. Rev. Lett.*, **22**: 1369–1371.

GODWIN, R. P. (1969): Synchrotron radiation as a light source, in G. Höhler (ed.), *Springer tracts in modern physics*, vol. 51, Springer-Verlag, Berlin.

HAENSEL, R., and KUNZ, C. (1967): Experiments with synchrotron radiation, *Z. Angew. Phys.*, **23**: 276–295.

——, ——, and SONNTAG, B. (1967): Absorption measurements of copper, silver, tin, gold and bismuth in the far ultraviolet, *Phys. Lett.*, **A25**: 205–206.

——, ——, SASAKI, T., and SONNTAG, B. (1968a): Absorption measurements of copper, silver, tin, gold, and bismuth in the far ultraviolet, *Appl. Opt.*, **7**: 301–306.

——, ——, and SONNTAG, B. (1968b): Measurement of photoabsorption of the lithium halides near the lithium K edge, *Phys. Rev. Lett.*, **20**: 262–264.

——, SASAKI, T., and SONNTAG, B. (1968c): Measurement of photoabsorption of the sodium halides near the sodium $L_{2,3}$ edge, *Phys. Rev. Lett.*, **20**: 1436–1438.

HAENSEL, R., KEITEL, G., SCHREIBER, P., and KUNZ, C. (1969a): Optical absorption of solid krypton and xenon in the far ultraviolet, *Phys. Rev.*, **188**: 1375–1380.

——, ——, KOCH, E. E., SKIBOWSKI, M., and SCHREIBER, P. (1969b): Reflection spectrum of solid argon in the vacuum ultraviolet, *Phys. Rev. Lett.*, **23**: 1160–1162.

——, RABE, P., and SONNTAG, B. (1970): Optical absorption of cerium, cerium oxide, praseodymium, praseodymium oxide, neodymium, neodymium oxide and samarium in the extreme ultraviolet, *Solid State Comm.*, **8**: 1845–1848.

HERICKHOFF, R. J., HANSON, W. F., ARAKAWA, E. T., and BIRKHOFF, R. D. (1965): Angular distribution and thickness dependence of transition radiation from thin aluminum foils, *Phys. Rev.*, **A139**: 1455–1458.

IVANENKO, D., and POMERANCHUK, I. (1944): On the maximal energy attainable in a betatron, *Dokl. Akad. Nauk (SSSR) (in Russian)*, **44**: 315–316; *Phys. Rev.*, **65**: 343.

—— and SOKOLOV, A. (1948): On the theory of the radiating electron, *Dokl. Akad. Nauk (USSR)*, **59**: 1551–1557.

JAEGLE, P., and MISSONI, G. (1966): Bulk absorption coefficient of gold in the wavelength region 26–120 A, *Compt. Rend., Acad. Sci. Paris*, **262**: 71–74.

JOOS, P. (1960): Measurements of the polarization of synchrotron radiation, *Phys. Rev. Lett.*, **4**: 558–559.

KEITEL, G. (1970): The optical properties of solid neon and argon in the photon energy region between 10 eV and 500 eV, DESY Internal Report F41-70/7.

KEY, P. J. (1970): Synchrotron radiation as a standard of spectral emission, *Metrologia*, **6**: 97–103.

KOROLEV, F. A., MARKOV, V. S., AKIMOV, E. M., and KULIKOV, O. F. (1956): Experimental investigations of the angular distribution and polarization of optical radiation from electrons in a synchrotron, *Dokl. Akad. Nauk (SSSR) (in Russian)*, **110**: 542–545.

KUNZ, A. B. (1969): Localized orbitals in polyatomic systems, *Phys. Stat. Sol.*, **36**: 301–309.

KUNZ, C., HAENSEL, R., and SONNTAG, B. (1968): Grazing-incidence vacuum-ultraviolet monochromator with fixed exit slit for use with distant sources, *J. Opt. Soc. Am.*, **58**: 1415.

LEMKE, D., and LABS, D. (1967): The synchrotron radiation of the 6-GeV DESY machine as a fundamental radiometric standard, *Appl. Opt.*, **6**: 1043–1048.

LIENARD, A. (1898): Electric and magnetic field produced when a point electric charge is put into motion arbitrarily, *L'Éclairage Élec.*, **16**: 5–14.

MADDEN, R. P., and CODLING, K. (1963): New autoionization atomic energy levels in He, Ne, and Ar, *Phys. Rev. Lett.*, **10**: 516–518.

—— and —— (1964): Recently discovered auto ionizing states of krypton and xenon in the λ 380–600 A region, *J. Opt. Soc. Am.*, **54**: 268–269.

—— and —— (1965): Two-electron excitation states in helium, *Astrophys. J.*, **141**: 364–375.

——, EDERER, D. L., and CODLING, K. (1967): Instrumental aspects of synchrotron xuv spectroscopy, *Appl. Opt.*, **6**: 31–38.

——, ——, and —— (1969): Resonances in the photoionization continuum of Ar I (20–150 eV), *Phys. Rev.*, **177**: 136–151.

MANSON, S. T., and COOPER, J. W. (1968): Photoionization in the soft x-ray range: Z dependence in a central-potential model, *Phys. Rev.*, **165**: 126–138.

MARR, G., and MUNROE, I. (1971): Light from electrons, *New Scientists and Science J.*, Feb. 4, 266–268.

MISSONI, G., and RUGGIERO, A. (1965): Characteristics of the radiation from the 1000-MeV synchrotron, *Lincei-Rend. Sc. fis, mat. e nat.*, **38**: 677–685.

MIYAKE, K. P., KATO, R., and YAMOSHITA, H. (1969): A new mounting of soft x-ray monochromator for synchrotron orbital radiation, *Sci. Light (Tokyo)*, **18**: 39–56.

NAKAMURA, M., SASANUMA, M., SATO, S., WATANABE, M., YAMASHITA, H., IGUCHI, Y., EJIRI, A., NAKAI, S., YAMAGUCHI, S., SAGAWA, T., NAKAI, Y., and OSHIO, T. (1968): Absorption structure near the $L_{2,3}$ edge of argon gas, *Phys. Rev. Lett.*, **21**: 1303–1305.

Ibid. (1969): Absorption structure near the K edge of the nitrogen molecule, *Phys. Rev.*, **178**: 80–82.

NEWSOM, G. H. (1971): Inner-shell absorption in the spectra of the alkaline earths. I: magnesium (Mg I), *Astrophys. J.*, **166**: 243–247.

PARRATT, L. G. (1959): Use of synchrotron orbit-radiation in x-ray physics, *Rev. Sci. Instr.*, **30**: 297–299.

PITZ, E. (1969): Absolute calibration of light sources in the vacuum ultraviolet by means of the synchrotron radiation of DESY, *Appl. Opt.*, **8**: 255–259.

RUBLOFF, G. W. (1971): Normal-incidence reflectance, optical properties, and electronic structure of Zn, *Phys. Rev.*, **B3**: 285–292.

———, FREEOUF, J., FRITZSCHE, H., and MURASE, K. (1971): Far ultraviolet reflectance spectra of ionic crystals, *Phys. Rev. Lett.*, **26**: 1317–1320.

RUDD, M. E., and SMITH, K. (1968): Energy spectra of auto-ionizing electrons in oxygen, *Phys. Rev.*, **169**: 79–84 and references therein.

SAGAWA, T., IGUCHI, Y., SASANUMA, M., NASU, T., YAMAGUCHI, S., FUJIWARA, S., NAKAMURA, M., EJIRI, A., MASUOKA, T., SASAKI, T., and OSHIO, T. (1966a): Soft x-ray absorption spectra of alkali halides. I: KCl and NaCl, *J. Phys. Soc. Japan*, **21**: 2587–2598.

———, ———, ———, EJIRI, A., FUJIWARA, S., YOKOTA, M., YAMAGUCHI, S., NAKAMURA, M., SASAKI, T., and OSHIO, T. (1966b): Soft x-ray absorption spectra of metals and alloys. I: Be, Al, Sb, Bi and Al-Mg alloys, *J. Phys. Soc. Japan*, **21**: 2602–2610.

SASAKI, T. (1965): Radiation from the electron synchrotron, *Oyo Buturi (Japanese J. Appl. Phys.)*, **34**: 231–234.

SCHOTT, G. A. (1907): On the radiation from groups of electrons, *Ann. Phys. (Leipzig)*, **24**: 635–660.

——— (1912): *Electromagnetic radiation*, chaps. 7 and 8, Cambridge University Press, London.

SCHWINGER, J. (1946): Electron radiation in high energy accelerators, *Phys. Rev.*, **70**: 798–799. See also reference to his calculations in E. M. McMillan (1945): Radiation from a group of electrons moving in a circular orbit, *Phys. Rev.*, **68**: 144–145; J. P. Blewett (1946): Radiation losses in the induction electron accelerator, *Phys. Rev.*, **69**: 87–95; L. I. Schiff (1946): Production of particle energies beyond 200 MeV, *Rev. Sci. Instr.*, **17**: 6–14.

SCHWINGER, J. (1949): On the classical radiation of accelerated electrons, *Phys. Rev.*, **75**: 1912–1925.

SKIBOWSKI, M., and STEINMANN, W. (1967): Normal-incidence monochromator for the vacuum ultraviolet radiation from an electron synchrotron, *J. Opt. Soc. Am.*, **57**: 112–113.

———, FEUERBACHER, B., STEINMANN, W., and GODWIN, R. P. (1968): Investigations of aluminum films with synchrotron radiation of wavelengths 500 to 1000 A. II: Polarization-dependent photoeffect, *Z. Physik*, **211**: 342–351.

SMITH, K. (1966): Resonant scattering of electrons by atomic systems, *Rept. Progr. Phys.*, **29**: 373–443.

SMITH, S. J. (1969): Survey on electron-atom collision experiments, in *Physics of the one- and two-electron atoms*, F. BOPP and H. KLEINPOPPEN (eds.), pp. 574–597, North-Holland Publishing Company, Amsterdam.

SOKOLOV, A. A., and TEMOV, I. M. (1968): *Synchrotron radiation*, English translation, E. Schmutzer (ed.), Academie-Verlag, Berlin.

STEINMANN, W., and SKIBOWSKI, M. (1966): Plasma resonance in the photoelectric yield of aluminum, *Phys. Rev. Lett.*, **16**: 989–990.

SUGAR, J. (1972): Potential barrier effects in photoabsorption. II: Interpretation of photoabsorption resonances in lanthanide metals at the $4d$-electron threshold, *Phys. Rev.*, **B5**: 1785–1792.

TOMBOULIAN, D. H., and HARTMAN, P. L. (1956): Spectral and angular distribution of ultraviolet radiation from the 300-MeV Cornell synchrotron, *Phys. Rev.*, **102**: 1423–1447.

VIDAL, C. R., and COOPER, J. (1969): Heat-pipe oven: A new, well-defined metal-vapor device for spectroscopic measurements, *J. Appl. Phys.*, **40**: 3370–3374.

WEISS, A. (1970): Private communication; see also Ederer et al. (1970).

ZIMKINA, T. M., FOMICHEV, V. A., GRIBOVSKII, S. A., and ZUKOVA, I. I. (1967*a*): Anomalies in the character of the x-ray absorption of rare earth elements of the lanthanide group, *Fiz. Tverd. Tela*, **9**: 1447–1449 [*Soviet Phys.-Solid State*, **9**: 1128–1130].

———, ———, ———, and ——— (1967*b*): Discrete absorption by $4d$ electrons of the lanthanide rare earths, *Fiz. Tverd. Tela*, **9**: 1490–1492 [*Soviet Phys.-Solid State*, **9**: 1163–1165].

8

X-RAY PHOTOELECTRON SPECTROSCOPY

Stig B. M. Hagström and Charles S. Fadley

I. INTRODUCTION

Our basic knowledge of the electronic structure of matter stems primarily from experiments based on the interaction of photons with electrons in bound states, and involves the emission and absorption processes of electromagnetic radiation over a wide energy range. One aspect of such studies has been x-ray physics where x-ray emission and absorption spectroscopy have played a vital role in increasing our understanding of electronic energy levels.

In recent years the photoelectron-spectroscopy technique has been developed to the degree that such experiments in many cases now give the most precise and accurate information on the electronic structure of gases, liquids, and solids. Both core and valence electrons are being studied with high accuracy. The technique is expanding very rapidly not only in physics but also in many other areas of scientific and technological importance.

The principle in photoelectron-spectroscopy experiments is to study the properties of electrons photoemitted from a sample by monoenergetic photons. There are basically three measurable properties of the photoelectron: the *kinetic-energy distribution* and its variation with photon energy, the *angular distribution* of photoelectron intensity (both with respect to the direction of the incident photon flux and some specified axes in the sample), and finally the *spin distribution* of the emitted photoelectrons. Most of the experiments performed to date have been concerned only with measurements of the kinetic-energy distribution.

The fundamental equation for the photoelectric process is based on the energy-conservation principle

$$hv = E^f(k) - E^i + E_{kin} \qquad (8\text{-}1)$$

where hv is the energy of the incident photon, $E^f(k)$ is the total energy of the final state with a hole in the kth subshell, E^i is the total energy of the initial state, and E_{kin} is the kinetic energy of the photoelectron expelled from the kth subshell. The photon energy thus determines the depth to which photoelectron spectroscopy can probe the electron energy levels. Due to the lack of readily available monoenergetic photon sources in the range between the vacuum ultraviolet and the soft x-ray region, photoelectron spectroscopy has developed in two distinct areas characterized by the energy of excitation. In *ultraviolet* (UV) photoelectron spectroscopy the upper limit of the photon energy is presently approximately 40 eV, while in the *x-ray* case the photon energy is typically about 1 keV. UV photoelectron spectroscopy is therefore restricted to the study of valence or quasi-valence electrons and has been applied extensively to studies of both gases and solids. X-ray photoelectron spectroscopy can be used to study both core- and valence-electron states (Siegbahn et al., 1967; Siegbahn et al., 1969). The x-ray photoelectron spectroscopy technique is often referred to as ESCA (electron spectroscopy for chemical analysis), thereby emphasizing its important chemical applications. To distinguish between the two excitation-energy regions, we will use the notations XPS for x-ray photoelectron spectroscopy and UPS for ultraviolet photoelectron spectroscopy. In this chapter we will deal primarily with the XPS technique.

Referring to equation (8-1), we note that the energy of the final state $E^f(k)$ will be reflected in the photoelectron spectrum. In the case of the final state being simply a hole left in the kth electron energy level, $E^f(k) - E^i$ is equal to the electron binding energy of the kth electron referred to the vacuum level of the sample

$$(\text{B.E.})_V = E^f(k) - E^i \qquad (8\text{-}2)$$

For solid samples, the Fermi level is usually a more convenient reference level, and the work function of the sample becomes involved in the energy equation. If the definition of E_{kin} is changed slightly to mean the kinetic energy as measured in the electron spectrometer, then we must account for the fact that the photoelectron will have adjusted to the work function of the spectrometer material ϕ_{sp}. The energy equation therefore becomes

$$hv = (\text{B.E.})_F + \phi_{sp} + E_{kin} \qquad (8\text{-}3)$$

Much of the successful development of photoelectron spectroscopy to its present importance can be ascribed to improved techniques in measuring E_{kin}. In the next paragraph we give the essential details of XPS experiments and the kind of information that can be obtained from them.

One of the objectives in the early development of the XPS technique was to improve the precision and accuracy of measurements of core-electron binding energies (see, for instance, Sokolowski, Nordling, and Siegbahn, 1957; Hagström, Nordling, and Siegbahn, 1965). Such measurements now constitute the basis for compilations of electron binding energies (see Chap. 2). For the case where an element is in monatomic gaseous form, these binding energies can be uniquely defined. However, if the element is studied as a constituent of a molecule or a solid, the valence-electron charge distribution will be changed, and this change will influence the core energy levels. Therefore, "chemical shifts" of binding energies may be observed in photoelectron spectra. These shifts represent an important source of information on the outer-electron structure of bonded atoms and constitute the basis for many important applications of XPS in, for example, chemistry. These aspects will be treated more fully in Section III.

Most photoelectron spectroscopic studies of the valence-electron levels have employed the UPS technique. However, x-ray photoelectron spectra from the outer electrons also give very valuable and complementary information. The possibility of studying the core levels and the valence electrons in a given sample is also of considerable advantage. A further advantage of XPS studies is that, in the case of solids, the final photoelectron state is high above the band-structure region in energy, and the photoelectron can thus be represented as a free electron. Such free-electron states introduce minimal final-state modulation of the observed photoelectron spectra. A discussion of valence-electron studies using XPS is given in Section IV.

The final state of the photoemission process is quite often more complex than is implied in the discussion above. It may very well exhibit various excitations which will be reflected in the photoelectron spectrum. This is best demonstrated in ultraviolet photoelectron spectra from gaseous molecules which clearly show the molecular vibrational and rotational levels. Furthermore, the interaction between the valence electrons and the core levels may be more complex than the spatially averaged Coulomb and exchange interactions which give rise to the chemical shifts mentioned above. For example, unpaired valence electrons will affect spin-up and spin-down core electrons unequally because of the exchange interaction and break their spin degeneracy. Thus, binding-energy *splittings* may be observed in photoelectron spectra. There are several other effects such as crystal-field splittings and multielectron processes that may add to the complexity of a photoelectron spectrum, thereby increasing the amount of information inherent in it. These effects are discussed in Section V.

In most XPS experiments to date, only the energy parameter has been measured. However, angular-distribution measurements have also been performed, and these are reviewed in Section VI. No spin-polarization measurements have as yet been performed on x-ray photoelectron spectra, although a few such measurements have been made on ultraviolet-excited photoelectrons (Busch, Campagna, and Siegmann, 1971).

In both gaseous and solid samples it is likely that many photoelectrons will be inelastically scattered before emerging into vacuum. These electrons will modify the appearance of each photoelectron peak on the low-kinetic-energy side. The elastically scattered photoelectrons in the primary peak come from a very thin surface layer. Relatively subtle effects on a sample surface will therefore influence photoelectron spectra, as discussed in Section VII.

Before discussing the various aspects of XPS studies, it may be worthwhile to outline briefly the most important steps in its development. The external photoelectric effect was first discovered by Hertz in 1887 and played an important role in the development of the idea of quantization of physical quantities. Such measurements constituted the crucial piece of experimental evidence in the conception of Einstein's photon theory of light. The principle of x-ray photoelectron spectroscopy was conceived as early as in 1913 by Robinson (Robinson and Rawlinson, 1914), who continued this work for approximately two decades. However, the poor resolution of electron spectrometers at that time prevented the extraction of most of the information contained in photoelectron spectra. It was not until high-resolution electron spectrometers were available that the method could yield unique information on the energy distribution of electrons in matter. The development of such spectrometers and the revitalization of x-ray photoelectron spectroscopy was mainly carried out by Siegbahn and his associates in Uppsala beginning in the middle 1950s (Nordling, Sokolowski, and Siegbahn, 1957; Sokolowski, Nordling, and Siegbahn, 1957). The primary aim of these measurements was to obtain more accurate electron binding energies of core electrons. The precision and accuracy of these measurements proved to be sufficient not only to revise the table of electron binding energies for the elements (Hagström, Nordling, and Siegbahn, 1965), but also to detect small chemical shifts in these binding energies (Nordling, Sokolowski, and Siegbahn, 1958). The establishment of the correlation between chemical oxidation state and electron-binding-energy shifts (Hagström, Nordling, and Siegbahn, 1964) was thus an important contribution in the use of x-ray photoelectron spectroscopy in chemistry. It was in connection with such chemical-shift work that the acronym "ESCA" was first used.

X-ray photoelectron spectroscopy has subsequently been utilized for work in a number of fields in various laboratories. The group at Oak Ridge has contributed considerably, especially to our understanding of multielectron processes in photoemission. The first XPS experiments on gaseous samples were also done there. Specific mention should also be made of the work on multiplet splittings primarily carried out by the group in Berkeley. A number of commercial instruments for XPS

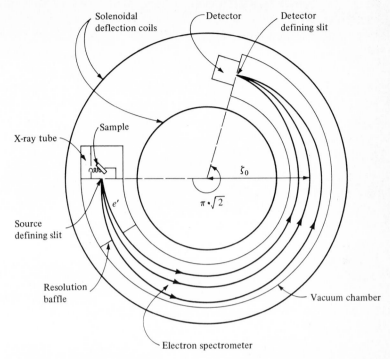

FIGURE 8-1

Schematic illustration of the instrumentation required for x-ray photoelectron experiments. The main parts are the excitation source, usually an x-ray tube; the sample arrangement; the electron-energy analyzer (here shown as a magnetic-deflection type); the detector system; and a data and control system (not shown).

work have recently been put on the market, and these represent a significant contribution to the development of instrumentation for photoelectron spectroscopy. Parallel to the development of the XPS technique, the evolution of UPS gave rise to a powerful tool for studying the band structure of solids (Berglund and Spicer, 1964) and the valence-electron orbitals of molecules (Turner et al., 1970).

II.   INSTRUMENTATION

The instrumentation required for XPS experiments (Fig. 8-1) can be divided into several parts: the source of the x radiation which is used to excite photoelectrons from the sample to be studied, the device maintaining the sample at whatever conditions are desired for the experiment, the electron spectrometer which measures the kinetic

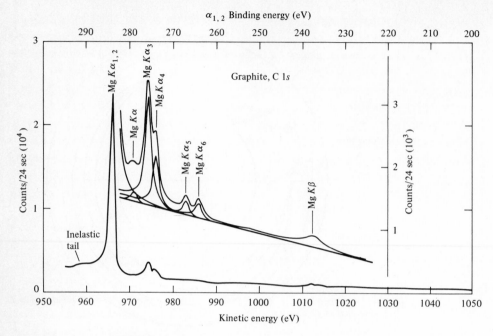

FIGURE 8-2
Photoelectron spectrum of C $1s$ electrons in graphite excited by Mg x rays showing the contributions from various lines in the incident Mg x-ray spectrum. (*After Fadley*, 1970.)

energies of the electrons emitted from the sample, a detector which detects these electrons after kinetic-energy measurement, and some system for controlling the spectrometer and recording data. We discuss these different aspects below.

A.   Radiation sources

The x-ray sources most commonly used in XPS studies are Mg and Al x-ray tubes of fairly standard design (Siegbahn et al., 1967; Fadley, 1970). These were first utilized in XPS studies by the Uppsala group (Siegbahn et al., 1967). The most intense x ray emanating from such an x-ray tube is the unresolved $K\alpha_{1,2}$ doublet, resulting from the $2p_{1/2,3/2} \rightarrow 1s$ transition in the metallic anode. For Mg and Al these x rays have the following energies: Mg $K\alpha_{1,2} - 1253.6$ eV and Al $K\alpha_{1,2} - 1486.6$ eV. Satellite x rays are also present, due to $2p \rightarrow 1s$ transitions in atoms which are doubly- or triply-ionized. These satellites are shown in Fig. 8-2. The most intense are the $K\alpha_3$ and $K\alpha_4$ x rays whose energies are about 10 eV above those of $K\alpha_{1,2}$ for both Mg and Al, and whose intensities are about 15 percent of $K\alpha_{1,2}$ (Fadley, 1970). They result in a characteristic double peak slightly above the strong $K\alpha_{1,2}$ photoelectron peak in kinetic energy. The $K\alpha_{5-8}$ satellites are also present at energies about 20 eV above $K\alpha_{1,2}$, but only with about 1 percent of its intensity. The $K\beta$ x rays resulting from valence-electron $\rightarrow 1s$ transitions are also present at ~ 60 eV above $K\alpha_{1,2}$ and with approximately 1 percent of its intensity. Thus, to a first approximation, the x-ray spectrum consists only of the very intense $K\alpha_{1,2}$ x ray, and most work has been based solely on an analysis of $K\alpha_{1,2}$ photoelectron peaks. However, in any study involving weak photoelectron peaks, or $K\alpha_{1,2}$ peaks which overlap with satellite peaks due to other electrons (Fadley and Shirley, 1970a), the nonmonochromatic character of the x-ray source must be taken into account. A final source of radiation from the x-ray tube is a continuous background of Bremsstrahlung, which is partially responsible for the high background observed in photoelectron spectra.

The natural line widths of the $K\alpha_{1,2}$ x rays set the present limit of instrumental resolution in XPS studies. The full width at half-maximum intensity (FWHM) values are about 0.8 eV for Mg $K\alpha_{1,2}$ and 0.9 eV for Al $K\alpha_{1,2}$ (Siegbahn et al., 1969). The $K\alpha_{1,2}$ line width should decrease with atomic number, because of the decrease of the $2p_{1/2} - 2p_{3/2}$ spin-orbit splitting and also because the $1s$ and $2p$ hole lifetimes should be longer. However, sodium does not make a particularly stable anode material (Siegbahn et al., 1967). Also, elements below neon cannot be used as the $2p$ levels become broad valence levels and the $K\alpha$ x rays are broadened accordingly. Recently, the Hewlett-Packard Company has developed a bent-crystal monochromator and dispersion-compensating electron-optical system (Siegbahn et al., 1967) which appears to permit a reduction of the photoelectron-peak line widths from Al $K\alpha_{1,2}$ to about 0.5 eV.

Other x-ray sources of lower energy have also been proposed for use in photoelectron spectroscopy by Krause (Krause, 1971; Krause and Wuilleumier, 1971). These make use of the $M\zeta$ transition ($4p_{3/2} \rightarrow 3d_{5/2}$) in the elements Y to Rh. The energies are in the range 132 to 260 eV, and the FWHM line widths vary from ~ 0.45 for Y $M\zeta$ to 4.0 eV for Rh $M\zeta$. Thus, these x rays are in the ultra-soft region, and would be capable of exciting photoelectrons from valence levels and outer core levels. They also furnish photon energies intermediate between those presently used in UPS and XPS studies, with line widths which can be larger or smaller than those of the common XPS x rays.

Another source of radiation in the x-ray or XUV range is the centripetal acceleration of high-energy electrons in such devices as synchrotrons or storage rings (Tomboulian and Hartman, 1956; Madden, Ederer, and Codling, 1967). As discussed in Chap. 7, such radiation sources have the advantage of an intense, continuous spectrum of radiation, from which a narrow energy band can be selected by a monochromator of some kind. Several existing facilities of this type have electron energies high enough to achieve reasonable intensities of photons with energies of a few hundred eV or even up to several thousand eV (Godwin, 1969). However, no photoelectron spectroscopy experiments have been performed with such sources as yet for photon energies higher than about 50 eV (Haensel et al., 1969).

B. Sample preparation

XPS samples may be in either gaseous, solid, or liquid form, although liquid-phase experiments are limited considerably by the need for keeping a vacuum of $\sim 10^{-4}$ torr or better in the electron spectrometer in order to prevent excessive electron scattering from gas-phase molecules. The sample arrangements utilized so far have thus dealt primarily with gaseous or solid samples in various environments and at various temperatures. A typical sample area for solid samples is a few mm by 1 cm, and a typical volume for gaseous samples is 1 cm^3.

Most experiments to date have been performed on room-temperature solids, for which sample mounting can be exceedingly simple. In many cases, finely powdered substances have been merely dusted on a sticky, Scotch-tape-like backing (Hagström, Nordling, and Siegbahn, 1964; Fadley et al., 1968). If problems are encountered with charging of an insulating sample, it can also be powdered and pressed into a fine metal mesh, thereby improving the electrical contact of the sample material with the spectrometer. Powdered samples can also be pressed into pellets (Delgass, Hughes, and Fadley, 1970). Metals can be studied as foils (Fadley and Shirley, 1970a) or as freshly evaporated films (Hedén, Löfgren, and Hagström, 1971).

Devices for raising or lowering the temperature of solid samples have also been used (Siegbahn et al., 1967; Fadley, 1970). Temperatures near that of liquid nitrogen have been achieved with cold-finger arrangements, although the usefulness of such devices for studying solids has been limited by the rapid freezing-out of surface impurities from the gas phase. Such devices have proved useful, however, in studying high-vapor-pressure solids or gases by condensing them at cryogenic temperatures (Siegbahn et al., 1967; Fadley, 1970); in such experiments, the sample surface can, in principle, be continuously replenished by providing a high vapor pressure of the desired compound around the cold finger. Frozen aqueous solutions have also been studied on such devices (Kramer and Klein, 1969). Solids have also been studied at temperatures as high as 1000°C, by means of specially constructed heaters which generate low, stray magnetic fields (Fadley, 1970; Fadley and Shirley, 1970a; Baer et al., 1970a). Studies of various transition metals have been made in which hydrogen was passed over high-temperature samples to reduce away surface oxide (Fadley and Shirley, 1968; Fadley, 1970; Fadley and Shirley, 1970a; Baer et al., 1970a). It is also clear from this discussion of studies of solids that the lack of ultra-high-vacuum conditions in the sample region has restricted development along certain lines, but suitable UHV systems are becoming available.

Most gas-phase experiments have been done at room temperature, although in certain instances, the gas actually in the sample chamber could be produced by heating a remote reservoir of high-vapor-pressure liquid or solid. The basic requirement for gas-phase studies is a chamber to contain the gas, with an x-ray-transparent window in one side and a small exit slit for photoelectrons in another (Siegbahn et al., 1969; Fadley, 1970). Some differential pumping at the exit slit may also be necessary in order to maintain a gas pressure in the chamber of at least 10^{-2} torr, while at the same time keeping the pressure in the spectrometer sufficiently low to minimize inelastic-scattering effects (Siegbahn et al., 1969). Rather large changes in photo-electron-peak positions and relative intensities as a function of chamber pressure have also been reported by Siegbahn et al. (1969), although the pressures in these measurements were as high as 1 torr. An oven for studying metals and other vaporizable solids in the gas phase has also been constructed by Fadley (1970).

C. Spectrometers

Several types of electron spectrometers are being used in XPS studies, including commercially available systems (Siegbahn et al., 1967; Fadley, Miner, and Hollander, 1969, 1971). Regardless of type, however, there are several basic characteristics that are a priori desirable in such a spectrometer (Fadley, Miner, and Hollander, 1969, 1972): (1) A resolution capability of $\Delta E_{kin}/E_{kin} \cong 0.01$ percent. This corresponds to 0.1 eV for 1000-eV electrons. Most XPS spectrometers are presently operated in the 0.02 to 0.04 percent range. (2) The highest possible efficiency. That is, the highest possible fraction of electrons leaving the sample should be energy-analyzed and detected at the same time. (3) Unrestricted physical access to the sample and detector regions. This permits a wide variety of excitation sources, sample arrangements, and detector systems to be used. (4) Ultra-high-vacuum capability for work on solid samples. As we have pointed out, the surface condition of solid samples often has important effects on photoelectron spectra (see also Section VII). (5) Ease of construction. (6) Relative insensitivity to external environment, particularly as regards the shielding of extraneous magnetic fields.

The resolution and efficiency of a spectrometer are of paramount importance. These variables are linked, since for operation at a lower resolution (higher $\Delta E_{kin}/E_{kin}$), a higher fraction of electrons can usually be energy-analyzed and detected. For operation at a given resolution, the overall efficiency E will be proportional to the following product (Fadley et al., 1972; Helmer and Weichert, 1968)

$$E \propto B \cdot A \cdot \Omega \cdot \delta E_{kin} \qquad (8\text{-}4)$$

where B is the intensity of the electron source for the energy analyzer in electrons per unit area-steradian, A the area of this source, Ω the solid angle over which electrons

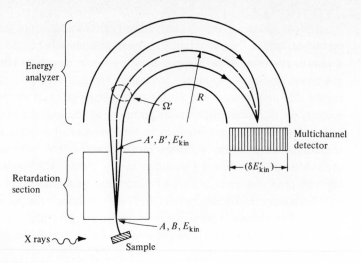

FIGURE 8-3
Schematic diagram of an XPS spectrometer utilizing retardation of electrons before energy analysis and a multichannel detector.

from the source are accepted into the energy analyzer, and δE_{kin} the range of electron energies which is analyzed at the same time, as, for example, by a multichannel detector (Fadley et al., 1972). The electron source is usually defined by an aperture in front of the photoemitting sample (Hagström, 1964); this aperture may be at the exit of a section used to retard the electrons in energy before analysis. If a multichannel detector is utilized, δE_{kin} may in principle be as large as 10 percent of E_{kin}, whereas the resolution ΔE_{kin} will be ~ 0.01 percent of E_{kin}. In this case, the detector would correspond to ~ 1000 channels. The notation used in this discussion is indicated in the schematic drawing in Fig. 8-3, where primes have been used on all quantities after the retarding section. This retarding section may be absent in many spectrometers.

The two types of energy analyzers most commonly used in XPS studies are based on the double-focusing magnetic field (Siegbahn et al., 1967; Fadley et al., 1972; Siegbahn, 1965) and the electrostatic field between two concentric spherical-sector electrodes (Siegbahn et al., 1967; Purcell, 1938). In both these fields, electrons are deflected in roughly circular orbits. For a given average orbit radius R, numerical calculations indicate that the product $A'\Omega'$ is roughly equal for these two fields (Siegbahn et al., 1967). Helmer and Weichert (1968) have also shown that, for this general class of energy analyzers, the product $A'\Omega'$ will be approximately given by

$$A'\Omega' \cong CR^2 \left(\frac{\Delta E_{kin}}{E'_{kin}}\right)^2 \qquad (8\text{-}5)$$

where C is a constant characteristic of the particular energy analyzer used, and E'_{kin} is the energy at which the electrons are analyzed (cf Fig. 8-3). In the absence of retardation, $E'_{kin} = E_{kin}$ and $A\Omega = A'\Omega'$. Furthermore, in connection with retardation before energy analysis, the source intensity will be given by (Helmer and Weichert, 1968)

$$B' = B \frac{E'_{kin}}{E_{kin}} \qquad (8\text{-}6)$$

where B' and E'_{kin} are the intensity and kinetic energy after retardation and B and E_{kin} are the intensity and kinetic energy before retardation. B and E_{kin} thus refer to electrons just as they leave the sample. Equation (8-6) is equivalent to Abbe's sine law or the Lagrange-Helmholtz equation (Helmer and Weichert, 1968; Siegbahn et al., 1967; Klemperer, 1953). For a fixed *absolute* energy resolution ΔE_{kin}, the relative resolution of the energy analyzer will thus be $\Delta E_{kin}/E'_{kin}$ [cf equation (8-5)]. Substituting equations (8-5) and (8-6) into equation (8-4) yields an efficiency *with retardation* of

$$E' \propto B'A'\Omega'(\delta E_{kin})'$$

$$E' \propto CR^2B \frac{E'_{kin}}{E_{kin}} \left(\frac{\Delta E_{kin}}{E'_{kin}}\right)^2 (\delta E_{kin})' \qquad (8\text{-}7)$$

where the prime on δE_{kin} denotes that detection is done after retardation. Thus, if photoelectrons are retarded so that $E_{kin}/E'_{kin} = N$ (where $N \geq 1$), the efficiency will be given by

$$E' \propto CR^2B(\Delta E_{kin})^2 \frac{1}{N} \frac{N^2}{E_{kin}^2} (\delta E_{kin})' \qquad (8\text{-}8)$$

$$E' \propto CR^2B(\Delta E_{kin})^2 \frac{N}{E_{kin}^2} (\delta E_{kin})' \qquad (8\text{-}9)$$

For a given source intensity B, initial kinetic energy E_{kin}, and absolute energy resolution ΔE_{kin}, the ratio of efficiencies with and without retardation will thus be given by

$$\frac{E'}{E} \cong \frac{N(\delta E_{kin})'}{\delta E_{kin}} \qquad (8\text{-}10)$$

If a single-channel detector is used, $(\delta E_{kin})' \cong \delta E_{kin} \lesssim \Delta E_{kin}$ so that

$$\frac{E'}{E} \cong N \qquad (8\text{-}11)$$

Thus, retardation to $E'_{kin} = E_{kin}/N$ yields an increase in efficiency by approximately a factor of N, provided only a single-channel detector is used. Equation (8-11) was first pointed out by Helmer and Weichert (1968). For the situation where a multi-channel detector is used, no simple relationship can be stated for the ratio of the maximum possible detector widths consistent with a given overall resolution $\Delta E_{kin}/E_{kin}$. However, recent theoretical studies of multichannel-detector operation by Fadley et al. (1972) seem to indicate that the ratios $(\delta E_{kin})'/E'_{kin}$ and $\delta E_{kin}/E_{kin}$ will be approximately equal:

$$\frac{(\delta E_{kin})'}{E'_{kin}} \cong \frac{\delta E_{kin}}{E_{kin}} \qquad (8\text{-}12)$$

so that

$$(\delta E_{kin})' \cong \frac{\delta E_{kin}}{N} \qquad (8\text{-}13)$$

Substitution of this equation in equation (8-10) yields

$$\frac{E'}{E} \cong 1 \qquad (8\text{-}14)$$

Thus, if a multichannel detector is included as part of the overall spectrometer system, the overall efficiencies should be approximately equal, with or without retardation. It is assumed throughout this discussion of multichannel detectors that the individual detector channels can be constructed so as to be compatible with an absolute resolution of ΔE_{kin}. This appears to be possible, for example, with the use of arrays of glass channel electron multipliers (Wiley and Hendee, 1962; Adams and Manley, 1967; Nilsson et al., 1970).

Having reviewed the general aspects of spectrometer efficiency, we shall now briefly discuss the specific types of spectrometers presently being used, indicating the most important positive and negative features of each. The first spectrometers to be used in x-ray photoelectron spectroscopy were of the magnetic double-focusing type first suggested by Svartholm and Siegbahn (1946) and utilized no retardation and single-channel detectors (Siegbahn et al., 1967; Fadley et al., 1968). Such a system is shown in Fig. 8-1. The double-focusing magnetic field achieves a high solid angle Ω by accurately analyzing electrons with finite departure angles from the sample in both radial and axial directions, bringing them to a focus at $\pi\sqrt{2}$ rad around the optic circle (see Fig. 8-1). The axial angles can also be several times larger than the radial angles. This magnetic field also possesses a "focal plane," along which electrons of different kinetic energies are distributed at unique radial positions (Lee-Whiting and Taylor, 1957; Siegbahn, 1965). It has recently been pointed out by Fadley, Miner, and Hollander (1969) that the full exploitation of this focal plane with a multichannel detector should yield efficiencies approximately 100 times higher than in single-channel operation. A new coil geometry for a double-focusing magnetic spectrometer with a multichannel detector has also been proposed (Fadley, Miner, and Hollander, 1969; Fadley et al., 1972). This geometry produces a field very close to the theoretical field form, and also has very unrestricted physical access to the sample and detector regions. It is superior in both these respects to previous magnetic double-focusing coil geometries (Fadley, Miner, and Hollander, 1969; Fadley et al., 1972). A weakness of any magnetic spectrometer, however, is that extraneous magnetic fields cannot be excluded by means of high-permeability Mumetal shielding. Helmholtz coils must be used, often with dynamic feedback control.

Electrostatic spectrometers based on spherical-sector electrodes without retardation are also being used in XPS studies (Siegbahn et al., 1967). The electron-optical properties of such spectrometers are similar to those of the double-focusing magnetic type and yield comparable solid angles at a given resolution (Siegbahn et al., 1967). No detailed investigations of the focal-plane properties of such spectrometers have been made, but some commercially available systems make use of a multichannel detector in conjunction with such an energy analyzer. Electrostatic spectrometers in general possess the advantages of being slightly simpler in construction and easier to shield from stray magnetic fields than magnetic spectrometers.

Some commercially available spectrometers make use of spherical-sector electrostatic analyzers with initial energy retardation. Aside from the possible efficiency advantages of such spectrometers in single-channel operation [equation (8-11)], the fact that the *relative* resolution of the analyzer is lower ($\Delta E_{kin}/E'_{kin}$ higher) means that both the construction and magnetic shielding are somewhat easier. It may be desirable for the analyzer to have higher relative resolutions, however, if it is also to be used for UPS studies utilizing very narrow excitation sources. The physical access to the source and detector regions is reasonably good for most spectrometers based on spherical-sector electrodes.

The experimental and theoretical properties of electrostatic spectrometers based on two concentric cylindrical electrodes have also been investigated. Such cylindrical-mirror spectrometers appear to give high intensities at moderate resolutions (~ 0.05 percent or greater), and Berkowitz (1971) has recently reported the use of such a system for XPS studies. Sar-el (1970) has compared the theoretical $A\Omega$ values of various electrostatic spectrometers as a function of resolution, and finds the cylindrical mirror to be superior even at resolutions better than 0.05 percent. The overall geometry of the electron trajectories in a cylindrical mirror appears to greatly restrict its use with multichannel detectors, however, and the physical access to the source region is somewhat limited.

The use of a spherical retarding grid spectrometer has also been proposed for use in XPS studies by Huchital and Rigden (1970). The principle of this spectrometer is distinct from those previously discussed in that it is nondispersive; that is, electrons are not electromagnetically deflected so as to spatially disperse them on the basis of energy. The total electron current is measured as a function of retarding voltage and differentiated with respect to this voltage to give a kinetic-energy distribution. In this recent design, some velocity-selection capability is added near the electron collector to improve signal-to-background ratios. This analyzer has been used at resolutions as high as 0.05 percent, although the signal-to-background ratios still appear to be rather high in comparison with spectra from conventional dispersive spectrometers.

Although some types of spectrometers may be easier to shield from the effects of stray magnetic fields, all will require some sort of shielding to operate with overall resolutions in the 0.01 percent range. It is thus essential to know what the maximum magnitudes of the extraneous magnetic fields can be in order to permit operation at a given resolution. Estimates of these magnitudes have recently been made by Fadley et al. (1972) for both homogeneous magnetic fields and constant magnetic-field gradients.

D. Control and data handling

Although the specific type of spectrometer control required may vary considerably with the type of energy analysis used, there are several desirable features in any control and data-handling system. The obvious minimum requirement is to be able to scan a given spectral region in kinetic energy and to output the electron kinetic-energy distribution. Beyond this, however, it is extremely useful to be able to scan several widely separated spectral regions without operator intervention and, in particular, to accumulate data for all spectra at the same time (Fadley, 1970). That is, if the spectra are denoted $1, 2, 3, \ldots, j$ and each scan is also numbered up to some maximum number m desired for all spectra, there are two possible modes in which the data may be accumulated (Fadley, 1970): scan 1, spectrum 1; scan 2, spectrum 1; ... scan m, spectrum 1; scan 1, spectrum 2; scan 2, spectrum 2; ... etc.; *or* scan 1, spectrum 1; scan 1, spectrum 2; ... scan 1, spectrum j; scan 2, spectrum 1; scan 2, spectrum 2; and so forth. The latter mode has the considerable advantage that any instrumental drifts or changes in sample composition will be more equally reflected in all spectra. This is useful, for example, when several core levels and the valence-electron levels of a given sample are being studied. The core levels may include those of suspected sample contaminants also (Fadley and Shirley, 1970a). The desirability of these rather sophisticated control and data-handling features has led to the use of small dedicated computers in connection with several XPS spectrometers (Fadley, 1970).

III. CORE-ELECTRON BINDING ENERGIES

Much of the work in the XPS field has been devoted to the accurate determination of core-electron binding energies, the reasons being both to obtain precise information on these physically important quantities and to use them as sensitive probes of the detailed environment of an atom. In this section we will review the general principles behind the method for obtaining binding energies from XPS data and discuss the way in which these energies are influenced by chemical surroundings.

A *binding energy* is defined as the energy required to remove an electron from the bound state to infinity. However, the binding energy of a core electron has a single, uniquely defined value only for a closed-shell element in the form of a monatomic gas. In all other cases either the final state is not uniquely defined or the core levels are perturbed by the environment. In the former case the corresponding photoelectron spectrum will exhibit multicomponent lines, as discussed in Section V, and in the latter case the lines will be shifted in energy. The two effects may also occur together.

Most binding-energy measurements have been performed on solids (Siegbahn et al., 1967; Hollander and Shirley, 1970). In the case of metallic samples it is convenient to use the Fermi level in the sample as a reference level for binding energies. This procedure avoids the inclusion of the work function of the sample in the energy-conservation equation, as is shown in equation (8-2). The sample work function is critically dependent on the physical condition of the sample surface, whereas the spectrometer work function should be relatively constant with time. As mentioned in Section II, the experimental arrangement is usually such that the sample is located behind an entrance slit to the spectrometer and in electrical contact with this slit. Before the photoelectrons enter the region of energy analysis, they will thus adjust their energy to the vacuum level of the inside walls of the spectrometer. The energy equation will therefore contain the work function of the spectrometer, as is illustrated in Fig. 8-4, where the simple case of a one-electron transition and a one-hole final state is used for illustration. A consideration of this figure yields equation (8-2).

In semiconductors and insulators the choice of a reference level for binding energies is more ambiguous. The Fermi level is located in the energy gap between the valence and conduction bands, and equilibrium between the sample and the spectrometer can be established only if enough charge carriers are present in the conduction band. Furthermore, surface effects will induce shifts of all electronic levels relative to the Fermi level. However, the bulk of experimental results accumulated to date on such materials indicates that the reference level remains relatively constant, even though its absolute location cannot be unambiguously determined (Sharma et al., 1971).

Since the Fermi level serves as the zero point in the binding-energy scale, one can also use this level as an internal calibration for spectra. It can be determined by recording the photoelectron spectrum from the band-structure region of a metal, as discussed in the following section. Convenient reference samples are thus Pd and Pt, for which the Fermi level is located at the sharp edge of the upper part of the d band (Fadley and Shirley, 1970a; Baer et al., 1970a).

In the case of nonmetallic samples the situation is more complex. Due to the continuous loss of electrons during the x-ray bombardment an insulating sample may

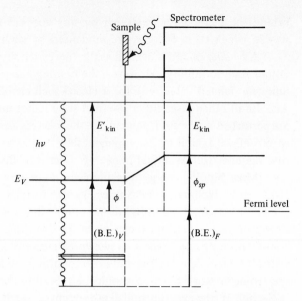

FIGURE 8-4
The effect of the contact potential between a metallic sample and the electron-energy analyzer. During energy analysis the electrons will have adjusted to the vacuum potential of the analyzer material, and thus it is the work function of the latter that enters into the energy-conservation equation.

become strongly charged and the photoelectron lines shifted toward lower kinetic energy. Such charging effects are in fact observed (Fadley et al., 1968) and can easily be misinterpreted as chemical shifts. However, the magnitudes of such charging effects are usually much less than expected, being only a few eV (Fadley et al., 1968). This reduction in magnitude as compared to the corresponding phenomena observed in beta-ray spectroscopy is probably due to the neutralizing cloud of low-energy electrons around the sample caused by photoemission and secondary emission from the sample and its surroundings. Also, the charging effect can be minimized by using very thin samples or possibly coating the sample with a thin conducting layer. It can sometimes be distinguished from genuine binding-energy shifts by observing the shift as a function of x-ray flux, or by applying a precisely known voltage to the sample and measuring the corresponding shift of the photoelectron line (Fadley, Geoffroy et al., 1969).

Solid samples will always produce inelastically scattered electrons which show up on the low-energy side of a photoelectron peak. However, the energy losses are usually spread over a very broad energy range, and in high-resolution spectra a peak of elastically scattered electrons can in most cases be easily distinguished from the "tail" of inelastically scattered electrons. The sample surface will strongly influence the shape of photoelectron spectra, not only because of inelastic-scattering effects, but also because of the presence of surface oxidation and surface states, for example (Hedén, Löfgren, and Hagström, 1971). These effects will be further elaborated in Section VII.

In order to establish binding energies on an absolute energy scale, the spectrometer must be calibrated. A procedure for this based on accurately known x-ray transition energies is described by Nordling (1959). A modification of this procedure has been proposed by Fadley, Geoffroy et al. (1969). In this scheme, an accurately measured potential difference is applied between a conducting sample and the entrance slit to the spectrometer. By recording photoelectron peaks with and without the potential difference applied and with the two possible signs of the potential difference, a set of three photoelectron peaks is obtained whose energy separations are determined precisely by the applied voltages. By applying the same technique to an insulating sample one can also obtain an indication of the electric-field penetration through the sample and therefore of possible charging effects.

Figure 8-5 shows a portion of the photoelectron spectrum from a gold sample (Siegbahn et al., 1967). The atomic levels $4f_{7/2}$ and $4f_{5/2}$ give rise to two well-separated peaks which can easily be located with an uncertainty of less than 0.1 eV. One can also see the band-structure part of the spectrum with the zero-point reference level at the Fermi level. In this way, it is possible to accurately determine all electronic levels provided that the excitation energy is large enough. In measuring deeper lying levels the line width will be increased by contributions from the natural line widths of the exciting x ray and of the level itself.

By combining XPS data on binding energies with x-ray emission line measurements, compilations of inner-atomic-energy levels have been made. The first such table was published in 1965 (Hagström, Nordling, and Siegbahn, 1965) and has been followed by several others (Bearden and Burr, 1967; Siegbahn et al., 1967). Most of the data in these tables refer to solids. Attempts have been made to convert these data to free atoms (Lotz, 1970). Since binding energies are influenced by the chemical surroundings, care should be exercised in using tabulated binding energies as they will depend on the chemical compound to which the tabulated values refer. The establishment of such tables of electron binding energies is further discussed in Chap. 2.

An XPS spectrum will also contain peaks originating from nonradiative transitions such as the Auger transitions. The energies of these lines are solely dependent on the energy levels in the sample and not on the energy of the incident radiation. This property can be used to distinguish between peaks connected to the photoelectric effect and Auger peaks. Auger spectroscopy is also an important source of information on electronic structure and has found a number of important applications, especially in characterizing the surface of solid samples (Palmberg, 1971). However, this type of spectroscopy will not be treated in this review.

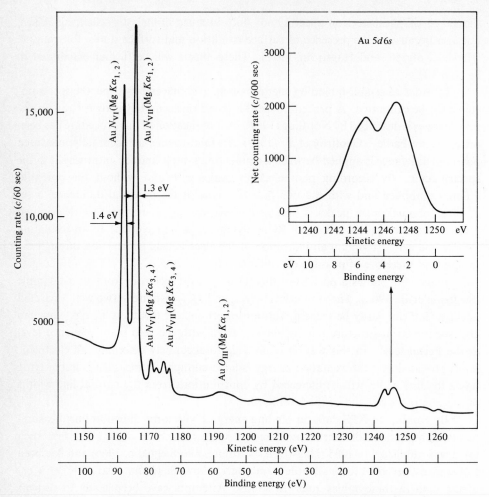

FIGURE 8-5
The photoelectron spectrum from a gold sample excited by Mg x rays. The core levels $4f_{7/2}$ and $4f_{5/2}$ give rise to well-defined peaks. The band-structure part of the spectrum is also clearly visible. (*After Siegbahn et al.*, 1967.)

It has been mentioned previously that the core-level energies of an atom are dependent on the total environment of the atom. This gives rise to chemical shifts in recorded photoelectron peaks. The conclusive evidence that such shifts are actually due to shifts of the core levels, rather than changes in the reference level, came from experiments on substances containing one type of atom in different chemical positions (Hagström, Nordling, and Siegbahn, 1964). Such a spectrum is shown in Fig. 8-6.

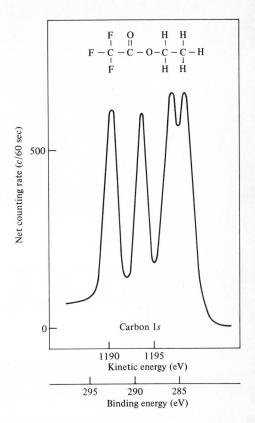

FIGURE 8-6
Chemical shifts of the 1s photoelectron
line of carbon in the chemical compound ethyl trifluoroacetate. (*After
Siegbahn et al.*, 1967.)

To gain a qualitative idea of the origin of such shifts, we can consider the simplest
model of a free ion which consists of core electrons inside a spherical charge shell of
valence electrons with a radius r. If the ion has been formed by changing the charge
on the spherical shell by an amount q, all core electrons will experience a change in
the uniform electrostatic potential from the valence electrons and thereby a change in
binding energy of q/r. If the charge q has been removed, all core-electron binding
energies will be increased by this amount. The binding energies will decrease by
q/r if charge q has been added.

In the actual case of an atom contained in a crystal or a molecule, one has to
take into account not only a more refined model for the charge transfer between the
atom and its surroundings but also the influence of all other atoms in the crystal or
molecule in the form of a crystal- or molecular-potential correction. There are also
other effects that influence the derivation of a relationship between charge and binding-
energy shifts. By taking these influences into account, it is often possible to establish
a correlation between measured shifts and the effective charge of an atom in a chemical

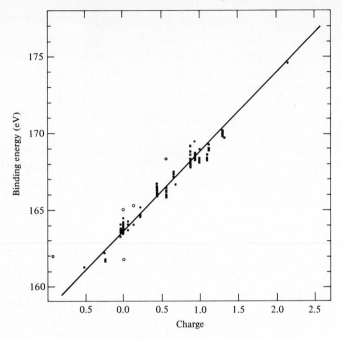

FIGURE 8-7
Correlation between measured binding energies of S $2p$ electrons and atomic charge for a number of chemical compounds. (*After Lindberg et al.*, 1970.)

compound (Siegbahn et al., 1967; Fadley et al., 1968; Jørgensen, 1971). Such a correlation is shown in Fig. 8-7. This application of XPS has found a broad range of applications in the field of chemistry and related areas.

A.   General review of theory

By using self-consistent-field methods it is possible to calculate accurately the binding energies of core electrons. The most general of these methods is the Hartree-Fock (HF) method. A modification of this method which simplifies the form of the exchange interaction is the Hartree-Fock-Slater method (HFS). By introducing adjustable parameters which are determined by minimizing the total energy, Rosén and Lindgren (1968) have been able to obtain good agreement between calculated and measured binding energies using the HFS method.

In calculating the chemical shifts of the core-electron binding energies of an atom between two compounds, one can utilize an energy cycle as outlined by Fadley et al. (1968). In this cycle, the photoemission process is broken up into the following three steps: (1) The atom A with its charge Z is removed from the lattice of compound X. (2) A core electron (i) is removed from the free ion A^Z. This step requires the energy $(B.E.)_V(A,i,Z)$ and results in the ion A^{Z+1}. (3) The ion A^{Z+1} is inserted back into the lattice. The shift in core-electron binding energy of atom A in two different compounds X and Y, $\Delta(B.E.)_V(A, i, X - Y)$, can then be written

$$\Delta(B.E.)_V(A, i, X - Y) = \Delta(B.E.)_V(A, i, Z - Z') + \Delta E_{latt} \qquad (8\text{-}15)$$

where the first term relates to the effective-charge change in the free ion and the second term is due to the difference in crystal potential. In actual cases these two terms are of the same order of magnitude and of opposite sign. Both must therefore be calculated with high accuracy to make meaningful comparisons with experimental data possible. Calculations of free-ion shifts show the same general trends as the simple spherical-charged-shell model (Fadley et al., 1968). In removing an outer electron all core levels are shifted by approximately the same amount (Fadley et al., 1968). This explains why chemical effects are not easily observable in x-ray emission spectra. The free-ion shifts are also smaller in magnitude than those obtained with a classical model. In the case of the rare earths, however, the removal of a $4f$ electron causes the core electrons to shift by different amounts, as is expected because of the core-like character of the $4f$ electrons (Fadley et al., 1968).

The calculation of ΔE_{latt} is similar to the calculation of the Madelung energy in ionic solids. A certain ambiguity arises in comparing calculated data with experimental results because it is only atoms in a thin surface layer that contribute to the photoelectron spectrum, and these may have a different crystal potential than bulk atoms, One has also to consider the effect of electron relaxation. Estimates of this effect show that it is of the order of 1 eV (Fadley et al., 1968).

A similar model to the one sketched above has been applied to the case of an atom in a molecule. The measured chemical shifts are fitted to the following expression (Siegbahn et al., 1969):

$$\Delta(B.E.) = kq_A + V_A + l \qquad (8\text{-}16)$$

where k and l are empirical constants determined from the fitting and q_A is the atomic charge on atom A; V_A is a molecular potential analogous to the crystal potential for a solid. In estimating the atomic charge one can use methods with a broad range of sophistication. Even such fairly empirical methods as the Pauling electronegativity concept (Pauling, 1960) appear to give useful chemical information for some systems (Siegbahn et al., 1969).

A procedure for correlating measured shifts in binding energies with thermodynamic data has also been made (Jolly, 1971). The central assumption of this model is that atomic cores that have the same charge are chemically equivalent. An element with an inner-shell vacancy is thus equivalent to the neutral element one step below in the periodic table. A good correlation is obtained between measured binding-energy shifts and chemical-reaction energies.

Thus, it seems clear that precise measurements of core-electron binding energies by means of x-ray photoelectron spectroscopy can yield considerable information of physical and chemical interest. Chemical shifts of these energies can be correlated with valence-electron structure, although the prediction of shifts to within ~ 1 eV may require the consideration of several important corrections and theoretical calculations of high precision.

IV. VALENCE-ELECTRON LEVELS IN SOLIDS AND GASES

Valence-electron structure is a fundamental concept in understanding the physical and chemical properties of matter, as it is the valence electrons that in many cases determine these properties. Much of the work in XPS on core-level energies is actually aimed at obtaining information about the valence electrons by means of their effects on the core levels (Siegbahn et al., 1967; Fadley, 1970). However, the valence-electron structure of both solids and gases can also be directly studied with the XPS technique, yielding unique information. Such studies will be reviewed in this section.

There are a number of experimental methods for studying the outer-electron structure of both solids and gases. Many of these involve the interaction of photons with matter. In optical studies, for example, one measures the absorption and emission as a function of photon energy. An unambiguous assignment of the corresponding electron transitions is usually very difficult to make since only the energy difference between the final and initial states is known. Furthermore such studies are limited to easily accessible spectral regions of the photons. In photoelectron spectroscopy, by contrast, the final state is directly measurable in terms of energy and sometimes also in terms of properties which are reflected in the angular- or spin-polarization distributions of the emitted electrons. In solid-state physics, there are also a number of methods to study the electronic properties in metals in the immediate vicinity of the Fermi surface, but relatively few which probe deeper into the energy bands.

Most studies of the valence-electron levels of both solids and gases by means of photoelectron spectroscopy have been done with the UPS technique (Eastman, 1971; Turner et al., 1970). As mentioned previously, the easily obtainable radiation sources have placed an upper limit of about 40 eV in photon energy for UPS measurements. This limit excludes the study of the bottom part of the electron bands in solids and deep-lying levels in gases with the UPS technique. There are also other considerations that make a study of the outer-electron levels with the XPS technique a meaningful undertaking. We will review the most important aspects of such measurements and will restrict ourselves to the XPS technique, but we shall make comparisons, where appropriate, with the UPS technique.

In the same way that an XPS spectrum reflects the core-electron energy levels it will also reflect the occupied part of the electron density of states in a solid. However, in the latter case there are several important contributions to the photoelectron spectrum that make it nontrivial to extract a true picture of the density of states (Fadley and Shirley, 1970a).

The energy distribution of excited electrons using a one-electron-transition model is given by

$$I(E_{\text{kin}}) \propto \int \overline{\sigma(E')}\rho(E')\rho(E' + h\nu)F(E')L(E - E')\, dE' \qquad (8\text{-}17)$$

where E' is the energy of the initial state and $E' + h\nu$ is that of the final state. The energy distribution is modulated both by the density of states at the initial energy $\rho(E')$ and at the final energy $\rho(E' + h\nu)$. The factor $\overline{\sigma(E')}$ is an average cross section for all states at the energy E', and $F(E')$ is the Fermi function. Instrumental and natural line-width contributions are gathered together in the function $L(E)$ and will be discussed below. In principle, there should also be an equation stating momentum conservation. However, due to the high photon energy employed in XPS measurements, the final state is to a very good approximation free-electron-like, thereby reducing the importance of the selection rules imposed by the momentum-conservation law. By the same argument, the density of states at the final energy is nearly constant. The structure in the energy-distribution curve is therefore mainly caused by variation in $\bar{\sigma}$ and ρ as a function of energy.

In escaping into vacuum, the photoexcited electrons may suffer inelastic collisions which will further modulate the spectrum. This effect can be corrected for in XPS by using a narrow core level in the sample (Fadley and Shirley, 1970a). The shape of the corresponding core photoelectron spectrum is determined by the function $L(E)$ and the energy-loss spectrum. By making reasonable assumptions about the form of $L(E)$, one can isolate the energy-loss spectrum and use this to correct the energy-band spectrum for inelastic-scattering events. Such a corrected spectrum is shown in Fig. 8-8.

The function $L(E)$ includes the following contributions to the line shape:

1 Natural line width of the excitation radiation. For Mg $K\alpha_{1,2}$ this amounts to about 0.8 eV. A reduction in this to about 0.5 eV can be obtained by using a crystal monochromator. The same reduction can be obtained by using the Y $M\zeta$ line (Krause, 1971).

2 Spectrometer resolution. This contribution has been discussed in Section II. It usually amounts to 0.2 to 0.6 eV.

3 Lifetime broadening of the final-hole state of the system.

The cross section $\sigma(E)$ can usually be considered to be roughly constant for states with the same symmetry (e.g., *d* bands or *s* bands). However, states with different

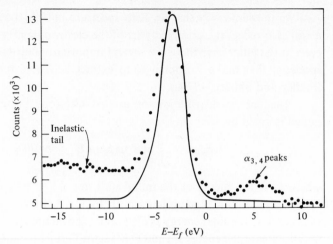

FIGURE 8-8
Valence-band photoelectron spectrum of Cu excited by Mg x rays. The full line represents the corrected spectrum obtained after allowance for the effects of inelastic scattering and Mg $K\alpha_{3,4}$ x rays, and the dots are the original data. (*After Fadley and Shirley, 1970a.*)

symmetries may show large differences. This assumption of constancy deserves further study, however.

In comparing the two photoemission techniques for studying the band structure of solids, we note the following main differences between UPS and XPS:

1 The resolution is usually better in UPS (0.2 to 0.3 eV) than in XPS (~ 1.0 eV), mainly due to the difference in line width between the exciting radiations.

2 In the XPS case, the final state is free-electron-like, while in the UPS case the final state is still influenced strongly by the periodic potential of the lattice giving rise to band-structure effects that impose selection rules on the transitions.

3 The high photon energy employed in XPS makes the entire band structure easily accessible for study. Furthermore, core levels can be used to correct for the inelastic scattering of photoemitted electrons. Core-level studies can also be used as an inherent analytical tool to study the chemical composition of the sample surface. The presence of oxygen can, for example, be monitored very sensitively (Fadley and Shirley, 1968; Hagström et al., 1971), and chemical shifts of contaminant peaks may even give some information about the chemical state of the surface.

4 The energy-distribution curve of photoemitted electrons is strongly dependent on the surface cleanliness. This seems to be more so in UPS than in XPS.

5   The cross section of certain types of valence electrons may be larger for x rays than for ultraviolet photons. For example, the cross section of f electrons has a maximum of several hundred eV above threshold. The $4f$ levels in the rare earths can therefore be advantageously studied by means of XPS (Hagström, 1971).

Systematic XPS studies of transition metals have been reported by Fadley and Shirley (1970a) and Baer et al. (1970a). The studies included $3d$, $4d$, and $5d$ elements. A summary of the results obtained by Fadley and Shirley is shown in Fig. 8-9. The spectra of the lighter elements appear relatively structureless, while the spin-orbit splitting is resolvable for the heavier elements, this effect being most clearly depicted for HgO. The filling of the d bands with increasing atomic number, and the narrowing of the d states into core-like levels, is also clearly indicated in this figure. The higher resolution obtainable in UPS is illustrated in Fig. 8-10, which shows a spectrum for Cu obtained with the UPS technique (Lindau and Hagström, 1971). The spectrum was obtained with a very narrow line width, 21.2-eV photon energy for excitation. However, even at this high photon energy the deviation of the final state from a free-electron-like state appears to be appreciable, as is demonstrated for Au in Fig. 8-11.

Figure 8-11a shows UPS spectra from gold obtained with the photon energies 10.2, 11.6, 16.8, 21.2, and 26.9 eV (Eastman, 1971). For comparison we also show two XPS spectra (Fig. 8-11b and c) from Au recorded with and without a crystal monochromator to narrow the x-ray line width (Hewlett-Packard). It is apparent from the figure that the structure in the UPS spectra is strongly dependent on the photon energy, indicating modulation effects of the density of states in the final state. One can also note that the largest similarity between the UPS and XPS spectra is obtained for the highest UV photon energy. This seems to be a general trend, and it thus appears that high-photon-energy UPS spectra and XPS spectra represent the best experimental picture of the density-of-states function $\rho(E)$.

Furthermore, we note from the spectra shown that the width of the band is not easily obtained from the UPS spectra. Even if the excitation photon energy is large enough to excite photoelectrons from the bottom of the band into the vacuum, the low-energy part of the photoelectron spectrum is smeared out by inelastically scattered electrons. If this effect is also present in XPS spectra, it can be corrected for by means of the recording of a sharp core level, as described earlier (Fadley and Shirley, 1970a). This particular feature of XPS is, of course, most valuable in the study of wide-band solids. For example, in a study on carbon in the form of graphite and diamond, occupied valence-band widths of 31 and 33 eV, respectively, were reported (Thomas et al., 1971).

The XPS technique has also been applied to the study of the outer levels in the

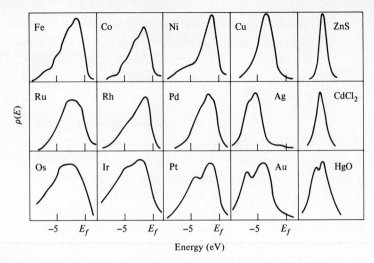

FIGURE 8-9
Summary of corrected valence-band photoelectron spectra of the 15 solids
studied by Fadley and Shirley (1970*a*).

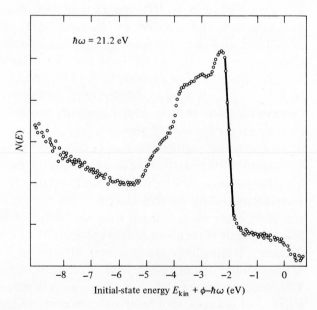

FIGURE 8-10
Photoelectron spectrum of Cu obtained with a photon energy of 21.2 eV by
Lindau and Hagström (1971).

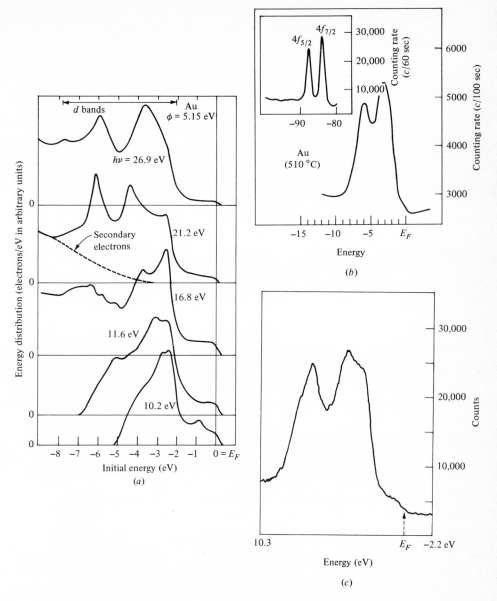

FIGURE 8-11

Photoemission spectra from the 5d electron band structure of gold. (a) UPS spectra obtained with photon energies 10.2, 11.6, 16.8, 21.2, and 26.9 eV. (*After Eastman*, 1971.) (b) The corresponding XPS spectrum obtained with unfiltered Mg x rays. (*After Baer et al., 1970a*.) (c) Finer details in the *d* bands resolved using crystal monochromatized x rays (Al *K*α). (*Obtained by Hewlett-Packard Company using 256 channels per 12.5 eV during 1 hour*.)

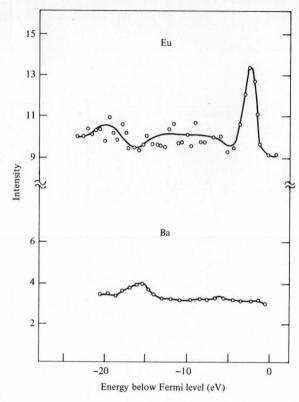

FIGURE 8-12
XPS spectra of Eu and Ba which are isoelectronic except for the 4f electrons.
The strong excitation of the 4f electrons with x rays is clearly seen in the figure.
(*After Brodén et al.,* 1971.)

rare-earth metals (Hagström, Hedén, and Löfgren, 1970). The interesting problem
with regard to the density of states in this series of metals relates to the 4f electrons.
Even though these electrons are energetically accessible to both optical and UPS
studies, it has proved difficult to obtain reliable information on them from such data,
due to the small oscillator strengths of the corresponding transitions. As mentioned
previously, the photoelectric cross section for these electrons increases with energy
to a maximum which is several hundred eV away from the threshold (Combet-
Farnoux, 1969). In XPS spectra, the 4f electrons are therefore strongly excited.
This is illustrated in Fig. 8-12, which shows the spectrum from Eu (Hedén, Löfgren,
and Hagström, 1971). For comparison the spectrum of Ba is also shown. This element
has the same electron and crystal structure as Eu except for the 4f electrons. Spectra
similar to those for Eu have been reported for the elements Gd, Yb, and Lu, all of

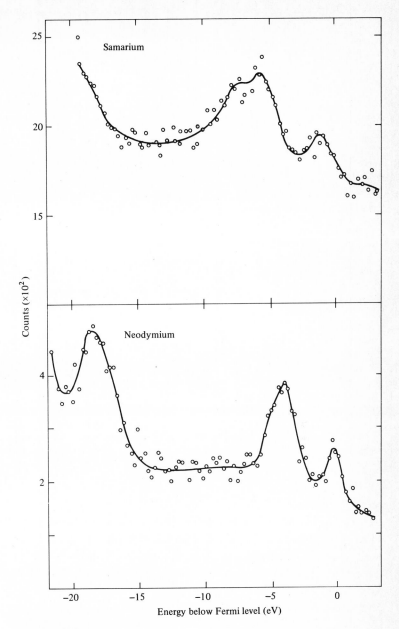

FIGURE 8-13
XPS spectra of Sm and Nd, showing the effect of multiplet splittings in the final
state. (*After Hagström et al.*, 1971.)

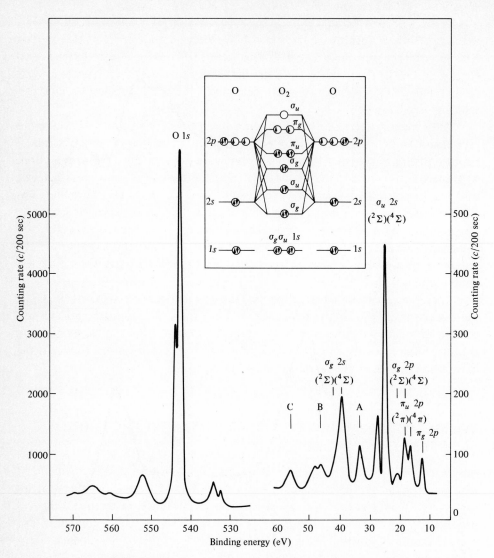

FIGURE 8-14
XPS spectrum of core and valence electrons in gaseous oxygen excited by Mg $K\alpha$ radiation. (*After Siegbahn et al.*, 1969.)

which have a filled or half-filled 4f shell (Hedén, Löfgren, and Hagström, 1971). For the elements which do not have these low-energy electron configurations, the recorded XPS spectra are more complicated, as illustrated in Fig. 8-13 (Hedén, Löfgren, and Hagström, 1971). The extended structure in this case has been interpreted as a result of the final states not being uniquely determined, but rather subject to multiplet splittings (see Section V).

Even though the resolution limitation in XPS as compared to UPS is more pronounced in photoelectron spectra from the valence electrons of gaseous samples, the high photon energy of XPS makes it possible to extract unique information from XPS spectra. Figure 8-14 shows the XPS spectrum for oxygen with an assignment of the molecular orbitals (Siegbahn et al., 1969). In this figure is also shown the core-electron spectrum with a multiplet splitting of the 1s level as explained in Section V. The lines marked A, B, and C are interpreted as being due to multielectron processes and energy-loss effects. It is obvious from the figure that the high energy of the x-ray photons employed is an advantage in reaching the molecular orbitals with the highest binding energies.

Since electrons with different angular momenta will have different cross sections, the relative intensities of the photoelectron lines from the molecular orbitals will contain information on their symmetry properties (Gelius, 1971). Similar information is also contained in the angular distribution of the photoelectrons. However, the latter field has not yet been explored to any extent.

V. MULTIPLET SPLITTINGS AND MULTIELECTRON PROCESSES

In this section, we shall discuss two distinct, but closely related, sources of additional structure in photoelectron spectra. Multiplet splittings of electron binding energies can be described in terms of a one-electron-transition model of the photoemission process. On the other hand, multielectron processes involve the excitation of two or more electrons during photoemission and represent a fundamentally different effect. Both these effects have the potential of adding considerably to the information inherent in a photoelectron spectrum.

The fundamental energy-conservation relation governing the photoemission process has been stated in equation (8-1). As we have noted, the binding energy of an electron from the kth subshell is by definition given by B.E.$(k) = E^f(k) - E^i$. (For the present discussion, a specification of the reference level for binding energies is unnecessary.) Thus, if more than one energy $E^f(k)$ is possible for the final state of the system with a hole in the kth subshell, the binding energy will be split into components. Each of these components can then be specified by the quantum numbers of the different final states. If E^i is assumed to be a unique energy for all measurements on a given system, the separations of the different binding energies will be independent of it and will be given simply by some difference $E^f(k)_1 - E^f(k)_2$.

Thus, the reference level for absolute binding energies is not critically important in the analysis of structure due to various final-state energies. In general, $E^f(k)$ may exhibit several values because of energy associated with rotational, vibrational, or electronic motion. Because of the present resolution limit of ~ 1.0 eV in XPS studies, only the effects of electronic motion have so far been detected. By contrast, the effects of vibrational excitations on $E^f(k)$ are very commonly seen in the analyses of higher resolution UPS spectra from the valence levels of gaseous molecules (Turner et al., 1970). In multiplet splittings, the different $E^f(k)$ values arise from a detailed consideration of the potential experienced by the electron ejected, including perhaps the coupling of the final-state hole to the valence electrons. The various $E^f(k)$ values occurring in connection with multielectron transitions involve the excitation of one or more additional electrons to different bound subshells or perhaps to free continuum states; in this case, therefore, the final states may differ in the quantum numbers of the additional excited electrons, plus any further differences introduced by multiplet effects. Thus, it is clear that the two types of structure are closely related. If a multielectron process involves the excitation of additional electrons into the continuum, $E^f(k)$ will also include the kinetic energies of these electrons.

Before proceeding to discuss these two effects in detail, we note several other sources of structure in photoelectron spectra which may be erroneously attributed to binding-energy splittings or multielectron processes (Fadley, 1971): (1) Auger-electron peaks: These are easily distinguished because they have a constant kinetic energy, regardless of excitation source. (2) Photoelectron peaks arising from weak satellite or impurity x rays: These will usually have fixed relative intensities and positions in comparison to the intense x-ray components and can also be detected by utilizing different x-ray sources (e.g., Al and Mg). (3) Inelastic-scattering processes in the photoelectron escape from the sample: These will cause almost identical low-kinetic-energy tails on each photoelectron peak, particularly for peaks close together in kinetic energy. (4) Chemical effects on electron binding energies caused by spurious chemical reaction in the sample (as, for example, at the surface of a solid sample): Such chemical effects are more difficult to distinguish, but observations of the positions, shapes, and relative intensities of core and valence photoelectron peaks for all the constituents of a sample will generally give a very good indication of its chemical state. Thus, with careful analysis of photoelectron spectra, it is generally possible to distinguish structure due to multiplet splittings and multielectron processes. It may not be possible, however, to separate structure due to the two effects, since multielectron processes often exhibit their own multiplet splittings.

A.   Multiplet splittings of electron binding energies

We shall here review the experimental and theoretical aspects of electron binding-energy *splittings*, as contrasted with the core-electron binding-energy *shifts* discussed in Section III. A binding-energy shift can be considered to be the result of a change in the spatially averaged Coulomb and exchange potentials exerted upon an electron by its environment. Because the potential has been averaged over the region of the electron, the initial and final states in the photoemission process can be considered in first approximation to be closed-shell systems with one hole present in the final state. In this model, therefore, the resulting binding energies for different core electrons are fully characterized by the usual one-electron quantum numbers for free atoms: n, l, and j. In describing valence-electron binding energies, on the other hand, the various electron-electron and electron-nuclear interactions must be considered in detail, with the result that quantum numbers appropriate to the symmetry of the total atomic ensemble must be used. For example, an atomic state may be specified by d^5 ^6S, a homonuclear diatomic molecular state by $(\sigma_g)^2$ $^1\Sigma_g$, or a state in an octahedral environment by $(t_{2g})^2$ T_1. Thus, multiplet splittings form an integral part of the study of photoelectron spectra from valence electrons (Turner et al., 1970). If similar considerations are applied to the description of core-electron states or, in particular, states with a single hole in a core level, the binding energies are found to be split into several components relative to a description in terms of n, l, and j alone. Such splittings we shall term "multiplet splittings," since they arise from the various electronic multiplet states occurring in atoms, molecules, and solids. Multiplet effects involving core-electron holes are very commonly used in discussing structure observed in x-ray emission spectra (Coster and Druyvesteyn, 1927; Nefedov, 1964a, 1964b, 1966; Ekstig et al., 1970) and Auger-electron spectra (Burhop, 1952; Asaad and Burhop, 1958; Stalherm et al., 1969). However, it is only recently that such effects were first discussed in connection with x-ray photoelectron spectra emanating from core levels (Novakov and Hollander, 1968, 1969; Hedman et al., 1969; Siegbahn et al., 1969; Fadley, Shirley et al., 1969; Fadley and Shirley, 1970b). The measurement and interpretation of such splittings have recently been reviewed in more detail by Fadley (1971).

As a simple illustration of one type of multiplet splitting found in XPS studies (Fadley, Shirley et al., 1969; Fadley and Shirley, 1970b), we consider photoemission from the 3s level of an Mn^{++} free ion, as shown in the left portion of Fig. 8-15. The ground state of this ion can be described in Russell-Saunders (L-S) coupling as $3d^5$ ^6S. In this state, the five 3d spins are coupled parallel. Upon ejecting a 3s electron, however, two final states may result: $3s3d^5$ ^5S or $3s3d^5$ ^7S. The difference between these two states is that in the ^5S state, the spin of the remaining 3s electron is coupled

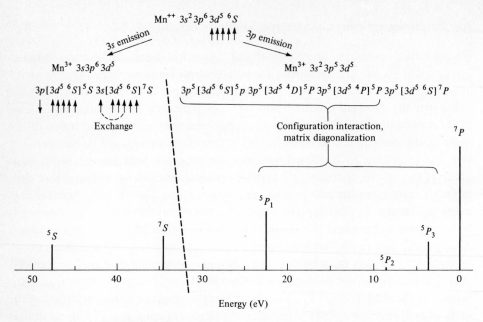

FIGURE 8-15
The various possible final-state multiplets arising from $3s$ and $3p$ photoemission from an Mn^{++} ion. The theoretical separations and relative intensities of these multiplets are also indicated. (*After Fadley, 1971.*)

antiparallel to those of the five $3d$ electrons, whereas in the ^7S state the $3s$ and $3d$ spins are coupled parallel. Because the exchange interaction acts only between electrons with parallel spins, the ^7S energy will be lowered relative to the ^5S energy because of $3s$-$3d$ exchange (Slater, 1960). The magnitude of this energy separation will be proportional to the $3s$-$3d$ exchange integral, and will be given by (Slater, 1960)

$$
\begin{aligned}
\Delta[\text{B.E.}(3s)] &= \Delta E^f(3s3d^5) \\
&= E^f(3s3d^5\ {}^5S) - E^f(3s3d^5\ {}^7S) \\
&= \tfrac{6}{5} \cdot e^2 \int_0^\infty \int_0^\infty \frac{r_<^2}{r_>^3} R_{3s}(r_1)R_{3d}(r_2)R_{3s}(r_2)R_{3d}(r_1)\,dr_1\,dr_2
\end{aligned}
\tag{8-18}
$$

where e is the electronic charge, $r_<$ and $r_>$ are chosen to be the smaller and larger of r_1 and r_2 in performing the integration, and $R_{3s}(r)/r$ and $R_{3d}(r)/r$ are the radial wavefunctions for $3s$ and $3d$ electrons. A calculation of the energy splitting in equation (8-18) for Mn^{3+} gives a value of $\Delta E^f(3s3d^5) \simeq 13$ eV (Fadley, Shirley et al., 1969; Fadley and Shirley, 1970*b*). A generalization of the relationship for the energy splitting in equation (8-18) shows that for a final-state configuration sl^n, where l can be any angular momentum, the energy separation $\Delta E^f(sl^n)$ between the two possible final spin states will be proportional to the multiplicity of the initial state (Slater, 1960). That is, if S is the total spin of the l^n configuration in the initial state, there will be two possible final states with spins of $S \pm \frac{1}{2}$, and energy separations equal to

$$\Delta[\text{B.E.}(s)] = \Delta E^f(sl^n) \propto (2S + 1) \qquad S \neq 0 \qquad (8\text{-}19)$$

or

$$\Delta[\text{B.E.}(s)] = \Delta E^f(sl^n) = 0 \qquad\qquad S = 0 \qquad (8\text{-}20)$$

Thus, such s-electron binding-energy splittings have the potential of directly measuring the spin properties of valence electrons. Similar splittings are expected for non-s electrons, although their interpretation is less straightforward (Fadley, Shirley et al., 1969; Fadley and Shirley, 1970b).

The first observation of s-electron binding-energy splittings analogous to those described by equations (8-19) and (8-20) were in gaseous, paramagnetic molecules (Hedman et al., 1969; Siegbahn et al., 1969). Hedman et al. (1969) found splittings as large as 1.5 eV in the $1s$ photoelectron spectra of the molecules NO and O_2. These results are shown in Fig. 8-16 along with an unsplit $1s$ spectrum from the diamagnetic molecule N_2. In each case, it can be shown that the observed energy splitting should be proportional to an exchange integral between the unfilled valence orbital and the $1s$ orbital of N or O (Siegbahn et al., 1969), in analogy with equation (8-18). Theoretical estimates of these splittings from molecular orbital calculations give values in reasonable agreement with experiment (Hedman et al., 1969). The observed intensity ratios of the peaks are very close to the ratios of the final-state degeneracies (Siegbahn et al., 1969), also in agreement with theory (Cox and Orchard, 1970).

As might be expected from the 13-eV splitting predicted for the $3s$ binding energy of Mn^{++}, rather large splittings have been observed in solid compounds containing $3d$ transition metals. Fadley, Shirley et al. (1969) first observed such splittings in Mn and Fe compounds. The magnitudes of these $3s$ splittings are as large as 7 eV (Fadley, Shirley et al., 1969; Fadley and Shirley, 1970b). Figures 8-17 and 8-18 show the x-ray photoelectron spectra from some of these compounds in the region corresponding to $3s$ and $3p$ emission (Fadley, Shirley et al., 1969; Fadley and Shirley, 1970b). Roughly, the left half of each spectrum is the $3s$ region, and doublets are observed there for all systems containing unpaired electrons (MnF_2, MnO, MnO_2, FeF_3, and Fe metal). The diamagnetic compounds $K_4Fe(CN)_6$ and $Na_4Fe(CN)_6$ exhibit no splitting of the $3s$ photoelectron peak. These results are thus qualitatively consistent with the predictions of equations (8-19) and (8-20). However, free-ion calculations for Mn^{++} based on several models yield $3s$ splittings 1.5 to 2.0 times as large as the observed values (Fadley, Shirley et al., 1969; Fadley and Shirley, 1970b). One likely cause for this discrepancy is spatial delocalization and spin-pairing of the $3d$ electrons in the formation of chemical bonds (Fadley, Shirley et al., 1969). The

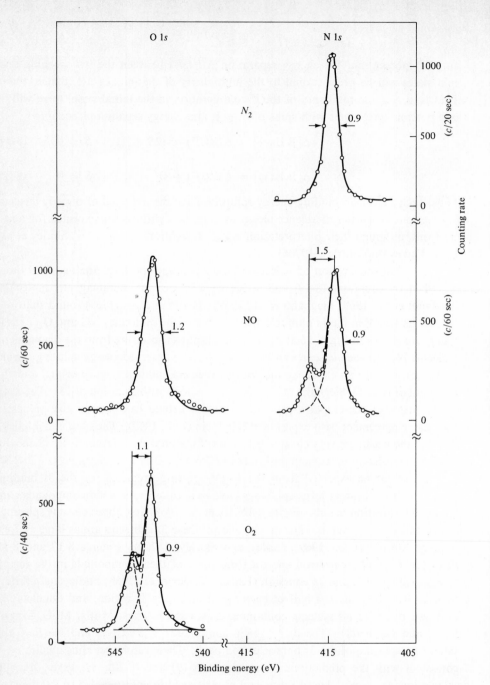

FIGURE 8-16
XPS spectra from $1s$ electrons of the gaseous molecules N_2, NO, and O_2. The peaks from the paramagnetic molecules NO and O_2 are split due to final-state multiplets. (*After Siegbahn et al., 1969.*)

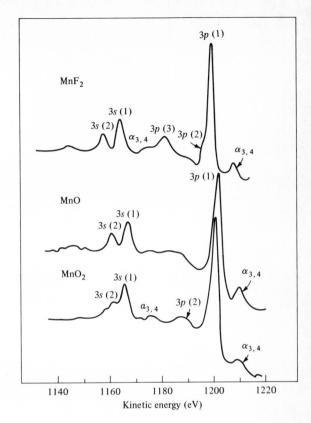

FIGURE 8-17
XPS spectra from the solid compounds MnF_2, MnO, and MnO_2 in the kinetic-energy region corresponding to ejection of Mn $3s$ and $3p$ electrons. (*After Fadley and Shirley*, 1970*b*.)

possible origins of this discrepancy are discussed in detail elsewhere (Fadley, 1971; Fadley, Shirley et al., 1969; Fadley and Shirley, 1970*b*). $3s$ binding-energy splittings or broadenings have also been observed in the ferromagnetic transition metals Fe, Co, and Ni by Fadley and Shirley (1970*b*).

Inspection of the $3p$ regions of the spectra in Figs. 8-17 and 8-18 indicates that they do not show the same simple doublet structure as that observed in the $3s$ region. The primary reason for this is that the orbital angular momentum of the $3p$ electron permits several possible L-S couplings within the $3p^5 3d^5$ final-state configuration (Fadley, Shirley et al., 1969; Fadley and Shirley, 1970*b*). These states are indicated on the right-hand side of Fig. 8-15. Only one ^7P state is possible, but three ^5P states can be formed by coupling $3p^5$ (which must couple to $L = 1$, $S = \frac{1}{2}$) with $3d^5$ (which can couple to $L = 0$, $S = 5/2$; $L = 1$, $S = 3/2$; or $L = 2$, $S = 3/2$) (Slater, 1960).

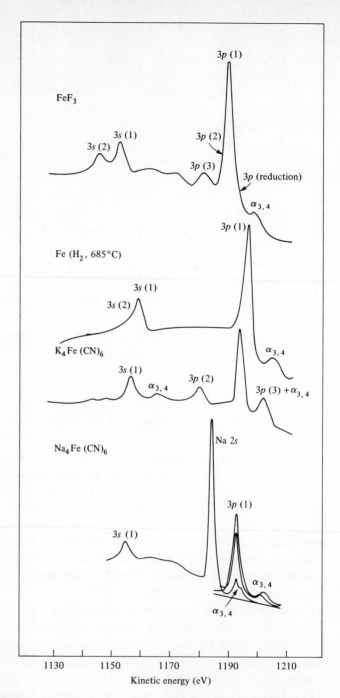

FIGURE 8-18
XPS spectra from solids containing Fe in the kinetic-energy region corresponding
to emission of Fe 3s and 3p electrons. (*After Fadley and Shirley, 1970b.*)

Thus, four final multiplet states are possible, and the actual ^5P states will be some linear combination of the three different $3p^53d^5$ coupling schemes (Fadley, Shirley et al., 1969; Fadley and Shirley, 1970b), derived most simply by a matrix-diagonalization procedure. Free-ion calculations of $3p$ splittings for Mn^{++} based on this model yield results in qualitative agreement with the experimental results for MnF_2 (Fadley, Shirley et al., 1969; Fadley and Shirley, 1970b). These are summarized at the bottom of Fig. 8-15. An important difference between s and non-s binding-energy splittings is that both Coulomb and exchange integrals are involved in determining the magnitudes of the splittings. Other complications inherent in the interpretation of splittings of non-s electrons are the presence of spin-orbit coupling and crystal-field effects (Fadley, 1971; Fadley, Shirley et al., 1969; Fadley and Shirley, 1970b).

As Fig. 8-15 indicates, the relative intensities of the various components in the $3s$ and $3p$ photoelectron spectra are also qualitatively consistent with theory, although certain discrepancies have been noted and discussed (Fadley, 1971; Fadley, Shirley et al., 1969; Fadley and Shirley, 1970b). Other core levels in $3d$ transition-metal atoms should also exhibit similar splittings, although the magnitudes will be reduced because of the decreased overlap between the core-electron wavefunction and the $3d$ wavefunction. The $2p$ levels in MnF_2 are found to be broadened by ~ 1.5 eV relative to low spin compounds, and this broadening is probably connected to multiplet effects (Fadley and Shirley, 1970b)

Multiplet-splitting measurements have also been made for a number of solid compounds containing atoms with unfilled valence shells (Novakov, 1971a; Carver et al., 1971; Helmer, 1971; Wertheim et al., 1971). Novakov (1971a) has measured $3s$ splittings in several $3d$-metal compounds. Carver et al. (1971) have measured s-electron splittings in a number of $3d$- and $4d$-metal compounds. Helmer (1971) has observed what appears to be multiplet effects on the Cr $2p$, Cr $3s$, and Cr $3p$ binding energies in Cr_2O_3. Similar multiplet splittings should also be observed in the rare-earth elements, where the partially filled $4f$ shell can couple in various ways to a core-electron hole (Fadley and Shirley, 1970b). Measurements on *gaseous* Eu indicate probable multiplet effects on the $4d$ photoelectron spectrum (Fadley and Shirley, 1970b). Recently, Wertheim et al. (1971) have reported multiplet-splitting measurements on solid rare-earth compounds.

Finally, we note that multiplet splittings may be important factors in the interpretation of valence-band photoelectron spectra from solids (Fadley and Shirley, 1970a). In particular, multiplet effects account very well for the changes in width observed in XPS spectra from the $4f$ electrons in various rare-earth metals (Hedén, Löfgren, and Hagström, 1971; Brodén, 1971). These effects are also consistent with a broadening observed in the $4f$ photoelectron peak from gaseous Eu (Fadley and Shirley, 1970b).

Core-electron binding-energy splittings were first observed by Novakov and

Hollander (1968; 1969) in x-ray photoelectron spectra from the core levels of the heavy metals Th, U, and Au in various solid compounds. Some of their results for Au compounds are shown in Fig. 8-19 (Novakov and Hollander, 1969). Splittings as large as 10 eV were observed in some solids (Novakov and Hollander, 1968). However, these splittings cannot be explained in terms of free-atom-like multiplets such as those discussed above. It has been proposed that they may be due to crystal-field splittings of the core-electron levels (Novakov and Hollander; 1969, Apai et al., 1969). Calculations based on first-order perturbation theory do not seem adequate to quantitatively explain the observed results (Apai et al., 1969). A theoretical formulation for interpreting such splittings has been proposed by Gupta, Rao, and Sen (1971), but no detailed analyses of the experimental data have as yet been made. More recently, Novakov (1971a) has measured the known 2-eV crystal-field splitting of the valence $3d$ levels in $CoSO_4$. Thus, it appears that crystal-field effects may be important considerations in the interpretation of the core-electron binding energies of certain systems. In contrast with free-atom-like multiplets, crystal-field effects should also be observable in closed-shell systems.

In comparison to chemical shifts of core-electron binding energies, multiplet splittings of these energies thus represent higher order effects yielding a different type of information. Chemical-shift measurements detect a change in the average potential experienced by an electron, whereas multiplet effects appear to have the capability of determining the valence-electron configuration of an atom or the detailed potential around it. The two types of measurements are thus complementary. Numerous applications of multiplet-splitting measurements seem possible in the study of the transition-series metals, the rare earths, the transuranium elements, and open-shell systems in general.

B.   Multielectron processes

Multielectron processes in connection with x-ray absorption were first studied in detail by Krause, Carlson, and coworkers (Krause et al., 1964; Krause, Carlson, and Dismukes, 1968; Carlson and Krause, 1965a, 1965b; Krause and Carlson, 1966, 1967; Carlson, Hunt, and Krause, 1966). In these studies, gaseous neon and argon were exposed to x rays having energies in a range from 270 eV to 1.5 keV. Measurements were then made of both the charge distributions of the resulting ions and the kinetic-energy distributions of the ejected photoelectrons. From these studies, it was concluded that two-electron and even three-electron transitions occur in photoabsorption, with total probabilities which may be as high as 20 percent for each absorbed photon. It was also pointed out that the probability for multielectron processes should increase from zero for a value of hv equal to the one-electron-transition binding energy to a constant value for hv equal to a few times this value.

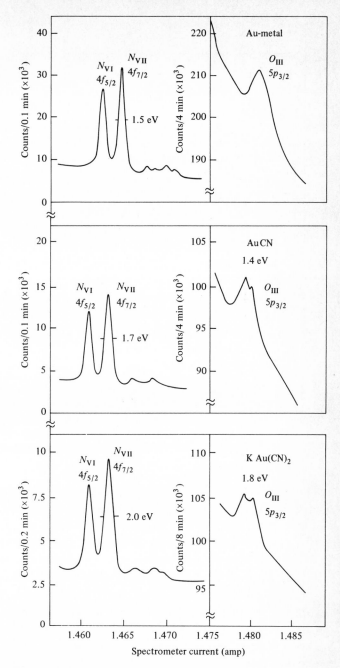

FIGURE 8-19
XPS spectra from several solids containing Au atoms. The splittings of the Au $5p_{3/2}$ peaks in AuCN and KAu(CN)$_2$ and the corresponding broadenings of the Au $4f_{7/2}$ peaks are believed to be due to crystal-field splittings. (*After Novakov and Hollander*, 1969.)

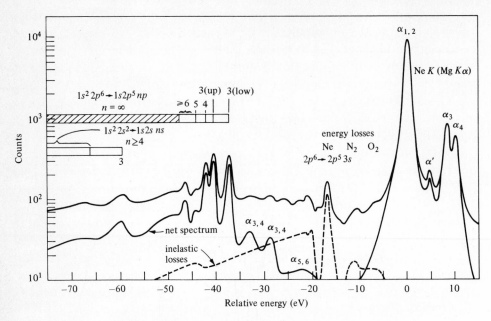

FIGURE 8-20

Photoelectron spectrum of Ne $1s$ electrons excited by Mg $K\alpha$ x rays. Energy given relative to principal peak, whose kinetic energy is 616 eV. Dotted line shows inelastic losses as determined in separate experiments. Net spectrum is derived by subtracting the background and contribution from inelastic scattering from the gross data. (*After Carlson, Krause, and Moddeman*, 1971.)

Thus, x-ray photoemission experiments with $h\nu \approx 1.5$ keV should exhibit a maximum effect from multielectron processes for electrons whose binding energies are less than approximately 500 eV. Theoretical predictions of the intensities of multielectron processes based on one-electron models give reasonable agreement with experiment if the electrons involved in the transition are in different subshells (e.g., the $1s$ and $2p$ electrons in neon) (Krause, Carlson, and Dismukes, 1968; Siegbahn et al., 1969; Carlson, Krause, and Moddeman, 1971). However, if the electrons are in the same subshell (e.g., the $2s$ and $2p$ electrons in neon), electron-correlation effects appear to increase the experimental multielectron intensities above those based on a one-electron model (Krause, Carlson, and Dismukes, 1968; Carlson, Krause, and Moddeman, 1971).

Higher resolution XPS spectra have been obtained recently for neon by Siegbahn et al. (1969) and for neon and helium by Carlson, Krause, and Moddeman (1971). A spectrum obtained in the most recent study on neon (Carlson, Krause, and Moddeman, 1971) is shown on a semilogarithmic plot in Fig. 8-20. The notations α_3, α_4, etc., indicate photoelectron peaks due to the various satellite x rays of Mg (themselves the result of multielectron excitations in the Mg anode of the x-ray tube). The dotted curve represents intensity due to inelastic scattering. The solid curve is thus the net spectrum due to elastically scattered photoelectrons. The two-electron transitions responsible for the observed spectrum intensity at relative energies less than ~ -36 eV are diagrammatically indicated in Fig. 8-20. The total two-electron intensity in this spectrum is estimated to be approximately 10 percent of that of the one-electron peak (Carlson, Krause, and Moddeman, 1971).

The theoretical assignment of peak positions and relative intensities in multi-electron spectra has usually been done with a model based on the "sudden approximation" (Krause et al., 1964; Krause, Carlson, and Dismukes, 1968; Carlson and Krause, 1965a, 1965b; Krause and Carlson, 1966, 1967; Carlson, Hunt, and Krause, 1966; Carlson, Krause, and Moddeman, 1971; Siegbahn et al., 1969; Åberg, 1967; Manne and Åberg, 1970). This model has also been used in the explanation of satellite x rays in terms of multielectron excitations (Åberg, 1967). A concise review of this model has been presented recently by Manne and Åberg (1970). The fundamental assumption is that the primary photoabsorption process which excites a core electron from the kth subshell into a continuum photoelectron state occurs so rapidly that the valence electrons do not have time to readjust to the change in the potential due to the core. The primary excitation is considered in this sense to be "sudden." A theoretical criterion for the validity of the sudden approximation is that (Åberg, 1967)

$$|E^f(k,k') - E^f(k)| \frac{\tau}{\hbar} \ll 1 \qquad (8\text{-}21)$$

where $E^f(k,k')$ is the total energy of a two-electron-transition final state involving electrons k and k' and τ is the time of transit of the k photoelectron past the k' subshell. This criterion has proved in practice to be overly restrictive, however (Åberg, 1967). A further assumption implicit in this treatment is that the one-electron orbitals describing the initial and final states do not change appreciably. The initial wavefunction of the N-electron system is denoted by $\Psi^i(N)$ and is expressed as an antisymmetrized product of the kth one-electron orbital and a remainder $\Psi_R(N-1)$. That is,

$$\Psi^i(N) = \mathbf{A}\phi_k\Psi_R(N-1) \qquad (8\text{-}22)$$

where $\mathbf{A}$ is an antisymmetrizing operator, and

$$H^i(N)\Psi^i(N) = E^i\Psi^i N \qquad (8\text{-}23)$$

where $H^i(N)$ is the N-electron Hamiltonian. Thus Ψ_R is *not* a wavefunction for the $N-1$ electron system. The actual final states will be described by wavefunctions $\Psi^f(N-1)_j \neq \Psi_R(N-1)$, such that

$$H^f(N-1)\Psi^f(N-1)_j = E^f(k)_j\Psi^f(N-1)_j \qquad (8\text{-}24)$$

It can then be shown (Manne and Åberg, 1970) that the probability of forming a given final state described by a wavefunction $\Psi^f(N - 1)_j$ will be given by

$$P_j = |\langle \Psi^f(N - 1)_j \,|\, \Psi_R(N - 1)\rangle|^2 \qquad (8\text{-}25)$$

This equation has been used in the most recent theoretical calculations of the relative intensities of the two-electron processes shown in Fig. 8-20 (Siegbahn et al., 1969). The form of equation (8-25) thus implies very strict "selection rules" between $\Psi^f(N - 1)_j$ and the somewhat fictitious state $\Psi_R(N - 1)$ (Siegbahn et al., 1969):

$$\Delta J = \Delta L = \Delta S = \Delta M_J = \Delta M_S = 0 \qquad (8\text{-}26)$$

That is, the multielectron processes are monopole transitions relative to $\Psi_R(N - 1)$. Manne and Åberg (1970) have also pointed out that a weighted-average binding energy over all one-electron and multielectron processes is precisely equal to a Koopman's theorem binding energy based on the one-electron energies from a Hartree-Fock calculation of the initial state.

It is clear from equation (8-26) that the various $L\text{-}S$ multiplets formed in the final state must be considered. For example, the peaks indicated as "3(low)" and "3(up)" in Fig. 8-20 are due to a multiplet splitting of the configuration-interaction type noted on the right-hand side of Fig. 8-15 for ^5P states. In the case of neon, two ^2S states are possible for the total configuration $1s2s^22p^53p$: one in which the $1s$ electron is coupled with $2s^22p^53p\ ^1S$ and one in which it is coupled with $2s^22p^53p\ ^3S$ (Siegbahn et al., 1969). Thus, there may be considerable interaction between multielectron processes and multiplet splittings.

If a one-electron orbital approximation is used in describing $\Psi^f(N - 1)_j$ and $\Psi_R(N - 1)$, the overall description of a two-electron transition is as indicated below:

$$\text{Overall transitions:} \qquad \Psi^i(N) \overset{h\nu}{\to} \Psi^f(N - 1)_j \cdot \phi_c \qquad (8\text{-}27)$$

$$\text{One-electron transitions:} \qquad \phi_k \to \phi_c \qquad (8\text{-}28)$$

$$\phi_{nl} \to \phi_{n'l'} \qquad (8\text{-}29)$$

where we have used ϕ_c to denote a continuum photoelectron state, ϕ_{nl} and $\phi_{n'l'}$ to denote the initial and final orbitals of the second electron, and where each index j will refer to specific values of n' and l'. If $\phi_{n'l'}$ is a bound orbital, the second electron is called a *shake-up electron*, whereas if it is a continuum state, it is termed a *shake-off electron* (Krause, Carlson, and Dismukes, 1968). The orbital approximation also yields a one-electron expression of equation (8-25) (Krause, Carlson, and Dismukes, 1968)

$$P_j = |\langle \phi_{n'l'} \,|\, \phi_{nl}\rangle|^2 \qquad (8\text{-}30)$$

and an approximate one-electron selection rule

$$l - l' = \Delta l = 0 \qquad (8\text{-}31)$$

Equations (8-30) and (8-31) have been used in the analysis of much experimental data

(Krause et al., 1964; Krause, Carlson, and Dismukes, 1968; Carlson and Krause, 1965a, 1965b; Krause and Carlson, 1966, 1967; Carlson, Hunt, and Krause, 1966; Carlson, Krause, and Moddeman, 1971), although equations (8-25) and (8-26) are more rigorously correct (Siegbahn et al., 1969).

Multielectron transitions also appear to be present in x-ray photoelectron spectra from gaseous molecules (Siegbahn et al., 1969; Carlson, Krause, and Moddeman, 1971) and solids (Fadley and Shirley, 1970b; Novakov, 1971b; Wertheim and Rosencwaig, 1971). Spectra from N_2 (Siegbahn et al., 1969; Carlson, Krause, and Moddeman, 1971), O_2 (Carlson, Krause, and Moddeman, 1971), and CO_2 (Carlson, Krause, and Moddeman, 1971) exhibit such structure with 10 to 15 percent probability. Multielectron satellite peaks have also been reported for potassium in inorganic salts (Fadley and Shirley, 1970b), for nickel and copper in oxides and halides (Novakov, 1971b), and for alkali metal and halogen atoms in the alkali halides (Wertheim and Rosencwaig, 1971). Photoelectron spectra from three potassium halides are shown in Fig. 8-21.

The effects of inelastic scattering are indicated by the dotted curve. The broad peaks observed at approximately 14 eV below the narrow peaks due to one-electron photoemission of K $3s$ electrons have been explained in terms of two-electron transitions connected with the photoemission of K $3p$ electrons (Wertheim and Rosencwaig, 1971). A broad two-electron peak similar to those in Fig. 8-21 has also been observed in $K_4Fe(CN)_6$ (Fadley and Shirley, 1970b). Wertheim and Rosencwaig (1971) suggest that these peaks are connected with final-state configurations of $3s^2 3p^4 4s$ and $3s^2 3p^4 3d$ and note that the position of the satellite peaks in energy can be approximately correlated with tabulated free-ion energies for these configurations of K^{++} (Moore, 1949). This interpretation satisfies the selection rule given in equation (8-26), but violates the one-electron selection rule given in equation (8-31). It is also inconsistent with the interpretation of two-electron peaks for Ar gas (Carlson, Krause, and Moddeman, 1971), for which the energies of the $\Delta l = 0$ configuration $3s^2 3p^4 4p$ (Moore, 1949) give excellent agreement with experiment. A check of the pertinent energy levels in K^{++} (Moore, 1949) makes it seem plausible that the two-electron structure observed in Fig. 8-21 could also be explained in terms of the configuration $3s^2 3p^4 4p$ satisfying both selection rules. Two-electron peaks with very high intensities have also been observed in connection with Rb $4s$ emission (Wertheim and Rosencwaig, 1971). These two-electron peaks are found to be more intense than the one-electron peaks. It should be noted in connection with both the K $3p$ and Rb $4s$ two-electron processes that the primary photoelectron is ejected from a subshell with the same principal quantum number as that of the excited electron. Thus, electron-correlation effects would be expected to increase the relative intensity of two-electron events, as has been discussed previously (Krause, Carlson, and Dismukes, 1968; Carlson, Krause, and Moddeman, 1971).

Novakov (1971b) has also observed very strong satellite peaks in connection

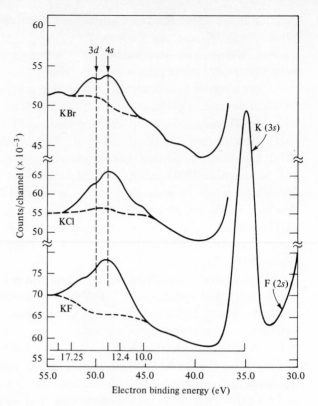

FIGURE 8-21
XPS spectra of K in the compounds KBr, KCl, and KF, showing multielectron satellite peaks on the low-energy side. (*After Wertheim and Rosencwaig*, 1971.)

with photoemission from the $2p$ levels of $3d$-transition-metal atoms in semiconductor compounds. Some of these results for CuS and Cu_2O are shown in Fig. 8-22. Data such as these have been correlated with the unoccupied valence-band structure as measured in optical absorption (Novakov, 1971*b*).

Although our discussion has been confined primarily to the observations and interpretations of two-electron transitions, it is entirely possible that three-electron transitions may occur (Krause et al., 1964; Krause, Carlson, and Dismukes, 1968). For example, in neon gas, triple ionization has been observed in connection with the x-ray excitation of an Ne $1s$ photoelectron (Krause, Carlson, and Dismukes, 1968). This triple ionization has a relative intensity of ~ 1 percent of all events and is attributed to two shake-off electrons from the $2s$ and $2p$ subshells (Krause, Carlson, and Dismukes, 1968). A relative intensity of ~ 1 percent for triple ionization is also con-

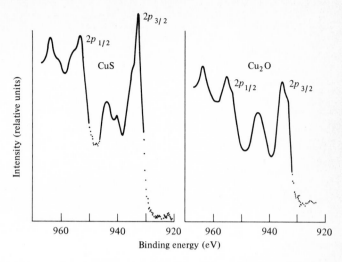

FIGURE 8-22
XPS spectra of Cu $2p_{1/2}$ and $2p_{3/2}$ electrons in the compounds CuS and Cu$_2$O.
The satellite structure on the low-energy side is interpreted as originating from
the simultaneous excitation of valence electrons from the filled bands into the
empty ones. (*After Novakov*, 1971*b*.)

sistent with the known ratios of the $\alpha_{1,2}$ and $\alpha_{5,6}$ satellite x rays of Mg, as can be
seen in Fig. 8-20.

Multielectron effects analogous to those we have been discussing have been
predicted to occur also in metals (Hedin and Lundqvist, 1969; Doniach, 1970). The
multielectron excitations for a metallic system are more complicated and may involve
the excitation of plasmons (Hedin and Lundqvist, 1969). No definite experimental
verifications of such effects have as yet been made, however (see also Chap. 5).

It is thus clear that multielectron transitions represent an important consideration in the interpretation of any x-ray photoelectron spectrum. Where a detailed
analysis of these transitions is possible, it may yield significant information about
electronic structure, particularly as to the excited states of valence electrons.

VI. PHOTOELECTRIC CROSS SECTIONS AND ANGULAR DISTRIBUTIONS

The photoelectric cross section represents the transition probability per unit time for
exciting a photoelectron from a subshell with a photon flux of a given energy. It
should thus be proportional to the intensity of an observed photoelectron peak.
However, this cross section should also be a function of the angle of electron emission

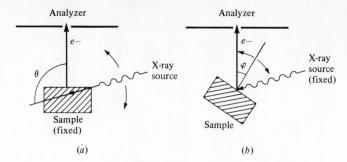

FIGURE 8-23
Schematic diagram indicating the two types of angular-distribution measurements that can be performed in photoemission experiments. (*a*) Angle between the incident radiation and the electron analyzer is varied. (*b*) Arrangement in which a single-crystal specimen is being rotated.

relative to the photon-propagation direction, so that measurements of the angular distribution of peak intensity are also of interest. A second type of angular-distribution measurement is also permitted if the sample has a preferred axis of orientation, as, for example, the axis perpendicular to the surface of a polycrystalline solid sample or the crystal axes of a single-crystal solid sample. The experimental arrangements for these types of angular-distribution measurements are indicated in Fig. 8-23. We shall first discuss atomic and molecular photoelectric cross sections and their angular dependence, and then review angular-distribution measurements from solids.

A.   Free atoms and molecules

The basic theory of photon absorption in the one-electron approximation has been reviewed by Bethe and Salpeter (1955). Bates (1946) and Cooper (1962) have discussed the application of this theory to many-electron atoms. Fano and Cooper (1968) have reviewed the experimental and theoretical data bearing on atomic photoabsorption near threshold, including a consideration of various many-body effects. More recently, Cooper and Manson (1969) have discussed the one-electron theory of angular distributions in photoemission from atoms. We shall briefly review the one-electron theory for free atoms, indicating its basic assumptions and predictions.

There are several simplifying assumptions which are usually made in calculating photoelectric cross sections: (1) The effect of the photon is treated as a perturbation. (2) The interaction between the initial one-electron orbital ϕ_k and the continuum photoelectron orbital ϕ_c is assumed to be described by an electric-dipole operator acting between ϕ_k and ϕ_c. This is a reasonable assumption if the photon wavelength is long compared to the spatial extent of ϕ_k, as is the case for Al $K\alpha$ or Mg $K\alpha$ x rays with $\lambda \sim 10$ A. This assumption is termed a "neglect-of-retardation effect." (3) If the wavefunction of a many-electron atom is described in terms of a Slater determinant, only a single one-electron orbital is assumed to change in this determinant during photoabsorption. That is, an electron is excited from ϕ_k to ϕ_c, with all other orbitals remaining the same. The orbital ϕ_c is thus associated with an energy E_{kin} [cf equation (8-1)].

Subject to the above assumptions, the cross section for a transition from the orbital k to the orbital c is given by

$$\sigma_{k,c}(E_{kin}) = \frac{4\pi\alpha a_0{}^2}{3} h\nu \left| \int \phi_k \bar{r} \phi_c \, dt \right|^2 \qquad (8\text{-}32)$$

where α is the fine-structure constant, a_0 is the Bohr radius, and the final factor is the square of a one-electron dipole matrix element between ϕ_k and ϕ_c. The dipole operator connecting ϕ_k and ϕ_c implies certain selection rules, with the most important being $\Delta l = \pm 1$. It is common to sum over the possible final states ϕ_c which can be reached by excitation of a given ϕ_k and also to average over the various orbitals ϕ_k in a given n,l subshell. Such summing and averaging yield the total subshell cross section as

$$\sigma_{nl}(E_{kin}) = \frac{4\pi\alpha a_0{}^2}{3} \left[E_{kin} + \text{B.E.}(nl)\right] \cdot \left[lR_{E_{kin}, l-1} + (l+1)R_{E_{kin}, l+1}\right] \qquad (8\text{-}33)$$

where B.E.(nl) is the binding energy of the nl subshell, $E_{kin} + \text{B.E.}(nl) = h\nu$, and $R_{E_{kin}, l\pm 1}$ are radial integrals common to all the one-electron dipole matrix elements between ϕ_k and ϕ_c. These radial integrals are given by

$$R_{E_{kin}, l\pm 1} = \int_0^\infty P_{nl}(r) r P_{E_{kin}, l\pm 1}(r) \, dr \qquad (8\text{-}34)$$

where $P_{nl}(r)$ is the radial part of the ϕ_k orbitals and $P_{E_{kin}, l\pm 1}$ is the radial part of the continuum ϕ_c orbitals. The $l+1$ and $l-1$ continuum orbitals are those allowed by the dipole selection rule.

Total subshell cross sections for photon energies relevant to XPS have been calculated by Bearden (1966), Rakavy and Ron (1967), Brysk and Zerby (1968), and Cooper and Manson (1969). Comparisons with experiment are often made through the total absorption coefficient, which consists primarily of sums over several subshell cross sections. Such comparisons yield reasonably good agreement between experiment and theory except near threshold where $h\nu \simeq \text{B.E.}(nl)$ (Rakavy and Ron, 1967; Brysk and Zerby, 1968). Cooper and Manson (1969) have also calculated relative subshell cross sections which compare favorably with the experimental values of Krause (1969).

In general, $\sigma_{nl}(E_{kin})$ will be a decreasing function of E_{kin} for large values of E_{kin} (i.e., for $h\nu$ well above threshold). However, large oscillations and even zeros

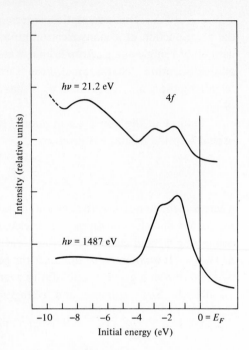

FIGURE 8-24
Photoemission spectra from Yb metal at different photon energies showing the increase in cross section for excitation of $4f$ electrons with x rays. (*After Brodén, Hagström, and Norris, 1970, and Hagström, Hedén, and Löfgren, 1970.*)

in the cross section may occur as $h\nu$ is increased above threshold (Cooper and Manson, 1969). Also $\sigma_{nl}(E_{kin})$ may be much different for different nl quantum numbers at the same E_{kin} or for the same nl quantum numbers at different E_{kin}. Such effects have already proved very useful in studying the valence-electron states in gases (Hamrin et al., 1968) and solids (Hedén, Löfgren, and Hagström, 1971; Eastman and Kuznietz, 1971). Molecular orbitals with $2s$ character are found to be emphasized in XPS studies, whereas orbitals with $2p$ character are more pronounced in UPS studies (Hamrin et al., 1968). Similarly, the $4f$ electrons in rare-earth metals are seen much more clearly in XPS spectra than in UPS spectra, as is shown in Fig. 8-24.

The total subshell cross section $\sigma_{nl}(E_{kin})$ represents an integration over all emission angles relative to the photon-propagation direction. The detailed dependence of photoelectron emission on angle can be expressed in terms of a differential cross section $d\sigma_{nl}(E_{kin})/d\Omega$. If retardation is again neglected, this differential cross section is given by

$$\frac{d\sigma_{nl}(E_{kin})}{d\Omega} = \frac{\sigma_{nl}(E_{kin})}{4\pi}\left[1 - \tfrac{1}{2}\beta(E_{kin})P_2(\cos\theta)\right] \qquad (8\text{-}35)$$

where $\beta(E_{kin})$ is termed the "asymmetry parameter," θ is the angle between the photon-propagation direction and the photoelectron-emission direction (cf Fig. 8-23), and $P_2(\cos\theta) = \tfrac{1}{2}(3\cos^2\theta - 1)$. $\beta(E_{kin})$ can be obtained theoretically from

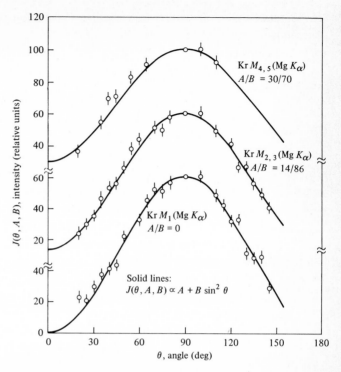

FIGURE 8-25
Angular distributions of 3s (M_1), 3p ($M_{2,3}$), and 3d ($M_{4,5}$) photoelectrons of Kr excited by Mg $K\alpha$ x rays. (*After Krause*, 1969.)

the radial integrals $R_{E_{kin},\, l\pm1}$ and certain continuum orbital phase shifts $\delta_{l\pm1}$. Thus equation (8-35) can be simplified to the form

$$\frac{d\sigma_{nl}(E_{kin})}{d\Omega} = A + B \sin^2 \theta \qquad (8\text{-}36)$$

A comparison between the function predicted by equation (8-36) and experimental results is shown in Fig. 8-25. The parameters A and B have, in this case, been empirically adjusted to give the best fit to data obtained for photoemission from the Kr 3s, Kr 3p, and Kr 3d levels with Mg $K\alpha$ x rays. The data are reasonably well described by equation (8-36). The small displacement of the experimental curve for Kr 3d toward smaller angles is probably due to retardation effects (Krause, 1969). These effects have been discussed in detail by Cooper and Manson (1969). Manson (1971) has also recently pointed out that for certain subshells, $\beta(E_{kin})$ may exhibit large oscillations with E_{kin}.

Similar total-cross-section and angular-distribution relations are obtained for photoemission from free molecules (Berkowitz, Erhardt, and Te Kaat, 1967; Thomas, 1971). Thomas (1971) has recently included the effects of final-state rotational and vibrational excitation in a theoretical calculation of the form of photoelectron angular distributions from molecules.

B. Solids

In solids the fundamental photoemission process is modified in the sense that the final state ϕ_c and perhaps the initial state ϕ_k must be described in terms of Bloch waves. The periodic potential in a crystalline solid can thus produce electron-diffraction-like effects on photoelectron angular distributions. In addition, the inelastic-scattering processes in solids lead to rather short escape depths for photoelectrons (see Section VII). These short escape depths in turn give rise to distinctly different behavior in the angular distributions of photoelectrons from "bulk" and "surface" atoms.

Electron-diffraction effects have been observed in x-ray photoelectron spectra from both sodium chloride single crystals (Siegbahn et al., 1970) and gold single crystals (Fadley and Bergström, 1971a; 1971b). The angular distributions of Au $4f$ electrons which have escaped with minimal inelastic scattering from a gold single crystal are shown in the upper portion of Fig. 8-26, along with the experimental geometry (cf Fig. 8-23b). In this particular experiment, the single crystal was rotated about an axis perpendicular to [111]. The peaks in these Au $4f$ angular distributions are believed to be associated with Bragg reflection of electrons from the planes indicated (Fadley and Bergström, 1971a; 1971b). Theoretical treatments of closely related diffraction or channeling processes indicate that the appearance of such angular distributions should depend on both the spatial distribution of the initial ϕ_k state and the spatial distribution and de Broglie wavelength of the continuum ϕ_c state (Hirsch and Howie, 1953; DeWames and Hall, 1968). Very similar angular-distribution anisotropies have recently been discussed theoretically (Mahan, 1970), and verified experimentally (Gerhardt and Dietz, 1971) for ultraviolet photoemission studies on single crystals.

The angular distribution shown in the lower portion of Fig. 8-26 is for C $1s$ photoelectrons from surface contaminants on a gold single crystal (Fadley and Bergström, 1971a; 1971b). The appearance of this angular distribution is much different from those for Au $4f$ photoelectrons from the same crystal. The C $1s$ distribution exhibits no structure and is relatively flat with angle, in contrast to the general decrease in intensity observed with increasing ϕ for Au $4f$ (note that ϕ increases to the left in this figure). As the penetration depth of elastically scattered Au $4f$ photoelectrons through a carbon-containing contaminant layer is probably of the order of 100

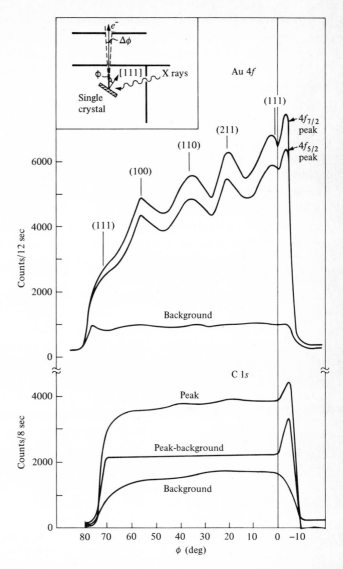

FIGURE 8-26
The angular distributions of Au 4f photoelectrons from a single crystal and C 1s photoelectrons from surface impurities. (*After Fadley and Bergström, 1971a.*)

A (see Section VII), the contaminant layer on which these measurements were made was probably much thinner than this. Thus, a very large fraction of the C 1s photoelectrons from the contaminant layer could escape from the surface without suffering any inelastic-scattering events. Therefore, the relative constancy of the C 1s angular distribution is not surprising. By contrast, the Au 4f photoelectrons are emitted from a surface layer of the crystal much thicker than 100 A, but a very large fraction of these will be inelastically scattered as they leave the crystal. The net result is that inelastic scattering should play an important role in determining the intensity of photoelectrons from "bulk" atoms, whereas photoelectrons from "surface" atoms should be minimally affected by inelastic scattering. Such considerations relate to angular-distribution measurements in that the average path length traversed by a "bulk" photoelectron in leaving a crystal such as that shown in Fig. 8-26 will increase as $1/\cos \phi$. Thus, it would be expected that at grazing angles of electron escape (high ϕ), the number of photoelectrons from bulk atoms would be much decreased. This is consistent with the general slope of the Au 4f angular distributions in Fig. 8-26, as well as with Au 4f angular-distribution measurements on polycrystalline gold samples (Fadley and Bergström, 1971a; 1971b). For the particular experimental arrangements shown in Fig. 8-26, it can also be shown that simple geometric effects should lead to an *increase* in surface-atom photoelectron intensities with increasing ϕ (Fadley and Bergström, 1971b). Thus, the angular distributions of photoelectrons from bulk and surface atoms are fundamentally different, and surface-atom photoemission should be much accented at low angles of electron escape (Fadley and Bergström, 1971a; 1971b).

In summary, measurements of photoelectron-peak intensities, as a function of incident photon energy, photoelectron emission angle relative to the photon-propagation direction, or photoelectron emission angle relative to the axes of a crystalline sample, appear capable of yielding considerable information about both the detailed dynamics of the photoemission process and the spatial distributions of the wavefunctions involved. An added benefit from angular-distribution measurements on solids is that it may be possible to semiquantitatively distinguish surface atoms.

VII.   INELASTIC SCATTERING AND SURFACE STUDIES

We have already indicated several instances where the effects of inelastic scattering on x-ray photoelectron spectra are important considerations. Inelastic scattering can be an important effect in photoelectron spectra from both gases (Carlson, Krause, and Moddeman, 1971; Siegbahn et al., 1969) and solids (Fadley and Shirley, 1970a; 1970b). The mechanisms of such scattering processes in solids are varied, but may involve phonon excitation, plasmon excitation, intraband transitions or interband

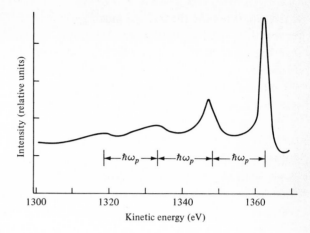

FIGURE 8-27
XPS spectrum of Al $2s_{1/2}$ electrons expelled by Al $K\alpha_{1,2}$ radiation. Several plasmon-loss peaks are observed.

transitions. The relative importance of the various contributions will vary from solid to solid (Swanson and Powell, 1968), but the overall scattering process can be described in terms of a complex dielectric constant (Swanson and Powell, 1968; Pines, 1963). As an example of a system where plasmon excitation is dominant, we show in Fig. 8-27 an XPS spectrum from aluminum (Hedén and Hagström, 1971) in which three plasma-loss peaks are clearly visible. In general, however, the loss processes involved in XPS spectra are not well characterized and must simply be viewed as possible sources of peak asymmetry and a low-kinetic-energy tail composed primarily of inelastically scattered electrons. An additional and very important effect of inelastic scattering is that photoelectrons which have *not* been inelastically scattered come from only a very thin surface layer of a sample. As it is only these electrons which contain information easily interpretable through equation (8-1), the photoelectron spectroscopy of solids is necessarily a quasi-surface technique.

In estimating the thickness of the surface layer responsible for elastically scattered photoelectrons, we are really interested in the fraction of photoelectrons excited at a given depth which escape from the surface without "significant" inelastic scattering. At the present time a significant loss could be defined as that resulting in a decrease in kinetic energy of more than approximately 0.5 eV or one-half of the present inherent line width of XPS measurements. This fraction will be a function of the perpendicular distance d from the surface at which photoelectrons are excited (Fadley and Shirley, 1970a), the angle ϕ at which they are emitted from the surface, and their kinetic energy E_{kin}. Thus, we can denote this fraction as $f(d,\phi,E_{kin})$. In general, the exciting x rays will penetrate the solid to much greater depths than those for inelastic electron escape (~ 2000 A for Mg $K\alpha$ in Au, for example). Thus, we may assume that there is a uniform excitation of photoelectrons in number per unit volume per second for all depths for which $f(d,\phi,E_{kin})$ is significantly above zero. If this is the case, the average depth from which photoelectrons are emitted becomes simply

$$\bar{d}(\phi,E_{kin}) = \frac{\int_0^\infty f(d',\phi,E_{kin}) \cdot d' \cdot dd'}{\int_0^\infty f(d',\phi,E_{kin})dd'} \qquad (8\text{-}37)$$

$\bar{d}(\phi,E_{kin})$ should decrease with increasing ϕ, as the total path length for escape will be given by $d/\cos\phi$, and the perpendicular component of momentum required to penetrate the surface barrier will also decrease as $\cos\phi$. Also, $\bar{d}(\phi,E_{kin})$ will probably decrease with E_{kin} (Delgass, Hughes, and Fadley, 1970; Henke, 1971), although little quantitative data are as yet available on this point. It has been pointed out that for fixed ϕ and E_{kin}, the functional form of f should be (Baer et al., 1970b)

$$f(d,\phi,E_{kin}) = e^{-d/\Lambda(\phi,E_{kin})} \qquad (8\text{-}38)$$

where $\Lambda(\phi,E_{kin})$ is an attenuation length. Substituting this form for $f(d,\phi,E_{kin})$ into equation (8-37) shows that

$$\bar{d}(\phi,E_{kin}) = \Lambda(\phi,E_{kin}) \qquad (8\text{-}39)$$

Thus, for the exponential relationship in equation (8-38), $\bar{d}(\phi,E_{kin})$ occurs at a depth where $f(d,\phi,E_{kin}) = 1/e \cong 0.36$.

The magnitudes of $\bar{d}(\phi,E_{kin})$ and the functional form of $f(d,\phi,E_{kin})$ are not well understood for electrons in the 200- to 1500- eV range. Two types of experiments have been performed in order to measure these quantities for XPS photoelectrons, however. In the first (Larsson et al., 1966; Siegbahn et al., 1967), stearic-acid layers of known thickness were labeled with halogen atoms. The photoelectron-peak intensities from carbon and halogen atoms were then measured as a function of total layer thickness. One conclusion resulting from these studies was that the average depth for photo-emission is less than 100 A (Larsson et al., 1966; Siegbahn et al., 1967). However, a further analysis of the data obtained in these studies for iodine-labeled stearic acid shows that the results are consistent with each stearic-acid layer having approximately a 70 percent transmission for I $3d_{5/2}$ photoelectrons. Since each layer is known to be 40 A thick (Larsson et al., 1966; Siegbahn et al., 1967), this transmission can be used in equation (8-38) to give an estimate for $\Lambda = \bar{d}$. The value so obtained is $\bar{d}(\phi \approx 45°$, $E_{kin} \simeq 850$ eV) $\simeq 110$ A for inelastic scattering in a loosely stacked hydrocarbon array. This value is thus slightly larger than previous estimates (Larsson et al., 1966; Siegbahn et al., 1967). More detailed recent measurements by Henke (1971) on stearic-acid layers indicate comparable transmissions. Similar studies have also been performed on chromium covered with gold layers of varying thickness by

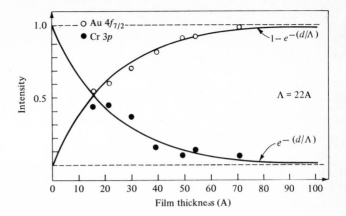

FIGURE 8-28
Core-level intensities obtained from samples consisting of thin gold films deposited on chromium substrates. (*After Baer et al., 1970b.*)

Baer et al. (1970*b*). The data obtained from these measurements are shown in Fig. 8-28. The curves in this figure were obtained by empirically fitting equations based on equation (8-38) to the experimental data. These curves yield a value of $\bar{d}(\phi \approx 45°$, $E_{kin} \simeq 1200 \text{ eV}) \simeq 22$ A for inelastic scattering in gold (Baer et al., 1970*b*). However, there appear to be systematic deviations of the experimental data from exponential curves for Au film thickness less than $\sim$35 A. These deviations could be due to non-uniform Au film deposition (island formation) and would lead to too high an estimate for $\bar{d} \cdot \bar{d}$ values in the range 15 to 25 A, thus appearing to be consistent with these data. A further comparison with experiment can be made by considering the range for electrons in a given material (Siegbahn et al., 1967). The range can be defined as the maximum penetration depth of electrons with energy E_{kin}, or equivalently, the depth d_{max} at which $f(d_{max}, \phi, E_{kin}) \approx 0.01$. If f is assumed to be exponential, this will occur at $d_{max} \simeq 4.5\bar{d}$. With a value of $\bar{d} = 20$ A for gold, an estimate of $d_{max}(\phi \approx 45°, E_{kin} \simeq 1200 \text{ eV}) \simeq 90$ A is thus obtained. This XPS-derived value can be compared with independent range measurements (Kanter and Sternglass, 1962), which indicate $d_{max}(\phi = 90°, E_{kin} \simeq 600 \text{ eV}) \simeq 55$ A for gold. These two values for d_{max} are thus in reasonable agreement, especially in view of the differences in kinetic energy involved. A similar estimate based on photoelectron spectra from stearic-acid layers yields $d_{max}(\phi \approx 45°, E_{kin} \simeq 850 \text{ eV}) \cong 500$ A. In summary, the data presently available indicate that $\bar{d}$ for x-ray photoelectrons is in the approximate range 15 to 110 A. The maximum penetration depth d_{max} should thus be in the range 70 to 500 A.

Few attempts have been made to measure the dependence of $\bar{d}$ on ϕ or E_{kin} (Henke, 1971). The ϕ dependence has been discussed qualitatively in Section VI and

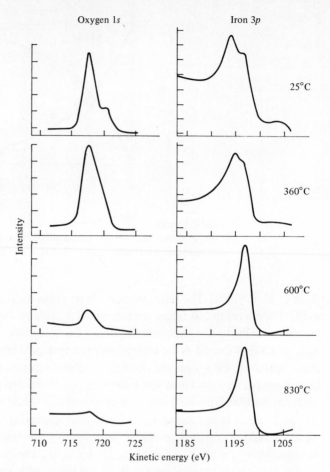

FIGURE 8-29
Oxygen 1s and iron 3p photoelectron peaks from metallic iron at various temperatures in a hydrogen atmosphere. (*After Fadley and Shirley*, 1970a.)

also by Fadley and Bergström (1971a; 1971b). For x-ray photoelectrons with energies of a few hundred eV or more, it is probably reasonable to set

$$\bar{d}(\phi, E_{kin}) \simeq \bar{d}(90°, E_{kin}) \cdot \cos \phi \qquad (8\text{-}40)$$

Based on this model, we thus expect a very thin surface layer to be emphasized in experiments at large ϕ values near 90° (Fadley and Bergström, 1971a; 1971b).

It is clear from the previous discussion that XPS measurements should be very sensitive to the surface condition of a solid sample. This can be an advantage if surface properties are of primary interest (Delgass, Hughes, and Fadley, 1970), or a

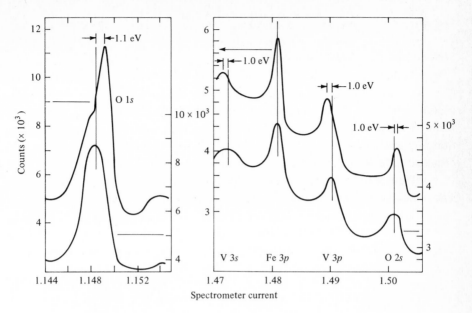

FIGURE 8-30
Partial photoelectron spectra from FeV_2O_4 before (lower curves) and after (upper curves) it was used as a catalyst for the dehydrogenation of cyclohexane at 425°C. (*After Delgass, Hughes, and Fadley*, 1970.)

disadvantage if the aim of a study is to measure bulk properties (Fadley and Shirley, 1970a, 1970b; Fadley et al., 1968). A graphic illustration of this surface sensitivity is shown in Fig. 8-29, where the oxide layer on metallic iron (~ 50 A thick) is readily observed in XPS spectra. Fadley and Shirley (1968, 1970b) have obtained photoelectron spectra from the O 1s and Fe 3p core levels of an iron sample which was heated in hydrogen to reduce away the surface oxide. The reduction of the oxide is most apparent in Fig. 8-29 as the disappearance of the O 1s photoelectron peak with increasing temperature. However, the Fe 3p photoelectron peak also exhibits two components, one at lower kinetic energy (an "oxide" peak) and one at higher kinetic energy (a "metal" peak). The oxide peak is also observed to disappear at high temperature. The O 1s peak shows structure also, with the shoulder at high kinetic energy probably being due to absorbed oxygen, and the main peak at lower kinetic energy to oxide oxygen (Fadley and Shirley, 1968; 1970b).

Several systems have been studied in an attempt to exploit this surface sensitivity. Delgass, Hughes, and Fadley (1970) have discussed the application of XPS to studies of heterogeneous catalysts, and have presented preliminary experimental results relevant to several catalytic systems. The effects of catalytic action on the XPS spectra of an FeV_2O_4 catalyst are shown in Fig. 8-30. The spectra were obtained before and after using the catalyst to dehydrogenate cyclohexane at 425°C. The V $3s$ and V $3p$ photoelectron peaks are observed to shift together by ~ 1.0 eV toward lower kinetic energy after dehydrogenation. On the other hand, the O $2s$ peak shifts by ~ 1 eV in the opposite direction, and the Fe $3p$ peak is not found to shift significantly. The O $1s$ peak changes in position and shape. Such observations may thus prove useful in the study of various catalytic systems (Delgass, Hughes, and Fadley, 1970). More recently, XPS has been used to study the surface of carbon fibers and the oxidation of graphite on different single-crystal faces (Barber et al., 1970), and the surface chemical states of a cobalt-molybdate catalyst (Miller et al., 1971). Thus, XPS seems very well suited to the study of many problems in surface physics and surface chemistry. The primary experimental limitation up to the present time has been the lack of ultra-high-vacuum capacity for precisely controlling the surface condition of the sample. However, XPS systems with capabilities in the 10^{-10}-torr range are now becoming available, and the number of surface applications should enlarge considerably.

REFERENCES

ÅBERG, T. (1967): Theory of x-ray satellites, *Phys. Rev.*, **156**: 35–41.

ADAMS, J., and MANLEY, B. W. (1967): The channel electron multiplier: A new radiation detector, *Philips Tech. Rev.*, **28**: 156–161.

APAI, G. R., DELGASS, W. N., HOLLANDER, J. M., NOVAKOV, T., and SHIRLEY, D. A. (1969): Splitting of gold $p_{3/2}$ core levels by chemical bonding, Lawrence Radiation Lab. Rept. UCRL-19530, 252–255.

ASAAD, W. N., and BURHOP, E. H. S. (1958): The K Auger spectrum, *Proc. Phys. Soc. London*, **71**: 369–382.

BAER, Y., HEDÉN, P. F., HEDMAN, J., KLASSON, M., NORDLING, C., and SIEGBAHN, K. (1970a): Band structure of transition metals studied by ESCA, *Physica Scripta*, **1**: 55–65.

———, ———, ———, ———, and NORDLING, C. (1970b): Determination of the electron escape depth in gold by means of ESCA, *Solid State Comm.*, **8**: 1479–1481.

BARBER, M., SWIFT, P., EVANS, E. L., and THOMAS, J. M. (1970): High energy photoelectron spectroscopic study of carbon fiber surfaces, *Nature*, **227**: 1131–1132.

BATES, D. R. (1946): Calculation of the cross section of neutral atoms and positive and negative ions towards the absorption of radiation in the continuum, *Monthly Notices Roy. Astron. Soc.*, **106**: 432–445.

BEARDEN, J. A. (1966): X-ray photoeffect cross sections in low- and medium-Z absorbers for the energy range 852 eV to 40 keV, *J. Appl. Phys.*, **37**: 1681–1692.

——— and BURR, A. F. (1967): Reevaluation of x-ray atomic energy levels, *Rev. Mod. Phys.*, **39**: 125–142.

BERGLUND, C. N., and SPICER, W. E. (1964): Photoemission studies of copper and silver: Experiment, *Phys. Rev.*, **A136**: 1044–1064.

BERKOWITZ, J. (1971): Photoelectron spectroscopic studies with a cylindrical-mirror analyzer, *Proc. Intern. Conf. Electron Spectry.*, pp. 391–399, North-Holland Publishing Company, Amsterdam.

——, ERHARDT, H., and TE KAAT, T. (1967): Spectra and angular distributions of photoelectrons from atoms and molecules, *Z. Physik*, **200**: 69–83.

BETHE, H. A., and SALPETER, E. E. (1955): Quantum mechanics of one- and two-electron systems, *Handb. Physik*, **35**: 88–436.

BRODÉN, G. (1971): Private communication.

——, HAGSTRÖM, S. B. M., HEDÉN, P. O., and NORRIS, C. (1971): Ultraviolet and x-ray photoemission from europium and barium, Third IMR Symposium, *Natl. Bur. Std., Spec. Publ. 323*, pp. 217–223.

——, ——, and NORRIS, C. (1970): *Phys. Rev. Lett.*, **24**: 1173–1174.

BRYSK, H., and ZERBY, C. D. (1968): Photoelectric cross sections in the keV range, *Phys. Rev.*, **171**: 292–298.

BURHOP, E. H. S. (1952): *The Auger effect and other radiationless transitions*, Cambridge University Press, London.

BUSCH, G., CAMPAGNA, M., and SIEGMANN, H. CH. (1971): Photoelectric detection of spin-polarized electron states in antiferromagnets, *Proc. Intern. Conf. Electron Spectry.*, pp. 827–834, North-Holland Publishing Company, Amsterdam.

CARLSON, T. A., HUNT, W. E., and KRAUSE, M. O. (1966): Relative abundances of ions formed as the result of inner-shell vacancies in atoms, *Phys. Rev.*, **151**: 41–47.

—— and KRAUSE, M. O. (1965a): Atomic readjustment to vacancies in the K and L shells of argon, *Phys. Rev.*, **A137**: 1655–1662.

—— and —— (1965b): Electron shake-off resulting from K-shell ionization in neon measured as a function of photoelectron velocity, *Phys. Rev.*, **A140**: 1057–1065.

——, ——, and MODDEMAN, W. E. (1971): Excitation accompanying photoionization in atoms and molecules and its relationship to electron correlation, *J. Phys.*, **C4**: 76–84.

CARVER, J. C., CARLSON, T. A., CAIN, L. C., and SCHWEITZER, G. K. (1971): Use of x-ray photoelectron spectroscopy to study bonding in transition metal salts by observation of multiplet splitting, *Proc. Intern. Conf. Electron Spectry.*, pp. 803–812, North-Holland Publishing Company, Amsterdam.

COMBET-FARNOUX, F. (1969): Photoionisation des atomes lourds: Étude théorique dans un modèle non relativiste à potentiel central, *J. Phys. (Paris), Colloq.*, **30**: 521–530.

COOPER, J. W. (1962): Photoionization from outer atomic subshells: A model study, *Phys. Rev.*, **128**: 681–693.

—— and MANSON, S. T. (1969): Photo-ionization in the soft x-ray range: Angular distributions of photoelectrons and interpretation in terms of subshell structure, *Phys. Rev.*, **177**: 157–163.

COSTER, D., and DRUYVESTEYN, M. J. (1927): Über die Satelliten der Röntgendiagrammlinien, *Z. Physik*, **40**: 765–774.

COX, P. A., and ORCHARD, F. A. (1970): On band intensities in the photoelectron spectra of open-shell molecules, *Chem. Phys. Lett.*, **7**: 273–275.

DELGASS, W. N., HUGHES, T. R., and FADLEY, C. S. (1970): X-ray photoelectron spectroscopy: A tool for research in catalysis, *Catalysis*, **4**: 179–220.

DEWAMES, R. E., and HALL, W. F. (1968): A wave mechanical description of electron and positron emission from crystals, *Acta Cryst.*, **A24**: 206–212.

DONIACH, S. (1970): Many-electron theory of nondirect transitions in the optical and photoemission spectra of metals, *Phys. Rev.*, **B2**: 3898–3905.

EASTMAN, D. E. (1971): Photoemission spectroscopy of metals, in E. PASSAGLIA (ed.), *Techniques of metals research VI*, chap. 10, Interscience Publishers, division of John Wiley & Sons, Inc., New York.

—— and KUZNIETZ, M. (1971): Energy-dependent photoemission intensities of "*f*" states in EuS, GdS, and US, *Phys. Rev. Lett.*, **26**: 846–850.

EKSTIG, B., KÄLLNE, E., NORELAND, E., and MANNE, R. (1970): Electron interaction in transition metal x-ray emission spectra, *Physica Scripta*, **2**: 38–44.

FADLEY, C. S. (1970): Core and valence electronic states studied with x-ray photoelectron spectroscopy, Lawrence Radiation Lab. Rept. UCRL-19535 (thesis).

—— (1971): Multiplet splittings in photoelectron spectra, *Proc. Intern. Conf. Electron Spectry.*, pp. 781–802, North-Holland Publishing Company, Amsterdam.

—— and BERGSTRÖM, S. A. L. (1971a): Angular distribution of photoelectrons from a metal single crystal, *Phys. Lett.*, **A35**: 375–376.

—— and —— (1971b): Angular distributions of photoelectrons from metal crystals, *Proc. Intern. Conf. Electron Spectry.*, pp. 233–243, North-Holland Publishing Company, Amsterdam.

——, GEOFFROY, G. L., HAGSTRÖM, S. B. M., and HOLLANDER, J. M. (1969): Direct voltage calibration of an electron spectrometer, *Nucl. Instr. Methods*, **68**: 177–178.

——, HAGSTRÖM, S. B. M., KLEIN, M. P., and SHIRLEY, D. A. (1968): Chemical effects on core-electron binding energies in iodine and europium, *J. Chem. Phys.*, **48**: 3779–3794.

——, HEALEY, R. N., HOLLANDER, J. M., and MINER, C. E. (1972): Design of a high-resolution, high-efficiency, magnetic spectrometer for electron spectroscopy, *J. Appl. Phys.*, **43**: 1085–1102.

——, MINER, C. E., and HOLLANDER, J. M. (1969): Design of a magnetic spectrometer for photoelectron spectroscopy, *Appl. Phys. Lett.*, **15**: 223–225.

—— and SHIRLEY, D. A. (1968): X-ray photoelectron spectroscopic study of iron, cobalt, nickel, copper, and platinum, *Phys. Rev. Lett.*, **21**: 980–983.

—— and —— (1970a): Electronic densities of states from x-ray photoelectron spectroscopy, *Natl. Bur. Std., J. Res.*, **A74**: 543–558.

—— and —— (1970b): Multiplet splitting of metal-atom electron binding energies, *Phys. Rev.*, **A2**: 1109–1120.

——, ——, FREEMAN, A. J., BAGUS, P. S., and MALLOW, J. V. (1969): Multiplet splitting of core-electron binding energies in transition-metal ions, *Phys. Rev. Lett.*, **23**: 1397–1401.

FANO, U., and COOPER, J. W. (1968): Spectral distribution of atomic oscillator strengths, *Rev. Mod. Phys.*, **40**: 441–507.

GELIUS, U. (1971): Molecular orbitals and line intensities in ESCA spectra, *Proc. Intern. Conf. Electron Spectry.*, pp. 311–334, North-Holland Publishing Company, Amsterdam.

GERHARDT, U., and DIETZ, E. (1971): Angular distribution of photoelectrons emitted from copper single crystals, *Phys. Rev. Lett.*, **26**: 1477–1480.

GODWIN, R. P. (1969): Synchrotron radiation as a light source, *Springer tracts in modern physics*, vol. 51, pp. 1–73.

GUPTA, R. P., RAO, B. K., and SEN, S. K. (1971): Quadrupole crystalline electric field shielding and antishielding factors at some rare-earth and heavy ions, *Phys. Rev.*, **A3**: 545–553.

HAENSEL, R., KEITEL, G., PETERS, G., SCHREIBER, P., SONNTAG, B., AND KUNZ, C. (1969): Photoemission measurement on NaCl in the photon energy range 32–50 eV, *Phys. Rev. Lett.*, **23**: 530–532.

HAGSTRÖM, S. (1964): Atomic level energies in hafnium, *Z. Physik*, **178**: 82–100.

———— (1971): X-ray photoemission studies of valence electronic levels in solids, *Proc. Intern. Conf. Electron Spectry.*, pp. 515–534, North-Holland Publishing Company, Amsterdam.

————, BRODÉN, G., HEDÉN, P. O., and LÖFGREN, H. (1971): X-ray and UV photoelectron spectra from outer electrons of some rare earths, *J. Phys.*, **C4**: 269–273.

————, HEDÉN, P. O., and LÖFGREN, H. (1970): Electron density of states in Yb metal as observed by x-ray photoemission, *Solid State Comm.*, **8**: 1245–1248.

————, NORDLING, C., and SIEGBAHN, K. (1964): Electron spectroscopic determination of the chemical valence state, *Z. Physik*, **178**: 439–444.

————, ————, and ———— (1965): Tables of electron binding energies and kinetic energy versus magnetic rigidity in K. SIEGBAHN (ed.), *Alpha-, beta- and gamma-ray spectroscopy*, Appendix 2, North-Holland Publishing Company, Amsterdam.

HAMRIN, K., JOHANSSON, G., GELIUS, U., FAHLMAN, A., NORDLING, C., and SIEGBAHN, K. (1968): Ionization energies in methane and ethane measured by means of ESCA, *Chem. Phys. Lett.*, **1**: 613–615.

HEDÉN, P. O., and HAGSTRÖM, S. B. M. (1971): Unpublished data.

————, LÖFGREN, H., and HAGSTRÖM, S. B. M. (1971): 4*f* electronic states in the metals Nd, Sm, Dy, and Er studied by x-ray photoemission, *Phys. Rev. Lett.*, **26**: 432–434.

HEDIN, L., and LUNDQVIST, S. (1969): Effects of electron-electron and electron-phonon interactions on the one-electron states of solids, *Solid State Phys.*, **23**: 1–181.

HEDMAN, J., HEDÉN, P. O., NORDLING, C., and SIEGBAHN, K. (1969): Energy splitting of core electron levels in paramagnetic molecules, *Phys. Lett.*, **A29**: 178–179.

HELMER, J. C. (1971): Private communication, to be published.

———— and WEICHERT, N. H. (1968): Enhancement of sensitivity in ESCA spectrometers, *Appl. Phys. Lett.*, **13**: 266–267.

HENKE, B. L. (1971): Measurement of primary electron interaction cross sections, *J. Phys.*, **C4**: 246–252.

HIRSCH, P. B., and HOWIE, A. (1953): *Electron microscopy of thin crystals*, Butterworth, London.

HOLLANDER, J. M., and SHIRLEY, D. A. (1970): Chemical information from photoelectron and conversion-electron spectroscopy, *Ann. Rev. Nucl. Sci.*, **20**: 435–466.

HUCHITAL, D. A., and RIGDEN, J. D. (1970): High-sensitivity electron spectrometer, *Appl. Phys. Lett.*, **16**: 348–351.

JORGENSEN, C. K. (1971): Photo-electron spectrometry of inorganic solids, *Chimia (Aarau)*, **25**: 213–222.

JOLLY, W. L. (1971): Thermochemical estimates of core-level binding energy shifts, *Proc. Intern. Conf. Electron Spectry.*, pp. 629–645, North-Holland Publishing Company, Amsterdam.

KANTER, H., and STERNGLASS, E. J. (1962): Interpretation of range measurements for kilovolt electrons in solids, *Phys. Rev.*, **126**: 620–626.

KLEMPERER, O. (1953): *Electron optics*, 2d ed., Cambridge University Press, London.

KRAMER, L. N., and KLEIN, M. P. (1969): Solute investigation by means of photoelectron spectroscopy, *J. Chem. Phys.*, **51**: 3620–3621.

KRAUSE, M. O. (1969): Photo-ionization of krypton between 300 and 1500 eV; relative subshell cross sections and angular distributions of photoelectrons, *Phys. Rev.*, **177**: 151–157.

——— (1971): The $M\zeta$ x-rays of Y to Rh in photoelectron spectrometry, *Chem. Phys. Lett.*, **10**: 65–69.

——— and CARLSON, T. A. (1966): Charge distributions of krypton ions following photo-ionization in the M shell, *Phys. Rev.*, **149**: 52–58.

——— and ——— (1967): Vacancy cascade in the reorganization of krypton ionized in an inner shell, *Phys. Rev.*, **158**: 18–24.

———, ———, and DISMUKES, R. D. (1968): Double electron ejection in the photoabsorption process, *Phys. Rev.*, **170**: 37–47.

———, VESTAL, M. L., JOHNSTON, W. H., and CARLSON, T. A. (1964): Readjustment of the neon atom ionized in the K shell by x-rays, *Phys. Rev.*, **A133**: 385–390.

——— and WUILLEUMIER, F. (1971): Energies of $M\zeta$ x-rays of Y to Mo, *Phys. Lett.*, **A35**: 341–342.

LARSSON, K., NORDLING, C., SIEGBAHN, K., and STENHAGEN, E. (1966): Photoelectron spectroscopy of fatty acid multilayers, *Acta Chem. Scand.*, **20**: 2880–2881.

LEE-WHITING, G. E., and TAYLOR, E. A. (1957): Higher-order focusing in the $\pi\sqrt{2}$ β-spectrometer, *Can. J. Phys.*, **35**: 1–15.

LINDAU, I., and HAGSTRÖM, S. B. M. (1971): High resolution electron energy analyzer at UHV conditions, *J. Phys.*, **C4**: 936–940.

LINDBERG, B. J., HAMRIN, K., JOHANSSON, G., GELIUS, U., FAHLMAN, A., NORDLING, C., and SIEGBAHN, K. (1970): Molecular spectroscopy by means of ESCA. II: Sulfur compounds; correlation of electron binding energy with structure, *Physica Scripta*, **1**: 286–298.

LOTZ, W. (1970): Electron binding energies in free atoms, *J. Opt. Soc. Am.*, **60**: 206–210.

MADDEN, R. P., EDERER, D. L., and CODLING, K. (1967): Instrumental aspects of synchrotron XUV spectroscopy, *Appl. Opt.*, **6**: 31–38.

MAHAN, G. D. (1970): Theory of photoemission in simple metals, *Phys. Rev.*, **B2**: 4334–4350.

MANNE, R., and ÅBERG, T. (1970): Koopmans' theorem for inner-shell ionization, *Chem. Phys. Lett.*, **7**: 282–284.

MANSON, S. T. (1971): Oscillations in the energy dependence of the angular distribution of photoelectrons, *Phys. Rev. Lett.*, **26**: 219–220.

MILLER, A. W., ATKINSON, W., BARBER, M., and SWIFT, P. J. (1971): *J. Catalysis*, **22**: 140–142.

MOORE, C. E. (1949): Atomic energy levels, *Natl. Bur. Std. (U.S.) Circ. 467*, vol. 1.

NEFEDOV, V. I. (1964a): Multiplet structure of the $K_{\alpha_{1,2}}$ lines of the transition elements, *Izv. Akad. Nauk SSSR, Ser. Fiz.*, **28**: 816–822. [English translation: *Bull. Acad. Sci. USSR, Phys. Ser.*, **28**: 724–730.]

———— (1964b): Multiplet structure of the $K_{\beta_1\beta'}$ lines of transition elements, *J. Struct. Chem. (USSR) (English Transl.)*, **5**: 603–604. Determination of the effective charge on an atom in a molecule from the shifts of the lines of the x-ray emission spectrum, *J. Struct. Chem. (USSR) (English Transl.)*, **5**: 605–607.

———— (1966): Multiplet structure of the $K_{\alpha_{1,2}}$ and $K_{\beta_1\beta'}$ lines in the x-ray spectra of iron compounds, *J. Struct. Chem. (USSR) (English Transl.)*, **7**: 672–677.

NILSSON, Ö., HASSELGREN, L., SIEGBAHN, K., BERG, S., ANDERSSON, L. P., and TOVE, P. A. (1970): Development of parallel plate channel multipliers for use in electron spectroscopy, *Nucl. Instr. Methods*, **84**: 301–306.

NORDLING, C. (1959): K and L energy levels in some fourth and fifth period elements, *Arkiv Fysik*, **15**: 397–429.

————, SOKOLOWSKI, E., and SIEGBAHN, K. (1957): Precision method for obtaining absolute values of atomic binding energies, *Phys. Rev.*, **105**: 1676–1677.

————, ————, and ———— (1958): Evidence of chemical shifts of the inner electronic levels in a metal relative to its oxides (Cu, Cu_2O, CuO), *Arkiv Fysik*, **13**: 483–500.

NOVAKOV, T. (1971a): Private communication.

———— (1971b): X-ray photoelectron spectroscopy of solids; evidence of band structure, *Phys. Rev.*, **B3**: 2693–2698.

———— and HOLLANDER, J. M. (1968): Spectroscopy of inner atomic levels: Electric field splitting of core $p_{3/2}$ levels in heavy atoms, *Phys. Rev. Lett.*, **21**: 1133–1136.

———— and ———— (1969): Splitting of atomic core $p_{3/2}$ levels observed by photoelectron spectroscopy, *Bull. Am. Phys. Soc.*, **14**: 524.

PALMBERG, P. W. (1971): Auger electron spectroscopy, *Proc. Intern. Conf. Electron Spectry.*, pp. 835–859, North-Holland Publishing Company, Amsterdam.

PAULING, L. (1960): *The nature of the chemical bond*, 3d ed., Cornell University Press, Ithaca, New York.

PINES, D. (1963): *Elementary excitations in solids*, W. A. Benjamin, Inc., New York.

PURCELL, E. M. (1938): The focusing of charged particles by a spherical condenser, *Phys. Rev.*, **54**: 818–826.

RAKAVY, G., and RON, A. (1967): Atomic photoeffect in the range $E_\gamma = 1$–2000 keV, *Phys. Rev.*, **159**: 50–56.

ROBINSON, H., and RAWLINSON, W. F. (1914): The magnetic spectrum of the β rays excited in metals by soft x-rays, *Phil. Mag.*, **28**: 277–281.

ROSÉN, A., and LINDGREN, I. (1968): Relativistic calculations of electron binding energies by a modified Hartree-Fock-Slater method, *Phys. Rev.*, **176**: 114–125.

SAR-EL, H. Z. (1967): Cylindrical capacitor as an analyzer. I. Nonrelativistic part, *Rev. Sci. Instr.*, **38**: 1210–1216.

———— (1970): Criterion for comparing analyzers, *Rev. Sci. Instr.*, **41**: 561–564.

SHARMA, J., STALEY, R. H., RIMSTIDT, J. D., FAIR, H. D., and GORA, T. F. (1971): Effect of doping on the x-ray photoelectron spectra of semiconductors, *Chem. Phys. Lett.*, **9**: 564–567.

SIEGBAHN, K. (1965): Beta-ray spectrometer theory and design; magnetic alpha-ray spectroscopy; high resolution spectroscopy, in K. SIEGBAHN (ed.), *Alpha-, beta- and gamma-ray spectroscopy*, chap. 3, North-Holland Publishing Company, Amsterdam.

———, GELIUS, U., SIEGBAHN, H., and OLSEN, E. (1970): Angular distribution of electrons in ESCA spectra from a single crystal, *Phys. Lett.*, **A32**: 221–222.

———, NORDLING, C., FAHLMAN, A., NORDBERG, R., HAMRIN, K., HEDMAN, J., JOHANSSON, G., BERGMARK, T., KARLSSON, S.-E., LINDGREN, I., and LINDBERG, B. (1967): *ESCA atomic, molecular and solid state structure studied by means of electron spectroscopy*, *Nova Acta Reg. Soc. Sci. Upsalien. Ser. IV*, vol. 20. Second revised edition in preparation, North-Holland Publishing Company, Amsterdam.

———, ———, JOHANSSON, G., HEDMAN, J., HEDÉN, P. F., HAMRIN, K., GELIUS, U., BERGMARK, T., WERME, L. O., MANNE, R., and BAER, Y. (1969): *ESCA applied to free molecules*, North-Holland Publishing Company, Amsterdam.

SLATER, J. C. (1960): *Quantum theory of atomic structure*, vol. II, McGraw-Hill Book Company, New York.

SOKOLOWSKI, E., NORDLING, C., and SIEGBAHN, K. (1957): Magnetic analysis of x-ray produced photo and Auger electrons, *Arkiv Fysik*, **12**: 301–318.

STALHERM, D., CLEFF, B., HILLIG, H., and MEHLHORN, W. (1969): Energies of excited states of doubly ionized molecules by means of Auger electron spectroscopy. Part I. Electronic states of N_2^{2+}, *Z. Naturforsch.*, **A24**: 1728–1733.

SVARTHOLM, N., and SIEGBAHN, K. (1946): An inhomogeneous ring-shaped magnetic field for two-directional focusing of electrons and its application to β-spectroscopy, *Arkiv Mat. Astr. Fysik*, **A33**: No. 21.

SWANSON, N., and POWELL, C. J. (1968): Excitation of *L*-shell electrons in Al and Al_2O_3 by 20-keV electrons, *Phys. Rev.*, **167**: 592–600.

THOMAS, I. L. (1971): Angular dependence of the vibrational and rotational excitations seen in photoelectron spectroscopy, *Phys. Rev.*, **A4**: 457–459.

THOMAS, J. M., EVANS, E. L., BARBER, M., and SWIFT, P. (1971): Determination of the occupancy of valence bands in graphite, diamond and less-ordered carbons by x-ray photoelectron spectroscopy, *Trans. Faraday Soc.* (in press).

TOMBOULIAN, D. H., and HARTMAN, P. L. (1956): Spectral and angular distribution of ultraviolet radiation from the 300-MeV Cornell synchrotron, *Phys. Rev.*, **102**: 1423–1447.

TURNER, D. W., BAKER, C., BAKER, A. D., and BRUNDLE, C. R. (1970): *Molecular photoelectron spectroscopy*, John Wiley & Sons, Inc., New York.

WERTHEIM, G. K., COHEN, R. L., ROSENCWAIG, A., and GUGGENHEIM, H. J. (1971): Multiplet splitting and two-electron excitation in the trivalent rare earths, *Proc. Intern. Conf. Electron Spectry.*, pp. 813–826, North-Holland Publishing Company, Amsterdam.

——— and ROSENCWAIG, A. (1971): Configuration interaction in the x-ray photoelectron spectra of alkali halides, *Phys. Rev. Lett.*, **26**: 1179–1182.

WILEY, W. C., and HENDEE, C. F. (1962): Electron multipliers utilizing continuous strip surfaces, *IRE Trans. Nucl. Sci.*, **9**: 103–106.

9

BONDING EFFECTS IN X-RAY SPECTRA

D. J. Nagel and W. L. Baun

I. INTRODUCTION

The preceding chapters have reviewed the tremendous amount of work in x-ray spectroscopy which has been done since Bragg's invention of the flat crystal spectrometer in 1913. The reason for this activity is, of course, the fact that the spacings between the electron-energy levels are equivalent to x-ray energies. Hence the x-ray region of the electromagnetic spectrum is best suited for study of energy levels in atoms, individually or bonded together in molecules and solids.

X-ray spectra are measured for two major reasons: the first is to obtain a fundamental understanding of atoms, molecules, and solids, and the second is the practical characterization of materials. Historically, early work was aimed at an understanding of matter, and later techniques were developed for practical measurements. In the case of x-ray line spectra, for example, the major concern in the 1920s and 1930s was determination of the separation of inner-electron energy levels. This basic work culminated in the x-ray energy-level tables of Bearden and Burr (1967) (see Appendix B). Later, in the 1940s, x-ray fluorescence and in the 1950s electron microprobe methods of using line spectra for measurement of elemental composition were developed (Birks, 1969; 1971).

While x-ray line spectra are uniquely suited for the study of inner-electron levels of atoms, all x-ray spectra and especially valence-band spectra are useful for the study of bonding between atoms. There is in progress a movement from basic work to application for valence-band and other x-ray spectra similar to that for line spectra.

However, development of the field of bonding effects is occurring later, and its time scale is somewhat more compressed. Also, the separation of basic and applied work concerning bonding effects in x-ray spectra is less clear because the fundamental understanding of spectra is only now growing rapidly, but already bonding effects are being put to practical use.

The assertion that rapid progress in the basic understanding of bonding effects in spectra is only now occurring requires discussion. After all, the first observation of chemical effects in x-ray spectra was made in an absorption experiment by Bergengren in 1920. However, only in the last 20 years or so have equipment advances made it possible to reproducibly measure numerous soft x-ray spectra which are relatively free of experimental factors such as self-absorption. Also, having good measured spectra is not sufficient for full understanding and utilization of bonding effects. It is necessary to compare the experimental spectra with the results of bonding theory to know the origin of spectral details and to make use of spectra. The requisite bonding theories have been developed over several decades, but only recently have they passed from the specialist in bonding theory to those whose interest is interpretation of x-ray spectra. For example, detailed calculations of valence-band emission spectra based on bonding theory began to appear only in the late 1960s. To date only a few valence-band spectra have been calculated, but interest in computing them and other x-ray spectra is increasing. Hence, both good experimental spectra and the needed theoretical tools are available and are being brought together now. Major progress in the understanding and use of those details of x-ray spectra which are affected by the bonding of the emitting atom will continue to result from this combination of experiment and theory.

We are concerned in this chapter with both the interpretation of bonding effects in x-ray spectra and their use for both fundamental and practical studies of molecules and solids. It is not possible to understand and use bonding effect wsithout extensive background on the relation of material properties and spectra. Electronic structure, the description of bonding, is central to understanding material properties as well as spectra. Hence, the connection between electronic structure and properties will be discussed in the next section. This relation provides the major utilitarian motivation for studying bonding effects and what can be learned from them about electronic structure. In the same section, the connection between electronic structure and spectra will be outlined to set the stage for the rest of the chapter. Further necessary background on the various types of spectra will then be summarized in Section III. Details of the relation of materials and spectra which are necessary for the interpretation of spectra will be covered in three closely related sections (IV to VI) on ground- and ionized-state electronic structure and the calculation of spectra. With this extensive prelude out of the way, examples of bonding effects and their use will be given for the different types of x-ray spectra and several kinds of materials in

Sections VII and VIII. Additional effects in x-ray spectra, and the relation of x-ray spectra to other experimental methods of probing electronic structure, will be mentioned in the concluding section.

The basic question underlying the material presented here is, What good is high-resolution x-ray spectroscopy for bonding studies and the characterization of materials, i.e., what bonding and other information can be gotten from x-ray spectra, and what must be done to extract this information? Our approach to this question is mainly through examples. We aim first to give a clear physical picture and an outline of the relevant theory for bonding effects in various spectra, and then to present illustrative examples. Accordingly, we have attempted to emphasize work, mostly recent, which shows the greatest advances in understanding and application. Thus this chapter is more of a status report than a complete review of past work. With apologies to authors whose work is not covered, however high its quality, it is hoped that our selection of illustrations will allow the larger framework of a unified approach to bonding effects in x-ray spectra to show clearly through the examples.

II. RELATION OF ELECTRONIC STRUCTURE TO A MATERIAL'S PROPERTIES AND X-RAY SPECTRA

We could turn immediately to the connection between electronic structure and spectra but it is useful first to pause and consider the reasons for wanting bonding information on a material. Certainly one major motivation is the desire to understand bonding itself, regardless of any further use of bonding information. But electronic-structure information is also required to understand and calculate a material's properties. Hence, since the properties of a material determine its uses, electronic structure is fundamental to the fullest use of materials.

The relationship between electronic structure and materials applications is outlined in Fig. 9-1. At the top are the basic "static" characteristics of a material. The types of atoms and their arrangement in a material determine and are prerequisite to the calculation of bonding features. We note, as an aside, that x rays are useful for all three aspects of a material's characterization: (1) x-ray line spectra provide a means to measure elemental composition, (2) x-ray diffraction is the most heavily used means to determine perfect- and defect-crystal structure, and (3) details of x-ray spectra measured with high resolution provide information on electronic structure.

In the middle of Fig. 9-1 are the "dynamic" characteristics of a material, those aspects which describe how a material will respond to external stimuli such as stresses, electric and magnetic fields, temperature gradients, etc. Perfect crystal properties, such as elastic constants and some optical properties, are determined directly by the character and strength of bonding. Defect-dependent properties, for example,

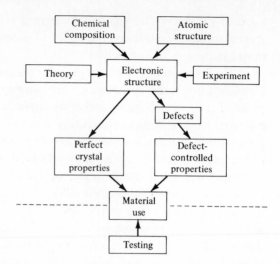

FIGURE 9-1
Schematic representation of the relation
of electronic structure to the properties
and use of materials.

mechanical strength and interdiffusion, depend on both electronic structure and the
history of a particular piece of material. Bonding determines which defects can exist
and their energies of formation and motion, while the defects which are actually
present and their arrangement are governed by previous handling of the material. The
fabrication, performance, and lifetime of a component depends on the properties and,
hence, the electronic structure of its constituent materials.

The basic approach to materials use depicted above the dashed line in Fig. 9-1
is an alternative to the time-honored empirical or "handbook" method of obtaining
values for practical properties which govern a particular material's application. The
empirical method is very useful because it alone can be applied to complex materials,
such as multicomponent alloys. But increasing understanding, better theoretical tools,
and the growing utility of experimental electronic-structure methods, such as x-ray
spectroscopy, are making the basic approach to the use of materials increasingly
viable. Hence bonding information is becoming of increasing practical as well as
basic use.

We now turn to the relation of electronic structure and spectra. Along the top
of Fig. 9-2 (Nagel, 1970) are the material characteristics and relations just discussed.
The figure is drawn to emphasize the relation of ground-state electronic structure to
spectral details via the electronic structure of the ionized states which are created
during most x-ray spectral measurements. As was mentioned, three successive sections
are devoted later to ground- and ionized-state electronic structure and to the calculation of spectral details. They give the primary link between electronic structure and
spectra. We will see that the intervention of an ionized state between the desired
electronic structure and most measured x-ray spectra is a significant but not insuper-

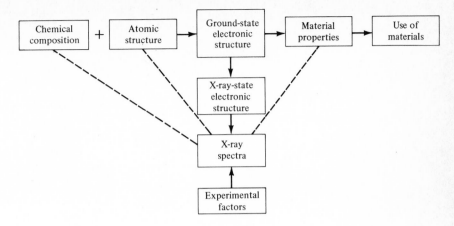

FIGURE 9-2
Relationship of material characteristics (top row) to the factors important in the generation and measurement of x-ray spectra (center column). Correlations between material characteristics and spectra are shown by dashed lines (Nagel, 1970).

able complication. Similarly, dependence of measured spectra on experimental factors introduces several problems (and a few advantages). We will give most attention to problems concerning the physics of x-ray generation, with some attention to transport effects which depend on particular experimental conditions, but with relatively little discussion of the measurement of spectra. Reviews of instrumentation are given elsewhere (Baun, 1968; Baun and Fischer, 1970), as well as in Chaps. 2 and 3. A subsection of this chapter (VIII A) will be devoted to extraction of electronic-structure information from spectra.

A second important type of connection between materials and spectra is indicated by the dashed lines in Fig. 9-2. Correlations between composition and atomic structure on one hand and the details of x-ray spectra on the other exist because of the mutual relation of these factors to the ground-state electronic structure. For example, the chemical compounds present in an unknown can be determined from the location and separation of peaks in valence-band x-ray spectra. Correlations of structural factors, such as bond length and coordination, with x-ray energies and spectral shapes have also been exhibited. Sometimes such correlations can be found empirically and used without much understanding of the bonding in a sample. The third dashed line in Fig. 9-2 indicates the fact that various valence-band spectral details can also be used to understand and calculate electronic, magnetic, and thermodynamic properties of materials. Examples of correlations between spectral details and material characteristics, plus examples of the use of spectra to study material properties, will

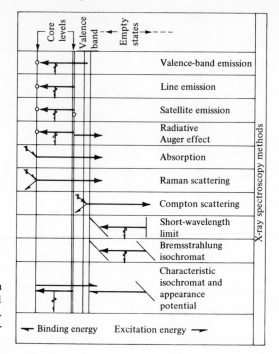

FIGURE 9-3
Binding- and excitation-energy diagram showing schematically the states probed in x-ray spectroscopy experiments. Open circles represent initial-state electron vacancies (Nagel, 1973).

be given in two subsections (VIII B and C) near the end of this chapter. By then we will have discussed the material necessary for an understanding of such relations.

Next we consider the different types of spectra which are known. The fine structure of different x-ray spectra is affected in different ways by bonding and therefore gives different information on the electronic structure.

III.   TYPES OF X-RAY SPECTRA AND BONDING EFFECTS

Emission and absorption spectra are the most familiar types of x-ray spectra and are of greatest use for the study of bonding effects. However, there are other less familiar types of x-ray spectra which will be increasingly useful for the study of bonding effects. Figure 9-3 summarizes the different types of x-ray spectra and indicates which states are involved in the measurement of each. Electron spectra (Siegbahn et al., 1967, 1969; Shirley, 1972) might be included in Fig. 9-3 because of their close relation to x-ray spectra. The information obtained from these two types of spectroscopy tends to be complementary. Much of the literature on electron spectroscopy is useful for interpretation of x-ray spectra because the physics of producing x-ray and electron

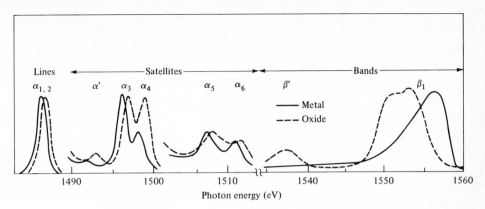

FIGURE 9-4
Al K line, satellite, and valence-band spectra from the metal and oxide (Dodd and Glen, 1968).

spectra is essentially the same. It is perhaps worth noting that, while electron spectroscopy has stimulated much useful theoretical effort, several of the concepts invoked to interpret electron spectra were first used to understand x-ray data, especially in the Russian literature. We will occasionally take advantage of the relations between x-ray and electron spectroscopy but will not be able to give separate consideration to electron spectra. See Chap. 8 for a review of electron spectra.

The first three kinds of x-ray spectra in Fig. 9-3 are all emission methods which involve only the filled electron-energy levels. Valence-band diagram spectra directly reflect the energies of bonding electrons, while diagram line spectra provide an indirect measure of the valence-electron distribution. Satellite spectra originate in multiply-ionized atoms[1] and may or may not involve the valence electrons. The satellite spectra associated, for instance, with the $K\alpha$ lines of third-row elements are intense, well-resolved, and sensitive to bonding. But the complex multiplet structure of satellite spectra makes it difficult to obtain bonding information from them. The Al K diagram and satellite spectra from aluminum metal and oxide are shown in Fig. 9-4. Several chemical effects in going from the metal to the oxide are evident and will be discussed later.

The remaining kinds of spectra in Fig. 9-3 are mostly probes of the empty states in a material. Weak radiative Auger spectra appear on the low-energy side of some line spectra because part of the available energy goes into excitation of an electron into an empty state, simultaneous with photon emission. Thus, they might

---

[1] Sometimes low-energy valence-band peaks and multiplet structure are referred to as "satellites," but in this chapter the term is reserved for emission features from multiply-ionized atoms.

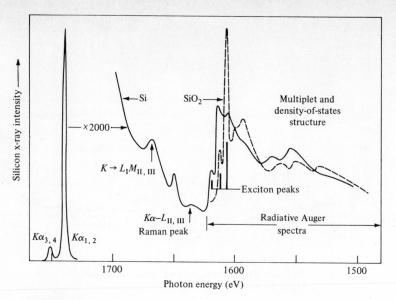

FIGURE 9-5
Si K radiative Auger spectra from Si and SiO_2 (Aberg and Utriainen, 1972).

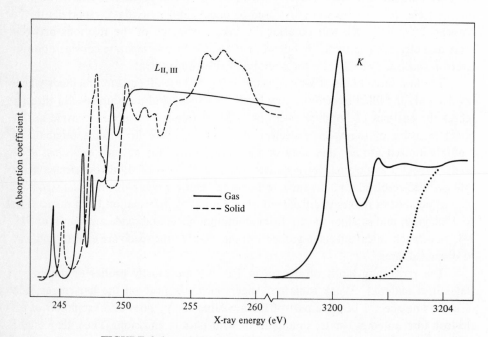

FIGURE 9-6
Absorption spectra of argon near the K edge (Parratt, 1939) and the L edges (Haensel et al., 1970). The dotted line in the K spectrum separates low-energy excitonic structure from the continuum.

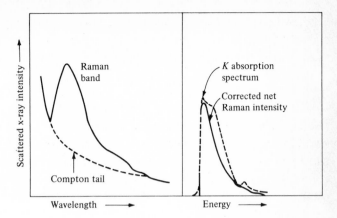

FIGURE 9-7
X-ray Raman and K absorption spectra of Be (Suzuki, 1967).

reflect the availability of empty states. As seen in Fig. 9-5, however, radiative Auger spectra are complicated by overlapping multiplet and exciton structure. X-ray absorption spectra also involve core-level-to-empty-state transitions as shown in Fig. 9-3, but they are usually free of multiplets. K and L x-ray absorption spectra of argon in Fig. 9-6 show the effects of excited-bound states below the edges. The empty states form a smooth continuum in the gaseous state but are formed into bands in the solid. The L spectra show this difference. Fine structure due to solid-state effects can extend to about 1 keV above the K edge of cubic materials. Absorption spectra, which are discussed in Chap. 6, can be obtained three ways: by the usual transmission technique, by ratioing emission intensities when absorption and emission spectra overlap (Fischer and Baun, 1967a; Liefeld, 1968), and by use of ultra-soft x-ray reflection data (Ershov, 1969).[1] In x-ray Raman scattering, a core electron is also excited into the empty bands as indicated in Fig. 9-3, but the spectrum of scattered rather than absorbed radiation is measured (Das Gupta, 1969). Similarly to the radiative Auger process, the photon carries off the energy not used in electron excitation. The correspondence between the x-ray Raman and absorption spectra of beryllium is shown in Fig. 9-7. X-ray Raman spectra are weak and hard to separate from Compton scattering, and are therefore more difficult to measure than absorption spectra. Compton scattering results in excitation of valence electrons and core electrons to high-energy bands. The spectrum of Compton-scattered radiation reflects the momentum distribution of bonding electrons more than the energy-band

[1] For a recent example of the use of photoelectric yield curves to obtain absorption spectra, see Gudat and Kunz (1972).

structure. Thus Compton spectra are different from the others in Fig. 9-3.[1] Since momentum and energy are related and the energy of valence electrons depends on bonding, the momentum distribution of valence electrons is also sensitive to bonding. Compton spectra are becoming increasingly useful for bonding studies for low-atomic-number, low-absorption materials.

Radiative Auger, x-ray absorption, and Raman spectra sample the empty states from the deeper core levels. In contrast to this, short-wavelength-limit (SWL) and Bremsstrahlung isochromat spectra both involve Bremsstrahlung emission accompanying downward transitions of incident electrons from the high-energy bands to states near the Fermi level. The SWL measurement requires a scan of photon energies at a fixed value of incident-electron energy, while in the Bremsstrahlung isochromat measurement the incident-electron energy is scanned at one spectrometer setting (Chap. 6). As for Compton spectra, the energies of SWL and Bremsstrahlung isochromat spectra are not characteristic of the emitting material, but they still yield electronic-structure information. Figure 9-8 demonstrates the similarity of SWL and Bremsstrahlung isochromat spectra from Ta which is due to the common final state involved in the two types of spectra. In the measurement of characteristic isochromats, the spectrometer is peaked on a line, and the incident energy is scanned through the threshold value. Characteristic isochromat spectra are therefore also called *excitation curves*. This method gives information on the availability of empty states since the number of core holes created, and hence the emitted line intensity, depends on the probability of core electrons finding empty states. Appearance-potential spectra are similar to characteristic isochromat spectra, but are measured nondispersively by modulation techniques which yield a derivative of the characteristic intensity. Such spectra from nickel are compared in Fig. 9-9, along with the appearance-potential spectrum of oxidized Ni.

We turn now from the classification of x-ray spectra to consideration of the types of bonding effects. For any given spectrum, bonding effects appear in the precise peak energies or spectral shape. These effects are often most easily noticed by comparing spectra from a series of materials in which one aspect of the bonding, e.g., phase or valence, is varied. Spectra are, of course, plots of emitted or absorbed intensity (I) versus photon energy ($\mathscr{E}$). Hence, bonding effects include changes in spectral shapes, such as the appearance or disappearance of peaks, changes of peak and integrated intensities (ΔI), and shifts ($\Delta \mathscr{E}$) in sufficiently resolved peaks. Although not commonly done, bonding effects can be highlighted by plotting a difference curve ΔI vs. $\mathscr{E}$. Narbutt (1955) has used this method.

----

[1] X-ray diffraction involves elastic scattering which gives the electron-density distributions, in comparison to inelastic Compton scattering which yields electron-momentum distributions. But x-ray diffraction involves measurement of spatial, not spectral, photon distributions and hence is excluded from consideration here.

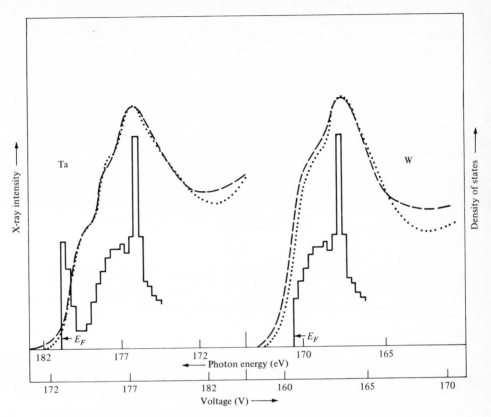

FIGURE 9-8
Short-wavelength limit (dashed) and Bremsstrahlung isochromat (dotted) spectra of tantalum and tungsten (Böhm and Ulmer, 1971) compared to the density of empty states (solid) (Mattheiss, 1970).

For all the emission and scattering spectra mentioned above, the measured intensity depends on the transport of the incident radiation through the sample, the cross sections for the interaction of interest, and the transport of excited or scattered radiation out of the sample. In general, it is only the relative cross sections for generation or scattering of the radiation to be measured which are of interest, since they directly reflect bonding effects. Matrix effects on radiation transport, such as absorption of emerging radiation, do indeed depend on the chemical makeup and bonding in the sample, but they are considered here to be distinct from bonding effects. Usually they do not provide incisive information on electronic structure and are hence viewed as unwanted and sometimes unavoidable experimental factors. Matrix effects can be critically important if they affect spectra in such a way as to make it difficult

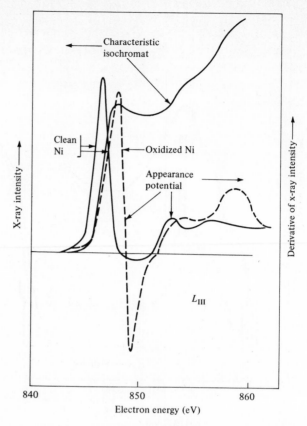

FIGURE 9-9
Characteristic isochromat spectrum (excitation curve) of nickel (Liefeld, 1968) compared with the appearance-potential spectra of clean and oxidized nickel (Houston and Park, 1972a).

to separate the desired electronic-structure effects from apparent bonding effects due to changes in the transport of incident or exciting radiation. If matrix effects influenced only the absolute intensity (the spectral scale factor), they would be of no bother since relative spectra are almost universally used in the study of bonding effects. The complications introduced by matrix effects stem from their influence on relative spectral shapes.

No examples of bonding effects are shown now because of our desire first to discuss the material necessary for the interpretation of bonding effects which is given in the following three sections. The next section on ground-state electronic structure is fundamental to the interpretation and use of x-ray spectra. Then the creation and electronic structure of ionized states is discussed. Examination of Fig. 9-3 shows that the various x-ray spectroscopic methods mentioned above have several features in common. In particular, the same ionized states appear often. Understanding one kind of ionized state can thus contribute to interpretation of different types of spectra. This is one reason for discussing ionized states by themselves. Then the use of calculated electronic structure to compute x-ray spectra for comparison with experiment is outlined.

We may see the motivation behind this sequence of sections by considering the equation for the energy $\mathscr{E}$ of an x-ray line

$$\mathscr{E} = E_2 - E_1 \qquad (9\text{-}1)$$

given as the difference of the binding energy ($E < 0$) of the inner orbital 1 and the less tightly bound orbital 2. Binding energy E is the difference in the total energy of an atom before and after ionization. Calculation of E thus requires knowledge of both the ground and ionized states. Sometimes, however, knowledge of the ground-state electronic structure alone is enough to interpret spectra. For instance, shifts in line energies $\mathscr{E}$ may be written

$$\Delta\mathscr{E} = \mathscr{E}^S - \mathscr{E}^R = (E_2{}^S - E_1{}^S) - (E_2{}^R - E_1{}^R)$$

$$= (E_2{}^S - E_2{}^R) - (E_1{}^S - E_1{}^R) = \Delta E_2 - \Delta E_1 \simeq \Delta\varepsilon_2 - \Delta\varepsilon_1 \qquad (9\text{-}2)$$

where S and R denote the sample and reference. Changes in binding energies can be approximated by changes in the ground-state eigenvalues ε for the orbitals involved if electron-relaxation and other effects cancel for the sample and reference (Schwartz et al., 1972). Also the ground state for filled and empty bands is used in calculating some spectra. Because of these factors and because it is fundamental to the understanding of bonding effects, ground-state electronic structure is considered at length. But the ground state alone is not sufficient to interpret spectra. The intensity of some spectra is determined by the probability of creation and the lifetimes of ionized states. Also, there are some effects reflected in spectra which do not appear until ionization. Multiplet splitting is one of the most prominent. We therefore need to consider the creation and history of ionized states, and their electronic structure.

Another way to view the following three sections is by relating them to the three distinct, but related, kinds of energies of interest in understanding bonding effects:

Energy	Section
Eigenvalues ε	IV. Ground-state electronic structure
Binding energy E	V. Ionized-state creation and electronic structure
X-ray energy $\mathscr{E}$	VI. Calculation of x-ray spectra

Also, of course, "ground state → excitation and possibly nonradiative readjustment → emission" is the time sequence of events in most x-ray measurements.

IV.   GROUND-STATE ELECTRONIC STRUCTURE

As shown in Fig. 9-2 and as we have emphasized, the ground-state electronic structure of a material is essential to understanding and calculating its properties and spectra. By ground state, we imply the absence of any ionization or electronic excitation. At ordinary temperatures, all materials contain thermal excitations (phonons), but they have relatively little influence on most spectra, and will be mentioned briefly in the last section.

The term "electronic structure" is commonly used, and we have not felt the need to define it before now. But perhaps it is worthwhile to spell out that is meant in this chapter by electronic structure. There are two important independent variables: $\mathbf{r}$, the position in ordinary "real" space, and $\mathbf{k}$, the wave number in quasi-momentum ($\hbar\mathbf{k}$) space ($\hbar$ is Planck's constant divided by 2π). The electronic structure of a material is usually obtained by solution of Schrödinger's equation using a potential defined in $\mathbf{r}$ space which contains nuclear and electronic contributions. It is described by the wavefunctions $\psi(\mathbf{r})$ and electron-density distribution $\rho(\mathbf{r}) = |\psi(\mathbf{r})|^2$ in real space, or equivalently, the wavefunctions $\chi(\mathbf{p})$ and momentum density $|\chi(\mathbf{p})|^2$ in momentum space. The energy distribution of valence electrons, that is band structure $E(k)$, and constant-energy surfaces are plotted in $\mathbf{k}$ space for solids. Electrons occupy the energy bands up to the Fermi level (energy) in conductors. The Fermi surface is the boundary in $\mathbf{k}$ space inside of which all states are filled with electrons. Both constant-energy surfaces and the density of states $N(E)$ are derived directly from the energy bands. The density of states gives the number of electron states per unit energy at the energy E. It is similar to degeneracy, which gives the number of states at the same energy in an atom or molecule.

In general, electronic-structure details are gotten from theory and checked by experiment. Except for peak separations which can be associated with energy-level differences, x-ray spectra alone do not usually yield information which is directly comparable to any of the different aspects of ground-state electronic structure. Rather, x-ray spectroscopy provides a test of calculated electronic structure, i.e., data for a comparison with certain integrals which involve the calculated electronic structure and which will be discussed (Altmann, 1968). Effective charges, related to line shifts, are $\mathbf{r}$-space integrals over electron density. And, we will see that x-ray spectra are given theoretically by integrals over $\mathbf{k}$ space or energy.

To understand and use x-ray spectra, it is necessary to have a clear picture of the various aspects of electronic structure and their relation to spectra. Hence it would be nice to give here several calculated examples of the various aspects of electronic structure mentioned above. But these would take us somewhat far afield, and we will generally content ourselves with only references to especially useful examples of computed electronic structure. The reader is encouraged to study these to become familiar with the different aspects of electronic structure, all of which enter in one way or another in the interpretation of x-ray spectra.

The potential has a central importance since in a calculational sense it determines the rest of the electronic structure. Most papers on solid-state electronic-structure calculations contain a discussion of the potential used. The problem of constructing potentials was discussed in detail at a recent conference (Marcus, Janak, and Williams, 1971). Reviews of pseudopotentials are given by Heine and his colleagues in volume 24 of the *Solid state physics* series (Ehrenreich, Seitz, and Turnbull, 1970). Potentials are directly useful for interpretation of x-ray and electron line shifts as well as for electronic-structure calculations.

Wavefunctions contain all information about the electronic structure of a material, and hence they are one of the most important aspects of electronic structure. Also, wavefunctions are used in thorough calculations of x-ray spectra; their extent determines the region in space which is sampled by x-ray spectroscopy, and their detailed shape determines transition probabilities. It is interesting and useful to compare (1) accurate and approximate wavefunctions, (2) wavefunctions for free atoms and for molecules or solids, and (3) wavefunctions for different states within the band structure of given material. For example, atomic wavefunctions calculated from the Schrödinger equation and tabulated by Herman and Skillman (1963) can be contrasted with approximate but analytical Slater-type orbitals (see, e.g., Cumper, 1966). Both these types of wavefunctions provide a chance to note the very limited extent of the inner-core wavefunctions. For example, K spectra for nickel probe only about 0.1 percent of the atomic volume. One comparison of free-atom and solid-state bonding-electron wavefunctions used for interpretation of chemical shifts in x-ray lines was given by Shubaev (1961). Numerous comparisons can now be made between the output of atomic programs and various band theory calculations for solids. For example, Cohen and Heine (1970) contrast Ge $4p$ free-atom wavefunctions with the p-like band states in solid Ge. Rigorous and approximate d-electron solid-state wavefunctions, and their variation across the $3d$ band of Fe, are given by Pettifor (1972). Many other examples of thorough and approximate atomic and solid-state wavefunctions could be given.

The overall electron density calculated from wavefunctions is dominated by the core electrons. Johnson and Smith (1972) recently presented charge densities for MnO_4^- calculated by the cluster method. Figure 9-10 gives valence-electron charge distributions for Ge, GaAs, and ZnSe, computed using the pseudopotential method. These clearly show bond formation and penetration of the core regions by the valence electrons. Decrease in the charge between atoms in going from Ge to ZnSe is evident. It is the change of the valence-electron distribution in the core region of atoms such as Zn and Se which alters the screening and binding energies of core electrons and leads to line shifts.

FIGURE 9-10
Theoretical valence-electron charge distributions showing contours of equal density (Walter and Cohen, 1971).

Figure 9-10 provides a good opportunity to consider the fact that electronic structure is decidedly different for the different kinds of bonding. In proceeding from metallic through covalent to ionic bonding, there is increasing localization of the valence electrons. In metals, the bonding electrons occupy all parts of the atomic volume although they are concentrated on the atoms. As illustrated in Fig. 9-10, valence electrons in covalent materials such as Ge form bonds between atoms and no longer occupy some regions. The structure of covalent materials is relatively open, with the bonding-electron charge density very dependent on direction. For ionic materials, exemplified by ZnSe in Fig. 9-10, the valence electrons are highly localized on the anion. The increasing removal of the outer electron(s) from the cation corresponds to ionization of a free atom and is, in some sense, the ultimate bonding effect. The relation of ionization and effects due to bonding will be exploited later.

We have already noted that effective charges can be given by an integral of the electron density over some region in **r** space. Effective charges so determined are not uniquely defined because of arbitrariness in the extent of the integration. In fact, while the concept of effective charges is very useful, numerical values depend on the means used to obtain them (Nefedov, 1964). Other methods besides **r**-space integrals yield effective charges. These include orbital population analysis (in molecular orbital theory, Mulliken, 1962), the use of electronegativities (Pauling, 1960; Phillips, 1970), and the use of the equation $\eta = Z - cn$, where η is the "effective coordination charge," Z the valence, c the degree of covalency, and n the coordination number (Ovsyannikova and Nasanova, 1970). Also, different experiments probe different aspects of a charge distribution in a solid. The Knight shift (in nuclear magnetic-resonance spectroscopy) and the isomer shift (in Mössbauer spectroscopy) sample only s-like electrons at the nucleus, while x-ray and electron line shifts are often sensitive to the entire electron density in the region of core orbitals. X-ray diffraction probes the electron distribution everywhere.

Momentum distributions can also be calculated using modern computer techniques. Epstein (1970) presents momentum-density maps in **p** space computed for several simple hydrocarbons. He also gives difference maps which highlight the effects of bond formation on momentum density. Data on electron momenta are available from positron annihilation experiments (Stewart and Roellig, 1967) as well as Compton scattering.

While wavefunctions and related charge and momentum densities are important aspects of electronic structure, the computation of electronic structure first yields the energy levels or bands. For atoms and molecules, energy levels are sharp, of course. The molecular levels contain a mixture of several atomic states due to the sharing of electrons in covalent bonds, as illustrated by the atomic and molecular orbital diagrams in books such as that by Ballhausen and Gray (1964). Molecular orbital theory is primarily designed for materials in which the bonding is mostly covalent, but it can also be used to describe both covalent aspects of bonding in metals and mixed co-valent-ionic bonding in materials which are mainly ionic.

Band theory is applicable to ordered materials with all types of electronic struc-ture. It is instructive to compare the energy bands for four materials close to each other in the periodic table, yet exhibiting different bonding types, namely Al, Si, KCl, and solid Ar. For Al (see references in Callaway, 1964, p. 177), the Fermi level cuts across bands so that empty states are available immediately above it, making Al a conductor. The bands are quite similar to those for nearly free electrons. In Si (Callaway, 1964, p. 158), an energy gap exists between filled and empty states, but it is not large and impurity states in the gap generally make Si a semiconductor. The filled and empty bands in Si can be associated with bonding and antibonding molec-ular orbitals (Weaire and Thorpe, 1971). KCl is an insulator because it has a larger energy gap (Fong and Cohen, 1969) with relatively little tendency to develop itinerant states within it. The bands in KCl are narrower (flatter) because there is less inter-action between electrons on different atoms (ions) than in the covalent case. Each band is clearly associated with an energy level in one of the ions. For solid argon, the valence band is very flat in accord with its atomic-like character (Mattheiss, 1964). In fact, core-level states in all materials are almost perfectly flat on an energy-band diagram because electrons in core states have essentially the same energy no matter what their "direction of motion" around the nucleus. The slight curvature of the filled bands in argon reflects the effects of the near-neighbor atoms on the outer elec-trons of a given atom. The higher energy empty bands in argon have more curvature because "hot" electrons moving in them temporarily sense the lattice and are dif-fracted in a manner similar to the mobile conduction electrons in metals. The high-energy band structure is reflected in the L absorption spectrum of solid argon shown in Fig. 9-6.

The topology of the Fermi surface of conductors is very important, of course, for discussion of their transport properties. Similar constant-energy surfaces in **k** space, below the Fermi surface for emission and above it for absorption, are also useful for understanding x-ray band spectra. States on such surfaces which have symmetry consistent with the selection rules are probed at each setting of the spectrom-eter. Harrison (1966) shows constant-energy surfaces for face-centered-cubic metals at energies equivalent to one, two, three, and four electrons per atom. For aluminum (valence 3), the surfaces corresponding to one and two electrons per atom would be sampled by an emission spectrum, while the surface for four electrons per atom would be probed by an absorption spectrum.

We have so far referred to examples concerning wavefunctions, electron density, energy levels, energy bands, and constant-energy surfaces. The local density of band states with the symmetry sampled by a particular core state is the last general aspect of electronic structure needed for understanding x-ray spectra. This doubly decomposed density of states, i.e., broken down according to spatial location and wavefunction symmetry, is available from band theory. We will discuss the origin and calculation of component densities of states further in Section VI.

This completes our review of examples of electronic structure. We next consider the theoretical methods in use for calculation of the details of bonding. The different techniques have been summarized in many textbooks, for example those of Slater (1963; 1965). All we can do here is consider some of the grosser features of the various methods and comment on their applicability and previous use for interpretation of spectra.

All bonding calculations are many-body problems, but most are done in the one-electron approximation. This assumes each electron can be treated independently if the effects of all electrons are lumped into an averaged potential. The exchange correction, which accounts for the fact that an electron does not act on itself, and the correlation correction, which concerns direct electron-electron interactions, are central problems in the calculation of electronic structure. However the potential is constructed, calculations may be carried to self-consistency to remove the sensitivity of the results to the initial choice of electron configuration.

The particular theory used for an electronic-structure calculation depends on several factors, the first of which is the state of the material, whether atomic, molecular, or solid. For each of these, there is a range of theories varying from quite approximate, but easily used, methods to the most thorough, self-consistent calculations which can be done only on the faster computers. Next the general type of material (rare gas, simple metal, transition metal, covalent compound, ionic compound, etc.) must be taken into account. Then the particular material requires attention, since the approach used may well involve quite specific approximations. In the rest of this section, we will consider classes of calculation methods applicable to atoms, molecules, and solids, with little attention to the type of material.

In the case of atoms, there are a few different approximate methods. One, the Thomas-Fermi statistical approach does not include consideration of the shell structure of the atom in its usual form, and so it finds little use in x-ray spectroscopy. The simplest method which separates the various orbitals involves the use of screening constants in wavefunctions of analytical form. Slater-type orbitals lack the inner nodes of the true wavefunctions but are useful for many atomic calculations. Screening constants are also used in hydrogen-like orbitals which have the proper radial behavior and yet are still easy to use. For these cases, the energy of a core level nl is given by a term-value equation such as

$$\varepsilon_{nl} = \frac{13.6}{n^2} Z_{nl}{}^2 = \frac{13.6}{n^2} (Z - \sigma_{nl})^2 \qquad (9\text{-}3)$$

where the screening constant is denoted σ_{nl} (Kuhn, 1962). This equation is a generalization of the nonrelativistic result for the hydrogen atom. Another term-value equation comes from Dirac's relativistic theory (Dirac, 1947):

$$\varepsilon_{nj} = mc^2(1 + (\alpha Z_{nj})^2\{n - j - \tfrac{1}{2} + [(j + \tfrac{1}{2})^2 - (\alpha Z_{nj})^2]^{1/2}\}^{-2})^{-1/2} \tag{9-4}$$

where α is the fine-structure constant and Z_{nj} the effective atomic number. Z_{nj} is the actual atomic number less a sum over occupied orbitals of the product of electron numbers and orbital screening constants. It is calculated from the matrix equation

$$Z_{nj} = (Z) - (S_{nj})(N_{nj}) + (S_{nj}) \tag{9-5}$$

where (N_{nj}) is the orbital-occupation matrix and (S_{nj}) is the screening matrix. Veigele, Stevenson, and Henry (1969) have written a screening-constant program based on the use of Z_{nj} in the Dirac equation.

The methods of atomic calculation mentioned so far do not require solution of Schrödinger's equation and hence do not yield accurate wavefunctions. Many thorough calculations of atomic wavefunctions have been made. One of these makes use of the Hartree-Fock scheme but treats exchange in the Slater free-electron-gas approximation. This procedure, carried to self-consistency in the Herman-Skillman program (1963), is one of the most widely used today. Self-consistent Hartree-Fock calculations without approximating exchange have been programmed by Froese (1963) and also find extensive application for the calculation of eigenvalues and wavefunctions. Clementi (1965) published wavefunctions for atoms and ions calculated by use of the Roothaan-Hartree-Fock method.

Almost all approaches for atoms are also applicable to free ions, although there are questions of convergence for self-consistent calculations for negative ions in some cases. The Veigele, Herman-Skillman, Froese, and Liberman programs are all useful for positive ions, even when the electron vacancies are distributed in an arbitrary manner in the inner levels.

If an ion is bound in a molecule or solid, it is not enough to consider only orbitals on the particular ion. Core-level energies in the ground state are also influenced by fields produced by neighboring ions. In a molecule, other ions can be treated as effective point charges (Siegbahn et al., 1969; Schwartz et al., 1972) in computing their influence on the energy of a localized orbital. In a more or less ionic solid, the effect of all other ions can be handled via a Madelung field. This field opposes the energy shifts resulting from ion formation. We will return to the subject of core-level energies and their shifts when discussing the calculation of x-ray line shifts.

Absorption and other spectra which directly involve electron bands rather than core levels cannot, of course, be treated with only the atomic and ionic electronic-structure methods discussed thus far. We now turn to the variety of methods for calculating bonding-electron structure. As for the atomic and ionic cases, there are many approaches available.

For molecules, the spectrum of methods is discussed by Pople (1965). There is in general a tradeoff between the number of electrons which can be treated and the thoroughness of the calculation. For the interpretation of x-ray spectra, methods of intermediate complexity and accuracy are commonly used. In recent years, approximate molecular orbital (MO) theory has become very popular for understanding x-ray spectra (Anderman and Whitehead, 1971). Reference is made to several types of MO theories by Schwartz et al. (1972).

Most applications of MO theory for interpreting x-ray spectra have not been to spectra from molecules but rather to the spectra of simple compounds in the solid state. This seems surprising. The energy bands of most solids seem to have so much curvature that a description in terms of energy levels (flat bands) as in MO theory sounds quite extreme. But there are two factors behind the applicability of MO theory. First, in many solids, particular atoms are found within rather isolated units having largely covalent bonds, such as S in SO_4^{--} tetrahedra. The bonding of these units to other atoms has little influence on the encapsulated atom. Second, even in solids where there is a rather uniform lattice of at least partly covalent bonds, the electronic structure in the core region of a given atom is primarily determined by the type, coordination, and distance of its nearest neighbors. Hence it is a good approximation to treat the atom and its neighbors as a small molecule whose levels are merely perturbed (and broadened) by interaction with the rest of the solid. This approximation has already proven useful for interpretation of spectra from ordered materials for which band calculations could be but were not performed. It is anticipated that some form of molecular or local orbital theory will be useful for spectra from disordered materials, such as semiconductor alloys and glasses for which the usual band calculations cannot be performed (see Dodd and Glen, 1970).

The use of MO theory for interpreting the valence-band and absorption spectra from solids has one drawback, at least in principle. While degeneracies and wavefunctions result from a MO calculation and can be used for estimating spectra, only degeneracy numbers, not densities of states, are available from MO theory. Full calculations of x-ray spectra based on band theory show that x-ray band spectra mimic the density-of-states curves, especially for transition metals and their alloys. Spectra estimated from MO theory are bar graphs, with the locations of the bars dictated by the energy levels (and hence not uniformly spaced). However, we will show some examples of x-ray spectra calculated using MO theory, and other examples of the use of MO theory to interpret spectra, even without calculations of spectra. Hence, the absence of the density-of-states curve in MO theory is not a practical handicap in some cases.

To this point we have mentioned some theories for calculation of the electronic structures of atoms, ions, and molecules which have proved useful for understanding x-ray spectra. The electronic structure of ordered solids can be computed by a wide variety of methods (Callaway, 1964; Marcus, Janak, and Williams, 1971). The end-point theories of the tight-binding and nearly free electron pictures are simple but not sufficient to interpret spectral details. Intermediate theories are most widely used. From these, in a manner similar to the various bond theories, additional more accurate information results but at the price of increased complexity and less general applicability. For simple metals, semiconductors, and compounds, the orthogonalized plane wave (OPW) and the related pseudopotential (PP) methods are most efficient, while for transition metals and their alloys and compounds the augmented plane wave (APW) and Korringa-Kohn-Rostoker (KKR) methods are generally used. The first three of these use a propagating-electron-wave picture, while the KKR method is based on scattering theory. Although the band-calculation schemes are quite complex, they are all reduced to computer programs which increasingly can be used with little attention to underlying theory or inner details of the program. The density of states and wavefunctions gotten from these programs provide the starting point for calculation of band spectra, which will be discussed two sections hence.

The basic theory for the electronic structure of disordered alloys is only now being developed. No present theory gives the electronic-structure information needed to calculate the spectra from a disordered alloy in the same detail as is available for ordered materials. However, progress has been made recently in the calculation of density-of-states curves for disordered covalent semiconductors (Weaire and Thorpe, 1971; Thorpe and Weaire, 1971a). For disordered metallic alloys, new theories such as the coherent potential approximation (CPA) and average t-matrix approximation (ATA) yield local densities of states which will be increasingly useful for interpretation of valence-band emission spectra of alloys (Schwartz et al., 1971; Stocks, Williams, and Faulkner, 1971). In the area of disordered materials, experimental x-ray spectra may provide a guide for the development of theory, or at least data against which emerging approaches can be tested.

New theoretical methods, some of which are applicable to disordered materials, are being developed to compute the electronic structure of surface regions (Haydock, Heine, and Kelly, 1972). It was found for nickel that the rms width of the d-band density of states decreases as the square root of the coordination number of the surface atoms (Haydock et al., 1972). Another new surface model, embracing both band-structure and many-body effects, has also been applied to nickel (Levin, Liebsch, and Bennemann, 1973). It predicts that the d band near the surface lies lower and is narrower than that in the bulk.

This completes our enumeration of electronic-structure methods. We have mentioned the classical approaches to bonding calculations. Nowadays, hybrids of these approaches tailored to specific types of materials are being used with increasing frequency. Also, the relation of bond and band theory is being explored. Carter (1971) has discussed extraction of density-of-states information from chemical bonding theory, while Heine and Weaire (1970) examined the treatment of covalency by band theory. The methods mentioned above are the bases for combined and comparative approaches and are sufficient for interpretation of x-ray spectra from most materials.

It should be mentioned, before passing from a discussion of the ground state to that of the ionized state in the next section, that the complex molecular orbital and band theory programs can also yield accurate values of core-level energies for calculation of level shifts even though such programs were developed primarily to calculate valence-electron structure. Shifts calculated from such programs reflect the potential in a molecule or solid and do not depend on ionic or other models. If one program can be applied to a series of compounds, core-level energies and shifts can be calculated. However, if different bonding programs must be used for the sample and reference material, there is a problem of matching energy scales to calculate a core-level shift. It is to avoid this problem and to save labor that the various less rigorous models for level shifts which we will discuss in Section VI are used.

V.   IONIZED-STATE CREATION AND ELECTRONIC STRUCTURE

Bonding theories used to understand x-ray spectra deal with the ground state even for those spectra involving ionization. Because of this and the fact that the ground state is the starting point for understanding properties, we have presented an extensive discussion of the ground state. But if we did not explicitly consider the creation and electronic structure of the ionized state, it would not be possible to interpret some valence-band, line, absorption, and Raman spectra, and all satellite and radiative Auger spectra. Core-level vacancies produce a greater disturbance than do valence-electron vacancies, and so we will mainly be concerned with their creation and effects. Rearrangement of core holes among different levels also influences some spectra, and must be considered. The creation and electronic structure of ionized states are discussed before bonding effects on both development and electronic structure of x-ray (core-ionized) states are summarized at the end of the section. This section is largely a description of the physics of the creation and rearrangement of ionized states.

It should be emphasized that the present discussion which treats x-ray states as separate entities—quasi-stationary states—is largely artificial. The creation, rearrangement, and decay of an ionized state is probabilistic, and also there are questions of the completeness of relaxation. But the idea of a quasi-stationary x-ray state with a lifetime τ equal to the $1/e$-time for decay is still a useful concept. If τ is

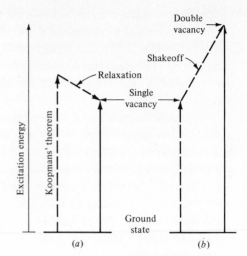

FIGURE 9-11
Ionization processes (schematic). (*a*)
Single vacancies in core levels; (*b*) double
vacancies in core levels.

sufficiently long for relaxation to occur, we can use the one-electron picture to discuss the x-ray state prior to emission.

Core-energy levels in the ground state are as sharp as solid-state effects allow. But, since x-ray states are not stationary states, broadening $\Delta E \sim \hbar/\tau$ occurs due to the finite lifetime τ. Such broadening is a most obvious result of creating ionized states and one of the major obstacles to studying the ground state by use of x-ray spectra.

X-ray states may be created in a variety of ways. Production of ionization in a core level depends on the cross section for photon absorption (fluorescence excitation) or Coulomb excitation (electronic and light-ion excitation). A review of relativistic calculations for these processes is given by Mohr (1968). Incident high-energy heavy ions produce multiple inner-shell vacancies by Pauli excitation, a process which can involve diabatic level crossing (Barat and Lichten, 1972) and is not entirely understood at present (Brandt and Laubert, 1970). Whatever the method of ionization, we find it useful to discuss separately single- and multiple-core-level-hole x-ray states.

Figure 9-11*a* shows a single-ionization process on an excitation-energy diagram. Although ionization is a single physical event, this process can be conceptually separated into two steps: (1) ionization with all other orbitals frozen, and (2) relaxation of the other orbitals in response to the vacancy. Physically, orbital relaxation may be pictured as causing acceleration of the outgoing electron, which removes some excitation energy that would otherwise reside in the ionized system (Davis et al., 1970). In a Hartree-Fock calculation for an atom, Koopmans' theorem states that, ignoring relaxation, the binding energy of an electron is precisely the negative of its eigenvalue.[1] Ionization or binding energies are correctly calculated by taking the difference of the total-system energies in the ground and ionized states. Lindgren (1967) compared K-shell binding energies calculated in the frozen-orbital approximation (method 1) and those calculated including relaxation (method 2) with experiment. The difference between the two methods, the relaxation energy, varies from about 13 eV for carbon to 52 eV for krypton. Gelius, Roos, and Siegbahn (1970) have discussed contributions to the relaxation energy which is about 10 eV for the sulfur $2p$ level. Changes in electron correlation account for only a few eV of this value, the rest being primarily due to loss of screening. Lindgren's method 2, while agreeing better with experiment than method 1, still gives an energy which is 32 eV too high for Kr $1s$ electrons, as an example. Such remaining discrepancies between method 2 and experiment may be due to the shortcomings of the Hartree-Fock method and neglect of solid-state effects (since the measurements were made on solids while the calculation is for free atoms).

Figure 9-11b shows the production of a state doubly ionized in the core. The additional vacancy may be produced by two mechanisms: (1) direct ejection of a second electron, and (2) shake-off of an additional electron due to the impulsive perturbation produced by the initial ionization. In electron shake-off, the outer electrons respond to the apparent increase in nuclear charge caused by the sudden loss of a core electron which partially screens them from the nucleus in the ground state.

Direct multiple ionization of core levels is most evident in the spectra produced by heavy-ion impact. In Fig. 9-12, for example, Al K satellite spectra due to emission from ions with as many as five additional L-shell vacancies are overall more intense than the Al $K\alpha_{1,2}$ line for which no extra vacancies are produced prior to emission. The ratio of Al K satellite to $K\alpha_{1,2}$ intensity decreases with increasing light-ion energy, as does the L-shell cross section, showing that double vacancies are also produced directly, even by proton and alpha impact (Knudson, Burkhalter, and Nagel, 1973). For electron excitation, it was shown experimentally that the $\alpha_{3,4}$ to $\alpha_{1,2}$ (satellite to parent line) intensity saturates for Al at about 3 times threshold (Baun and Fischer, 1964). The relative contribution of direct ionization and shake-off with electron excitation near threshold is not yet known. Direct double ionization by photoabsorption near threshold has been discussed by Sachenko and Burtsev (1967) for third- and fourth-row elements. They calculated that the dual (K- and L-shell) ionization should rise rapidly and attain the value given by the impulse approximation within about 20% above threshold. Data on photoionization of neon given by Krause (1971) are consistent with this prediction.

For third-row elements, the ionization cross-section ratio σ_L/σ_K decreases with

---

[1] Note, however, that this is not true if the Slater exchange approximation is used, as in the Herman-Skillman program (Lindgren, 1965),

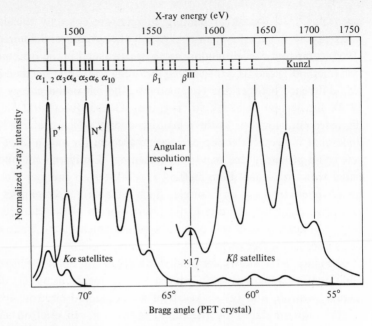

FIGURE 9-12
Al K spectra produced by impact of 5-MeV protons and nitrogen ions on Al metal (Knudson et al., 1971). Energies of strong peaks (labeled solid lines) and weak peaks (dashed lines) measured by Kunzl (1936) with electron excitation are indicated.

energy above the K threshold for electrons and ions and is approximately constant for photoionization. Hence, at least for particle excitation, only the shake-off mechanism remains at higher energies. The probability for multiple ionization by shake-off has been calculated in the impulse approximation for elements in the vicinity of the third row (Aberg, 1969). The theory predicts that, for both electron and photon excitation, the K-plus-L ionization probability attains a constant value at about twice threshold. The ratios of integrated intensities $K\alpha_{3,4}/K\alpha_{1,2}$ calculated by Aberg agree well with the experimental values measured in the region of satellite saturation for fluorine through calcium. This ratio is primarily sensitive to the relative probability of double-KL and single-K ionizations. The agreement supports the shake-off hypothesis. No difference between satellite spectral shapes and relative intensities excited by x rays and electrons was detected in a test done using the Si $K\alpha_{3,4}$ spectrum, which further supports the shake-off picture (Graeffe et al., 1969). Heinle and Faessler (1969) found Si satellite relative intensity 60 percent higher for electron excitation compared to photoionization. Herglotz (1953) found a 150 percent enhancement with electrons for Cr. Apparently the "enhancement" in both these cases was due to use of x-ray energies close to threshold which did not allow attainment of full shake-off values (Graeffe et al., 1969).

Whether one or more initial core-level vacancies is produced, they can in many cases move nonradiatively from the original level to one less tightly bound (less excited). The energy thus available is carried off by excitation of another electron into some empty level. In the Auger process, a more highly excited ionized state decays by an electron transition between shells and ejection of a core or valence electron, for example, $K \rightarrow L_I L_{II,III}$. In a Coster-Kronig transition, an electron moves between subshells, and a valence electron is excited, e.g., $L_I \rightarrow L_{II,III} V$. These well-known processes are mentioned here to emphasize the fact that a measured x-ray intensity depends heavily on the competition between radiative and nonradiative processes (besides excitation conditions, matrix effects, and other experimental factors). Nonradiative processes which occur to change the ionized state prior to emission are reflected prominently in satellite spectra, as we will illustrate later.

Computation of the likelihood of producing a particular ionized state, considering both initial and subsequent processes as well as transport effects, is complex. Increasing attention is being given to the processes for producing core-level vacancies (Fink et al., 1973) and the required yields (Bambynek et al., 1972). Aberg (1973) has recently given a discussion of direct and vacancy-cascade processes for production of ionized states.

The electronic structure of an ionized state, however produced, is similar to that for the ground state, readjusted to take the missing electron(s) into account. But there are other aspects which are peculiar to the electronic structure of an ionized state, and we now summarize them, starting with the effect of core holes on core levels. Just as two electrons outside a closed shell interact to produce multiplet levels, the angular momenta of two (or more) holes inside a closed shell are coupled to produce similar level splitting.[1] Even with only a single core ionization, multiplet splittings can arise from the interaction of the core vacancy with partially filled outer shells, e.g., d shells in the transition metals. Multiplets arising in this way produce the asymmetry and low-energy structure, as well as some of the shift observed in transition-metal "line" spectra. A useful review of multiplet structure in $3d$ and $4d$ line spectra has been given by Finster, Leonhardt, and Meisel (1971). In Fig. 9-13 we show an example of the sensitivity of such a multiplet structure to the number of unpaired d electrons in Fe. If there are two or more core-electron vacancies, their interaction produces the multiplet structure such as appears in satellite spectra. This will be seen when we discuss the calculation of satellite spectra.

Turning now to the effect of core holes on the band states, we consider states below, within and above the filled band structure. The presence of a core vacancy

---

[1] We are not discussing the spin-orbit splittings such as $2p_{1/2}$ and $2p_{3/2}$ which give the $K\alpha_{1,2}$ splitting.

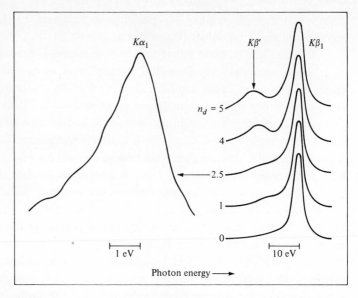

FIGURE 9-13
Multiplet structure in iron $K\alpha$ (Priest, 1971) and $K\beta$ (Nefedov, 1966*b*) line spectra as a function of n_d, the number of $3d$ electrons on the iron atom in various compounds.

in a metal is similar to an impurity with atomic number 1 greater than the metal, both producing an attractive potential. Friedel (1952) has considered the possibility that the attractive vacancy potential is strong enough to form a bound state below the conduction band in a metal in order to screen the core hole or impurity. Ashcroft (1968) emphasized the possibility that copious ionization in the surface of the sample in an x-ray spectroscopy experiment may alter the states within the valence band and prevent an x-ray measurement from reflecting ground-state characteristics. This question is not entirely settled, although good agreement of measured spectra with calculations based on ground state implies that core-level ionization effects do not heavily alter the electronic structure. Burkhalter et al. (1972) have studied Al $K\beta$ valence-band satellite spectra from the metal and Al_2O_3 excited by high-energy, heavy-ion impact. The metal-to-oxide shift changes from negative to positive and then decreases toward zero as the number of L-shell vacancies increases from zero to five. The decrease for high-ionization states indicates that the normal ground-state band structure has been upset by the large charge on the Al ion. Even without such heavy ionization, a nonuniform response (shift) of states within the filled band due to creation of a core hole also seems possible, especially in transition and rare-earth metals where the d or f electrons are quite different spatially from the s and p electrons.

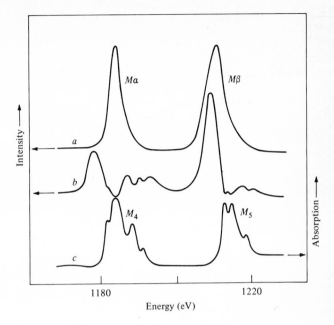

FIGURE 9-14
Gadolinium M spectra (Fischer and Baun, 1967a) with a negligible and b heavy self-absorption. The absorption spectra c were obtained from the ratio a/b of the emission spectra.

Recently, Stern and Sayers (1973) have discussed the response of bonding electrons to creation of a single core hole. Molecular orbital calculations, including the presence of a core vacancy, are now receiving attention also (Bagus, 1973; Gianturco, 1973).

Excitonic states can appear above the filled band in the band gap of an insulator (Cauchois and Mott, 1949; Parratt, 1959; Barinskii and Vainshtein, 1959). Structure due to excitons appears near the band edge of the radiative Auger spectra shown in Fig. 9-5 and absorption spectra in Fig. 9-6. The similarity between excitons in a solid and excited-but-bound states of an atom in a gas is evident from Fig. 9-6. Resonances, the occurrence of emission and absorption spectra at the same energy, are common in optical spectra but rare in the x-ray region. Recently, Lavilla (1972a) observed emission of resonant x rays due to the decay of bound, excited states in gaseous N_2. Resonance-like emission and absorption also have been found in rare-earth M spectra (Fischer and Baun, 1967a; Bonnelle and Karnatak, 1971). An example is shown in Fig. 9-14. Here the overlap of emission and absorption spectra is due to multiplet splittings caused by core-level vacancies (Sugar, 1972).

So much for the effects of core-level states on electronic structure. Missing valence electrons can be expected to have negligible effect on the band structure of conductors because of electron mobility. Best (1971) has discussed polarization of ionic crystals due to a valence hole.[1] Ruffa (1972) used valence-hole effects on the final state for emission from an insulator, SiO_2, to estimate the O $K\alpha$ spectrum on a relative basis. But such spectra can also be interpreted using the ground-state bonding-electron structure. The importance of valence-electron holes ("broken bonds") for interpreting x-ray spectra has not been conclusively established yet.

Finally, we consider the effects of bonding on the creation, rearrangement, and electronic structure of ionized states. Although the x-ray states produced depend indirectly on bonding through matrix effects on the exciting radiation, at an incident energy 2 or more times greater than threshold, the core-ionization probability is not directly sensitive to bonding. The availability and nature of empty states into which a core electron is ejected are important only for threshold excitation of a quasi-stationary x-ray state. In that case, the final state of the electron is just above the Fermi level where the band structure is quite sensitive to bonding. A sensitivity of shake-off to bonding might also be expected since the additional electron departs from the atom with little energy. For Mg, Al, and Si, the $K\alpha_{3,4}/K\alpha_{1,2}$ ratios are 10 percent greater for the oxides than the pure elements (Aberg, 1968), but this may be due to differences in transition probabilities, in addition to any sensitivity of shake-off to bonding. Core-level vacancy rearrangements can be sensitive to bonding when they directly involve the valence electrons. For example, the fewer the bonding electrons, the less likely a Coster-Kronig transition which involves valence-electron excitation. Also, for Coster-Kronig processes as for threshold excitation, an electron is ejected into bands near the Fermi level. We also note that lifetime broadening due to Auger processes within the valence band depends on electron mobility, and therefore it too should be sensitive to bonding. Lastly, since an ionized state can be considered to be a relaxed and sometimes split ground state, and since the ground state is sensitive to bonding, certainly the ionized state contains bonding information. Were this not so, the x-ray spectra involving ionization would not be useful for electronic studies! In short, bonding has only indirect effects on the ionization process except near threshold; it does sometimes affect nonradiative rearrangements; it should affect valence-band lifetime broadening; and it largely determines the ionized-state electronic structure because of the relation of that state to the ground state.

Often a particular ionized state has one dominant decay mode, and so it is almost uniquely associated with one kind of spectrum. However, as we saw when considering the various types of spectra (Fig. 9-3), the same x-ray state can enter into several x-ray processes. Now we are ready to go on to the use of ground- and ionized-state electronic-structure information to calculate the different kinds of spectra.

---

[1] Deslattes (1963) gave an early discussion of final-state polarization effects in KCl.

VI. CALCULATION OF X-RAY SPECTRA

We already noted that spectra are curves of intensity, sometimes expressed as an absorption coefficient, as a function of energy. In the past two sections we were largely concerned with energies; now we go on to consider spectral intensities and shapes as well as energies. The requirements for spectral calculations (formulas and input electronic-structure information) will be outlined for different kinds of spectra with examples of calculated x-ray spectra given to illustrate present capabilities. The discussion will parallel the enumeration of different spectral types which was given earlier (see Fig. 9-3).

In the last section we emphasized quasi-stationary x-ray states as a useful though incomplete concept. Here we will give a similar emphasis to the one-electron picture. While many-electron effects are not entirely separable from one-electron effects, the grosser aspects of most x-ray spectra, and the details of some, can be understood without recourse to the many-electron problem. This section is related to the viewpoint of Chap. 4, in contrast to the many-body approach of Chap. 5.

Having just said that we are now mainly concerned with the one-electron model, we must admit that initial and final-state lifetime broadening effects are required in a full spectral calculation. And, such effects have a many-electron nature since the lifetime of a quasi-stationary state and the associated uncertainty-principle width depend on the disturbed system of electrons. The lifetimes of $1s$-hole states in light elements (below sulfur) have been found sensitive to the chemical environment in electron spectral measurements (Shaw and Thomas, 1972; Friedman, Hudis, and Perlman, 1972). The lifetime of a conduction band electron due to an Auger process has been well studied (e.g., Rooke, 1968a). Recently, attention has been given to broadening due to the limited lifetime of an excited ("hot") electron against electron-electron or electron-plasmon scattering (McCaffrey, Nagel, and Papaconstantopoulos, 1973). Fortunately, these lifetime widths, along with the spectrometer window, can be put into a one-electron spectral computation at the end.

Earlier we alluded to problems created by matrix effects on the incident radiation during penetration of the target and on the generated radiation while leaving the target. Although we are primarily concerned with calculation of spectral shapes and shifts due to bonding effects, transport (matrix) effects on emission spectra cannot in general be ignored. For some spectra they can be minimized, but for others they are unavoidable. As a rough rule, transport effects can be lessened by making measurements near threshold (Liefeld, 1968). This has two advantages. First, it ensures that the incident radiation does not suffer appreciable energy losses prior to exciting the radiation of interest. Then the distribution of electron energies and energy dependence of the ionization or Bremsstrahlung cross sections are not important. Away from threshold it is usually difficult to separate the fundamental emission processes of interest from transport effects. We will see this later for short-wavelength limit, isochromat, and appearance-potential spectra. Blokhin, Demekhin, and Shveitser (1962) presented a method to correct emission band spectra for self-absorption which considered both characteristic and continuum radiation. A second advantage of threshold measurements is the fact that the emitted radiation comes from near the surface of the sample so that self-absorption is minimal. Figure 9-14 shows the M spectra of Gd taken under conditions of negligible and heavy self-absorption obtained by varying the electron energy and take-off angle. The ratio of two such curves provides a good replica of the absorption spectrum (Liefeld, 1968). Extraction of absorption data from emission spectra is the major (and essentially the only) advantage of transport effects for high-resolution spectroscopic studies.

As an example of spectra which depend on the electron distribution in both energy and depth and which reflect energy-dependent cross sections, we mention Bremsstrahlung continua (Brown, 1973). Such spectra are not used to study the electronic structure of materials, but they nicely illustrate the importance of electron transport. The electron energy distribution and cross sections are reflected in the overall extent of the spectra. These factors and self-absorption determine the spectral shape computed by Brown. The different effective depths of x-ray production at different incident-electron energies are shown in his work by changes in absorption-edge jumps.

There have been very few computations of x-ray lines or band spectra which take into account all aspects of the excitation, emission, and self-absorption process. One recent example is analysis of the L line and satellite spectra of Zr by Krause, Wuilleumier, and Nestor (1972). In this section, we will be concerned primarily with the basic emission, scattering, or absorption process for each type of spectrum. Only limited attention will be given to transport effects such as characteristic energy losses.

Now we begin presentation and discussion of the equations used to compute various types of x-ray spectra. Valence-band emission intensities and absorption spectra may be calculated in the dipole approximation from

$$I(\mathscr{E}) \text{ or } \mu(\mathscr{E}) \sim \int_{\substack{E=\text{constant}}} \frac{d^2k}{|\nabla_{\mathbf{k}}E|} \left[\int_{\text{all } \mathbf{r}} \psi_f^*(\mathbf{r})\mathbf{r}\psi_i(\mathbf{r}) \, d^3r \right]^2 \qquad (9\text{-}6)$$

where ψ_i is a band state or molecular orbital for valence-band emission spectra or a core state for line spectra, and ψ_f is the final (core) state. The spatial integral squared in equation (9-6) is the transition probability. Rooke (1968a) and Harrison (1968) have emphasized the importance of transition probabilities for x-ray spectra. The angular part of the spatial integral is the source of the selection rules. The density of states $N(E) = 1/|\nabla_{\mathbf{k}}E|$ may be decomposed in accordance with the allowed transitions for particular spectra. That is, the angular part of the spatial integral may be effectively applied to the density of states, leaving the **k**-space integral over the component density of states and the remaining radial part of the **r**-space integral. The radial integral is, of course, what gives x-ray spectroscopy its spatial selectivity, and requires consideration of the local density of states in an alloy or compound. Spectra from alloys have been discussed in terms of two-band models (Rooke, 1969) because the core levels of constituent atoms sample the bonding-electron distribution at different sites. X-ray spectra probe a larger region around the nucleus than the Knight shift in nuclear magnetic resonance or the isotope shift in Mössbauer spectroscopy, both of which are limited to the immediate region of the nucleus. Also, different x-ray series involve core orbitals of varying extent in space. In short, transition probabilities are important in calculating band spectra because they dictate the double decomposition of the density of states according to final-state symmetry and the region in **r** space overlapping the core orbital. Component density of states for nickel are given by McCaffrey, Nagel, and Popaconstantopoulos (1974).

There have been several calculations of valence-band emission spectra using the results of band theory for ordered materials in the past few years.[1] See Chaps. 4 and 5. Most of these have been for pure elements and were based on the results of APW band calculations. Goodings and Harris (1969), Dobbyn et al. (1970), and Smrcka (1971) present details of the use of APW results to compute spectra. Recently Conklin and Schwarz (1972) used a self-consistent APW calculation for the compound TiC to calculate both titanium and carbon valence-band spectra. Results of the OPW method have also been used to compute spectra. Blokhin, Sachenko, and Nikiforov (1972) took a single orthogonalized plane wave to compute integrated K valence-band–line-intensity ratios for the $3d$ transition metals. A complete spectral calculation based on the OPW method was done by Klima (1970) for Si and Ge. Figure 9-15 shows the K and L results for Si. Here, as is common, the computed spectra before broadening are much sharper than in the experiment. Comparison between the as-calculated spectra and experimental spectra demonstrates the devastating effect that lifetime and spectrometer broadening have on x-ray spectra. In general, the higher the x-ray energy, the worse the smearing. After broadening, the computed spectra in Fig. 9-15 agree well with experiment.

So far there have been no full spectral calculations published for disordered materials. However, Gyoffry and Stott (1971; 1973) have developed a one-electron formalism using partially averaged Green's functions with muffin-tin potentials to calculate valence-band spectra from random substitutional alloys. X-ray intensity is given in terms of two factors: a transition probability independent of the environment of the emitting atom and an ensemble-averaged matching term which contains

---

[1] The earliest x-ray band calculation based on band theory and known to the authors is that of Nikiforov (1961) for Fe L_{III}.

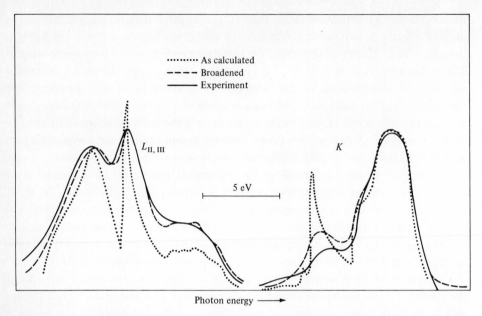

FIGURE 9-15
Valence-band spectra of silicon calculated from band theory (Klima, 1970) compared to the measurements of Wiech (1968) for the $L_{II,III}$ and Läuger (1968) for the K spectra.

all the information about the atomic types and the arrangement. Various approximations used in this formalism correspond to the different basic electronic-structure theories for disordered alloys which were mentioned earlier. Babonov and Sokolov (1971) are also developing a method for calculating band spectra for disordered alloys using averaged Green's functions. The x-ray spectrum for a particular element in an alloy is given in terms of the local density of states for that component. Jacobs (1969) did a model calculation for the Mg-Al system which reproduced the trends in the $L_{II,III}$ band spectra with alloying. Virtual-bound or resonance states have been invoked to interpret Al alloy spectra (Fabian, 1970), but no spectral calculations based on this concept seem to have been made yet.

Molecular orbital theory has been used in many instances to interpret spectra from solids (Best, 1966; Nefedov, 1967; Glen and Dodd, 1968; Urch, 1970). An example from Fischer's (1972) work will be discussed later when we consider what electronic-structure information can be obtained from valence-band spectra. Molecular orbital theory has also been used for detailed calculations of spectra from molecules. Figure 9-16 shows recent examples from the work of Klasson and Manne (1972). For molecular spectra, orbital occupancy replaces the density of states in equation (9-6). Each computed peak was broadened in order to compare theory with

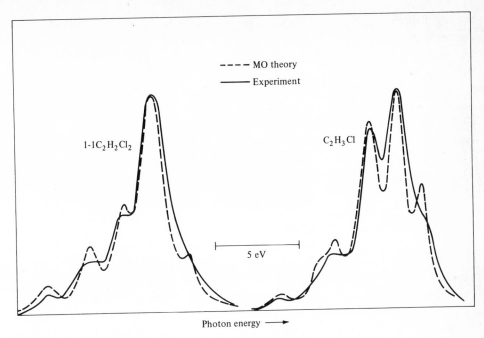

FIGURE 9-16
Chlorine valence-band spectra obtained from molecular orbital theory (Klasson and Manne, 1972) compared to experiment (Gilberg, 1970).

experiment. Here, as in Fig. 9-15 for spectra computed from band theory, the number and intensities of the component peaks agree well with experiment. (In this case peak positions were obtained from photoelectron data.) Recently, in a new type of calculation, Kortela and Manne (1973) have used molecular orbital theory to compute energy bands and x-ray valence-band spectra of graphite and diamond.

Relative intensities of the diagram lines in a spectral series have been computed from atomic theory (Scofield, 1969), using an equation similar to (9-6), but they do not yield bonding information. Nor are absolute line energies useful for determining electronic structure because of difficulties in accurately calculating x-ray energies. However, both shifts in going from one material to another and the shapes of line and satellite spectra do give information on electronic structure. We will review shift calculations before considering the use of multiplet theory to calculate shifts and shapes for both lines and satellites.

The shift in energy of an x-ray line due to changes in bonding is determined by self-atom and neighbor-atom effects in addition to multiplet splitting. That is, there is a change in energy due to loss or gain of valence electrons on the atom of interest and a change due to alteration of the field caused by neighboring charges.

We begin discussing line shifts by considering quantities related to changes in x-ray line energies. Then self-atom (free-ion) effects, which are related to electronic screening, will be considered before reviewing neighbor-atom effects. Since self-atom effects are most important in determining line shifts, the free-ion model is often used.

We mentioned earlier, following equation (9-2), that a line shift $\Delta\mathscr{E}$, which is the difference in binding-energy shifts ΔE, can be approximated by a difference in level shifts $\Delta\varepsilon$ if relaxation and other effects in the sample and reference material cancel. It is the relation of level shifts to changes in effective charges Δq and potentials ΔV which allows us to connect line shifts to aspects of the bonding in one material relative to another. Schematically, we have for the difference between two materials:

$$\Delta\mathscr{E} \leftrightarrow \Delta\,(\Delta E) \leftrightarrow \Delta\,(\Delta\varepsilon) \leftrightarrow \Delta q \leftrightarrow \Delta V$$

It is useful to examine the theoretical and experimental methods by which each quantity in this sequence can be obtained. X-ray line and binding-energy shifts can be calculated from the same theories, but binding-energy changes are obtained by the measurement of electron rather than x-ray spectra. Eigenvalue shifts $\Delta\varepsilon$ are available only from theory. Effective charges are obtained by a wide variety of means, some as simple as taking a fraction of the oxidation number. We noted several means of obtaining charges earlier, including theoretical integrals over the electron density, orbital population analyses, and the use of electronegativities, as well as measurement of Knight and isomer shifts and x-ray diffraction. Changes in potential are usually calculated directly, but they can sometimes be obtained from effective charges by use of point-ion models. In addition to this array of experimental as well as theoretical methods, the various calculations can be performed using a wide variety of models and methods with many different approximations concerning, for example, construction of the potential, the number of electrons treated, and neglect of orbital overlap. A vast array of correlations among the various quantities obtained in different ways is available in the literature. Sometimes theoretical quantities are compared with each other and sometimes with experiment. In the following, we will use the free-ion picture and discuss models first for shifts $\Delta\varepsilon$ in energy levels and then for shifts $\Delta\mathscr{E}$ in line energies.

Charge redistribution attendant to changes in bonding is equated to formation or alteration of the charge on an ion in the free-ion model. Changes in the screening of the core levels by the inner parts of outer-electron wavefunctions, such as those shown in Fig. 9-10, result in a change in core-electron binding energy. This ionic model for core-level shifts is widely used, not only for interpreting and using x-ray line shifts but also for calculating that part of shifts in satellite and absorption spectra which is due to changes in the screening of core electrons by outer orbitals. We present free-ion-level and x-ray-shift methods in order of increasing complexity, beginning with techniques which yield the ground-state energy of a core level for an atom in one material relative to another reference material.

In the simplest approach to the free-ion model, the derivative of the term-value equation (9-3) with respect to σ yields

$$\Delta\varepsilon_{nl} = \frac{27.2}{n^2}(Z - \sigma_{nl})\,\Delta\sigma_{nl} \qquad (9\text{-}7)$$

This was used by Varley (1956) and Shubaev (1960). The change in screening can be calculated from the assumed charge transfer using tables of screening constants. Similar use could be made of Veigele, Stevenson, and Henry's approach (1969), where Z_{nj} changes with bonding.

Somewhat more sophisticated approaches to level-shift computations using the free-ion model still avoid solution of Schrödinger's equation but require wavefunctions. Nefedov (1966a) has written level shifts in terms of the Coulomb (F) and exchange (G) integrals over the occupied orbitals (Slater, 1960). For example, the 1s-level shift due to loss of one ns electron is given by

$$\Delta\varepsilon_{1s} = F^0(1s\ ns) - \tfrac{1}{2}G^0(1s\ ns) \qquad (9\text{-}8)$$

Shubaev (1963) calculated 1s-level shifts in terms of the changes in potential ΔV between materials

$$\Delta\varepsilon_{1s} = \int_0^\infty P_{1s}{}^2(r)\,\Delta V(r)\,dr \qquad (9\text{-}9)$$

where $P_{1s}(r)$ is the radial wavefunction for the 1s level. The required potential change is related to the difference in electron distribution $\Delta\rho$

$$\Delta V(r) = \frac{1}{r}\int_0^r \Delta\rho(r)\,dr + \int_r^\infty \frac{\Delta\rho(r)}{r}\,dr \qquad (9\text{-}10)$$

Shubaev (1961) also used the derivative of Schrödinger's equation to obtain an expression for level shifts in terms of differences in the radial wavefunctions $P_{nl}(r)$ and screening constants. Alder, Bauer, and Raff (1972) have computed line shifts in terms of the change in the radial electron density due to compound formation.

The most thorough free-ion level shift calculations are made with one of the self-consistent-field programs which we mentioned earlier. These include the Herman-Skillman (1963) and Froese (1963) programs.

We turn now to free-ion equations relating x-ray line shifts to changes in various aspects of the emitting materials. Karalnik (1956) was one of the first to consider shifts in terms of changes in electron screening. A very simple empirical equation for estimating x-ray energy shifts is based on equating shifts in x-ray energy per unit change in outer-valence-electron screening to the observed difference in x-ray energy for a unit change of nuclear charge ($Z \to Z + 1$) (Karalnik, 1957):

$$\frac{(\mathscr{E}^S - \mathscr{E}^R)}{(\sigma^S - \sigma^R)} = \frac{\Delta\mathscr{E}}{\Delta\sigma} = \frac{\mathscr{E}_Z - \mathscr{E}_{Z+1}}{1} \qquad (9\text{-}11)$$

The left-hand side of this equation concerns the change in outer (electron) screening, while the right-hand side is for the change in inner (nuclear) screening.

Much attention has been paid to relations between shifts and effective charges (Barinskii and Nefedow, 1969; Leonhardt and Meisel, 1970). Equations relating these quantities allow computation of charge values from observed shifts. $K\alpha$ line shifts for third-row elements have been calculated as a function of the degree of ionization from a self-consistent free-ion model (Clementi, 1965). It is possible to use a linear relation between charge and line shift because Δq seldom exceeds 2. Hence, $\Delta \mathscr{E} = K \Delta q$, where Δq is the charge transfer. Theoretical values of K range from 0.33 for Mg to 1.32 for S in units of eV per charge unit (Shubaev, 1963). Line shifts for the $3d$ transition metals depend on the participation of $4s$ and $3d$ electrons in the bonding (Karalnik, 1957). Hence, Leonhardt and Meisel (1970) wrote

$$\Delta \mathscr{E}_{K\alpha} = \Delta q_{4s}\alpha_{4s} + \Delta q_{3d}\alpha_{3d}$$
$$\Delta \mathscr{E}_{K\beta} = \Delta q_{4s}\beta_{4s} + \Delta q_{3d}\beta_{3d} \qquad (9\text{-}12)$$

Another linear relation between shifts and charges is due to Sumbaev (1970):

$$\Delta \mathscr{E} = \mathscr{E}^S - \mathscr{E}^R = (i_S - i_R) \sum_l m_l C_l \qquad (9\text{-}13)$$

where the charge is now expressed as ionicity i times valence m, here decomposed according to the angular momentum of the contributing orbital. C_l is given in eV per l subshell electron. The equation due to Nefedow (1962),

$$\Delta \mathscr{E} = \mathscr{E}^S - \mathscr{E}^R = a\mathscr{E} \sum_i^q \sqrt{I_i} \qquad (9\text{-}14)$$

where i labels the outer electrons, q of which are missing, and a is a constant, is essentially a linear relation between the shift and charge. The ionization potential I_i of an outer electron is related to the square of a sum including the effective charge of the ion. In equations (9-12) through (9-14) the proportionality constants can be calculated or determined by fitting data.

All the equations (9-7) through (9-14) provide means to calculate level and x-ray line shifts due to self-atom effects only. We now expand our view to include neighbor-atom effects also. Other-atom influences on x-ray line positions are small because core orbitals both are spatially compact and tend to respond similarly to fields.

A simple electrostatic model taking into account both self-ion and neighbor-ion effects was developed for interpretation of binding-energy shifts as measured by electron spectroscopy (Siegbahn et al., 1967):

$$\Delta E_i = \frac{\Delta q_i}{r} - \sum_j \frac{\Delta q_j}{R_{ij}} \qquad (9\text{-}15)$$

Here, (q_i), (q_j) is the charge on atoms (i), (j), r is the valence-electron shell radius and R_{ij} is the distance from atom i to atom j. The second term, a point-ion model, gives the field contribution which can be viewed as a correction to the free-ion model (the first term) to account for the fact that electrons were transferred only to neighboring atoms and not to infinity as in the free-ion model.

An empirical equation, similar to (9-15), was used by Siegbahn et al. (1969) to correlate binding-energy shifts in molecules as measured by electron spectroscopy:

$$\Delta E_i = k \, \Delta q_i + V + l \qquad (9\text{-}16)$$

where $V = \sum_j (\Delta q_j / R_{ij})$. Both k and l are adjustable parameters. Schwartz and his colleagues (Schwartz, 1970; Schwartz, Switalski, and Stronski, 1972) presented a generalization of the point-ion model for molecules. In their model, core-level shifts are related to changes in the potential (rather than charge) which are calculated from the full valence-electron distribution.

The self-atom plus neighbor-atom models developed to interpret shifts in electron spectra (binding energies) cannot be used as they stand to calculate x-ray line shifts since there is no distinction between different core orbitals in the models. Use of the models for x-ray line shifts would require attention to the different spatial extent of the two core orbitals relative to the valence-electron orbitals and the non-uniform crystal field. The models can be used as they are to estimate that part of a valence-band emission peak shift or absorption-edge shift which is due to chemical shift in the core level.

Shifts in the positions of satellites may be obtained from the same considerations as line spectral shifts. By use of an ionic model which included multiplet structure, it was shown by Demekhin and Sachenko (1967a) that the K satellites of third-row elements shift more than the parent $K\alpha_{1,2}$ lines in going from the elements to compounds (i.e., in removing M-shell electrons). This occurs because in compounds the atoms are already effectively ionized, and it agrees with the observed trend. The crystal field in compounds may also have some influence on satellite locations and structure via its influence on the multiplet structure.

Earlier we noted that multiplet structure develops when a core-level hole interacts with either normally present vacancies or other vacancies due to ionization. Multiplet structure appears in line, satellite, and radiative Auger spectra. In general, given a hole configuration, multiplet structure can be calculated using the methods of theoretical atomic spectroscopy (Slater, 1960). Cowan (1968) has reduced multiplet calculations to a routine for free atoms.

Ekstig et al. (1970) have used multiplet theory to calculate the shape of $K\beta_1\beta'$ line spectra from $3d$ transition elements due to interaction of the ionization hole with d-electron vacancies. The theoretical spectra are not in good agreement with measurements such as those shown in Fig. 9-13. However, the calculations do indicate that the observed low-energy structure is due to multiplet structure. The lack of close agreement may be due to the neglect of solid-state effects.

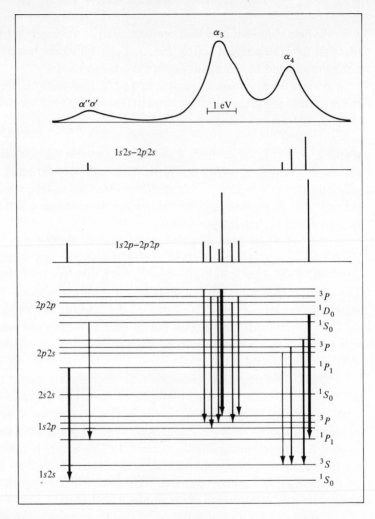

FIGURE 9-17
Multiplet-level diagram (Sawada, Taniguchi, and Nakamura, 1969), and theoretical component positions and intensities compared to the measured Si K satellite spectrum (Demekhin and Sachenko, 1967a).

Interactions between multiple-ionization vacancies produce rich structure in satellite spectra. There have been many measurements of satellite spectra. Those of Parratt (1936) are especially interesting. Relatively few calculations of x-ray satellite structure are available. Detailed calculations of the energies and intensities of satellite components for third-row elements have been made by Sureau (1971). Figure 9-17 shows the multiplet-level diagram with allowed transitions and the computed, un-broadened $K\alpha'\alpha_{3,4}$ intensity compared to experiment for Si. Combinations of Coulomb and exchange integrals determine the line positions, while level degeneracies and transition probabilities give the relative intensities of the component lines. The theoretical structure is in good agreement with experiment. Hartmann, Papula, and Strehl (1971) recently calculated the $K\alpha$ satellites of phosphorus which also agree well with experiment. We have already mentioned that the impulse approximation predicts correctly the relative integrated satellite intensities for third-row elements excited by x rays or electrons. Overall, the electron shake-off picture, plus an ionic model with multiplet interactions, yields the integrated intensity, position, and shape of third-row K-series satellites which agree well with experiment. Differences between satellite spectra from elements and compounds will be discussed in the next section.

The same models useful for calculating satellite spectra are also applicable to radiative Auger spectra. For such spectra (see Fig. 9-4), the intensity is governed by the probability for simultaneous emission of a lower energy x ray and an electron, relative to the probability for normal $K\alpha_{1,2}$ emission. Relative radiative Auger intensities were computed by Aberg (1971) for $K \to L^2$ ($9 \leq Z \leq 22$) and $K \to M^2$ ($17 \leq Z \leq 22$) transitions. These were based on consideration of both electron shake-off and configuration interaction in the final state. The shake-off is due to the rapid change in potential caused by the electron transition from the L shell to a K-shell vacancy during the radiative Auger process. Aberg ascribed the overestimation of radiative Auger probabilities for the lower atomic numbers to ignoring configuration interaction in the final state when shake-off occurs, although it was included when there is no shake-off. There have been no detailed calculations of radiative Auger spectra yet. Such a computation would have to account for the presence of excitonic and multiplet structure in these spectra.

Turning now to the calculation of absorption spectra, we find that the situation, at least for molecules and solids, is not as well developed as for computation of emission spectra. Because there has been much study of x-ray absorption by atoms and because it is useful background, we first briefly consider atomic absorption before reviewing work on molecules and solids.

Calculation of high-energy (core-level) absorption by free atoms is relatively straightforward because of the hydrogen-like nature of core states and the near-plane-wave nature of the photoelectron state. References to work in this area are given by Hubbel (1970). For soft and ultra-soft absorption by electrons in shallow levels, the situation is more complex. The potential experienced by outer electrons in the ground state is no longer simply hydrogenic, and the low-energy photoelectron is distorted by the potential of the absorbing atom. Delayed oscillator strength because of centrifugal barriers due to the nature of the potential results in broad but strong maxima in the absorption coefficient. Inclusion of electron correlation in calculations of ultra-soft x-ray absorption is necessary to get agreement with experiment. Reviews in this area are given by Fano and Cooper (1968) and Zimkina and Gribovskii (1971).

X-ray absorption by free molecules is similar to that by atoms, especially at higher energies, but shows departures from atomic behavior near threshold. The excited-bound states of molecules have different energies than those in atoms. Further, they are split by Stark (electric-field) effects due to neighboring atoms (Barinskii and Malyukov, 1962) and by hybridization effects (Barinskii and Malyukov, 1963). Another effect, not present in atomic absorption, is the suppression near threshold of absorption by positive ions due to the repulsive field set up by nearby negative ions. The potential barrier leads to quasi-stationary states and narrow absorption peaks in the continuum. Dehmer (1972) gave a recent evidence for this effect which was discussed earlier by Nefedov (1970a).

Absorption of x-rays by condensed matter is more complex than that by free atoms and molecules. Most atomic and molecular features are retained, and additional effects arise. In highly perfect crystals, reduced absorption occurs due to anomalous transmission of x rays (the Borrman effect). This is reviewed by Batterman and Cole (1964). It does not yield information on electronic structure. Other many-electron solid-state effects which do depend on bonding occur near an absorption edge. Anion effects on cation absorption near threshold appear in ionic solids (Nefedov, 1970b). In metals, effects of the core hole occur right at the edge. Some of the peaks within a few volts of absorption edges in metals are attributed to plasmon excitation which is more probable near the edge where the ejected electron has little kinetic energy and can more effectively couple to the conduction electron plasma (see, e.g., Mande and Joshi, 1969).

Calculations of absorption spectra can be somewhat artificially separated into three groups: edge shifts, near-edge (Kossel) structure, and extended-fine (Kronig) structure. Unlike x-ray line shifts, which require consideration of two similar (core) levels, absorption-edge shifts involve two qualitatively different orbitals, one core and the other the lowest energy empty level. The core level is influenced by both own-ion and neighboring-ion (crystal-field) effects as in equation (9-15). The normally empty state is influenced by many direct-bonding effects. It may be an excitonic or continuum level, and in many cases it involves atomic (multiplet) or many-electron effects. Interpretation of absorption-edge chemical shifts has much in common with study of line shifts in electron spectra, especially for materials like metals with a well-defined Fermi level. An extensive theoretical discussion of absorption-edge shifts is given by Barinskii and Nefedow (1969). In a recent treatment, Dey and Agarwal (1971) relate shifts to the valence and ionicity of the absorbing atom in a compound.

Aside from such edge effects, x-ray absorption in solids can be treated in the one-electron picture in which transitions from core levels to **k**-dependent Bloch states are computed. There have been many comparisons of band energies and total-state densities with absorption as well as emission spectra, but little attention has been given to actually computing spectra. A partial calculation was made for Ni within 20 eV of the K edge by Irkhin (1961) using the Slater-Koster method.

McCaffrey, Nagel, and Papaconstantopoulos (1974) based a calculation of K, L_{III}, and $M_{II,III}$ edge structure of nickel on the results of an APW calculation. Transition probabilities were taken into account by a symmetry decomposition of the density of states to account for the angular part (which gives the selection rules) and a separate calculation of the smoothly varying radial part. Particular attention was given to the hot electron contribution to the final-state broadening. The complete L_{III} spectrum is compared with experiment in Fig. 9-18. Fair agreement between theoretical and measured spectra was also found for the K edge but not the $M_{II,III}$ edges. Apparently the M-edge structure is determined primarily by atomic effects.

A thorough band-theory calculation of extended absorption fine structure for harder radiation has yet to be made. The many theories for interpreting extended fine structure were reviewed by Azároff (1963 and in Chap. 6 of this book). They generally yield only the energies at which fine-structure peaks and valleys occur and do not give the full shape of the absorption curve.

One-electron calculations of absorption spectra make no distinction between various regions above the absorption edge. Hence the question, why, as in Chap. 6, are the near-edge (Kossel) and extended (Kronig) fine structure treated separately? The answer has several facets. Excitonic structure, as shown in Fig. 9-6, appears below the edge of insulators, and other many-body effects also appear at and near the edge in metals. Even within the one-electron-band picture, the edge region may be distinguished from the extended region by a decreasing importance of electron exchange away from the edge. This is still an open question in band theory. And finally, the "hot"-electron mean free path and lifetime increase with energy away from the edge and wash out the fine structure which may already be less distinct since high-energy electrons are nearly free. These features and variations are responsible for differences in near-edge and extended fine structure.

We saw in Fig. 9-7 that the experimental absorption and x-ray Raman spectra of Be are similar. The theoretical relation between these two spectral types has also been examined in the dipole approximation (Mizuno and Ohmura, 1967). It was found that both the absorption and Raman-scattering probabilities are proportional to the same function of energy for polycrystalline samples. The differential K x-ray Raman cross section for scattering of radiation $\hbar\omega_0$ at angles θ and ϕ is

$$W(\hbar\omega,\theta,\phi) \sim (1 + \cos^2 \theta) \sin^2 \frac{\theta}{2} \, \omega_0{}^2 t(\omega_0 - \omega) \qquad (9\text{-}17)$$

while for the absorption probability,

$$W_{abs}(\hbar\omega) \sim \omega t(\omega) \qquad (9\text{-}18)$$

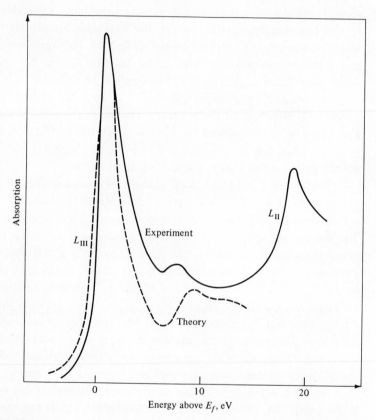

FIGURE 9-18
Theoretical absorption-edge structure for nickel (McCaffrey, Nagel, and Papaconstantopoulos, 1974) compared to the measured spectrum (Bonnelle, 1966).

The function t appearing in both equations (9-17) and (9-18) is essentially the same expression as in equation (9-6) and includes both the density states and transition probability. In another approach not making the dipole approximation (Kuriyama and Alexandropoulos, 1971), the Raman inelastic scattering and absorption probabilities were found to be

$$W_{\text{scatt}}(k,k') \sim (\mathbf{s} \cdot \mathbf{s}')^2 \iint d^4x \ d^4y \ e^{-i(k-k')(x-y)} \langle \Delta\rho(x) \cdot \Delta\rho(y) \rangle \qquad (9\text{-}19)$$

and

$$W_{\text{abs}}(k) \sim \iint d^4x \ d^4y \ e^{-ik(x-y)} \langle \Delta j_s(x) \cdot \Delta j_s(y) \rangle \qquad (9\text{-}20)$$

where $\mathbf{k}$ and $\mathbf{k}'$ are the x-ray wave vectors, $\mathbf{s}$ and $\mathbf{s}'$ the polarization vectors, $\mathbf{j}$ the many-electron current, and ρ the electron density in the target material. These equations show that the probabilities are double Fourier transforms of the current and density correlation, respectively, which in turn are related by charge conservation. Doniach, Platzman, and Yue (1971) computed the Raman scattering from lithium as a function of angle. They predict that the near-edge many-body effects are strongly dependent on the momentum transfer which varies with the scattering angle.

The profile of Compton-scattered x rays is given by

$$I(q) = \frac{1}{2} \int_{|q|}^{\infty} \frac{I(p)}{p} \, dp \qquad (9\text{-}21)$$

where $I(p)$ is the linear momentum distribution

$$I(p) = \int \chi(\mathbf{p})\chi(\mathbf{p})p^2 \, d\omega \qquad (9\text{-}22)$$

with $d\omega$ denoting an element of solid angle (Weiss, 1966). Momentum wavefunctions are related to $\psi(\mathbf{r})$ by

$$\chi(\mathbf{p}) = \frac{1}{(2\pi)^{3/2}} \int \exp(-i\mathbf{p} \cdot \mathbf{r})\psi(\mathbf{r}) \, d^3r \qquad (9\text{-}23)$$

The momentum parameter is $q = mc \, \Delta\lambda/(\lambda_0 \sin \theta)$, where $\Delta\lambda$ is the free-electron Compton shift for the scattering of incident radiation of wavelength λ_0 through angle θ (Weiss, 1966). The momentum distribution [equation (9-22)] can be derived from experimental $I(q)$ by use of the inverse transform (derivative) of equation (9-21). The above and related equations have been used to calculate Compton profiles for atoms (Phillips and Weiss, 1968), molecules (Epstein, 1970), and solids (De Cicco, 1969). Subtraction of accurately calculated core-electron contributions from measured Compton profiles is necessary to derive information or the valence-electron momentum density. Recent discussions of this problem are given by Currat and coworkers (Currat, De Cicco, and Kaplow, 1971; Currat, De Cicco, and Weiss, 1971).

Calculation capabilities for the remaining types of spectra shown in Fig. 9-3—

short-wavelength limit, isochromat, and appearance potential—are not well developed. There may be several reasons for this. Not as many measurements of these kinds of spectra have been made compared to other types, especially band spectra. Relatively few of these have been for compounds, so that not much attention has been given to chemical shifts. Also, calculation of SWL, isochromat, and appearance-potential spectra require attention to both band-structure (density-of-states) and electron-transport aspects of generating these types of spectra.

Equations for calculations of SWL spectra apply equally well to Bremsstrahlung isochromat spectra. The similarity of these spectra is shown in Fig. 9-8. Böhm and Ulmer (1969) summarized the theory for these spectra which includes the density-of-states but not the electron-transport part of the problem. They give the equation

$$I(E,\hbar\omega) \sim [1 - f(E)] \int\int \frac{P(\mathbf{k'},\mathbf{k}) \, dS' \, dS}{|\nabla_{\mathbf{k}}E|} \qquad (9\text{-}24)$$

where energy E is measured from the bottom of the conduction band, $f(E)$ is the Fermi distribution function, and $\mathbf{k'}$, $\mathbf{k}$ are the initial and final states. Assuming a constant Bremsstrahlung transition probability $P(\mathbf{k'}, \mathbf{k})$ and an isotropic distribution of initial plane-wave states, equation (9-24) reduces to

$$I(E,\hbar\omega) \sim [1 - f(E)] \int \frac{dS}{|\nabla_{\mathbf{k}}E|} = [1 - f(E)]N(E) \qquad (9\text{-}25)$$

where $N(E)$ is the density of states. Hence, SWL and Bremsstrahlung isochromat spectra are proportional to the density of empty states and independent of the measured photon energy in the simplified theory. This was demonstrated experimentally near threshold for Bremsstrahlung isochromat spectra by Böhm and Ulmer (1971). However, away from threshold, the spectra do depend on the incident-electron (and therefore the measured photon) energy.

A formalism for calculating SWL and Bremsstrahlung isochromat spectra which includes electron-transport effects has been outlined by Nagel (1973). Ignoring sample self-absorption, the central equation is

$$I(E_0,\hbar\omega) = \int_0^{E_0 - \hbar\omega} f(E_0 - \hbar\omega - \varepsilon)Q_B(\hbar\omega + \varepsilon)N(\varepsilon) \, d\varepsilon \qquad (9\text{-}26)$$

where E_0 is the incident-electron energy, f is the electron-distribution function, and Q_B is the Bremsstrahlung cross section [analogous to P in equation (9-24)]. The new ingredient, the distribution function, takes into account the quasi-continuous and discrete energy losses which electrons suffer prior to Bremsstrahlung emission. If the density of states is relatively featureless, as for simple metals, the emitted intensity will be sensitive mainly to electron-transport effects such as characteristic losses (Keiser, 1973). It remains to be seen if equation (9-26) will reproduce observed SWL and Bremsstrahlung isochromat spectra above 10 or 20 volts beyond threshold.

Presently, density-of-states information can be obtained from such spectra only within a narrow region above threshold. Use of equation (9-26) may broaden this range.

For characteristic isochromat and appearance-potential spectra, the incident and excited core electrons both wind up in states near the Fermi level for threshold excitation. Hence, these spectra involve a self-convolution of the density of states. When electron transport is ignored but the ionization cross section Q_i which may vary significantly near threshold is included, the intensity of a characteristic isochromat spectrum (prior to broadening) is given by Nagel (1973):

$$I(E_0) = \int_0^{E_0 - E_i} Q_i(E_0)N(\varepsilon)N(E_0 - E_i - \varepsilon) \, d\varepsilon \tag{9-27}$$

where E_i is the ionization energy of the core level. If the electron-energy distribution is also included, then

$$I(E_0) = \int_0^{E_0 - E_i} N(\varepsilon)$$

$$\times \left[\int_0^{E_0 - E_i - \varepsilon} f(\mathscr{E})Q_i(E_0 - \mathscr{E})N(E_0 - E_i - \varepsilon - \mathscr{E}) \, d\mathscr{E} \right] d\varepsilon \tag{9-28}$$

This equation has not been used to calculate characteristic isochromat spectra, but Nagel and Criss (1973) did compute isochromat spectra using equation (9-27) given above. Such an approach, due to Dev and Brinkman (1970), can be explained with reference to Fig. 9-19. Figure 9-19a schematically represents a high density of empty states near E_F, e.g., for a transition metal. Its self-convolution according to equation (9-27), shown in Fig. 9-19b, is the intensity that would be measured by the characteristic isochromat method. We already noted that an appearance-potential spectrum is the derivative of a characteristic isochromat spectrum (see Fig. 9-9). The derivative is indicated in Fig. 9-19c. Hence an appearance-potential spectrum, being the derivative of a convoluted density of states, is doubly removed from the density of states, which is often desired for comparison with band theory. However, such a spectrum more nearly resembles $N(E)$ than the characteristic isochromat spectrum.

The central importance of the local density of states for computation of x-ray spectra should be emphasized. Although this is clear for valence-band and absorption spectra of alloys, it also holds for the spectra from pure metals because of the radial transition probability in equation (9-6). Furthermore, short-wavelength-limit, isochromat, and appearance-potential spectra should also be sensitive to the local state density because of the close-encounter nature of Bremsstrahlung or ionization processes involved in these spectra (Nagel et al., 1973). Calculations based on local densities of states and comparisons with experiment are needed to test this possibility.

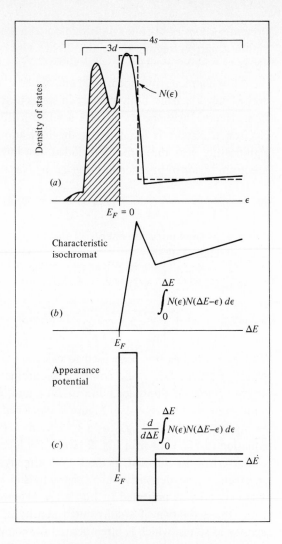

FIGURE 9-19
(*a*) Schematic density of states; (*b*) characteristic isochromat; and (*c*) appearance-potential spectrum for a $3d$ transition metal (Houston and Park, 1971).

In the introduction to this section, we said that the discussion would be given mainly in terms of the one-electron picture. This general aim could not be followed rigidly for spectra which reflect many-electron processes such as core-hole effects, electron shake-off, nonradiative rearrangements, and plasmon peaks. For all the spectral types discussed, the one-electron approach, as outlined in this section, is indeed a very useful tool. However, most of the spectra do exhibit important many-body effects. Core-hole effects (edge singularities) appear in band, Raman, characteristic isochromat, and appearance-potential spectra from conductors. Auger and electron-scattering lifetime broadening influences all spectra involving bands. Asymmetries may occur in some lines due to many-electron influences. Creation of multiple vacancies largely determines satellite and radiative Auger spectra. Plasmon production influences emission and absorption-band, Raman, SWL, isochromat, and appearance-potential spectra. Not only do many-body effects alter the spectra which can be calculated from one-electron theory, but also some spectral features cannot be calculated at all from one-electron theory. Distinct low-energy structure due to many-body effects occurs in a few cases. For example, separate plasmon satellites occur below some valence-band spectra (Rooke, 1963). Also, separate peaks appear in some scattered x-ray spectra because of plasmon creation in a Raman-like excitation of valence electrons (Priftis, Theadossiou, and Alexopoulos, 1968; Tanokura, Hirota, and Suzuki, 1969; Koumelis, Leventouri, and Alexopoulos, 1971). Distinct high-energy structure attributed to plasmon annihilation has been measured in Auger spectra (Matthew and Watts, 1971), but Löfgren and Wallden (1973) argue that multiple ionization is the source of this structure.

Chapter 5 discusses many-body effects both in a thorough, many-body approach to spectral theory and in relation to the question of modifying one-electron theory to include many-electron effects. Little explicit discussion of bonding effects on spectra calculated from many-body theory will be found there or in the literature. For example, most band-spectra calculations up to now which were based on a many-electron approach have concentrated on simple metals using the nearly free-electron picture in which there is no consideration of bonding details. Pseudopotentials have been included in many-body calculations to properly account for the core hole. These potentials reflect the bonding electrons, of course. But no many-body calculation which yielded both band-structure and many-body features of a spectrum is known to the authors. It seems that knowledge of bonding effects on several of the many-electron processes important in x-ray spectroscopy is not yet well developed.

VII. EXAMPLES OF BONDING EFFECTS

So far we have reviewed the different types of x-ray spectra, given examples of them, and discussed how to calculate them from bonding, and to an extent, from electron-transport theory. Now we will give a few examples of bonding effects on different kinds of spectra, still keeping the order in Fig. 9-3. The motivation here is similar to that which led to giving separate attention to ionized states. Just as a given ionized state can be involved in different types of spectra, so a particular bonding effect can have more than one use. While the past three sections were mainly concerned with going from electronic structure to spectra, now and in the next section the emphasis is reversed. We first exhibit different kinds of bonding effects (in this section) to augment those already seen in some of the earlier figures, and then we discuss what information on bonding and material characteristics the various bonding effects can yield (in the following section).

Bonding is determined by the state and composition of a particular material. Hence bonding effects can be studied as a function of state or composition. Also, the environment of the emitting atom is relevant insofar as the electronic states in a material in a particular state are sensitive to temperature and pressure changes or the application of fields. We will discuss such factors in the concluding section. In this section, the examples will concern state and composition changes, and orientation and excitation effects.

The largest bonding effects occur in going from a free atom to the atom bound in a molecule or solid. Molecular orbital levels or band structure develops, and large line shifts occur relative to those in going from one compound to another. Only a few high-resolution measurements have been made on the same material in both gaseous and condensed states. We saw one of them—the L absorption spectrum of argon—in Fig. 9-6. Mattson and Ehlert (1968) have presented spectra of CO_2 and benzene in both gaseous and solid targets. Studies of spectra from gases, even without comparable spectra from solids, are especially useful as tests of molecular orbital theory. A nice example is the study of SF_6 by Lavilla (1972b).

Dissolution of a molecular or ionic material also alters its x-ray spectra. Emission spectra from solutions, which can be measured by fluorescence, change little if the emitting anion is in the center of an anion away from the solvent. Its immediate environment is little different than in the solid (Anderman and Whitehead, 1971). Absorption spectra, on the other hand, are significantly altered by dissolution since, as we discussed, the fine structure depends on the states into which the photoelectron is ejected. The extended fine structure is more washed out in solutions than in crystalline solids, because of the lack of long-range order. See Bhide and Bhat (1968), for example, and the discussion in Chap. 6. Significant smearing of extended fine structure is also observed in absorption spectra of disordered semiconductors (Sayers, Stern, and Lytle, 1971).

The other examples given in this section will all have to do with the solid state. Spectral differences from element to compound are usually largest, although major compound-to-compound changes do occur, especially for valence-band spectra.

Bonding effects have commonly been studied for a series of compounds in which one variable, such as the valence, elements present, or geometry (coordination and bond distance), changes while, if possible, the others are kept constant. Such measurements are experimentally convenient and allow interpretation of changes even in the absence of bonding calculations on the same materials. Figure 9-20 shows the Ti L_{III} band spectra from Ti compounds as a function of the anion valence (Fischer, 1973). The absorption spectra were obtained by measuring the high-energy satellites at high and low incident-electron energy and ratioing the results, as in Fig. 9-14. Component peaks obtained with a curve resolver are shown under these spectra. Several regularities can be seen. As the valence of the anion increases from

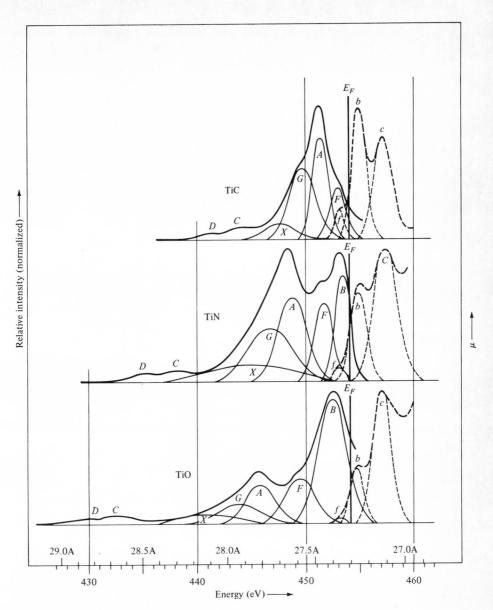

FIGURE 9-20
Titanium L_{III} emission and absorption spectra from the indicated compounds (Fischer, 1973) stripped into component curves.

TiO to TiC, more $3d$ electrons participate in the bonding and are no longer localized on the Ti atoms. The decrease in $3d$ occupancy is reflected in the decrease in peak B of the emission spectra. Simultaneously, peak b in the absorption spectra increases since it reflects the empty $3d$ states. We will say more later about how such bonding information is gotten and used when discussing Fig. 9-22. Another systematic variation in Fig. 9-20 is the decreasing separation of peaks C and D from the main band (peaks G, A, and F) with increasing valence. This occurs because C and D are mainly associated with the $2s$ level of the anion whose $2p$ level contributes to G, A, and F. A similar variation for compounds of third-row elements can be used for chemical analysis (we see this later in Fig. 9-27).

Wiech and Zöpf (1973) gave examples of valence-band emission spectra from intermetallic compounds, in relation to the crystal structure of the emitting compound, as measured with a grating spectrometer. Fabian (1970) presented similar band spectra as a function of alloy composition. Many high-quality measurements of soft and ultra-soft x-ray valence-band spectra are becoming available. In addition to flat single- and two-crystal and grating spectrometer measurements, curved single-crystal spectrometers in an electron microprobe yield high-resolution spectra with good efficiency. Valence-band measurements made across an Al-Cu diffusion couple containing several intermetallic phases (Baun and Solomon, unpublished) agree well with data taken on macroscopic samples with a flat crystal instrument.

The bonding effects on band spectra just mentioned are, in general, due to major orbital overlap (covalency). Best (1971) observed the orbital structure of molecular units, near cations within highly ionic materials, in the cation valence-band emission spectra. For example, hydrated $CaSO_4$ exhibited a "cross-over" structure in the $K\beta$ spectrum which did not occur in the anhydrous salt.

The shapes of line spectra also change with bonding, as we saw in Fig. 9-13 for the Fe $K\beta_1\beta'$ spectrum from a series of compounds. The intensity of β' relative to β_1 increases about a factor of 2 in going from Fe to Fe_2O_3. The shape of even $K\alpha_{1,2}$ lines is affected by changes in the multiplet structure with bonding. Asymmetry of $K\alpha_{1,2}$ lines is sensitive to the number of d electrons and hence to magnetic properties of first-transition-series elements. This will be discussed in Subsection VIII C. A more extreme example of multiplet effects on line spectra appears in the Cr $K\alpha_{1,2}$ from Cr_2O_3 which shows a split peak while the same line from Cr metal is unsplit (Meisel and Nefedow, 1962; Nigavekar and Bergwall, 1969). Other chemical effects on the shape of line spectra are discussed by Tsutsumi and Nakamori (1968) and Finster, Leonhardt, and Meisel (1971).

Many line-shift measurements have been made in order to calculate effective charges using the theory of level and line shifts which we discussed earlier. X-ray line shifts are usually about one order of magnitude smaller than shifts in electron-spectra lines (tenths of 1 eV rather than about 1 to 10 eV). However, x-ray shifts can sometimes be measured with better precision than electron shifts by a similar factor (Ramqvist, 1971). X-ray-shift measurements are free of problems associated with the Fermi level as a reference which occur in the interpretation of electron line shifts (Watson, Hudis, and Perlman, 1971).

In a manner similar to Fischer's parametric study of changes in valence-band spectra, Blokhin and Shubaev (1962) studied shifts in the Ti K lines as a function of valence, anion, and neighboring geometry. Alteration of geometry has the smallest effect on the line spectra. Shifts tend to become smaller with increasing atomic number since the outer electrons penetrate the K and L shell less at high atomic number. Barinskii and Nefedow (1969) present extensive tables of measured shifts.

Satellite spectra are sensitive to bonding when the availability of empty states influences the creation or rearrangement of the multiply-ionized state and when the valence electrons are involved in Coster-Kronig transitions which precede x-ray emission. Both these factors may enter into an explanation of the observation that satellite intensities in third-row elements are somewhat more intense relative to $K\alpha_{1,2}$ for oxides than for the pure elements (Aberg, 1968). An example of valence electrons being excited in Coster-Kronig processes is shown in Fig. 9-21 for the Al $K\alpha'$ and $\alpha_{3,4}$ satellites from Al and Al_2O_3. The peaks arising from the multiplet transitions in Fig. 9-17 are indicated. Those due to decay of the initial $1s2s$ vacancy configuration are shaded. Because there are fewer M-shell electrons available on the Al ions in Al_2O_3 than in Al, and because it is the M-shell electrons which are excited in a Coster-Kronig process, the probability of a $1s2s \rightarrow 1s2p$ Coster-Kronig transition is lower in Al_2O_3. Hence the intensity due to the $1s2s$ configuration is higher (Demekhin and Sachenko, 1967b). The observation that the Al $K\alpha_3/\alpha_4$ intensity ratio from Al alloys is sensitive to composition for some alloying elements but not others, as shown by Fischer and Baun (1967b), is not yet understood. When this variation is understood, it may be possible to extract useful electronic information about alloys from the slope of the intensity-ratio curves as a function of composition.

An example of chemical effects on Si $K\alpha_{1,2}$ radiative Auger spectra was given in Fig. 9-5. The radiative Auger edge occurs at a lower photon energy in SiO_2 than in Si because the electron requires more energy for excitation across the band gap in SiO_2. Differences in the exciton structure and multiplet structure are also evident, there being less excitonic structure in the semiconductor Si than in the insulator SiO_2. As for satellite spectra, little use of radiative Auger spectra has been made so far.

Numerous absorption spectra have been measured because large bonding effects can be observed, as discussed in Chap. 6. We will cite only one more example of absorption-spectra chemical effects. Barinskii and Malyukov (1964) measured the Si K edge from sodium thiosulfate, a compound containing two S atoms in non-equivalent positions. Bonding differences between the two are enough to separate the absorption edges for each atom. This result is similar to those often observed in x-ray photoelectron spectroscopy where atoms bound differently yield separate peaks

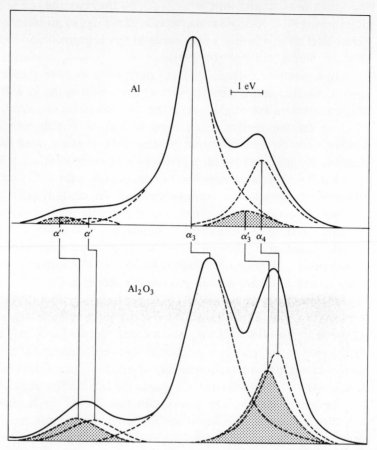

FIGURE 9-21
Al $K\alpha$ satellites from the metal and oxide (Sawada, Taniguchi, and Nakamura, 1969). Components associated with the initial-vacancy configuration $1s2s$ are shaded. Compare with Fig. 9-17.

(Siegbahn et al., 1967). We are reminded of the possibility that the derivative of an absorption spectrum might be useful for determining chemical shifts, similar to electron spectra.

There has been little work on chemical changes in Raman- and Compton-scattering spectra. We expect that Raman spectra will change in a manner similar to absorption spectra. Several Compton measurements have been done on pure elements, and scattering data from compounds are being obtained with increasing frequency. In measurements on ionic compounds, the experimental profile agrees better with an average of scattering from the component ions rather than from the neutral atoms (e.g., Fukamachi and Hosoya, 1970). This is not surprising, but it is a step in the development of Compton scattering as a sensitive tool to test bonding calculations. More recent measurements of scattering from small molecules are yielding good agreement with molecular orbital theory (e.g., Inkinen, Halonen, and Manninen, 1971).

Bonding effects in Bremsstrahlung isochromat spectra are illustrated by the work of Eggs and Ulmer (1968) on alloys, which is discussed in Chap. 6, Section V. As the electron-to-atom ratio increases in going from Rh, through Pb, to Ag, the bands fill and the isochromat peak decreases. These data illustrate the possibility of obtaining changes in the Fermi-level density of states, free of any many-body effects, if care is taken to account for broadening effects. This possibility will be discussed when we consider what bonding information can be obtained from x-ray spectra in the next section. Johansson (1960a; 1960b) has presented isochromat curves for Cr and Cr_2O_3. Change in the appearance-potential spectrum of Ni in going to the oxide is seen in Fig. 9-9. A higher bombarding-electron energy is needed because of the band gap in the oxide.

Bonding in single crystals of most materials is highly anisotropic. This has suggested study of bonding effects as a function of crystal orientation, and in relation to the polarization of incident or emitted x rays. Brümmer and Dräger (1973) and Feser et al. (1973) have studied the valence-band emission of single-crystal graphite. In the latter study, polarized (synchrotron) radiation was used to fluoresce the graphite, and anisotropic K-band emission was found. In an absorption study of single-crystal Fe_2O_3, Brümmer, Dräger, and Starke (1971) used diffraction at 45° to obtain polarized radiation.

Finally, we mention the sensitivity of bonding effects to the method of excitation for band and line spectra. Burkhalter et al. (1972) found that the chemical effects for Al K spectra from Al and Al_2O_3 are different for ion compared to x-ray or electron excitation. These effects, which are not merely a matter of self-absorption, may be due to target ion recoil effects.

We have now reviewed the needed theory, and given examples of what is possible experimentally. As indicated in Fig. 9-2, there are two kinds of relations between x-ray spectra and materials: the more or less direct relation between spectra and ground-state electronic structure, and correlations between spectra and the composition, structure, and properties of materials. The next subsection demonstrates the payoff from putting theory and experiment together to study electronic structure. After discussing that we will present further applications of bonding effects for the practical characterization of materials, and for the study of their properties. We note that some studies involve consideration of several aspects of particular materials. For example, Koster and Rieck (1970) recently studied iron $K\beta$ spectra in relation to valence (electronic structure), crystallography (atomic structure) and resistivity (one property).

VIII.   USE OF BONDING EFFECTS

A.   Electronic structure

We have put off discussion of the use of various kinds of spectra and bonding effects because of the important fact that no one spectrum or even type of spectrum suffices for a really thorough x-ray study of the bonding in a given material. Examples of the methods and results of bonding studies based on x-ray spectra will now be given.

In general, there are two kinds of uses of x-ray spectra details for electronic-structure studies. The first is to test the accuracy of electronic-structure calculations by using results of the bonding theory to calculate a spectrum for comparison with experiment (Nikiforov and Blokhin, 1963). Examples of calculated spectra were given for Si in Fig. 9-15. If such good agreement is attained, the comparison serves as a verification of that part of the computed electronic structure which is probed by the particular spectrum. Comparison with different spectra involving various core levels which overlap the bonding-electron distribution to a greater or lesser extent further tests the bonding theory results. If the comparison is not satisfactory, then some aspect of the bonding calculation, such as the starting potential, can be altered to bring the theoretical spectra into correspondence with experiment. For example, the location of peaks in valence-band emission spectra is a sensitive function of the electron-exchange multiplier used in a band calculation (Papaconstantopoulos and Nagel, 1971). Similarly, x-ray spectra can distinguish between atomic configurations (e.g., $4s^1$ or $4s^2$ for $3d$ transition metals) used for construction of the potential in a non-self-consistent band calculation. Fourier coefficients of the pseudopotential have also been chosen for Al on the basis of best agreement between calculated and experimental valence-band spectra. The coefficients determined from x-ray data compared well with those obtained by other means (Rooke, 1968b).

In the second type of use of spectra for bonding studies, some aspect of electronic structure is extracted more or less directly from x-ray data. We noted earlier that electronic structure involves the characteristics of bonding electrons in energy-momentum and in real-space regions. The first examples below demonstrate how density-of-states and energy-level information can be obtained from valence-band spectra. Obtaining bond type, overlap, and electron concentration from band spectra is also illustrated. Then the use of line shifts to obtain effective charges is discussed. Next the use of all x-ray band and line spectra information to obtain a rounded picture of bonding is emphasized. Finally, a test of electronegativity theory and the extraction of momenta distributions from Compton profiles are mentioned.

We saw earlier that the transition probability constrains x-ray band spectra to sampling only a local density of states $N_l(E)$ of a particular symmetry l satisfying the selection rules. Put another way, band spectra probe at least part of the density of states (or, in the viewpoint mentioned earlier in this section, spectra provide tests of calculated $N_l(E)$ in such cases). Many examples are now available. Recently, Thorpe and Weaire (1971b) predicted that the silicon $L_{II,\,III}$ spectrum which has two sharp peaks in crystalline Si should exhibit a smearing together of the peaks in amorphous Si. The predicted variation, based on their model in which $N_s(E)$ loses its sharp structure with disordering, was observed by Wiech (1973). This constituted an important verification of one aspect of the model which distinguishes those parts of $N_l(E)$ which are due to short- and long-range order. We anticipate that x-ray spectra will increasingly be used to test theories of disordered materials (see, e.g., Stocks, Williams, and Faulkner, 1971).

There have been many comparisons of x-ray spectra with calculated total density-of-states curves. Wiech (1968) noted that the sum of different valence-band spectra (i.e., K and L band spectra from silicon) approximates the total density of states. Kurmaev and Nemnonov (1971) used both band spectra from both components in V_3Si to obtain a curve which agreed well with the calculated total density of states.

One aspect of the density of states, namely $N(E_F)$, is of special interest because of its relation to electronic specific heat and magnetic susceptibility. It is not easy to determine $N(E_F)$, even on a relative basis, from x-ray band spectra because of lifetime and spectrometer broadening. As discussed in Chap. 5, many-body effects due to the core hole appear at the band edge. This at least complicates, and may preclude, x-ray determination of $N(E_F)$. Nevertheless, a theoretically expected correlation between x-ray emission intensity at the band edge and changes in the Knight shift was found in aluminum alloys (Bennett et al., 1971). Bremsstrahlung isochromat and short-wavelength-limit spectra, which involve no core hole, should be free of the associated many-body effects. Eggs and Ulmer (1968) obtained $N(E_F)$ for Rh-Pd-Ag alloys from the isochromat spectra and compared it to $N(E_F)$ from specific-heat measurements. The discrepancy between the two curves is large and was ascribed to many-body enhancement of the specific heat by electron-phonon interactions and spin fluctuations. The isochromat spectra yield the "bare" density of states, which is directly related to one-electron band calculations. However, the indicated enhancement is anomalously large. Recent work using temperature dependence of isochromat spectra (Merz, 1973) to determine $N(E_F)$ indicates that the values of $N(E_F)$ initially obtained from isochromats are too low. This implies more reasonable enhancement factors.

The ability to determine $N(E)$ from x-ray spectra points to an ability to determine energy-band or energy-level separations, since the latter influence $N(E)$. For example, Rooke (1968a) determined the singular points in the band structure of Al from x-ray spectra (see Chap. 4). Figure 9-22 shows how Fischer (1972) used the emission and absorption band spectra of TiO to determine the spacing between molecular orbital energy levels. The number and types of levels are available from

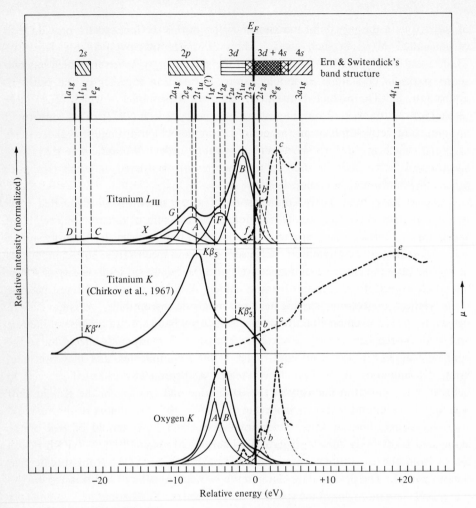

FIGURE 9-22
Molecular orbital energies obtained from the stripped emission and absorption spectra of TiO (Fischer, 1973) compared to the energy bands calculated by Ern and Switendick (1965).

group theory without the use of molecular orbital theory. Using stripped spectra, as shown, in conjunction with the x-ray selection rules allows unique association of spectral peaks and levels, all of which are taken into account. The result, in this case, is 14 energy differences. These separations could be used to evaluate atomic orbital contributions and some of the overlap integrals in a reversed molecular orbital calculation. The normal *ab initio* bonding calculation in these materials, using estimated overlap integrals to compute level spacings, is very sensitive to the assumptions made, and results vary widely. Ligand-field-splitting parameters can also be obtained from this matching of spectra to molecular orbital levels. Usually this parameter is obtained for a given material from optical absorption measurements which can link only the highest filled and lowest empty levels. X-ray spectra have the advantage of spanning a wide energy range near the Fermi level with sufficient resolution to accurately locate the desired energy levels. As seen in Fig. 9-22, the molecular orbital spacings obtained from x-ray spectra agree well with band calculations.

Other information is also available from Fischer's method. The widths of the component peaks give an indication of solid-state broadening effects. Their areas can be used to determine the number of electrons participating in the bonding if transition probabilities are assumed constant. For Cr_2O_3, for example, it was found that there are 3.5 $3d$ electrons on the Cr atom. Also, two distinct types of bonding—metal-metal covalent and metal-oxygen π bands—were found for corundum-structure materials such as Cr_2O_3.

There have been several other x-ray determinations of the nature of specific bonds besides Fischer's work. We already mentioned the difference between sulfur atoms in the thiosulfate radical. Blokhin, Shubaev, and Gorskii (1964) used S $K\alpha_{1,2}$ shifts to decide between chemically possible, alternate structures for methylene blue, a chelate polymer, and compounds of the thiocyanate group SCN. In the last instance, the possible structures were —S≡C=N— and —S—C≡N. The $K\alpha$ line shifts of S indicated that the second alternative is the actual structure for two of three compounds investigated. A similar study of sulfur bonding was performed by Shubaev et al. (1967). Bonding investigations such as this are now commonly carried out using shifts in electron spectral lines (Siegbahn et al., 1967; 1969). However, x-ray absorption-edge-shift data continue to be of use for studying the nature of bonds around a particular atom (Agarwal and Verma, 1970). Also, it has been shown recently (Stern and Sayers, 1973) that information on shielding around impurities can be deduced from the extended absorption fine structure.

Determinations of the overall type of bonding for materials with uniform bonding (in contrast to specific bonds in materials with different kinds of bonds) have also been made using valence-band spectra. The key to this work is that orbital overlap determines both the degree of covalency and the x-ray intensity. The greater the overlap (covalency), the larger the radial integral in equation (9-6). Urch (1970) and Aberg et al. (1970) have calculated the intensity of x-ray cross transitions. Comparison of their results with experiment was useful for both spectral interpretation and assessment of orbital mixing. Marshall et al. (1969) discussed the overlap of d electrons from neighboring sites with Al orbitals in aluminum alloys. Since the intensity of valence-band spectra indicates the spatial extent of particular wavefunctions, x-ray spectra should also prove useful for determining the contribution of covalent bonding in intermetallic alloys with open structures (e.g., Laves phases).

We just noted that use can be made of the dependency of valence-band x-ray intensity on orbital overlap. Since the intensity is also sensitive to the number of available electrons (the density-of-states factor), valence-band spectra can be used to determine electron concentrations (effective charges) in the region of a core orbital. Several bonding studies based on this situation have been made. Shubaev (1959; 1961) was among the first to use valence-band intensities to determine electron populations. Austin and Adelson (1970) used $K\beta_5$ intensities to determine the bonding in transition-metal germanides. They, and earlier authors (Korsunskii and Genkin, 1964; Menshikov and Nemnonov, 1963), discussed the use of x-ray spectra to determine the relative number of itinerant (collective or metallic) and localized (covalent) electrons. Most recently, Wenger, Bürri, and Steinemann (1971) took the ratio of different valence-band spectra intensities to determine the charge transfer in alloys of Al with transition metals as a function of composition. The change in the number of d electrons with alloying varied up to 1.6 for Co-Al.

Electron distributions in alloys can also be gotten from spectra not involving valence electrons. Use of K absorption-edge spectra to obtain information on changes in the number of d electrons on alloying transition metals (Azároff and Brown, 1973) is discussed in Chap. 6. Ulmer (1973) has developed models for using Bremsstrahlung isochromat spectra to determine whether charge exchange occurs upon alloying. He finds that data from the Rh-Pd system fits the charge-transfer model while Rh-Ni spectra indicate little charge exchange (minimum polarity).

Absorption spectra have been heavily used to determine effective charges for materials other than alloys. For example, Kirichok and Karalnik (1967) used edge shifts to study the valence of atoms in spinel- and garnet-type ferrites. Barinskii and Nadzhakov (1960) used their method of obtaining effective charges from K edge shifts for a variety of materials. Other examples of the use of absorption spectra to determine effective charges are given in the book by Barinskii and Nefedow (1969).

We now examine the use of line shifts to get effective charges. We just saw that recent work shows how valence-band, absorption, and isochromat spectra can yield effective charges. Earlier, such information was obtained primarily from line x-ray shifts. Now, of course, it is possible to use both electron and x-ray line shifts to obtain effective charges. While shifts in electron spectra give the direction of total charge transfer, we recall that x-ray line shifts can be used for the $3d$ transition metals to determine which orbitals ($3d$ or $4s$) contribute the transferred charge. Shubaev and Chechin (1964) pointed out that the $K\alpha_{1,2}$ shift essentially measures d-electron redistribution, while the $K\beta_{1,3}$ shift is sensitive to both s- and d-electron densities. $K\alpha$ and $K\beta$ line shifts were used with equation (9-12) by Leonhardt and Meisel (1970) to obtain $3d$ and $4s$ orbital populations in Cr and Fe compounds. The results were compared with the results of self-consistent charge and molecular orbital calculations. The computed charges on the metal ions have the same trend, but are only one-half as big as those derived from spectral shifts.

Effective charges obtained from $K\alpha$ line shifts have been determined for many compounds of elements in the third row of the periodic table as well as the first transition series. Gianturco and Coulson (1968) used S $K\alpha$ shifts from several molecules to compute effective charges and dipole moments. It is interesting to compare effective charges determined from shifts in x-ray lines with formal charges. This is done for S and Cl compounds (Leonhardt and Meisel, 1970), and it was found that the effective charges are approximately one-third of the formal charge in this case and rarely exceed 2. Reference is made to the book by Barinskii and Nefedow (1969) for discussion of other work on line shifts and effective charges.

Ramqvist and coworkers (1969) indicate that shifts of lines involving loosely bound core orbitals are not reliable for determining effective charges. This is due to direct bonding effects on the position and width of orbitals within less than about 30 volts of the valence band. Shifts in x-ray lines from the inner levels of heavy atoms are especially attractive for effective-charge determination. In that case, direct bonding effects are nil, both orbitals are small compared to the major valence-electron distribution, and the same crystal field is experienced by both orbitals.

Chemical shifts in high-energy x-ray lines are small, but they are measurable. Sumbaev (1970) and coworkers have measured $K\alpha_1$ shifts in the range $32 \leq Z \leq 74$ using a Cauchois transmission spectrometer. Some of the shifts were as small as a few millielectronvolts. All data were correlated by the use of equation (9-13) in which the valence m was assumed to be a single number for each element and the ionicity of the pure-element reference was set equal to zero. Thus, for the second transition series,

$$\frac{\Delta\mathscr{E}}{i} = xC_{5sp} - (m - x)C_{4d} \qquad (9\text{-}29)$$

where x is the number of sp electrons. In Fig. 9-23, equation (9-29) is fit to the measured $K\alpha_1$ shifts for oxides of the elements indicated. Pauling's electronegativity scale was used to calculate the ionicity i of the compounds measured. It is seen that participation of s and p electrons in the bonding (Cu-Se and Ag-Te) leads to positive shifts while d-electron (Y-Mo, La-W) and f-electron (rare-earths) bonding gives negative shifts. There is a single s electron in the series Rb-Mo, while there are two sp electrons in Ba-W. For the fourth through sixth rows of the periodic table it was found that $C_{sp} \simeq +80 \pm 10$ meV, $C_d \simeq -115 \pm 10$ meV, and $C_{4f} = -488 \pm 22$ meV for Eu and -426 ± 36 meV for Yb. Theoretical values of these constants agree with the values obtained from Fig. 9-23 to within $\pm 30\%$ in all but one instance. This agreement provides justification for the approach and a check on Pauling's

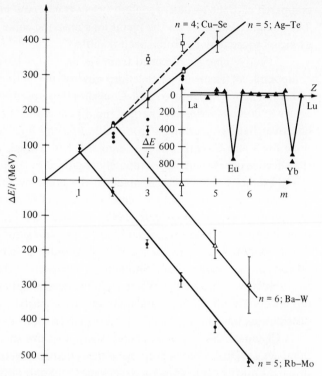

FIGURE 9-23
$K\alpha_1$ energy differences ΔE between elements and their simple compounds, divided by ionicity i, versus valence m for rows 4 to 6 of the periodic table and atomic number Z for the rare earths (Sumbaev, 1970).

ionicity scale,[1] although the sensitivity of this test is not clear. Sumbaev makes the interesting comment that conditions for the applicability of this simple approach to effective charges, namely, high-energy lines from medium- and high-atomic-number elements, are just those conditions for which it was once thought that there would be no chemical effects.

We note that, because of the sensitivity of x-ray line shapes to electron occupation numbers, as illustrated in Fig. 9-13, line shapes as well as shifts can be used, in principle, to obtain orbital occupancy values. Such an approach can yield information

[1] Shubaev (1966) also used x-ray line shifts to test Pauling's electronegativity scale. Good agreement between charges calculated from the experiment and from the electronegativity difference was obtained over an electronegativity-difference range of 0.3 to 2.2.

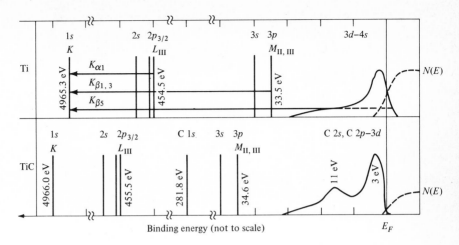

FIGURE 9-24
Core-level and valence-band energies for Ti and TiC obtained from x-ray and electron spectra (Ramqvist, 1971).

specifically on d electrons, but it has not been as widely used as line-shift measurements.

It seems logical to combine the use of valence-band spectra to determine electron-energy distributions with line shifts to determine electron spatial (charge) distributions, in order to obtain the fullest information on bonding available from x-ray spectra. This has been done by Ramqvist et al. (1969) for refractory-metal compounds. They have also used shifts in electron spectral lines to obtain changes in binding energies (relative to the Fermi level). Figure 9-24 shows the electronic structure obtained for Ti and TiC. Clementi's calculations (1965) and the $K\alpha_{1,2}$ line shift of -0.33 eV indicate a loss of 0.4 $3d$ electrons from Ti in bond formation. Only recently, with the combined use of x-ray and electron line and band spectra has it been possible to put the electronic structure of materials on a uniform diagram such as Fig. 9-24.

We now briefly consider the use of spectra to determine electron-momenta distributions, in contrast to the energy and spatial electron distributions just reviewed. We saw that a Compton profile can be transformed using the derivative of equation (9-21) to yield the momenta distribution. Results for Li, as well as Be and B, are shown in Fig. 9-25, plotted against a variable z which gives the component of initial electron momentum in the direction of the momentum transferred from the moving electron to the photon during the Compton-scattering process. The calculated momentum distributions from the core electrons have been subtracted, leaving only the valence-momentum distributions in Fig. 9-25. A sharp drop corresponding to the Fermi momentum is evident. Such data can provide a test of the wavefunctions

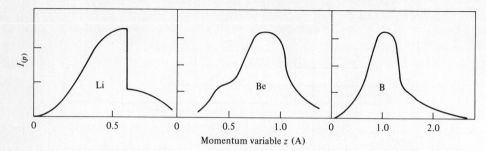

FIGURE 9-25
Valence-electron-momentum distributions obtained from measured Compton profiles (Phillips and Weiss, 1968).

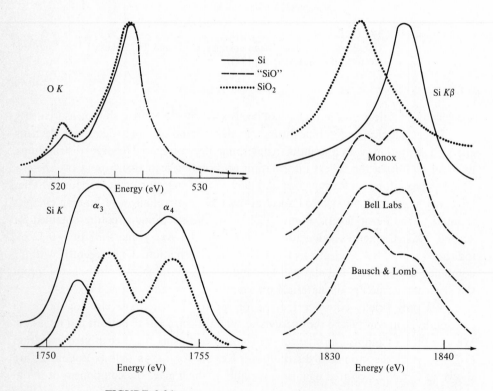

FIGURE 9-26
Si valence-band (White and Roy, 1964) and satellite spectra and oxygen valence-band spectra (Baun and Solomon, 1971) from Si, SiO, and SiO_2.

obtained from band calculations through use of equations (9-21) to (9-23). Interesting comparisons between electron-momenta data obtained from Compton scattering and positron annihilation are anticipated.

B.   Composition and atomic structure

The major uses of x-ray spectra for bonding studies were just reviewed. These exploit the close relation between spectra and electronic structure shown in Fig. 9-2. In this section we are concerned with the less direct correlations between spectra (mainly band) and the chemical composition and atomic structure of materials. Composition and structure are useful characteristics of a material to know because they give information on chemical and physical reactions which involve the material, and because they indicate how a material will respond during use, if relations between these characteristics and properties are known.

Elemental composition is available, of course, from the intensities of x-ray lines. But, often it is necessary to know what compounds are present in an unknown, not just which elements. Because the details of valence-band spectra are sensitive to the surroundings of the emitting atom, the compounds present in a sample can be determined from such data as spectral shapes, peak separations, and peak shifts.

Figure 9-26 shows Si and O K valence-band spectra and $K\alpha_{3,4}$ satellite spectra from a material with nominal composition SiO, plus spectra from Si and SiO_2. It was desired to determine whether the material was actually SiO or an intimate mixture of Si and SiO_2. Diffraction patterns could not yield an answer. The Si $K\beta$ and $K\alpha_{3,4}$ spectra from SiO appear to be a superposition of the spectra from Si and SiO_2, and the O K spectrum is the same from SiO as SiO_2, indicating a mixture.

The shapes of valence-band spectra vary so widely that sometimes they are essentially a fingerprint of a particular compound. This fact is useful for analysis where the number of possible compounds is small and they have distinct spectra. Examples include the analysis of corrosion and oxidation products.

Urch (1970) has pointed out that the energy separation of $K\beta_1$ and $K\beta'$ peaks (see Fig. 9-4) from third-row elements in compounds with second-row anions is characteristic of the anion, as illustrated in Fig. 9-27. This occurs because the separation of $K\beta'$ from the main band is set mainly by the anion $2s$-$2p$ separation, which also is the situation for peaks C and D in the Ti compound spectra shown in Fig. 9-20. Use of Fig. 9-27 allows determination of the anion from measurement of the more intense cation spectra which is in a more convenient wavelength region. A composition-sensitive separation of two peaks in the Al $K\beta$ spectra from Al-Cu alloys has also been found and used to perform quantitative analyses on Al-Cu thin films (Baun, 1969).

Transition metals form vacancy-interstitial compounds such as TiO_x

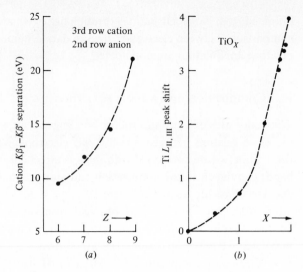

FIGURE 9-27
Empirical calibration curves relating the emitting compound to valence-band spectra. (*a*) Peak separations (Urch, 1970); (*b*) peak positions (Fischer and Baun, 1968).

$(0 \leq x \leq 2)$ for which the peak position of the Ti $L_{II, III}$ spectrum is related to x as shown in Fig. 9-27 (Fischer and Baun, 1968). This curve can be used for analysis of TiO$_x$ samples, with greatest sensitivity attainable in the range $1 \leq x \leq 2$. Holliday (1971) has plotted a curve similar to Fig. 9-27 for Ti $L_{II, III}$ from TiN$_x$ $(0 \leq x \leq 1)$. Krause, Savanick, and White (1970) used the peak position of the O K band to construct a calibration curve for characterization of oxides of Ti, Mn, and Fe. Also, O K and Al $K\beta$ peak positions were used by Gigl, Savanick, and White (1970) for measurement of corrosion layers on Al. Known oxides, oxyhydroxides, and hydroxides provided standards to plot the calibration curve.

Atomic arrangements have been determined from both x-ray line shifts and valence-band spectra details. Blokhin and Shubaev (1962) studied the effects of changes in coordination and bond distance (as well as valence and ligand type) on Ti $K\alpha_{1,2}$, $K\beta_1$, and $K\beta_5$ spectra from several Ti compounds. Shifts due to changes in atomic arrangement were less than 0.2 eV. Such shifts are due to changes in the valence-electron density within the core orbital region as the structure is altered (Shubaev, 1961). A recent discussion of the relation of Al $K\alpha_{1,2}$ line shifts and crystal structure was given by Läuger (1971).

White and Gibbs (1967; 1969) have studied Al and Si $K\beta$ valence-band shifts from minerals as a function of coordination and bond distance. The Al $K\beta$ energy increased by about 2 eV in going from fourfold to sixfold coordination and, for octahedral coordination, increased by about 1 eV for an increase in Al-O distance of

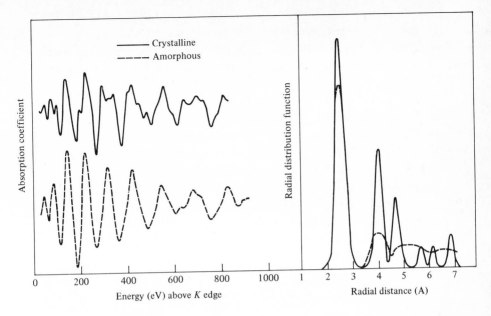

FIGURE 9-28
Extended absorption fine structure and radial distribution functions for crystalline and amorphous germanium (Sayers, Stern, and Lytle, 1971).

0.05 A. Both these trends can be understood in terms of decreasing valence-electron binding and increasing x-ray energy as the coordination or bond distance is increased. Subtleties of atomic geometry are also available from valence-band spectra. Klasson and Manne (1972) showed slightly different Cl $K\beta$ spectra (both experimental and theoretical) from cis- and trans-$C_2H_2Cl_2$.

Distances between atoms can also be deduced from absorption-edge extended fine structure. The nearly free-electron theory for high-energy bands predicts that, for the same crystal structure, the position of a peak should vary as the inverse square of the interatomic distance. This behavior is well verified, at least for peaks removed from the absorption edge. Interatomic distances determined by the use of absorption fine structure are in fair agreement with the results of crystal-structure determination by x-ray diffraction (e.g., Bhide and Bhat, 1969). It is not clear how much of the deviation from a perfect correspondence to the diffraction results is due to potential (band-structure) effects and how much is due to the approximate theories used to extract distances from the fine-structure positions. A thorough calculation of extended absorption structure based on band theory should answer this question.

A recent, more extensive use of absorption extended fine structure is shown in Fig. 9-28. The absorption spectra from crystalline and amorphous Ge shown in Fig. 9-28a were measured by Sayers, Stern, and Lytle (1971) and Fourier-analyzed to obtain the radial distribution function shown in Fig. 9-28b. Comparison of this distribution obtained from x-ray spectroscopy with one obtained from the conventional x-ray scattering experiment is needed to evaluate the new absorption technique.

The examples mentioned so far in this section have concerned links between line or band spectra and atomic arrangement. As a final illustration, we mention work by Merz and Ulmer (1971) which showed the sensitivity of Bremsstrahlung isochromat spectra to bcc-hcp phase changes in Ti, Zr, and Hf. Differences in the observed spectra are due to the different density of (empty) states for the two structures.

Several of the examples discussed in this subsection essentially concern calibration curves which relate the material characteristic of interest (composition or structure) to some aspect of an x-ray spectrum (e.g., peak separation or position). Figure 9-27 is one of these. It should be possible to use bonding theory to produce corresponding theoretical relations for the cases where structural and spectral characteristics of materials are correlated.

C.   Material properties

Earlier we emphasized the importance of electronic structure because of its relation to basic material properties. Some properties are so directly related to bonding that various aspects of electronic structure available from x-ray spectra can be used to understand and calculate properties. We will cite a few of them in this section.

Aside from entropy terms, heats of formation are determined mainly by the valence-electron spatial and energy distributions. Ramqvist (1971) has used the effective charge of Ti in TiC, as obtained from the Ti $K\alpha_1$ shift, to calculate the electrostatic (Madelung) energy. The answer almost exactly equals the heat of formation of TiC, showing that the electrostatic energy is what gives TiC its increased thermal stability relative to the pure elements.

Heats of formation were calculated directly from valence-band spectra shifts for some simple compounds by Das Gupta (1950). Dodd and Glen (1970) have used Al and Si $K\beta$ peak shifts to compute the decrease in Al-O and Si-O bond energies with increasing silica content in lithia-aluminosilicate glasses.

Valence-band peak positions can also be correlated with electrical conductivity, as shown for aluminum compounds by Baun and Fischer (1965). Metallic materials yield band spectra at higher energies than insulators. In a manner similar to the variation of peak positions with coordination and bond distance, this correlation can be understood in terms of variation in the binding energy of valence electrons. For insulators, the valence band (bonding orbitals) is more tightly bound, the energy gap (bonding to anti-bonding-orbital separation) greater, and the conductivity less. Separations between emission and absorption spectra also yield energy gaps directly, if care is taken to separate out broadening and satellite effects. Careful consideration was given to the edge widths of valence-band spectra by Gale, Catterall, and Trotter (1969). They found a correlation between the Mg $L_{II, III}$ edge width and the electrical resistivity of Mg-Cd alloys.

Widths of valence-band spectra from simple metals can be used to calculate plasmon energies if the low-energy tail is correctly subtracted from the measured spectra. The energies obtained from spectra are about 5% greater than plasmon energies determined by other means (Nagel, 1970). Plasmon energies are more directly available from the separation between valence-band and plasmon-satellite spectra (Rooke, 1963) and between coherent and plasmon Raman peaks in scattered-x-ray spectra (Priftis, Theodossiou, and Alexopoulos, 1968). Mande and Joshi (1969) used K absorption spectra of transition metals to determine plasmon energies, but there is some question concerning whether the peaks they used are due to plasmon or band-structure effects (McCaffrey, Nagel, and Papaconstantopoulos, 1974). Characteristic loss values can also be determined from Bremsstrahlung isochromat (e.g., Claus and Ulmer, 1965) and appearance-potential (Houston and Park, 1972b) measurements.

We noted earlier that the Fermi-level density of states $N(E_F)$ can sometimes also be obtained from x-ray spectra. This quantity is related to the electronic specific heat and magnetic susceptibility, in addition to the Knight shift already mentioned. A high $N(E_F)$ value is desired in a superconductor so that it has a high transition temperature. Valence-band x-ray spectra may be useful to screen prospective candidate superconducting materials in the search for high transition temperatures.

Excitonic states in the energy gap above E_F appear in radiative Auger and, especially, in x-ray absorption spectra (Fig. 9-6). The spacing of exciton peaks can be used to calculate the dielectric constant of a material. Wide separation of exciton peaks implies a lower dielectric constant (less polarizable medium).

Finally, we will briefly discuss some relations between magnetic properties and x-ray spectra. Early work in this area involved the measurement of K x-ray spectra from Cu and Mn in Heusler alloys (Cu_2MnAl) (Vainshtein and Kotliar, 1956). The $K\alpha_{1,2}$ asymmetry index was found to increase markedly as temperature was increased across the Curie point near 320°C. The results showed that the Cu atoms participate in the magnetic behavior of this material. Later Shubaev (1960) used valence-band spectra data to show that the Curie transition, and also the appearance of ferromagnetism as a function of composition in Heusler alloys, implies that the d-electron character of Cu is lower and Mn is greater in the ferromagnetic state.

A close relation between $K\alpha_1$ line asymmetry and magnetic moment for $3d$ transition elements was shown by Nemnonov and Kolobova (1958). The basis for this relation is the multiplet structure due to exchange interaction between the core hole and partially filled $3d$ level. The number of unpaired d electrons determines both the multiplet structure and the magnetic moment. Menshikov, Nemnonov, and Mischenko (1962) point out that since the multiplet structure also affects the line width and position, these spectral details should also be considered, e.g., in magnetic studies.

Adelson and Austin (1969a) used Fe $K\beta_5$ valence-band spectra from a series of iron germanides to study bonding and magnetism in these materials. They found that the Fe $K\beta_5$ intensities from the germanides, relative to pure Fe, could be used to calculate magnetic moments which agreed very well with the measured moments. The changes in intensity and magnetic character were attributed to changes in orbitals which occur due to different coordination in the Fe-Ge compounds. These authors also found a correlation between $K\beta_5$ band-intensity ratios and the magnetic moment of the pure $3d$ transition elements (Adelson and Austin, 1969b).

As a final example concerning the relation of x-ray spectra, electron structure, and properties, we cite a thorough study of refractory carbides by Ramqvist (1969). Attention was given to properties such as hardness and wetting angles, which are related to bond strength and heats of formation, as well as to resistivity which is more directly related to electronic structure.

IX.   DISCUSSION

We will recapitulate a few of the highlights and mention some problems and prospects in this section. Clearly the central theme of this chapter is the use as well as the understanding (interpretation) of all kinds of x-ray spectra. The relatively small number of examples given here was intended to give a picture of recent progress and the current status in the use of bonding effects. We note that the examples, which were chosen to illustrate various aspects of the interpretation and use of x-ray spectra, involved a wide range of materials. This emphasizes the breadth of x-ray spectroscopy, especially in relation to many other methods for studying materials.

An earlier review, written from a viewpoint similar to this chapter, was limited to valence-band emission spectra (Nagel, 1970). Here we have been concerned with all of the several kinds of x-ray spectra. It is clear that valence electrons influence all spectra to some extent, even the core-level transitions in heavy elements. Such indirect bonding effects are useful, but direct effects on spectra involving valence electrons or empty bands have greatest utility for both basic and practical work. As the examples showed, valence-band emission and absorption spectra are most used and most useful. Some of the other spectral types, such as radiative Auger and x-ray Raman spectra, await the use of high-intensity sources for a good evaluation of their utility.

We saw that many aspects of electronic structure can be tested against or extracted from x-ray spectra. Generally, this requires the use of theory, for example, atomic theory to calculate the $\Delta E - \Delta q$ curve for line-shift work, or bonding theory for full calculation of band spectra. Even Fischer's method (1973) for extracting level separations and other information from valence-band spectra without use of bonding theory calculations requires an initial group theoretical calculation to determine which molecular orbitals are expected.

It was also demonstrated that characteristics (composition and structure) and properties of materials can be obtained primarily from valence-band spectra. Sometimes empirical calibration curves suffice for this use of spectra, but an understanding of materials and x-ray theory is still desirable for such work. The use of spectra for materials characterization seems somewhat less important than the use of spectra for bonding studies. This is due to the more limited applicability of relations used for obtaining composition and structure information, and because, in comparison to other methods for bonding studies, x-ray spectroscopy offers some unique possibilities.

Consideration of remaining problems in x-ray spectroscopy is an important guide to future work. Each spectral type has its own peculiar difficulties, e.g., overlapping band and multiplet structure in radiative Auger spectra. Here we will consider the more general problem areas. We noted in connection with Fig. 9-2 that intervention of an ionized state and experimental influences on spectra are the two major obstacles to relating spectra and the ground state of a material.

There are problems associated with excitation of ionized states (e.g., by heavy-ion impact) and with x-ray emission, scattering, and absorption (e.g., many-body effects). These and related problems are being studied experimentally and theoretically. In particular, coincidence methods borrowed from nuclear physics are proving useful for studying excitation effects (Kessel, 1969; Briand et al., 1971). Also, calculation of spectra from one-electron band theory and the use of two-variable (energy and momentum or angle) x-ray-scattering experiments (Doniach, Platzman, and Yue, 1971) should allow separation of many-body effects out of the spectra of conductors.

General experimental problems include reproducibility of spectra measured in different laboratories, intensity, and resolution. The first of these problems is mainly a nuisance and not a basic difficulty. Low intensity is, to some extent, a technological problem, although anode or sample heating (and decomposition for some materials) sets the ultimate limit on intensity. The problem of resolution in the x-ray region seems more fundamental since it applies to all materials and since lifetime broadening cannot be avoided even though high available intensity may allow use of a very narrow spectrometer window. Two approaches to improving resolution are available. The first is mathematical and involves treatment (unfolding) of spectra measured in a conventional manner. The second requires a different experimental approach.

Many methods have been used to deconvolute lifetimes and spectrometer functions out of band spectra, with resultant resolution approaching 0.1 eV even for intermediate x-ray wavelengths (see, e.g., Nikiforov, 1960). Figure 9-29 shows the results of using a recently developed unfolding program on the valence-band spectra of silicon from SiO_2. We also saw in Figs. 9-20 and 9-22 how curve resolvers can be used to strip spectra into simple component curves which can be easily unfolded from the broadening functions.

For optical spectroscopy, the use of modulation and difference techniques in the past decade has led to an increase in resolution and the availability of new experimental information. A similar trend already beginning in the x-ray region is discussed by Nagel (1973). There are three basic components in a spectroscopy experiment: the exciting radiation, the sample, and the spectrometer. Any of these can be modulated (oscillated) to provide a reference for phase-sensitive detection of the measured x rays. Variation of the excitation or spectrometer serves to increase the resolution, while sample modulation brings out new information since the electronic structure of the sample is altered. Modulation procedures yield a derivative spectrum, as we saw in the case of appearance-potential spectroscopy. In that instance, the energy of the bombarding electrons was modulated. An x-ray experiment in which the sample (pure Cu) was modulated with a stress was performed by Willens et al. (1969) who plotted the difference between Cu $L_{II, III}$ spectra for stressed and unstressed conditions. The difference spectrum shows greater detail than the original spectra and can be matched to the calculated band structure of copper. No x-ray experiments in which some spectrometer component was modulated are known to the authors.

Earlier we noted that there are bonding effects due to a change of state as well as a change of chemical composition. Change in the environment of the emitting sample is another kind of variable which causes additional effects in x-ray spectra. Application of stress in the modulation experiment just discussed is one example. Similarly, temperature changes have been shown to have effects on spectra, especially absorption spectra (see Chap. 6). Bonding may enter temperature effects by its influence on the phonon spectrum, where changes cause the alteration of some spectral details (see Chap. 5). McAlister (1969) recently included phonon effects in a calculation of the Li K band spectrum. Besides stress and temperature changes, application of electric and magnetic fields during x-ray spectral measurements may also bring out new features of the bonding in some materials. High fields are required, however (Barinskii and Malyukov, 1962).

Many experimental techniques compete with the use of bonding effects in x-ray spectra for electronic-structure studies and measurement of the composition, structure, and properties of materials. For example, there are dozens of ways to determine chemical composition besides the use of fine details of x-ray spectra, including prominently the measurement of x-ray line intensities (Birks, 1969; 1971). And, certainly x-ray diffraction is the most powerful and widely used means of struc-

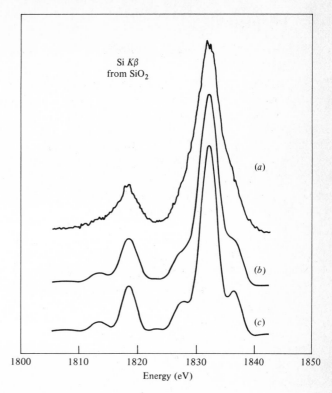

FIGURE 9-29
Si $K\beta$ (from Colby, 1972) (*a*) as measured; (*b*) spectrometer window only removed; (*c*) spectrometer and core-level broadening removed.

ture determination. Also, direct measurement of properties is usually the most efficient means of obtaining their values. Nonetheless, sometimes it is best to use x-ray spectra for measurement of a material's characteristics and for determination of its properties. Hence, it is important to develop this capability and know its strengths and weaknesses relative to alternate methods.

It is in the area of electronic-structure studies that x-ray methods are strongest compared to other techniques. The wide energy range probed in x-ray measurements, the availability of selection rules, and the relative simplicity of spectra involving core levels are qualities peculiar to x-ray spectra. It is anticipated that x-ray spectroscopy will grow to be one of the major experimental techniques used for bonding studies.

REFERENCES

ABERG, T. (1968): X-ray satellite intensities in the sudden approximation, *Phys. Lett.*, **A26**: 515–516.

—— (1969): Multiple excitation of a many-electron system by photon and electron impact in the sudden approximation, *Ann. Acad. Scientiarum Fennicae, Ser. A, VI, Physica*, no. 388, 1–46.

—— (1971): Theory of the radiative Auger effect, *Phys. Rev.*, **4A**: 1735–1740.

—— (1973): X-ray satellites and their interpretation, in A. FAESSLER (ed.), *Proceedings of international symposium on x-ray spectra and electronic structure of matter*, München, 18–22 Sept. 1972.

——, GRAEFFE, G., UTRIAINEN, J., and LINKOAHO, M. (1970): The x-ray K-emission spectrum of sodium and of fluorine in some alkali halides, *J. Phys. Chem.*, no. 3, 1112–1119.

—— and UTRIAINEN, J. (1971): The influence of chemical bonding on the low-energy $K\alpha$ spectrum of silicon, *J. Phys. (Paris)*, **Suppl. 10**: C4-295–C4-300.

ADELSON, E., and AUSTIN, A. E. (1969a): X-ray spectroscopic studies of bonding in iron germanides, *Advan. X-Ray Anal.*, **12**: 506–517.

—— and —— (1969b): Dependence of x-ray transitions from the conduction band upon non-localized $3d$ electrons, *Solid State Comm.*, **7**: 1819–1820.

AGARWAL, B. K., and VERMA, L. P. (1970): A rule for chemical shifts in x-ray absorption edges, *J. Phys. C. (Sol. St. Phys.)*, **3**: 535–537.

ALDER, K., BAUR, G., and RAFF, U. (1972): The chemical energy shift of K x-rays, *Helv. Phys. Acta*, **45**: 765–710.

ALTMANN, S. L. (1968): A theoretical summing-up, in D. J. FABIAN (ed.), *Soft x-ray band spectra and the electronic structures of metals and materials*, pp. 367–368, Academic Press, Inc., London.

ANDERMANN, G., and WHITEHEAD, H. C. (1971): Molecular x-ray spectra of sulphur and chlorine bearing substances, *Advan. X-Ray Anal.*, **14**: 453–486.

ASHCROFT, N. W. (1968): Density of states in simple metals and the soft x-ray spectrum, in D. J. FABIAN (ed.), *Soft x-ray band spectra and the electronic structure of metals and materials*, pp. 249–264, Academic Press, Inc., London.

AUSTIN, A. E., and ADELSON, E. (1970): X-ray spectroscopic studies of bonding in transition metal germanides, *J. Solid State Chem.*, **1**: 229–236.

AZÁROFF, L. V. (1963): Theory of extended fine structure in x-ray absorption edges, *Rev. Mod. Phys.*, **35**: 1012–1022.

—— and BROWN, R. S. (1973): Electronic structure of Cu-Ni-Zn ternary alloys. (See Fabian and Watson, 1973.)

BABANOV, YU. A., and SOKOLOV, O. B. (1971): Theory of the absorption spectra of soft x-rays in metals. I. One electron approximation (in Russian), *Fiz. Metal. i Metalloved.*, **31**: 675–684; *Phys. Metals Metallogr. (USSR) (English Transl.)*, **31**: 1–10.

BAGUS, P. S. (1973): Ab initio computation of molecular structure: Excited states involving x-ray phenomena, in A. FAESSLER (ed.), *Proceedings of international symposium on x-ray spectra and electronic structure of matter*, München, 18–22 Sept. 1972.

BALLHAUSEN, C. J., and GRAY, H. B. (1964): *Molecular orbital theory*, W. A. Benjamin, Inc., New York.

BAMBYNEK, W., CRASEMANN, B., FINK, R. W., FREUND, H. U., MARK, H., SWIFT, C. D., PRICE, R. E., and RAO, P. V. (1972): X-ray fluorescence yields, Auger, and Coster-Kronig transition probabilities, *Rev. Mod. Phys.*, **44**: 716–813.

BARAT, M., and LICHTEN, W. (1972): Extension of the electron promotion model to asymmetric atomic collisions, *Phys. Rev.*, **6A**: 211–229.

BARINSKII, R. L., and MALYUKOV, B. A. (1962): Interpretation of the *K* absorption spectra of sulphur in molecules and crystals in terms of the Stark effect in excitons, *Bull. Acad. Sci. USSR, Phys. Ser. (English Transl.)*, **26**: 412–418.

—— and —— (1963): The Stark effect and hybridization in *K* absorption spectra, *Bull. Acad. Sci. USSR, Phys. Ser. (English Transl.)*, **27**: 360–367.

—— and —— (1964): A study of two different states of sulphur in a thiosulphate by x-ray absorption spectroscopy, *J. Struct. Chem. (USSR) (English Transl.)*, **5**: 558–560.

—— and NADZHAKOV, E. G. (1960): Calculation of the charge of atoms in molecules from the *K* absorption spectra, *Bull. Acad. Sci. USSR, Phys. Ser. (English Transl.)*, **24**: 419–425.

—— and NEFEDOW, W. I. (1969): *Röntgenspektroskopische Bestimmung der Atomladungen in Molekülen*, Akademische Verlagsgesellschaft Geest & Portig K.-G., Leipzig.

—— and VAINSHTEIN, E. (1959): Collective interaction of electrons in crystals and its manifestation in the x-ray absorption spectra of atoms in polar crystals, *Bull. Acad. Sci. USSR, Phys. Ser. (English Transl.)*, **23**: 566–571.

BATTERMAN, B. W., and COLE, H. (1964): Dynamical diffraction of x-rays by perfect crystals, *Rev. Mod. Phys.*, **36**: 681–717.

BAUN, W. L. (1968): Instrumentation, spectral characteristics, and applications of soft x-ray spectroscopy, *Appl. Spectry. Rev.*, **1**: 379–432.

—— (1969): Al *K* x-ray emission fine features for characterizing Al-Cu films, *J. Appl. Phys.*, **40**: 4211–4112.

—— and FISCHER, D. W. (1964): High energy α satellites in the aluminum *K* x-ray emission spectrum, *Phys. Lett.*, **13**: 36–37.

—— and —— (1965): The effect of valence and coordination on *K* series diagram and nondiagram lines of Mg, Al, and Si, *Advan. X-Ray Anal.*, **8**: 371–383.

—— and —— (1970): Soft x-ray spectroscopy as related to inorganic chemistry, in C. N. RAO and J. R. FERRARO, *Spectroscopy in inorganic chemistry*, vol. 1, pp. 209–246, Academic Press, Inc., New York.

—— and SOLOMON, J. S. (1971): Characterization of SiO using fine features of x-ray emission *K* spectra, in *Proc. Sixth Natl. Conf. Electron Probe Anal.*, Electron Probe Analysis Society of America, Pittsburgh, 45A–45B.

BEARDEN, J. A., and BURR, A. F. (1967): Re-evaluation of x-ray atomic energy levels, *Rev. Mod. Phys.*, **39**: 125–142.

BENNETT, L. H., MCALISTER, A. J., CUTHILL, J. R., and DOBBYN, R. C. (1971): Correlation of changes in Knight shift and soft x-ray emission edge height upon alloying, in *Electronic density of states*, Natl. Bur. Std. Spec. Publ. **323**: 665–670.

BERGENGREN, J. (1920): Über die Röntgenabsorption des Phosphors, *Z. Physik*, **3**: 247–249.

BEST, P. E. (1966): Electronic structure of MnO_4^-, CrO_4^{2-}, and VO_4^{3-} ions from the metal K x-ray spectra, *J. Chem. Phys.*, **44**: 3248–3253.

——— (1971): Cation K x-ray emission from ionic crystals, *J. Chem. Phys.*, **54**: 1512–1514.

BHIDE, V. G., and BHAT, N. V. (1968): X-ray K absorption edge of yttrium in some yttrium compounds, *J. Chem. Phys.*, **48**: 3103–3108.

——— and ——— (1969): Fine structure in x-ray K absorption spectrum of yttrium, *J. Chem. Phys.*, **50**: 42–46.

BIRKS, L. S. (1969): *X-ray spectrochemical analysis*, 2d ed., Interscience Publishers, division of John Wiley & Sons, Inc., New York.

——— (1971): *Electron probe microanalysis*, 2d ed., Interscience Publishers, division of John Wiley & Sons, Inc., New York.

BLOKHIN, M. A., DEMEKHIN, V. F., and SHVEITSER, I. G. (1962): On correction of x-ray emission spectra for self-absorption, *Bull. Acad. Sci. USSR, Phys. Ser.* (*English Transl.*), **26**: 419–422.

———, SACHENKO, V. D., and NIKIFOROV, I. Y. (1971): On the nature of x-ray emission bands of the transition metals, *J. Phys.* (*Paris*), **Suppl. 10**: C4-211–C4-213.

——— and SHUBAEV, A. T. (1962): Concerning the influence of chemical bonds on the x-ray emission spectrum of titanium, *Bull. Acad. Sci. USSR, Phys. Ser.* (*English Transl.*), **26**: 429–432.

———, ———, and GORSKII, V. V. (1964): X-ray spectroscopic investigations of the chemical bond in sulphur compounds, *Bull. Acad. Sci. USSR, Phys. Ser.* (*English Transl.*), **28**: 709–712.

BÖHM, G., and ULMER, K. (1969): Energie-abhängigkeit der Isochromatenstruktur von Wolfram im Energiebereich von 0.15 bis 6 keV, *Z. Physik*, **228**: 473–488.

——— and ——— (1971): Highly resolved x-ray structures and the electronic structure of tantalum and tungsten, *J. Phys.* (*Paris*), **Suppl. 10**: C4-241–C4-245.

BONNELLE, C. (1966): Contribution a l'etude des etaux de transition du premier groupe, due cuirve et de leurs oxydes par spectroscopie X dans le domaine, *Ann. Phys.*, **1**: 439–481.

——— and KARNATAK, R. (1971): Distribution des états f dans les métaux et les oxydes de terres rares, *J. Phys.* (*Paris*), **Suppl. 10**: C4-230–C4-235.

BRANDT, W., and LAUBERT, R. (1970): Pauli excitation of atoms in collision, *Phys. Rev. Lett.*, **24**: 1037–1040.

BRIAND, J. P., CHEVALLIER, P., TAVERNIER, M., and ROZET, J. P. (1971): Observations of K hypersatellites and KL satellites in the x-ray spectrum of doubly-ionized gallium, *Phys. Rev. Lett.*, **27**: 777–779.

BROWN, D. B. (1973): *Calculation of absolute x-ray Bremsstrahlung intensities* (to be published).

BRÜMMER, O., and DRÄGER, G. (1973): Zur Orientienungsabhängigkeit der von Einkristallen emittierten langwelligen Röntgenbanden, in A. FAESSLER (ed.), *Proceedings of international symposium on x-ray spectra and electronic structure of matter*, München, 18–22 Sept. 1972.

———, ———, and STARKE, W. (1971): On the mechanism of 3d transitions in K-absorption spectra of transition-metal compounds, *J. Phys.* (*Paris*), **Suppl. 10**: C4-169–C4-171.

BURKHALTER, P. G., KNUDSON, A. R., NAGEL, D. J., and DUNNING, K. L. (1972): Chemical effects on ion-excited Al K x-ray spectra, *Phys. Rev.*, **6A**: 2093–2101.

CALLAWAY, J. (1964): *Energy band theory*, vol. 16, Pure and Applied Physics series, Academic Press, Inc., New York.

CARTER, F. L. (1971): On deriving density of states information from chemical bond considerations, in *Electronic density of states*, Natl. Bur. Std. Spec. Publ. **323**: 385–406.

CAUCHOIS, Y., and MOTT, N. F. (1949): The interpretation of x-ray absorption spectra of solids, *Phil. Mag.*, **40**: 1260–1269.

CHIRKOV, V. I., BLOKHIN, M. A., and VAINSHTEIN, E. E. (1967): Study of the x-ray K spectra of titanium in its nitride and carbide, *Sov. Phys. Solid State (English Transl.)*, **9**: 873–877.

CLAUS, H., and ULMER, K. (1965): Untersuchung der Zustandsdichte und der charakteristischen Energie verluste von Ir, Rh, Pt und Pd mit Isochromatenmessunger, *Z. Physik*, **185**: 139–154.

CLEMENTI, E. (1965): Tables of atomic functions, *IBM J. Res. Devel. Suppl.*, **9**: 2–19.

COHEN, M. L., and HEINE, V. (1970): The fitting of pseudopotentials to experimental data and their subsequent application, in *Solid state physics*, vol. 24, pp. 37–248, Academic Press, Inc., New York.

COLBY, J. (1972): Computer evaluation of electron microprobe x-ray spectral emission data, in SHIRODA *et al.* (1972).

CONKLIN, J. B., and SCHWARZ, K. (1972): The soft x-ray spectra of TiC, *J. Phys. (Paris)*, **Suppl. 8**: 76–94.

COWAN, R. D. (1968): Theoretical calculation of atomic spectra using digital computers, *J. Opt. Soc. Am.*, **58**: 808–818.

CUMPER, C. W. N. (1966): *Wave mechanics for chemists*, Academic Press, Inc., New York.

CURRAT, R., DE CICCO, P. D., and KAPLOW, R. (1971): Compton scattering and electron momentum density in beryllium, *Phys. Rev.*, **B3**: 243–251.

———, ———, and WEISS, R. J. (1971): Impulse approximation in Compton scattering, *Phys. Rev.*, **B4**: 4256–4261.

DAS GUPTA, K. (1950): The soft x-ray valence band spectra and the heat of formation of chemical compounds and alloys, *Phys. Rev.*, **80**: 281–282.

——— (1969): Modified x-ray scattering close to the Bragg spectrum, in *X-ray spectra and electronic structure of matter*, vol. 2, pp. 193–200, Institute of Metal Physics, Academy of Sciences of the Ukrainian SSR, Kiev.

DAVIS, D. W., HOLLANDER, J. M., SHIRLEY, D. A., and THOMAS, T. D. (1970): Comparison of core-level binding energy shifts in molecules with predictions based on Koopmans' theorem, *J. Chem. Phys.*, **52**: 3295–3296.

DE CICCO, P. D. (1969): Compton scattering of x-rays in solids, M.I.T. Solid State and Molecular Theory Group Semiannual Progress Report, no. 71, 24–40.

DEHMER, J. L. (1972): Evidence of effective potential barriers in x-ray absorption spectra of molecules, *J. Chem. Phys.*, **56**: 4496–4504.

DEMEKHIN, V. F., and SACHENKO, V. P. (1967a): Spectral position of $K\alpha$ satellites, *Bull. Acad. Sci. USSR, Phys. Ser. (English Transl.)*, **31**: 913–920.

DEMEKHIN, V. F. and SACHENKO, V. P. (1967b): Relative intensities of $K\alpha$ satellites and chemical bonding, *Bull. Acad. Sci. USSR, Phys. Ser.* (*English Transl.*), **31**: 921–925.

DESLATTES, R. D. (1963): $K\beta$ emission spectra of argon and KCl I, *Phys. Rev.*, **133**: A390–A398.

DEV, B., and BRINKMAN, H. (1970): Determination of electron binding energies of surface atoms, *Ned. Tijdschr. Vac.*, **8**: 176–184.

DEY, A. K., and AGARWAL, B. K. (1971): A relation between the chemical shifts of L_{III} absorption edges of Hg, Tl, Pb and Bi in binary compounds, *Lett. Nuovo Cimento*, **1**: 803–806.

DIRAC, P. A. M. (1947): *The principles of quantum mechanics*, 3d ed., Clarendon Press, Oxford.

DOBBYN, R. C., WILLIAMS, M. L., CUTHILL, J. R., and MCALISTER, A. J. (1970): Occupied band structure of Cu: Soft x-ray spectrum and comparison with other deep-band probes, *Phys. Rev.*, **B2**: 1563–1575.

DODD, C. G., and GLEN, G. L. (1968): Chemical bonding studies of silicates and oxides by x-ray K-emission spectroscopy, *J. Appl. Phys.*, **39**: 5377–5384.

——— and ——— (1970): Studies of chemical bonding in glasses by x-ray emission spectroscopy, *J. Am. Ceram. Soc.*, **53**: 322–325.

DONIACH, S., PLATZMAN, P. M., and YUE, J. T. (1971): X-ray Raman scattering in metals, *Phys. Rev.*, **B4**: 3345–3350.

EGGS, J., and ULMER, K. (1968): Zur Zustandsdichte der Übergangsmetalle. III. Die Legierungsreihe Rhodium-Palladium-Silber, *Z. Physik*, **213**: 293–315.

EHRENREICH, H., SEITZ, F., and TURNBULL, D. (eds.) (1970): *Solid state physics*, vol. 24, Academic Press, Inc., New York.

EKSTIG, B., KÄLLNE, E., NORELAND, E., and MANNE, R. (1970): Electron interaction in transition metal x-ray emission spectra, *Physica Scripta*, **2**: 38–44.

EPSTEIN, I. R. (1970): Molecular momentum distributions and Compton profiles. II. Localized orbital transferability and hydrocarbons, *J. Chem. Phys.*, **53**: 4425–4436.

ERN, V., and SWITENDICK, A. C. (1965): Electronic band structure of TiC, TiN, and TiO, *Phys. Rev.*, **A137**: 1927–1936.

ERSHOV, O. A. (1969): Reflection spectra and determination of optical constants in the ultrasoft x-ray region (in Russian), in *X-ray spectra and electronic structure of matter*, vol. 2, pp. 131–139, Institute of Metal Physics, Academy of Sciences of the Ukrainian SSR, Kiev.

FABIAN, D. J. (1970): Soft x-ray emission and electronic structure of alloys, *Mater. Res. Bull.*, **5**: 591–606.

——— and WATSON, L. M. (eds.) (1973): *Band structure spectroscopy of metals and alloys*, Academic Press, Inc., London.

FANO, U., and COOPER, J. W. (1968): Spectral distribution of atomic oscillator strengths, *Rev. Mod. Phys.*, **40**: 441–507.

FAUST, A. (1972): Berechnung der Röntgenspekteren von Lithium, Ph.D. thesis, Universitat München.

FESER, K., MÜLLER, J., FAESSLER, A., and WIECH, G. (1973): Studies of emission spectra in the soft x-ray range with fluorescence excitation using synchrotron radiation, in A. FAESSLER (ed.), *Proceedings of international symposium on x-ray spectra and electronic structure of matter*, München, 18–22 Sept. 1972.

FINK, R. W., MANSON, S. T., PALMS, J. M., and RAO, V. P. (eds.) (1973): *Proceedings of the international conference on inner shell ionization phenomena and future applications*, Atlanta, 17–22 April 1972, U.S. Atomic Energy Commission Conf.-720404, Oak Ridge.

FINSTER, J., LEONHARDT, G., and MEISEL, A. (1971): On the shape and width of the main lines of x-ray *K* emission of 3*d* and 4*d* elements, *J. Phys. (Paris)*, **Suppl. 10**: C4-218–C4-224.

FISCHER, D. W. (1973): Use of soft x-ray band spectra for determining valence conduction band structure in transition metal compounds. (See Fabian and Watson, 1973.)

—— and BAUN, W. L. (1967*a*): Self-absorption effects in the soft x-ray *Mα* and *Mβ* emission spectra of rare earth elements, *J. Appl. Phys.*, **38**: 4830–4836.

—— and —— (1967*b*): The effects of electronic structure and interatomic bonding on the soft x-ray Al *K* emission spectrum from aluminum binary alloys, *Advan. X-Ray Anal.*, **10**: 374–388.

—— and —— (1968): Band structure and the titanium $L_{2,3}$ x-ray emission and absorption spectra from pure metal, oxides, nitride, carbide and boride, *J. Appl. Phys.*, **39**: 4757–4776.

FONG, C. Y., and COHEN, M. L. (1969): Pseudopotential calculation of the optical constants of NaCl and KCl, *Phys. Rev.*, **185**: 1168–1176.

FRIEDEL, J. (1952): The distribution of electrons round impurities in monovalent metals, *Phil. Mag.*, **43**: 153–189.

FRIEDMAN, R. M., HUDIS, J., and PERLMAN, M. L. (1972): Chemical effects on line widths observed in photoelectron spectroscopy, *Phys. Rev. Lett.*, **29**: 692–695.

FROESE, C. (1963): Numerical solution of the Hartree-Fock equations, *Can. J. Phys.*, **41**: 1895–1910.

FUKAMACHI, T., and HOSOYA, S. (1970): Electron state in NaF studied by Compton scattering measurement, *J. Phys. Soc. Japan*, **29**: 736–745.

GALE, B., CATTERALL, J. A., and TROTTER, J. (1969): Soft x-ray $L_{2,3}$ emission edge-breadth in ordered and disordered Mg_3Cd, *Phil. Mag.*, **20**: 79–87.

GELIUS, U., ROOS, B., and SIEGBAHN, P. (1970): Ab initio MO SCF calculations of ESCA shifts in sulphur-containing molecules, *Chem. Phys. Lett.*, **4**: 471–475.

GIANTURCO, F. A. (1973): Inner-electron excitation processes in sulfur hexafluoride from ab initio MO-LCAO wavefunctions, in A. FAESSLER (ed.), *Proceedings of international symposium on x-ray spectra and electronic structure of matter*, München, 18–22 Sept. 1972.

—— and COULSON, C. A. (1968): Inner-electron binding energy and chemical bonding in sulphur, *Mol. Phys.*, **14**: 223–232.

GIGL, P. D., SAVANICK, G. A., and WHITE, E. W. (1970): Characterization of corrosion layers on aluminum by shifts in the aluminum and oxygen x-ray emission bands, *J. Electrochem. Soc.*, **117**: 15–17.

GILBERG, E. (1970): Das *K*-Röntgenemission-spektrum des chlors in freien Molekülen, *Z. Physik*, **236**: 21–44.

GLEN, G. L., and DODD, C. G. (1968): Use of molecular orbital theory to interpret x-ray *K*-absorption spectral data, *J. Appl. Phys.*, **39**: 5372–5377.

GOODINGS, D. A., and HARRIS, R. (1969): Calculations of x-ray emission bands of copper using augmented plane wave Bloch functions, *J. Phys. Chem*, **2:** 1808–1816.

GRAEFFE, G., SIIVOLA, J., UTRIAINEN, J., LINKOAHO, M., and ABERG, T. (1969): X-ray $K\alpha$ satellite spectra in primary and secondary excitation, *Phys. Lett.*, **A29:** 464–465.

GUDAT, W., and KUNZ, C. (1972): Close similarity between photoelectric yield and photo-absorption spectra in the soft x-ray range, *Phys. Rev. Lett.*, **29:** 169–172.

GYOFFRY, B. L., and STOTT, M. J. (1971): Soft x-ray emission from metals and alloys, *Solid State Comm.*, **9:** 613–617.

—— and —— (1973): One-electron theory of soft x-ray emission from alloys. (See Fabian and Watson, 1973.)

HAENSEL, R., KEITEL, G., KOSUCH, N., NIELSEN, U., and SCHREIBER, P. (1970): Optical absorption of solid neon and solid argon in the soft x-ray region, *Deut. Electron. Synchrotron* (Hamburg) *Rept.* 70/47.

HARRISON, W. A. (1966): *Pseudopotentials in the theory of metals*, W. A. Benjamin, Inc., New York.

—— (1968): Electronic structure and soft x-ray band spectra, in D. J. FABIAN (ed.), *Soft x-ray band spectra and the electronic structures of metals and materials*, pp. 227–248, Academic Press, Inc., London.

HARTMANN, H., PAPULA, L., and STREHL, W. (1971): Quantenmechanische Behandlung der $K\alpha$-Satellitenstruktur des Phosphors, *Theoret. Chim. Acta (Berlin)*, **21:** 261–266.

HAYDOCK, R., HEINE, V., and KELLY, M. J. (1972): Electronic structure based on the local atomic environment for tight-binding bands, *J. Phys. C.*, **5:** 2845–2858.

——, ——, ——, and PENDRY, J. B. (1972): Electronic density of states at transition metal surfaces, *Phys. Rev. Lett.*, **29:** 868–871.

HEINE, V., and WEAIRE, D. (1970): Pseudopotential theory of cohesion and structure, in H. EHRENREICH, F. SEITZ, and D. TURNBULL, *Solid state physics*, vol. 24, pp. 249–463, Academic Press, Inc., New York.

HEINLE, W., and FAESSLER, A. (1969): Experimental evidence against an x-ray satellite theory based upon the sudden approximation only, *Phys. Lett.*, **A28:** 783–784.

HERGLOTZ, H. (1953): Sitzungsber. *Österr. Ak. Wiss.*, **162:** 235.

HERMAN, F., and SKILLMAN, S. (1963): *Atomic structure calculations*, Prentice-Hall, Inc., Englewood Cliffs, N.J.

HOLLIDAY, J. (1971): The electronic properties of titanium interstitial and intermetallic compounds from soft x-ray spectroscopy, *J. Phys. Chem. Solids*, **32:** 1825–1834.

HOUSTON, J. E., and PARK, R. L. (1971): The effect of oxygen on the soft x-ray appearance potential spectrum of chromium, *J. Chem. Phys.*, **55:** 4601–4606.

—— and —— (1972a): The effect of oxygen on the soft x-ray appearance potential spectra of 3d transition metal surfaces. (See Shirley, 1972.)

—— and —— (1972b): Experimental evidence for strong plasmon coupling in the soft x-ray appearance potential spectrum of graphite, *Solid State Comm.*, **10:** 91–94.

HUBBEL, J. H. (1970): X-ray absorption 75 years later, *Phys. Bull.*, **21:** 353–357.

INKINEN, O., HALONEN, V., and MANNINEN, S. (1971): Compton profiles of hexane and decane, *Chem. Phys. Lett.*, **9:** 639–641.

IRKHIN, YU. P. (1961): The theory of the x-ray K absorption spectra in transition metals, *Phys. Metals Metallogr. (USSR) (English Transl.)*, **11**(1): 9–21.

JACOBS, R. L. (1969): The soft x-ray spectra of concentrated binary alloys, *Phys. Lett.*, **A30**: 523–524.

JOHANNSON, P. (1960a): An experimental investigation of x-ray excitation states in solids, *Arkiv Fysik*, **18**: 289–303.

—— (1960b): On the structure near the short wavelength limit of the continuous x-ray spectrum, *Arkiv Fysik*, **18**: 329–338.

JOHNSON, K. H., and SMITH, F. C. (1972): Chemical bonding of a molecular transition-metal ion in a crystalline environment, *Phys. Rev.*, **5B**: 831–843.

KÄLLNE, E. (1973): X-ray emission measurement on TiNi, TiCo, and TiFe, in A. FAESSLER (ed.), *Proceedings of international symposium on x-ray spectra and electronic structure of matter*, München, 18–22 Sept. 1972.

KARALNIK, S. M. (1956): X-ray spectra and interatomic bonds in alloys, *Bull. Acad. Sci. USSR, Phys. Ser. (English Transl.)*, **20**: 739–742.

—— (1957): External screening and the fine structure of x-ray spectra, *Bull. Acad. Sci. USSR, Phys. Ser. (English Transl.)*, **21**: 1432–1439.

KEISER, J. (1973): A new transition process concerning the generation of Bremsstrahlung in solids. (See Fabian and Watson, 1973.)

KESSEL, Q. C. (1969): Coincidence measurements, in MCDANIEL, E. W., and MCDOWELL, M. R. C. (eds.), *Case studies in atomic collision physics I*, pp. 401–462, North-Holland Publishing Company, Amsterdam.

KIRICHOK, P. P., and KARALNIK, S. M. (1967): X-ray spectroscopic study of spinel and garnet type ferrites of different composition and in different states, *Bull. Acad. Sci. USSR, Phys. Ser. (English Transl.)*, **31**: 1043–1046.

KLASSON, M., and MANNE, R. (1972): Molecular orbital interpretation of x-ray emission and photoelectron spectra. III. Chloroethylenes. (See Shirley, 1972, pp. 471–486.)

KLIMA, J. (1970): Calculation of the soft x-ray emission spectra of silicon and germanium, *J. Phys. Chem.*, **3**: 70–85.

KNUDSON, A. R., BURKHALTER, P. G., and NAGEL, D. J. (1973): Multiple inner shell vacancy production by light ion bombardment. (See Fink et al., 1973, pp. 1675–1681.)

——, NAGEL, D. J., BURKHALTER, P. G., and DUNNING, K. L. (1971): Aluminum x-ray satellite enhancement by ion-impact excitation, *Phys. Rev. Lett.*, **26**: 1149–1152.

KORSUNSKII, M. I., and GENKIN, YA. E. (1964): X-ray emission bands and the magnetic properties of niobium, *Bull. Acad. Sci. USSR, Phys. Ser. (English Transl.)*, **27**: 740–744.

KORTELA, E. K., and MANNE, R. (1973): Interpretation of x-ray emission band spectra of solids by semi-empirical energy band calculations, in A. FAESSLER (ed.), *Proceedings of international symposium on x-ray spectra and electronic structure of matter*, München, 18–22 Sept. 1972.

KOSTER, A. S., and RIECK, G. D. (1970): Determination of valence and coordination of iron in oxidic compounds by means of the iron x-ray fluorescence emission spectrum, *J. Phys. Chem. Solids*, **31**: 2505–2510.

KOUMELIS, C., LEVENTOURI, D., and ALEXOPOULOS, K. (1971): Plasmon excitation in colloidal graphite produced by x-rays, *Phys. Stat. Sol.*, **46**: K89–K91.

KRAUSE, H. B., SAVANICK, G. A., and WHITE, E. W. (1970): Oxygen x-ray emission band shifts applied to characterization of transition metal oxide surface layers, *J. Electrochem. Soc.*, **117**: 557–558.

KRAUSE, M. O. (1971): Rearrangement of inner shell ionized atoms, *J. Phys. (Paris)*, **Suppl. 10**: C4-67–C4-75.

———, WUILLEUMIER, F., and NESTOR, C. W. (1972): Interpretation of the L x-ray emission spectrum of Zr, *Phys. Rev.*, **6A**: 871–879.

KUHN, H. G. (1962): *Atomic spectra*, Academic Press, Inc., New York.

KUNZL, V. (1936): Über die K-Serie von Al, Mg und Na, *Z. Physik*, **99**: 481–491.

KURIYAMA, M., and ALEXANDROPOULOS, N. G. (1971): On the relationship between x-ray inelastic scattering and absorption spectra, *J. Phys. Soc. Japan*, **31**: 561–562.

KURMAEV, E. Z., and NEMNONOV, S. A. (1971): Density-of-states curve for V_3Si built up from experimental x-ray data, *Phys. Stat. Sol.*, **43**: K49–K53.

LÄUGER, K. (1968): Ph.D. thesis, Ludwig-Maximilians-Universität, Munich.

——— (1971): Über den Einfluss der Bindungsart under der Kristallstruktur auf das $K\alpha_{1,2}$-Röntgenspektrum von Aluminum, *J. Phys. Chem. Solids*, **32**: 609–622.

LAVILLA, R. E. (1972a): The $K\alpha$ emission spectrum of gaseous N_2, *J. Chem. Phys.*, **56**: 2345–2349.

——— (1972b): The sulfur K and L and fluorine K x-ray emission and absorption spectra of gaseous SF_6, *J. Chem. Phys.*, **57**: 899–909.

LEONHARDT, G., and MEISEL, A. (1970): Determination of effective atomic charges from the chemical shifts of x-ray emission lines, *J. Chem. Phys.*, **52**: 6189–6198.

LEVIN, K., LIEBSCH, A., and BENNMANN, K. H. (1973): Simple model for the electronic density of states near transition-metal surfaces: Application to ferromagnetic Ni, *Phys. Rev. B*, **B7**: 3066–3073.

LIEFELD, R. J. (1968): Soft x-ray emission spectra at threshold excitation, in D. J. FABIAN (ed.), *Soft x-ray band spectra and the electronic structures of metals and materials*, pp. 133–149, Academic Press, Inc., London.

LINDGREN, I. (1965): A note on the Hartree-Fock-Slater approximation, *Phys. Lett.*, **19**: 382–383.

——— (1967): Comparison between theoretical and experimental binding energies, in *Electron spectroscopy for chemical analysis*, Appendix 2, Almqvist & Wiksell, Stockholm.

LÖFGREN, H., and WALLDEN, L. (1973): High energy satellites in Auger spectra from metals, *Sol. St. Comm.*, **12**: 19–21.

MANDÉ, C., and JOSHI, N. V. (1969): X-ray spectroscopic study of plasmon energy in transition metals, in *X-ray spectra and electronic structure of matter*, vol. 1, pp. 57–64, Institute of Metal Physics, Academy of Sciences of the Ukrainian SSR, Kiev.

MARCUS, P. M., JANAK, J. F., and WILLIAMS, A. R. (eds.) (1971): *Computational methods in band theory*, vol. 1, Plenum Press, New York.

MARSHALL, C. A. W., WATSON, L. M., LINDSAY, G. M., ROOKE, G. A., and FABIAN, D. J. (1969): Interpretation of soft x-ray emission spectra of aluminum-silver alloys, *Phys. Lett.*, **A28**: 579–580.

MATTHEISS, L. F. (1964): Energy bands for solid argon, *Phys. Rev.*, **A133**: 1399–1403.

MATTHEISS, L. F. (1970): Electronic structure of niobium and tantalum, *Phys. Rev.*, **B1**: 373–380.

MATTHEW, J. A. D., and WATTS, C. M. K. (1971): The mechanism of plasmon gain in Auger spectra, *Phys. Lett.*, **A37**: 239–240.

MATTSON, R. A., and EHLERT, R. C. (1968): Carbon characteristic x-rays from gaseous compounds, *J. Chem. Phys.*, **48**: 5465–5470.

MCALISTER, A. J. (1969): Calculation of the soft x-ray *K*-emission and absorption spectra of metallic Li, *Phys. Rev.*, **186**: 595–599.

MCCAFFREY, J. W., NAGEL, D. J., and PAPACONSTANTOPOULOS, D. A. (1974): Calculated x-ray absorption structure of nickel (to be published).

MEISEL, A., and NEFEDOW, W. (1962): Einfluss der chemischen Bindung auf Form und Breite von Röntgenemissionslinien, *Z. Physik. Chem.* (*Leipzig*), **219**: 194–204.

MENSHIKOV, A. Z., and NEMNONOV, S. A. (1963): Influence of the chemical bonds on the valence state of chromium in different compounds, *Bull. Acad. Sci. USSR, Phys. Ser.* (*English Transl.*), **27**: 402–410.

———, ———, and MISCHENKO, L. B. (1962): Effect of the Chemical bond on the energy levels L_2 and L_3 of the Chromium atom, *Phys. Metals Metallogr.* (*USSR*) (*English Transl.*), **14** (3): 54–56.

MERZ, H. (1973): On temperature dependence of Bremsstrahlung isochromats. (See Fabian and Watson, 1973.)

——— and ULMER, K. (1971): Isochromat spectroscopic investigation of the hcp-bcc phase transformation in Ti, Zr and Hf, *J. Phys.* (*Paris*), **Suppl. 10**: C4-334–C4-337.

MIZUNO, Y., and OHMURA, Y. (1967): Theory of x-ray Raman scattering, *J. Phys. Soc. Japan*, **22**: 445–449.

MOHR, C. B. O. (1968): Relativistic inner shell ionization, in D. R. BATES and I. ESTERMAN (eds.), *Advances in atomic and molecular physics*, **4**: 221–236, Academic Press, Inc., New York.

MULLIKEN, R. S. (1962): Criteria for construction of good self-consistent-field molecular orbital wave functions, and the significance of LCAO-MO population analysis, *J. Chem. Phys.*, **36**: 3428–3439.

NAGEL, D. J. (1970): Interpretation of valence band x-ray spectra, *Advan. X-Ray Anal.*, **13**: 182–236.

——— (1973): X-ray probes of vacant energy bands. (See Fabian and Watson, 1973.)

——— and CRISS, J. W. (1973): Calculated appearance potential spectra, *Sol. St. Comm.* (to be published).

———, PAPACONSTANTOPOULOS, D. A., MCCAFFREY, J. W., and CRISS, J. W. (1973): Calculated x-ray band spectra, in A. FAESSLER (ed.), *Proceedings of international symposium on x-ray spectra and electronic structure of matter*, München, 18–22 Sept. 1972.

NARBUTT, K. I. (1955): On the structure of x-ray emission lines of ions in solution, *Bull. Acad. Sci. USSR, Phys. Ser.* (*English Transl.*), **20**: 107–110.

NEFEDOW, W. I. [NEFEDOV, V. I.] (1962): Bestimmung von Atomladungen in Molekülen mit Hilfe der Röntgenemissionsspektren, *Phys. Stat. Sol.*, **2**: 904–922.

NEFEDOV, V. I. (1964): Determination of the effective charge on an atom in a molecule from shifts of the lines of the x-ray spectrum, *J. Struct. Chem.* (*USSR*) (*English Transl.*), **5**: 605–607.

NEFEDOV, V. I. (1966*a*): Effective charges of the atoms in compounds according to x-ray emission spectra and the Mössbauer effect, *J. Struct. Chem. (USSR) (English Transl.)*, **7**: 518–523.

———— (1966*b*): Multiplet structure of the $K\alpha_{1,2}$ and $K\beta$-β' lines in the x-ray spectra of iron compounds, *J. Struct. Chem. (USSR) (English Transl.)*, **7**: 672–677.

———— (1967): Electronic structure of molecules according to x-ray data, *J. Struct. Chem. (USSR) (English Transl.)*, **8**: 919–923.

———— (1970*a*): Quasistationary states in x-ray absorption spectra of chemical compounds, *J. Struct. Chem. (USSR) (English Transl.)*, **11**: 272–276.

———— (1970*b*): X-ray absorption spectra of ionic crystals, *J. Struct. Chem. (USSR) (English Transl.)*, **11**: 277–282.

NEMNONOV, S. A., and KOLOBOVA, K. M. (1958): Connexion between certain x-ray and magnetic characteristics of iron base alloys, *Phys. Metals Metallog. (USSR) (English Transl.)*, (3)**6**: 82–88.

NIGAVEKAR, A. S., and BERGWALL, S. (1969): X-ray investigation of the $K\alpha$ doublet of chromium and its halides, *J. Phys.*, **B2**: 507–513.

NIKIFOROV, I. YA. (1960): Optimum correction of x-ray spectra for dispersive distortion, *Bull. Acad. Sci. USSR, Phys. Ser. (English Transl.)*, **24**: 394–397.

———— (1961): The shape of conduction bands of iron, *Phys. Metals Metallogr. (English Transl.)*, **11**: 110–106.

———— and BLOKHIN, M. A. (1963): Concerning the shape of the $K\beta_5$ emission band of iron. II. Transition probability as a function of energy, *Bull. Acad. Sci. USSR, Phys. Ser. (English Transl.)*, **27**: 323–327.

OVSYANNIKOVA, I. A., and NASANOVA, L. I. (1970): L_3 x-ray absorption spectra of praesodymium and cerium and effective coordination charges, *J. Struct. Chem. (USSR) (English Transl.)*, **11**: 505–506.

PAPACONSTANTOPOULOS, D. A., and NAGEL, D. J. (1971): Band structure and Fermi surfaces of ordered intermetallic compounds TiFe, TiCo and TiNi, *Intern. J. Quantum Chem.*, **5S**: 515–526.

PARRATT, L. G. (1936): $K\alpha$ satellite lines, *Phys. Rev.*, **50**: 1–15.

———— (1939): X-ray resonance absorption lines in the argon K spectrum, *Phys. Rev.*, **56**: 295–297.

———— (1959): Electronic band structure of solids by x-ray spectroscopy, *Rev. Mod. Phys.*, **31**: 616–645.

PAULING, L. (1960): *The nature of the chemical bond*, Cornell University Press, Ithaca, N.Y.

PEASE, D. M. (1973): Experimental test of theories of extended x-ray absorption edge fine structure in metals, *Phys. Rev.* **B7**: 3568–3572.

PETTIFOR, D. (1972): Transition metal wave functions: A first principles resonant TB approach, in *Proc. Conf. Calculation Electron. Struct. Ordered Disordered Solids* (to be published).

PHILLIPS, J. C. (1970): Ionicity of the chemical bond in crystals, *Rev. Mod. Phys.*, **42**: 317–356.

PHILLIPS, W. C., and WEISS, R. J. (1968): X-ray determination of electron momenta in Li, Be, B, Na, Mg, Al and LiF, *Phys. Rev.*, **171**: 790–800.

POPLE, J. A. (1965): Two-dimensional chart of quantum chemistry, *J. Chem. Phys. Suppl.*, **43**: S229.

PRIEST, J. F. (1971): Observation of fine structure in iron and chromium $K\alpha_{1,2}$ lines using a high-resolution three-crystal spectrometer, *J. Appl. Phys.*, **42**: 4750–4751.

PRIFTIS, G., THEODOSSIOU, A., and ALEXOPOULOS, K. (1968): Plasmon observation in x-ray scattering, *Phys. Lett.*, **A27**: 577–579.

RAMQVIST, L. (1969): Preparation, properties and electronic structure of refractory carbides and related compounds, *Jernkontorets Ann.*, **153**: 159–179.

——— (1971): Electronic structure of cubic refractory carbides, *J. Appl. Phys.*, **42**: 2113–2120.

———, EKSTIG, B., KÄLLNE, E., NORELAND, E., and MANNE, R. (1969): X-ray study of inner level shifts and band structure in TiC and related compounds, *J. Phys. Chem. Solids*, **30**: 1849–1860.

ROOKE, G. A. (1963): Plasmon satellites of soft x-ray emission spectra, *Phys. Lett.*, **3**: 234–236.

——— (1968a): Interpretation of aluminum x-ray band spectra. I. Intensity distribution, *J. Phys. Chem.*, **1**: 767–775.

——— (1968b): Interpretation of aluminum x-ray band spectra. II. Determination of effective potentials from $L_{2,3}$ emission spectra, *J. Phys. Chem.*, **1**: 776–783.

——— (1969): The interpretation of x-ray band spectra from disordered alloys of light metals, in *X-ray spectra and electronic structure of matter*, vol. 2, pp. 64–71, Institute of Metal Physics, Academy of Sciences of the Ukrainian SSR, Kiev.

RUFFA, A. R. (1972): A theoretical model for x-ray emission and absorption, bond breaking, and point defect production in SiO_2, *J. Appl. Phys.*, **43**: 4263–4265.

SACHENKO, V. P., and BURTSEV, E. V. (1967): Probability for multiple ionization of atoms under photon excitation, *Bull. Acad. Sci. USSR, Phys. Ser. (English Transl.)*, **31**: 980–984.

SAWADA, M., TANIGUCHI, K., and NAKAMURA, H. (1969): Relation between the x-ray $K\alpha$ non-diagram lines and their electronic conductivities, in *X-ray spectra and electronic structure of matter*, vol. 2, pp. 122–130, Institute of Metal Physics, Academy of Sciences of the Ukrainian SSR, Kiev.

SCHWARTZ, L., BROUERS, F., VEDYAYEV, A. V., and EHRENREICH, H. (1971): Comparison of the average-t-matrix and coherent-potential approximations in substitutional alloys, *Phys. Rev.*, **B4**: 3383–3392.

SCHWARTZ, M. E. (1970): Correlation of 1s binding energy with the average quantum mechanical potential at a nucleus, *Chem. Phys. Lett.*, **6**: 631–636.

———, SWITALSKI, J. D., and STRONSKI, R. E. (1972): Core-level binding energy shifts from molecular orbital theory. (See Shirley, 1972, pp. 605–627.)

SCOFIELD, J. H. (1969): Radiative decay rates of vacancies in K and L shells, *Phys. Rev.*, **179**: 9–16.

SHAW, R. W., and THOMAS, T. D. (1972): Chemical effects on the lifetime of 1s-hole states, *Phys. Rev. Lett.*, **29**: 689–692.

SHIRLEY, D. A. (ed.) (1972): *Electron spectroscopy*, North-Holland Publishing Company, Amsterdam.

SHIRODA, G., KOHRA, K., and ICHINOKAWA, T. (eds.) (1972): *Proceedings of the sixth international conference on x-ray optics and microanalysis*, University of Tokyo Press, Tokyo.

SHUBAEV, A. T. (1959): X-ray investigation of compounds with the perovskite structure, *Bull. Acad. Sci. USSR, Phys. Ser.* (*English Transl.*), **23**: 551–554.

—— (1960): Concerning interpretation of x-ray spectra, *Bull. Acad. Sci. USSR, Phys. Ser.* (*English Transl.*), **24**: 434–437.

—— (1961): Influence of the chemical bond on the energy and intensity of the x-ray lines of atoms in compounds, *Bull. Acad. Sci. USSR, Phys. Ser.* (*English Transl.*), **25**: 998–1001.

—— (1963): Determination of the charge of ions in compounds of period 2 elements from x-ray emission spectra, *Bull. Acad. Sci. USSR, Phys. Ser.* (*English Transl.*), **27**: 667–673.

—— (1966): Röntgenspektroskopie Prüfung der Grundlagen der Electronegativitäts-theorie, in *Röntgenspektren und Chemische Bindung*, pp. 325–332, Physikalisch-chemisches Institut, der Karl-Marx Universität, Leipzig.

—— and CHECHIN, G. M. (1964): Interpretation of the shifts of the K series lines of the transition elements; wave functions for three atomic configurations of titanium, *Bull. Acad. Sci. USSR, Phys. Ser.* (*English Transl.*), **38**: 838–842.

——, LANDYSHEV, A. V., BELOUSOV, A. K., and BAIDA, G. S. (1967): Chemical shifts of $K\alpha_{1,2}$ lines of sulphur in some compunds, *Bull. Acad. Sci. USSR, Phys. Ser.* (*English Transl.*), **31**: 911–913.

SIEGBAHN, K., NORDLING, C., FAHLMAN, A., NORDBERG, R., HAMRIN, K., HEDMAN, J., JOHANSSON, G., BERGMARK, T., KARLSSON, S. E., LINDGREN, I., and LINDBERG, B. (1967): *ESCA atomic, molecular and solid state structure studied by means of electron spectroscopy*, Almqvist & Wiksell, Stockholm.

——, ——, JOHANSSON, G., HEDMAN, J., HEDÉN, P. F., HAMRIN, K., GELIUS, U., BERGMARK, T., WERME, L. O., MANNE, R., and BAER, Y. (1969): *ESCA applied to free molecules*, North-Holland Publishing Company, Amsterdam.

SHIRLEY, D. A. (ed.) (1972): *Electron spectroscopy*, North-Holland Publishing Company, Amsterdam, 895–901.

SLATER, J. C. (1960): *Quantum theory of atomic structure*, vols. 1 and 2, McGraw-Hill Book Company, New York.

—— (1963): *Quantum theory of molecules and solids*, vol. 1, McGraw-Hill Book Company, New York.

—— (1965): *Quantum theory of molecules and solids*, vol. 2, McGraw-Hill Book Company, New York.

SMRCKA, L. (1971): Calculation of soft x-ray emission spectra of aluminum by APW method, *Czech. J. Phys.*, **B21**: 683–692.

SOMMER, G., VOLKOV, V. F., BLOKHIN, M. A., and NIKIFOROV, I. YA. (1970): Calculation of the shape of the L_{III} and M_{III} x-ray bands of chromium, *Phys. Met. Metallogr.* (English transl.), **30**: 233–237.

STERN, E. A., and SAYERS, D. E. (1973): Shielding of impurities as measured by extended x-ray-absorption fine structure, *Phys. Rev. Lett.*, **30**: 174–177.

STEWART, A. T., and ROELLIG, L. O. (eds.) (1967): *Positron annihilation*, Academic Press, Inc., New York.

STOCKS, G. M., WILLIAMS, R. W., and FAULKNER, J. S. (1971): Densities of states in paramagnetic Cu-Ni alloys, *Phys. Rev.*, **B4**: 4390–4405.

SUGAR, J. (1972): Interpretation of photoabsorption in the vicinity of the $3d$ edges of La, Er, and Tm, *Phys. Rev.*, **6A**: 1764–1767.

SUMBAEV, O. I. (1970): The effect of the chemical shift of the x-ray $K\alpha$ lines in heavy atoms; systematization of experimental data and comparison with theory, *Soviet Phys. JETP (English Transl.)*, **30**: 927–933.

SUREAU, A. (1971): Configuration interaction in atoms having several open shells; identification of the $K\alpha$ satellites of light elements, *J. Phys. (Paris)*, **Suppl. 10**: C4-105–C4-114.

SUZUKI, T. (1967): X-ray Raman scattering: Experiment I, *J. Phys. Soc. Japan*, **22**: 1139–1150.

TANOKURA, A., HIROTA, N., and SUZUKI, T. (1969): X-ray plasmon scattering, *J. Phys. Soc. Japan*, **27**: 515.

THORPE, M. F., and WEAIRE, D. (1971a): Electronic properties of an amorphous solid. II. Further aspects of the theory, *Phys. Rev.*, **B4**: 3518–3527.

—— and —— (1971b): Electronic density of states of amorphous Si and Ge, *Phys. Rev. Lett.*, **27**: 1581–1584.

TSUTSUMI, K., and NAKAMORI, H. (1968): X-ray K emission spectra of chromium in various chromium compounds, *J. Phys. Soc. Japan*, **25**: 1418–1432.

ULMER, K. (1973): Isochromat spectra of alloys. (See Fabian and Watson, 1973.)

URCH, D. (1970): The origin and intensities of low energy satellite lines in x-ray emission spectra: A molecular orbital interpretation, *J. Phys. Chem.*, **3**: 1275–1291.

VAINSHTEIN, E. E., and KOTLIAR, B. I. (1956): X-ray emission spectra of Mn and Cu in a Heusler alloy at temperatures around the Curie point, *Soviet Phys. "Doklady" (English Transl.)*, **1**: 527–529.

VARLEY, J. H. O. (1956): Effect of alloying on the energy levels of inner electronic shells, *Nature*, **178**: 939–940.

VEIGELE, W. J., STEVENSON, D. E., and HENRY, E. M. (1969): Screening constants and transition energies, *J. Chem. Phys.*, **50**: 5404–5407.

WALTER, J. P., and COHEN, M. L. (1971): Pseudopotential calculations of electronic charge densities in seven semiconductors, *Phys. Rev.*, **B4**: 1877–1892.

WATSON, R. E., HUDIS, J., and PERLMAN, M. L. (1971): Charge flow and d compensation in gold alloys, *Phys. Rev.*, **B4**: 4139–4144.

WEAIRE, D., and THORPE, M. F. (1971): Electronic properties of an amorphous solid. I. A simple tight-binding theory, *Phys. Rev.*, **B4**: 2508–2520.

WEISS, R. J. (1966): *X-ray determination of electron distributions*, North-Holland Publishing Company, Amsterdam.

WENGER, A., BÜRRI, G., and STEINEMANN, S. (1971): Charge transfer in alloys Mn-, Fe-, Co-, Ni-, Cu-Al measured by soft x-ray spectroscopy, *Solid State Comm.*, **9**: 1125–1128.

WHITE, E. W., and GIBBS, G. V. (1967): Structural and chemical effects on the Si $K\beta$ x-ray line for silicates, *Am. Mineralogist*, **52**: 985–993.

WHITE, E. W., and GIBBS, G. V. (1969): Structural and chemical effects on the Al $K\beta$ x-ray emission band among aluminum-containing silicates and aluminum oxides, *Am. Mineralogist*, **54**: 931–936.

———— and ROY, R. (1964): Silicon valence in SiO films studied by x-ray emission, *Solid State Comm.*, **2**: 151–152.

WIECH, G. (1968): Soft x-ray emission spectra and the valence-band structure of beryllium, aluminum, silicon and some silicon compounds, in D. J. FABIAN (ed.), *Soft x-ray band spectra and the electronic structures of metals and materials*, pp. 59–70, Academic Press, Inc., London.

———— (1973): X-ray emission bands of amorphous silicon. (See Fabian and Watson, 1973.)

———— and ZÖPF, E. (1973): Electronic properties of aluminum and silicon intermetallic compounds from x-ray spectroscopy. (See Fabian and Watson, 1973.)

WILLENS, R. H., SCHREIBER, H., BUEHLER, E., and BRASEN, D. (1969): Piezo soft-x-ray effect, *Phys. Rev. Lett.*, **23**: 413–416.

ZIMKINA, T. M., and GRIBOVSKII, S. A. (1971): The features of photoionization absorption in the ultrasoft x-ray region, *J. Phys. (Paris)*, **Suppl. 10**: C4-282–C4-289.

APPENDIX A

X-RAY WAVELENGTH TABLE

The selected wavelength values below are taken from the Bearden[1,2] table. In a comprehensive study during the period 1962–1964 Bearden and coworkers made a survey of virtually all previously published work, including some 2700 emission lines and absorption edges. As far as possible, necessary corrections were applied to put all the data on a basis consistent with the primary and secondary standards described in Section VII of Chap. 2. Where two or more measurements of comparable accuracy were available, a weighted average was usually employed. The results were expressed in terms of A* units (see Subsection VII C of Chap. 2); a probable error was given for each wavelength. It should be noted that *references to the original data, the type of corrections applied (if any), and other details are given only in the full report.*[1]

The table below gives values for the $K\alpha_1$, $K\beta_1$, $L\alpha_1$, $L\beta_1$, and $M\alpha_1$ lines; it excludes most cases where the upper electronic level occurs in the valence or conduction bands. The $K\alpha_1$ line is, of course, the most intense line of the K series. The $K\alpha_2$, the other component of this doublet, is second in intensity. It is not listed here; however, except for elements lighter than 26 Fe, it is about 0.004 to 0.005 A* above the $K\alpha_1$ value. The $K\beta_1$ ranks third in intensity within the K series. The $L\alpha_1$ and $L\beta_1$ lines represent, respectively, the first and second most intense lines of the L series, while the $M\alpha_1$ is the strongest line of the M series.

A thorough recheck of the measurements of the secondary standards (Subsection VII D of Chap. 2) is currently in progress. It now appears that one of the crystals used in the original work contained significant imperfections and that some of the wavelength ratios may have to be revised by one or two probable errors.

Furthermore at the time the table was first published the best values for the conversion factor Λ^* between A* units and angstroms was unity (although Λ^* has always been an

---

[1] J. A. Bearden (1964): *X-ray wavelengths*, NYO-10586, Clearinghouse for Federal Scientific and Technical Information, U.S. Dept. of Commerce, Springfield, Va.

[2] J. A. Bearden (1967): *Rev. Mod. Phys.*, **39**: 78.

experimental quantity involving an appropriate probable error). Hence at that time the numerical wavelength values were the same on either basis; only the probable errors differed. As indicated in Subsection VIII F of Chap. 2, the current value for Λ^* differs significantly from unity so that the two units must be carefully distinguished in precision work.

In the present tabulation the Bearden values are quoted without revision; these should be considered as on the A* scale. Accuracies are generally indicated by the number of significant figures.

In the table which follows, values which represent unresolved doublets ($K\alpha_{1,2}$, $K\beta_{1,2}$, $L\alpha_{1,2}$, or $M\alpha_{1,2}$) are indicated by a **D**. Interpolated values, which were given for a few cases where direct measurements were considered inaccurate or were nonexistent, are denoted by an asterisk. The wavelength of the W $K\alpha_1$ line is fixed by the definition of the A* unit; this value is indicated by an **S**.

Table A-1   WAVELENGTH (A*)

Atomic number and element	$K\alpha_1$ (KL_{III})	$K\beta_1$ (KM_{III})	$L\alpha_1$ $(L_{III}M_V)$	$L\beta_1$ $(L_{II}M_{IV})$	$M\alpha_1$ $(M_V N_{VII})$
11. Na	11.9101 **D**				
12. Mg	9.8900 **D**				
13. Al	8.33934				
14. Si	7.12542				
15. P	6.157*				
16. S	5.37216				
17. Cl	4.7278				
18. A	4.19180				
19. K	3.7414	3.4539 **D**			
20. Ca	3.35839	3.0897 **D**			
21. Sc	3.0309*	2.7796 **D**			
22. Ti	2.74851	2.51391 **D**			
23. V	2.50356	2.28440 **D**			
24. Cr	2.28970	2.08487 **D**			
25. Mn	2.101820	1.91021 **D**			
26. Fe	1.936042	1.75661 **D**			
27. Co	1.788965	1.62079 **D**			
28. Ni	1.657910	1.500135 **D**			
29. Cu	1.540562	1.392218 **D**	13.336 **D**	13.053	
30. Zn	1.435155	1.29525 **D**	12.254 **D**	11.983	
31. Ga	1.340083	1.20789	11.292 **D**	11.023	
32. Ge	1.254054	1.12894	10.4361 **D**	10.175	
33. As	1.17588	1.05730	9.6709 **D**	9.4141	
34. Se	1.10477	0.99218	8.9900 **D**	8.7358	
35. Br	1.03974	0.93279	8.3746 **D**	8.1251	
36. Kr	0.9801	0.8785	7.817 **D***	7.576*	
37. Rb	0.925553	0.82868	7.3183	7.0759	
38. Sr	0.87526	0.78292	6.8628	6.6239	
39. Y	0.82884	0.74072	6.4488	6.2120	
40. Zr	0.78593	0.70173	6.0705	5.8360	
41. Nb	0.74620	0.66576	5.7243	5.4923	
42. Mo	0.709300	0.632288	5.40655	5.17708	
43. Tc	0.67502*	0.60130*	5.1148*	4.8873*	
44. Ru	0.643083	0.572482	4.84575	4.62058	

Table A-1   WAVELENGTH (A*)   (continued)

Atomic number and element	$K\alpha_1$ (KL_{III})	$K\beta_1$ (KM_{III})	$L\alpha_1$ $(L_{III}M_V)$	$L\beta_1$ $(L_{II}M_{IV})$	$M\alpha_1$ $(M_V N_{VII})$
45. Rh	0.613279	0.545605	4.59743	4.37414	
46. Pd	0.585448	0.520520	4.36767	4.14622	
47. Ag	0.5594075	0.497069	4.15443	3.93473	
48. Cd	0.535010	0.475105	3.95635	3.73823	
49. In	0.512113	0.454545	3.77192	3.55531	
50. Sn	0.490599	0.435236	3.59994	3.38487	
51. Sb	0.470354	0.417085	3.43941	3.22567	
52. Te	0.451295	0.399995	3.28920	3.07677	
53. I	0.433318	0.383905	3.14860	2.93744	
54. Xe	0.41634*	0.36872*	3.0166*		
55. Cs	0.400290	0.354364	2.8924	2.6837	
56. Ba	0.385111	0.340811	2.77595	2.56821	
57. La	0.370737	0.327983	2.66570	2.45891	
58. Ce	0.357092	0.315816	2.5615	2.3561	
59. Pr	0.344140	0.304261	2.4630	2.2588	
60. Nd	0.331846	0.293299	2.3704	2.1669	
61. Pm	0.320160	0.28290*	2.2822	2.0797	
62. Sm	0.309040	0.27301	2.1998	1.99806	
63. Eu	0.298446	0.263577	2.1209	1.9203	
64. Gd	0.288353	0.25460	2.0468	1.8468	
65. Tb	0.278724	0.24608	1.9765	1.7768	
66. Dy	0.269533	0.23788	1.90881	1.71062	
67. Ho	0.260756	0.23012	1.8450	1.6475	
68. Er	0.252365	0.22266	1.78425	1.5873	
69. Tm	0.244338	0.21556	1.7268*	1.5304	
70. Yb	0.236655	0.20884	1.67189	1.47565	8.149 D
71. Lu	0.229298	0.20231*	1.61951	1.42359	7.840 D
72. Hf	0.222227	0.19607*	1.56958	1.37410	7.539 D
73. Ta	0.215497	0.190089	1.52197	1.32698	7.252 D
74. W	0.2090100 S	0.184374	1.47639	1.281809	6.983
75. Re	0.202781	0.178880	1.43290	1.23858	6.729 D
76. Os	0.196794	0.173611	1.39121	1.19727	6.490 D
77. Ir	0.191047	0.168542	1.35128	1.15781	6.262
78. Pt	0.185511	0.163675	1.31304	1.11990	6.047
79. Au	0.180195	0.158982	1.27640	1.08353	5.840
80. Hg	0.175068	0.154487	1.24120	1.04868	5.6476*
81. Tl	0.170136	0.150142	1.20739	1.01513	5.460
82. Pb	0.165376	0.145970	1.17501	0.98291	5.286
83. Bi	0.160789	0.141948	1.14386	0.951978	5.118
84. Po	0.15636*	0.13807*	1.11386	0.9220	
85. At	0.15210*	0.13432*	1.08500*	0.89349*	
86. Rn	0.14798*	0.13069*	1.05723*	0.86605*	
87. Fr	0.14399*	0.12719*	1.03049	0.83940*	
88. Ra	0.14014*	0.12382*	1.00473	0.81375	
89. Ac	0.136417*	0.12055*	0.97993*	0.78903*	
90. Th	0.132813	0.117396	0.95600	0.765210	4.1381
91. Pa	0.129325*	0.114345*	0.93284	0.74232	4.022
92. U	0.125947	0.111394	0.910639	0.719984	3.910
93. Np			0.889128	0.698478	
94. Pu			0.86830	0.67772	
95. Am			0.848187	0.657655	

*Interpolated values.

APPENDIX B

X-RAY ENERGY LEVELS

The table which follows gives a selected set of x-ray energy levels, based on the analysis of Bearden and Burr.[1,2] Values (in eV) are listed for the K, L_I, and M_I levels. These figures indicate the order of magnitude for other levels in the same shells and also represent the excitation voltages needed to produce the respective series. Only normally filled electronic shells are included in this tabulation.

The principle of the Bearden and Burr least-squares analysis is briefly outlined in Section IX of Chap. 2. As indicated there, the results rest chiefly on two sources: (1) emission line data as compiled by Bearden (see Appendix A), and (2) photoelectron measurements of K. Siegbahn and collaborators. *Full details of the analysis are found only in the original report*;[1] these include energy-level differences and adjusted wavelength values for all possible transitions, together with probable errors for the various output data.

Expressing results in eV requires an appropriate conversion factor, which is given by equation (2-216) in Subsection IX A of Chap. 2 as $(1.239\,837 \times 10^4 \text{ V} \cdot \text{A}^*) \pm 6$ ppm. This is 22 ppm higher than the value employed by Bearden and Burr. Hence the values below have been raised by 22 ppm above the original figures. This procedure is rigorously correct insofar as the energies depend on the emission lines and a reasonable approximation insofar as they are derived from photoelectron measurements. Except for this multiplication of the Bearden and Burr results by a constant factor, no other revision has been made.

No probable errors are given here. However, the accuracy of the data may be inferred from the number of significant figures quoted.

[1] J. A. Bearden and A. F. Burr (1965): *Atomic energy levels*, NYO-2543-1, Clearinghouse for Federal Scientific and Technical Information, U.S. Dept. of Commerce, Springfield, Va.

[2] J. A. Bearden and A. F. Burr (1967): *Rev. Mod. Phys.*, **39**: 125.

Table B-1 X-RAY ENERGY LEVELS (eV)

Atomic number and element	K	Level L_I	M_I
3. Li	54.75		
4. Be	111.0		
5. B	188.0		
6. C	283.8		
7. N	401.6		
8. O	532.0	23.7	
9. F	685.4	31*	
10. Ne	866.9	45*	
11. Na	1072.1	63.3	
12. Mg	1305.0	89.4	
13. Al	1559.6	117.7	
14. Si	1838.9	148.7	
15. P	2145.5	189.3	
16. S	2472.1	229.2	
17. Cl	2822.5	270.2	17.5
18. A	3203.0	320	25.3
19. K	3607.5	377.1	33.9
20. Ca	4038.2	437.8	43.7
21. Sc	4492.9	500.4	53.8
22. Ti	4966.5	563.7	60.3
23. V	5465.2	628.2	66.5
24. Cr	5989.3	694.6	74.1
25. Mn	6539.1	769.0	83.9
26. Fe	7112.2	846.1	92.9
27. Co	7709.1	925.6	100.7
28. Ni	8333.0	1008.1	111.8
29. Cu	8979.1	1096.1	119.8
30. Zn	9658.8	1193.6	135.9
31. Ga	10367.3	1297.7	158.1
32. Ge	11103.3	1414.3	180.0
33. As	11867.0	1526.5	203.5
34. Se	12658.1	1653.9	231.5
35. Br	13474.0	1782.0	256.5
36. Kr	14325.9	1921.0	
37. Rb	15200.0	2065.1	322.1
38. Sr	16105.0	2216.3	357.5
39. Y	17038.8	2372.6	394.6
40. Zr	17998.0	2531.7	430.3
41. Nb	18986.0	2697.8	468.4
42. Mo	19999.9	2865.6	504.6
43. Tc	21044.5	3042.6	
44. Ru	22117.7	3224.1	585.0
45. Rh	23220.4	3412.0	627.1
46. Pd	24350.8	3604.4	669.9
47. Ag	25514.6	3805.9	717.5
48. Cd	26711.8	4018.1	770.2
49. In	27940.5	4237.6	825.6
50. Sn	29200.7	4464.8	883.8
51. Sb	30491.9	4698.4	943.7
52. Te	31814.5	4939.3	1006.0
53. I	33170.1	5188.2	1072.1
54. Xe	34562.2	5452.9	

*Interpolated values.

Table B-1   X-RAY ENERGY LEVELS (eV)   (continued)

Atomic number and element	K	Level L_I	M_I
55. Cs	35985.4	5714.4	1217.1
56. Ba	37441.4	5988.9	1292.8
57. La	38925.5	6266.4	1361.3
58. Ce	40443.9	6548.9	1434.6
59. Pr	41991.5	6835.0	1511.0
60. Nd	43569.9	7126.2	1575.3
61. Pm	45185.0	7428.1	
62. Sm	46835.2	7737.0	1722.8
63. Eu	48520.1	8052.2	1800.0
64. Gd	50240.2	8375.8	1880.8
65. Tb	51996.8	8708.2	1967.5
66. Dy	53789.7	9046.0	2046.8
67. Ho	55618.9	9394.4	2128.3
68. Er	57486.8	9751.5	2206.5
69. Tm	59390.9	10115.9	2306.9
70. Yb	61333.6	10486.6	2398.2
71. Lu	63315.2	10870.6	2491.3
72. Hf	65352.2	11270.9	2601.0
73. Ta	67417.9	11681.8	2708.1
74. W	69526.5	12100.1	2819.7
75. Re	71678.0	12527.0	2931.8
76. Os	73872.4	12968.3	3048.6
77. Ir	76112.7	13418.8	3173.8
78. Pt	78396.5	13880.2	3296.1
79. Au	80726.7	14353.1	3425.0
80. Hg	83104.1	14839.6	3561.7
81. Tl	85532.3	15347.0	3704.2
82. Pb	88006.4	15861.1	3850.8
83. Bi	90527.9	16387.9	3999.2
84. Po	93107.0	16939.7	4149.5
85. At	95732.0	17493	4317*
86. Rn	98406	18049	4482*
87. Fr	101139	18639	4652*
88. Ra	103924.2	19237.1	4822.1
89. Ac	106757.6	19840	5002*
90. Th	109653.3	20472.6	5182.4
91. Pa	112603.9	21105.1	5367.0
92. U	115608.6	21757.9	5548.1
93. Np	118681	22427.3	5723.3
94. Pu	121821	23097.7	5933.0
95. Am	125030	23773.4	6120.6
96. Cm	128223	24461	6288*

*Interpolated values.

INDEXES

NAME INDEX

Page numbers in *italic* indicate bibliographic references.

SUBJECT INDEX

BELMONT COLLEGE LIBRARY